COLLEGE MISERICORDIA LIBRARY
RECEIVED
AUG 29 1983
DALLAS, PENNSYLVANIA

D1715376

TREATISE ON ANALYTICAL CHEMISTRY

PART I

THEORY AND PRACTICE

SECOND EDITION

TREATISE ON ANALYTICAL CHEMISTRY

PART I

THEORY AND PRACTICE

SECOND EDITION

VOLUME 10

Edited by PHILIP J. ELVING

Department of Chemistry, University of Michigan

Associate Editor: MAURICE M. BURSEY

Department of Chemistry, University of North Carolina

Editor Emeritus: I.M. KOLTHOFF

School of Chemistry, University of Minnesota

AN INTERSCIENCE® PUBLICATION

JOHN WILEY & SONS New York—Chichester—Brisbane—Toronto—Singapore

***Library of Congress Cataloging in Publication Data*:**

(Revised for pt. 1, v. 10)

Kolthoff, Izaak Maurits, 1894–
Treatise on analytical chemistry.

"An Interscience publication."
Includes bibliographies and indexes.
CONTENTS: pt. 1. Theory and practice.
1. Chemistry, Analytic. I. Elving, Philip Juliber, 1913– joint author. II. Title.

QD75.2.K64 1978 543 78-1707
ISBN 0-471-89688-8 (pt. 1, v. 10)

Printed in the United States of America

10 9 8 7 6 5 4 3 2 1

TREATISE ON ANALYTICAL CHEMISTRY

PART I

THEORY AND PRACTICE

VOLUME 10:

SECTION I

Magnetic Field and Related Methods of Analysis *Chapters 1–6*

AUTHORS OF VOLUME 10

ALLEN J. BARD
RICHARD H. COX
HANS G. FITZKY
IRA GOLDBERG
GEORGE C. LEVY
DONALD E. LEYDEN
ANTHONY LOMBARDO
A. E. MORGAN
MICHAEL J. RUIZ
JOHN G. STEVENS
H. W. WERNER

Authors of Volume 10

Dr. Allen J. Bard

Department of Chemistry, The University of Texas, Austin, Texas   Chapter 3

Dr. Maurice M. Bursey

Department of Chemistry, University of North Carolina, Chapel Hill, North Carolina

Dr. Richard Cox

Philip Morris USA Research Center P.O. Box 26583 Richmond, Virginia   Chapter 1

Dr. Hans G. Fitzky

Bereich Angewante Physik, Fa. Bayer AG, 509 Leverkusen, West Germany   Chapter 4

Dr. Ira Goldberg

Rockwell International, 1049 Camino Dos Rios, Thousand Oaks, California Chapter 3

Dr. George C. Levy

Department of Chemistry, Syracuse University, Syracuse, New York Chapter 2

Dr. Donald E. Leyden

Department of Chemistry, Colorado State University, Fort Collins, Colorado Chapter 1

Dr. Anthony Lombardo

Florida Atlantic University College of Science, Department of Chemistry, Boca Raton, Florida Chapter 2

Dr. A.E. Morgan

Philips Research Laboratories Sunnyvale, c/o Signetics Corporation, 811 E. Argnes, Sunnyvale, California Chapter 5

Dr. Michael J. Ruiz

Department of Chemistry, University of North Carolina, Asheville, North Carolina Chapter 6

Dr. John G. Stevens

Philips Research Laboratories, Gloeilampenfabrieken, Eindhoven, The Netherlands Chapter 6

Dr. H.W. Werner

Department of Chemistry, University of North Carolina, Asheville, North Carolina Chapter 5

Foreword

The division of chapters between Volumes 10 and 11, which will contain the section on "Magnetic Field and Related Methods of Analysis," was dictated in part by unforeseen delays in the preparation of some of the individual chapters, for example, authors who originally undertook the preparation of chapters were not able to fulfill their commitments and other experts had to be enlisted to prepare those chapters. In all fairness to the authors whose manuscripts were ready for publication, and to the users of the *Treatise*, it was decided to proceed with the present volume.

Volume 11 will contain, *inter alia*, chapters on NMR of nuclei other than ^{1}H and ^{13}C, and organic and spark source mass spectroscopy.

Preface to the Second Edition of the Treatise

In the mid 1950s, the plan ripened to edit a "Treatise on Analytical Chemistry" with the objective of presenting a comprehensive treatment of the theoretical fundamentals of analytical chemistry and their implementation (Part I) as well as of the practice of inorganic and organic analysis (Part II); an introduction to the utilization of analytical chemistry in industry (Part III) was also considered. Before starting this ambitious undertaking, the editors discussed it with many colleagues who were experts in the theory and/or practice of analytical chemistry. The uniform reaction was most skeptical; it was not thought possible to do justice to the many facets of analytical chemistry. Over several years, the editors spent days and weeks in discussion in order to define not only the aims and objectives of the Treatise but, more specifically, the order of presentation of the many topics in the form of a table of contents and the tentative scope of each chapter. In 1959, Volume 1 of Part I was published. The reviews of this volume and of the many other volumes of Part I as well as of those of Parts II and III have been uniformly favorable, and the first edition has become recognized as a contribution of classical value.

Even though analytical chemistry still has the same objectives as in the 1950s or even a century ago, the practice of analytical chemistry has been greatly expanded. Classically, qualitative and quantitative analysis have been practiced mainly as "solution chemistry." Since the 1950s, "solution analysis" has involved to an ever increasing extent physicochemical and physical methods of analysis, and automated analysis is finding more and more application, for example, its extensive utilization in clinical analysis and production control. The accomplishments resulting from automation are recognized even by laymen, who marvel at the knowledge gained by automated instruments in the analysis of the surfaces of the moon and of Mars. The computer is playing an ever increasing role in analysis and particularly in analytical research. This revolutionary development of analytical methodology is catalyzed by the demands made on analytical chemists, not only industrially and academically but also by society. Analytical chemistry has always played an important role in the development of inorganic, organic, and physical chemistry and biochemistry, as well as in that of other areas of the natural sciences such as mineralogy and geochemistry. In recent years, analytical chemistry—often of a rather sophisticated nature—has become increasingly important in the medical and biological sciences, as well as in the solving of such social problems as environmental pollution, the tracing of toxins, and the dating of art and archaeological objects, to mention only a few. In the area of atmospheric science, ozone reactivity and persistence in the stratosphere is presently a topic of great priority; extensive analysis is required both for monitoring atmospheric constituents and for investigating model systems.

One example of the increasing demands being made on analytical chemists is the growing need for speciation in characterizing chemical species. For example, in reporting that lake water contains dissolved mercury, it is necessary to report in which oxidation state it is present, whether as an inorganic salt or complex, or in an organic form and in which form.

As a result of the more or less revolutionary developments in analytical chemistry, portions of the first edition of the Treatise are becoming—and, to some extent, have become—out-of-date, and a revised, more up-to-date edition must take its place. In recognition of the extensive development and because of the increased specialization of analytical chemists, the editors have fortunately secured for the new edition the cooperation of experts as coeditors for various specific fields.

In essence, it is the objective of the second edition of the Treatise, as it was of the first edition (whose preface follows this one), to do justice to the theory and practice of contemporary analytical chemistry. It is a revision of Part I, which mirrors the development of analytical chemistry. Like the first edition, the second edition is not an extensive textbook; it attempts to present a thorough introduction to the methods of analytical chemistry and to provide the background for detailed evaluation of each topic.

Minneapolis, Minnesota I. M. KOLTHOFF
Ann Arbor, Michigan P. J. ELVING

Preface to the First Edition of the Treatise

The aims and objectives of this Treatise are to present a concise, critical, comprehensive, and systematic, but not exhaustive, treatment of all aspects of classical and modern analytical chemistry. The Treatise is designed to be a valuable source of information to all analytical chemists, to stimulate fundamental research in pure and applied analytical chemistry, and to illustrate the close relationship between academic and industrial analytical chemistry.

The general level sought in the Treatise is such that, while it may be profitably read by the chemist with the background equivalent to a bachelor's degree, it will at the same time be a guide to the advanced and experienced chemist—be he in industry or university—in the solution of his problems in analytical chemistry, whether of a routine or of a research character.

The progress and development of analytical chemistry during most of the first half of this century has generally been satisfactorily covered in modern textbooks and monographs. However, during the last fifteen or twenty years, there has been a tremendous expansion of analytical chemistry. Many new nuclear, subatomic, atomic, and molecular properties have been discovered, several of which have already found analytical application. In the development of techniques for measuring these and also the more classical properties, the revolutionary progress in the field of instrumentation has played a tremendous role.

It has been difficult, if not impossible, for anyone to digest this expansion of analytical chemistry. One of the objectives of the present Treatise is not only to describe these new properties, their measurement, and their analytical applicability, but also to classify them within the framework of the older classifications of analytical chemistry.

Theory and practice of analytical chemistry are closely interwoven. In solving an analytical chemical problem, a thorough understanding of the theory of analytical chemistry and of the fundamentals of its techniques, combined with a knowledge of and practical experience with chemical and physical methods, is essential. The Treatise as a whole is intended to be a unified, critical, and stimulating treatment of the theory of analytical chemistry, of our knowledge of analytically useful properties, of the theoretical and practical fundamentals of the techniques for their measurement, and of the ways in which they are applied to solving specific analytical problems. To achieve this purpose, the Treatise is divided into three parts: I, analytical chemistry and its methods; II, analytical chemistry of the elements; and III, the analytical chemistry of industrial materials.

Each chapter in Part I of the Treatise illustrates how analytical chemistry draws on the fundamentals of chemistry as well as on those of other sciences; it stresses for its particular topic the fundamental theoretical basis insofar as it affects the analytical approach, the methodology and practical fundamentals used both for the development of analytical methods and for their implementation for analytical service, and the critical factors in their application to both organic and inorganic materials. In general, the practical discussion is confined to fundamentals and to the analytical interpretation of the results obtained. Obviously then, the Treatise does not intend to take the place of the great number of existing and exhaustive monographs on specific subjects, but its intent is to serve as an introduction and guide to the efficient utilization of these specialized monographs. The emphasis is on the analytical significance of properties and of their measurement. In order to accomplish the above aims, the editors have invited authors who are not only recognized experts for the particular topics, but who are also personally acquainted with and vitally interested in the analytical applications. Only in this way the Treatise attain the analytical flavor which is one of its principal objectives.

Part II is intended to be very specific and to review critically the analytical chemistry of the elements. Each chapter, written by experts in the field, contains in addition to a critical and concise treatment of its subject, critically selected procedures for the determination of the element in its various forms. The same critical treatment is contemplated for Part III. Enough information is presented to enable the analyst both to analyze and to evaluate a product.

The response in connection with the preparation of the Treatise from all colleagues has been most enthusiastic and gratifying to the editors. It is obvious that it would have been impossible to accomplish the aims and objectives cited in this preface without the wholehearted cooperation of the large number of distinguished authors whose work appears in this and future volumes of the Treatise. To them and to our many friends who have encouraged us we express our sincere appreciation and gratitude. In particular, considering that the Treatise aims to cover all of the aspects of analytical chemistry, the editors have found it desirable to solicit the advice of some colleagues in the preparation of certain sections of the various parts of the Treatise. They would like at this time to acknowledge their indebtedness to Professor Ernest B. Sandell of the University of Minnesota for his interest and active cooperation in the organizing and detailed planning of the Treatise.

Minneapolis, Minnesota I. M. KOLTHOFF
Ann Arbor, Michigan P. J. ELVING

Acknowledgment

In view of the wide scope of the Treatise, it has been considered essential to have the advice and aid of experts in various areas of analytical chemistry. For the section on "Magnetic Field and Related Methods of Analysis," the editor has been fortunate to have the cooperation of Dr. Maurice M. Bursey of the University of North Carolina as Associate Editor; his collaboration is acknowledged with gratitude.

P. J. E

PART I. THEORY AND PRACTICE

CONTENTS VOLUME 10

SECTION I. Magnetic Field and Related Methods of Analysis

TREATISE ON ANALYTICAL CHEMISTRY

PART I

THEORY AND PRACTICE

SECOND EDITION

SECTION I: Magnetic Field and Related Methods of Analysis

Part 1
Section I

Chapter 1

NUCLEAR MAGNETIC RESONANCE: PRINCIPLES AND ^{1}H SPECTRA

BY RICHARD H. COX* AND DONALD E. LEYDEN,
Department of Chemistry, University of Georgia, Athens, Georgia

Contents

*Present address: Philip Morris USA Research Center, Richmond, Virginia.

I. INTRODUCTION

Nuclear magnetic resonance (NMR) spectroscopy is probably the single most important physical tool available to the chemist today. Since the only requirement for a nucleus to give an NMR spectrum is that the nucleus possess a nonzero magnetic moment, there is hardly an organic or inorganic compound that is not amenable to study by NMR. It is probably safe to say that an NMR spectrum of at least one, if not more, nucleus is obtained on each new compound isolated today. Scientists have come to rely on the information provided by NMR to such an extent that it is not uncommon to find several NMR spectrometers within a given laboratory. These spectrometers range from the routine-type instrument to the more sophisticated, research-type instrument. The developments in NMR instrumentation have continued to decrease the sample size requirement for an NMR spectrum and, hence, to increase the areas for application of NMR spectroscopy.

Like most spectroscopic techniques, NMR had its beginning in the earlier experiments of various physicists. The concept that certain nuclei possess a magnetic moment was proposed by Pauli in 1924 (176) to account for the hyperfine structure observed in atomic spectral lines. Molecular beam experiments by Stern and Gerlach (229,88) demonstrated that the component of electron magnetic moments is quantized in a magnetic field. Later refinements of these experiments permitted the measurement of the proton magnetic moment (72,86). Gouy balance-type experiments by Lasarew and Schubnikow in 1937 demonstrated nuclear paramagnetism and showed that the thermal equilibrium of nuclear spins in a magnetic field is rapidly established even at low temperatures (138).

In the strictest sense, the first NMR experiment was made in 1939 by Rabi and coworkers (194). While passing a beam of hydrogen molecules through a magnetic field, radio-frequency electromagnetic energy was applied to the molecules in the magnetic field. The frequency of the electromagnetic irradiation was varied until at a certain point energy was absorbed by the beam of hydrogen molecules, causing a slight deflection of the beam and a decrease in the number of molecules reaching the detector.

Although the early experiments were restricted to molecular beams, it had been pointed out that it should be possible to observe the absorption of energy in a magnetic field using samples in the solid, liquid, and gaseous states (93). Several attempts were made in the early 1940s to detect NMR absorption in solids. However, it was not until late 1945 that two groups first observed NMR absorption in experiments using bulk materials. Purcell, Torrey, and Pound (192) at Harvard observed proton NMR absorption in solid paraffin wax and Bloch, Hansen, and Packard (29) at Stanford observed the proton absorption in liquid water. The Nobel prize was awarded to these workers in 1952 for their discovery.

The potential for applications of NMR in chemistry dates to 1951 with the experiments of Arnold, Dharmatti, and Packard (16). In an examination of the NMR spectra of some alcohols, they showed that different signals could be observed for the chemically different types of protons. Since 1951, developments in both the theory and experimental aspects of NMR have been rapid. Major advances have been made in the quality of NMR spectra, in the ease in obtaining NMR spectra, and in improving the sensitivity of the NMR technique. One of the major improvements has been the development of magnets capable of operation at high-field strengths with a uniform, homogeneous field. This has led to improved sensitivity and has eased the problem of spectral interpretation. The development of low-priced minicomputers in recent years has led to the present-day pulsed Fourier transform NMR spectrometers that are becoming routine instruments in many laboratories.

Although data on molecular motion within a crystal and internuclear distances have been obtained from the spectra of solid samples, NMR spectra of the solid state is usually considered under the topic of wide-line NMR and will not be discussed here. We shall be concerned here primarily with high-resolution proton NMR spectroscopy where the observed spectral line widths approach the theoretical limits and are established by inhomogeneities in the magnetic field.

Unlike other forms of spectroscopy, where one often obtains either too little information or more information than can be readily interpreted, the interpretation of NMR spectra in terms of the fundamental parameters is straightforward in most cases. The parameters obtained from NMR spectra (chemical shifts, coupling constants, intensities, and relaxation times) provide the necessary information for solving a wide variety of problems in the chemical and biological sciences. The major application of NMR spectroscopy has been in structure elucidations. Using the chemical shifts and coupling constants obtained from a spectrum, considerable progress can usually be made towards determining the structure of an unknown compound through the use of empirical correlations established with known compounds. In many instances, it has been found that certain groupings of nuclei always give rise to a characteristic absorption pattern. This often allows one to distinguish isomers immediately. Two examples are shown in Fig. 1.1 and 1.2. A *para*-disubstituted benzene will in most cases give an absorption pattern similar to that in Fig. 1.1*c*, which appears on first inspection to contain two doublets in the low-field, aromatic region of the spectrum. Similarly, the large doublet and smaller septet in Fig. 1.2*b* are characteristic of the isopropyl group.

Other applications of NMR include the quantitative determination of the percentage composition of a complex mixture. Typical examples might be the determination of the percent deuteration at a particular site in a molecule, product distribution in a reaction mixture, or the analysis of a complex drug mixture. For this technique to be successful, one must first assign a peak in the spectrum to each individual component. Then the percentage composition of the mixture may be calculated from the known number of protons giving rise to each peak and the relative area under the peak. Another application of quantitative analysis using NMR is the determination of the percent hydrogen of an unknown compound. A known amount of a reference compound is added to the solution and the percent hydrogen of the unknown compound may be calculated from the relative areas under the peaks.

During recent years, the application of NMR to problems involving time-dependent phenomena has been increasing. In some cases, through the use of NMR it may prove possible to obtain reaction rates by carrying out the reaction in the probe of the spectrometer. This requires that one follow as a function of time either the disappearance of a peak in the spectrum due to a reactant or the appearance of a peak due to a product. One recent application of this technique has been the study of chemically induced dynamic nuclear polarization (CIDNP). Other examples of NMR studies of time-dependent phenomena include conformational analysis, rotational isomerism, restricted rotation, and fast chemical exchange—all of which may be investigated while the chemical system is at equilibrium.

One of the major limitations of NMR for analytical applications is the inherent insensitivity compared with other spectroscopic techniques. For normal continuous-wave spectrometers, somewhere between 5–50 mg of sample (depending on the molecular weight) are required to obtain a proton spectrum in a single scan. Thus, restrictions may be imposed by the solubility and/or the availability of the sample. Part of these difficulties may be overcome by using time-averaging techniques, smaller sample volumes, and spectrometers operating at higher magnetic fields. The availability of Fourier transform accessories has been of tremendous value in obtaining spectra on small samples. In those cases where a spectrum cannot be

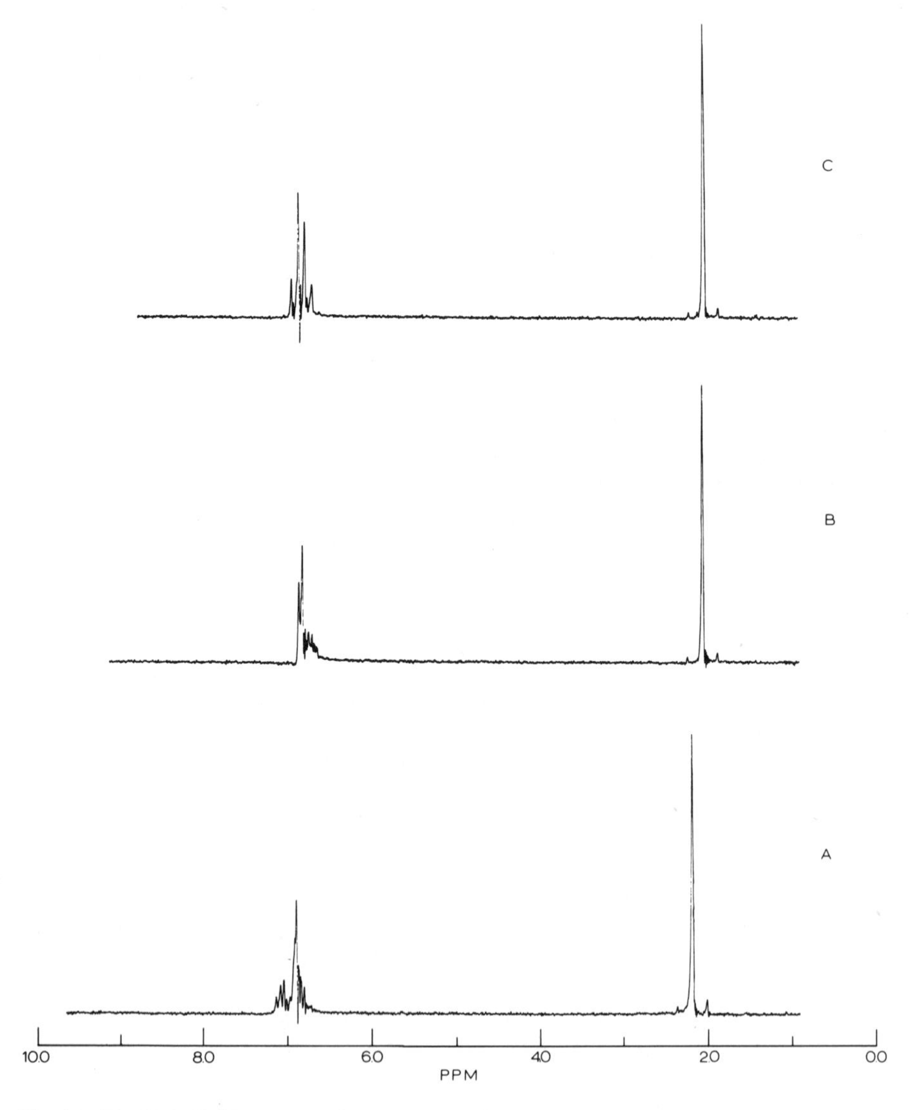

Fig. 1.1. The 100-MHz ^1H spectrum of (*a*) *o*-chlorotoluene; (*b*) *m*-chlorotoluene; and (*c*) *p*-chlorotoluene.

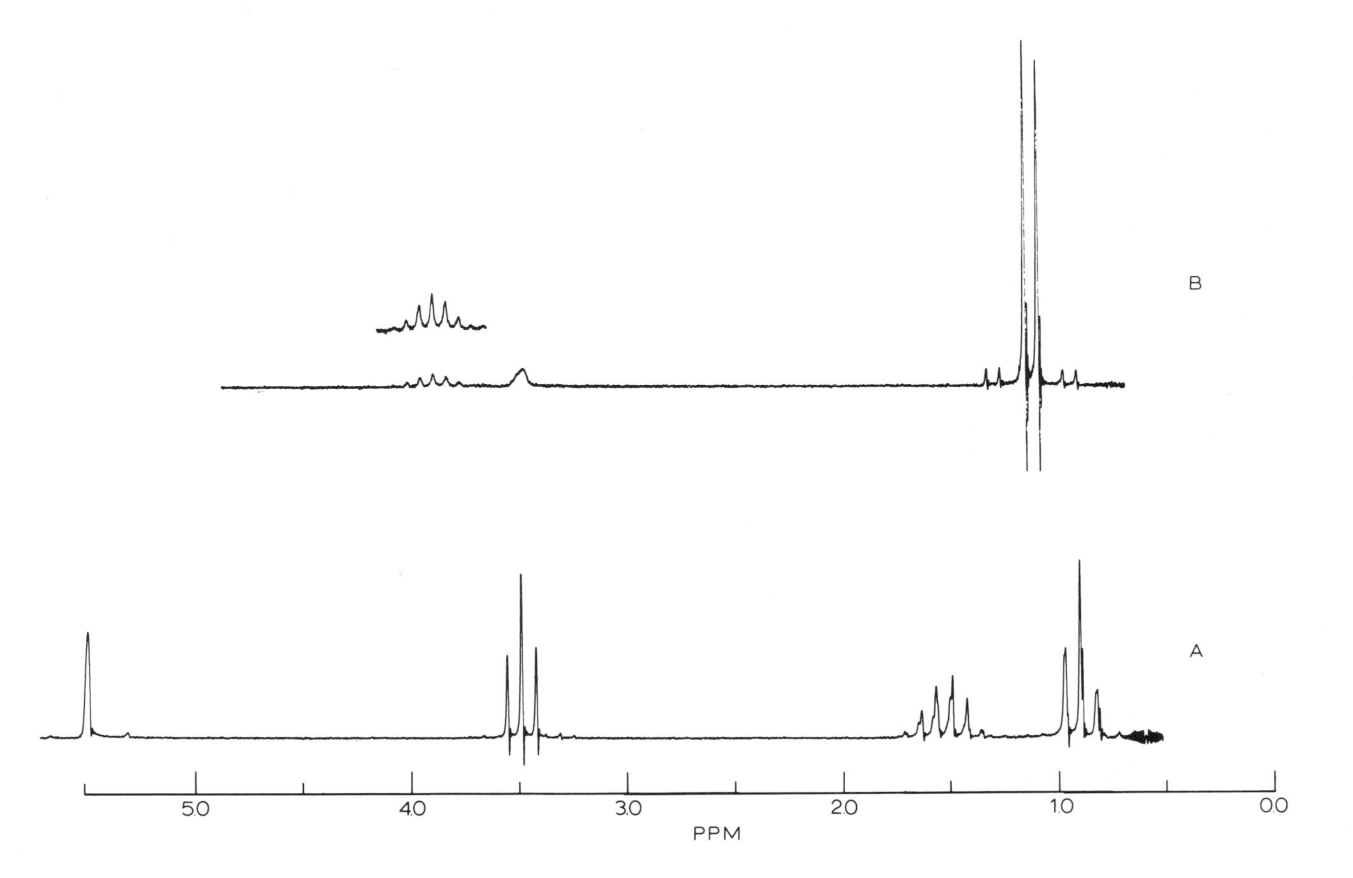

Fig. 1.2. The 100-MHz ^{1}H spectrum of (*a*) *n*-propyl alcohol; and (*b*) *iso*-propyl alcohol.

obtained in a single scan, time considerations may become important due either to the long time required to obtain the spectrum or to problems with decomposition of the sample.

Another problem often encountered in the application of NMR to structural elucidations of complex molecules is the overlap of absorptions in the spectrum. If a molecule contains several chemically similar types of protons, the absorptions of each of the protons may overlap to such an extent that it is not possible to obtain useful information from that region of the spectrum. In some cases, these problems may be alleviated by obtaining the spectrum at higher magnetic fields, by obtaining the spectrum in various solvents, and through the use of lanthanide shift reagents.

As there are variations in symbol conventions for magnetic resonance spectroscopy, a brief description of those used here is given. Two methods have been used previously for reporting proton chemical shifts: the tau, τ, and delta, δ, scales. To be consistent, we shall use only the δ scale in this chapter. Furthermore, low-field shifts will be reported as positive and high-field shifts as negative values with respect to the reference at δ 0.0. The unit for frequency will be the Hertz, Hz. As is now standard practice, spectra will be presented using the convention of increasing field from left to right.

NMR spectroscopists and chemists have become accustomed to writing $\overline{\mathbf{H}}$ for the magnetic field vector. Actually, $\overline{\mathbf{H}}$ is the magnetic field intensity vector and $\overline{\mathbf{B}}$ is the magnetic induction field vector. As long as electromagnetic units are used, $\overline{\mathbf{B}}$ and $\overline{\mathbf{H}}$ may be interchanged, although the observable magnetic properties depend on $\overline{\mathbf{B}}$ rather than $\overline{\mathbf{H}}$. Equations presented in this text will use $\overline{\mathbf{B}}$ for consistency in referring to the magnetic field. The electromagnetic unit, the gauss, is used for the magnetic induction field, $\overline{\mathbf{B}}$.

II. PRINCIPLES OF NUCLEAR MAGNETIC RESONANCE SPECTROMETRY

The treatment of the principles of magnetic resonance spectrometry will be limited in that we shall present the results of the theory and concentrate on providing a physical picture of the NMR phenomenon. Several excellent accounts of the theory of NMR are available, and those interested in a more detailed account of the theory are referred to these texts (46,73,159,185).

A. NUCLEAR MAGNETIC PROPERTIES

Certain nuclei, when placed in a magnetic field, behave as if they were spinning charged particles. Nuclei that possess this property have angular momentum, p. From quantum mechanics, the maximum observable component of the angular momentum is quantized and must be an integral or half-integral multiple of $\hbar$ (Planck's constant h divided by 2π). Furthermore, only certain states of p are allowed. Defining I as the spin quantum number, the maximum observable component of p is I. The angular momentum thus has $2I + 1$ states ($-I$, $-I + 1$, ...$I - 1$, I).

The spinning nucleus generates a magnetic moment, μ, that is parallel to and proportional to p. This relationship is given by equation 1,

$$\mu = \gamma p \tag{1}$$

where γ is the magnetogyric ratio and has different values for different nuclei. The magnetic moment is also quantized. The maximum observable component of μ has values of $m\mu/I$, where m is the magnetic quantum number and may have the values

$$m = -I, -I + 1, \ldots.., I - 1, I \tag{2}$$

A third magnetic property of nuclei related to the spin number I is the electric quadrupole moment, Q. The electric quadrupole moment is a measure of the nonsphericity of the electric charge distribution within the nucleus. Nuclei with $I = 1/2$ do not have electric quadrupole moments. However, nuclei with $I \geqslant 1$ often show line broadening along with a loss of certain information from the interaction of the nuclear magnetic moment with the electric quadrupole moment.

The only requirement for a nucleus to give an NMR absorption is that the nucleus possess a magnetic moment. All experiments to date indicate that the magnetic moment is zero if $I = 0$. Thus, nuclei with $I \neq 0$ give rise to NMR absorption. Rules for nuclear spins have been summarized in terms of the mass number, A, and the charge number, Z, as follows (185):

1. The nuclear spin is half-integral if the mass number A is odd.
2. The nuclear spin is integral if the mass number A is even and the atomic number Z is odd.
3. The nuclear spin is zero if both the mass number A and the atomic number Z are even.

Included in the first category with $I = 1/2$ are the important nuclei ^{1}H, ^{13}C, ^{19}F, and ^{31}P. Some nuclei with $I = 3/2$ are ^{7}Li, ^{11}B, ^{23}Na, and ^{35}Cl. The two major nuclei in the second category with $I = 1$ are ^{2}H (deuterium, D) and ^{14}N. Examples of nuclei with $I = 0$ are ^{12}C, ^{16}O, and ^{32}S. The results of rule 3 are important from the standpoint of the spectra of organic molecules. A compound containing only C, H, and O will give a proton spectrum without any complications due to interactions of the proton spins with the spins of the other nuclei in the molecule, as ^{13}C is present in low abundance. However, in some cases the proton spectrum either does not provide the desired information or the information is ambiguous and a carbon spectrum is needed.

B. RESONANCE PHENOMENON

Both classical and quantum mechanics have been used to treat the NMR experiment theoretically (185). Although both treatments give identical results, it is convenient to use parts of each method in describing the NMR experiment. The classical approach is better suited to derive a physical picture for the NMR experiment, whereas the quantum mechanics approach relates the NMR experiment to the absorption of energy during a transition between two energy states similar to other forms of spectroscopy. We shall discuss the classical mechanics approach first.

When a nucleus of spin $I \neq 0$ is placed in a magnetic field B_0, the magnetic moment takes up the allowed orientations along which the component of the magnetic moment may have values. For a spin 1/2 nucleus, the possible orientations in terms of the magnetic quantum number are $m = +1/2$ and $-1/2$ as shown in Fig. 1.3. The interaction between μ and B_0 results in a torque acting on μ that tends to tip it towards B_0. However, since the nucleus is spinning, instead of tipping μ towards B_0, the torque causes μ to precess about the magnetic field B_0, analogous to the way in which a gyroscope precesses about the earth's magnetic field. (It should be pointed out that in the absence of an applied field B_0, the components of μ are randomly distributed, since the earth's magnetic field is not strong enough to interact with μ.) The precessional frequency of μ about B_0 is given by the Larmor equation

$$\nu_0 = \frac{\gamma}{2\pi} B_0 \tag{3}$$

where ν_0 is the precessional frequency in cycles/sec. As can be seen from equation 3, the precessional frequency is dependent on both B_0 and γ and, thus, is different for each type of nucleus.

If a second, smaller magnetic field B_1 is applied in the $x - y$ plane (Fig. 1.3), and rotating in the same direction as μ, interactions between B_1 and μ occur. So long as B_1 is rotating at some frequency ν other than the Larmor frequency ν_0, the effect of B_1 on μ results in slight oscillations of the angle between μ and B_0. When B_1 is rotating at a frequency $\nu = \nu_0$, μ will feel the effects of both B_1 and B_0 and will exhibit large oscillations in the angle between μ and B_0, such that the direction of μ with respect to B_0 changes. Energy is absorbed by the nucleus and we speak of the magnetic moment μ as having "flipped" from one orientation in the magnetic field to the other. The absorption of energy when $\nu = \nu_0$ is the nuclear magnetic resonance (NMR) phenomenon.

In an actual experiment, the rotating field B_1 is obtained by passing a current through a coil at a frequency ν. This generates a magnetic field that is linearly polarized along the x axis and may be thought of as resulting from two equal fields rotating in opposite directions in the $x - y$ plane (Fig. 1.4). The resultant field is therefore 2 B_1 along the x axis.

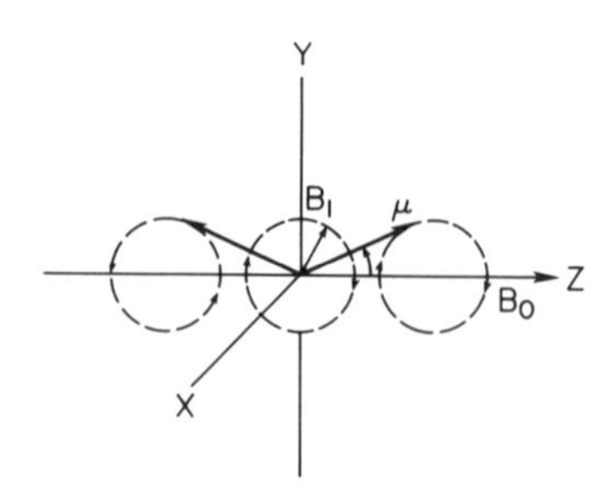

Fig. 1.3. The two orientations of the magnetic moments of a proton in a magnetic field.

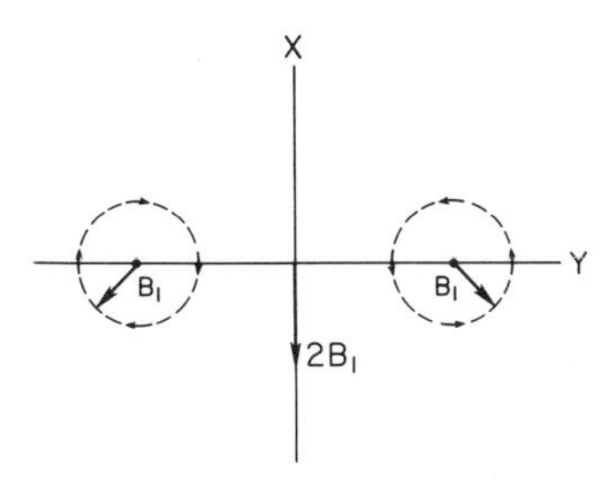

Fig. 1.4. An illustration of the two components of the rotating external magnetic field B_1.

According to the Larmor equation 3, the resonant condition depends upon the magnetogyric ratio, γ, the strength of the applied field, B_0, and the frequency of irradiation ν. Commercial NMR spectrometers usually employ a magnetic field of approximately either 14,092 or 23,487 gauss. The frequency of irradiation necessary for the resonant condition falls in the ratio frequency range (4–100 MHz) and overlaps with the standard FM band. In Table 1.I are given the frequencies of absorption and other properties important in observing an NMR spectrum for several nuclei.

TABLE 1.I
NMR Properties of Selected Nuclei

Isotope	I	NMR frequency in a 23,487 gauss field (MHz)	Natural abundance (%)	Relative sensitivity at constant field atnatural isotopic abundance
^{1}H	1/2	100.00	99.985	1.00
^{2}H	1	15.35	.015	.00000145
^{7}Li	3/2	38.86	92.58	.27123
^{11}B	3/2	32.08	80.42	.133
^{13}C	1/2	25.14	1.1	.00018
^{14}N	1	7.22	99.63	.001
^{15}N	1/2	10.13	0.37	.000004
^{17}O	5/2	13.56	.037	.00001
^{19}F	1/2	94.08	100	.833
^{23}Na	3/2	26.45	100	.0925
^{27}Al	5/2	26.06	100	.206
^{29}Si	1/2	19.86	4.70	.00037
^{31}P	1/2	40.48	100	.066
^{35}Cl	3/2	9.79	75.53	.0035
^{119}Sn	1/2	37.27	8.58	.0044
^{195}Pt	1/2	21.50	33.8	.0034
^{199}Hg	1/2	17.83	16.84	.00019
^{207}Pb	1/2	20.92	22.6	.002

In the quantum mechanical treatment of the NMR experiment, the interaction between the applied field and the magnetic moment appears in the Hamiltonian operator $\mathcal{H}$ as

$$\mathcal{H} = -\gamma\hbar\bar{B}_0 \cdot \bar{I} \tag{4}$$

Similarly, the interaction between B_1 and μ appears as another term in the Hamiltonian as

$$\mathcal{H}' = 2\mu_x B_1 \cos 2\pi\nu t = 2\gamma\hbar B_1 I_x \cos 2\pi\nu t \tag{5}$$

Solution of the Hamiltonian yields a discrete set of $2I + 1$ energy levels for the system

$$E_m = -\gamma\hbar m B_0 \tag{6}$$

where m has values of $-I, -I + 1, \ldots I - 1, I$. These energy levels correspond to the possible orientations of the magnetic moment with respect to B_0. For the case of a proton where $I = 1/2$, there are two energy levels as shown in Fig. 1.5 corresponding to the orientation of μ either aligned or opposed to the applied field B_0. The separation between the energy levels is linearly related to the applied field B_0.

Transitions between the two energy levels occur when the energy of B_1 corresponds to the energy difference between the two states. The probability of a transition per unit time is given by time-dependent perturbation theory as

$$P_{mm'} = \gamma^2 B_1^2 I(m \mid I_x \mid m')I^2 \delta(\nu_{mm'} - \nu) \tag{7}$$

where B_1 is the magnitude of the field applied in the x direction, $(m \mid I_x \mid m')I^2$ is the matrix element of the nuclear-spin operator in the x direction, δ is the Dirac delta function, and $\nu_{mm'} - \nu$ is the frequency corresponding to the energy difference between the two states m and m'. One important result of equation 7 is the selection rule that transitions occur only between energy states that differ in m by ± 1.

A second result of equation 7 is that the Dirac delta function vanishes for values of the frequency ν unless ν is exactly the frequency separation between the two states m

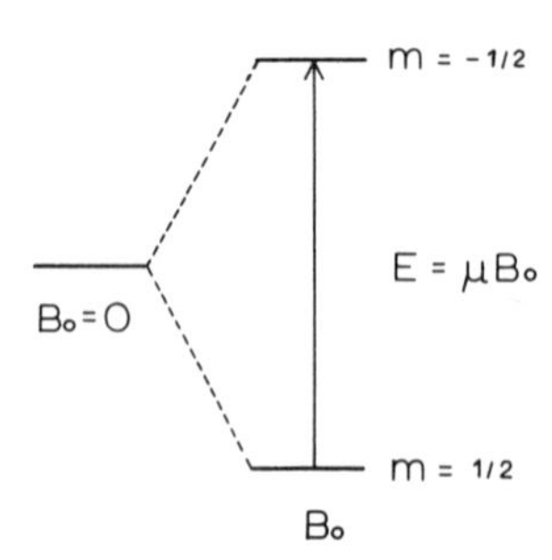

Fig. 1.5. The separation between the energy levels of a proton when placed into a magnetic field.

and m'. When $\nu = \nu_{mm'}$, an infinitely sharp absorption line is predicted. In actual practice, the NMR absorption usually approximates a Lorentzian line shape and the Dirac delta function is replaced with a line shape function $g(\nu)$ which is dependent on the frequency such that

$$\int_0^\infty g(\nu)dr = 1 \tag{8}$$

This substitution is necessary to reproduce the broadened lines normally found in NMR spectra.

Upon further consideration of the time-dependent perturbation theory and equation 7, one might raise the question as to why is it possible to observe an NMR transition. The probability of spontaneous absorption and emission is negligible. Furthermore, the probabilities of induced absorption and induced emission are equal. Therefore, the only way in which it would be possible to observe a transition would be if there exists a difference in the populations of the energy levels.

When a sample of nuclei is placed into a magnetic field B_0, the magnetic moments of the nuclei become orientated in the magnetic field according to the allowed energy levels. For a system of identical $I = 1/2$ nuclei, there will be two allowed energy levels corresponding to the orientation of μ either aligned with or opposed to B_0. At 23,487 gauss, the separation between the energy levels of a sample of protons is of the order of millicalories. The tendency for the nuclei to populate the lower energy level is opposed by thermal motions which tend to equalize the populations. The equilibrium population of the two energy levels at some temperature T is given by the Boltzmann equation

$$\frac{n_+}{n_-} = \exp(-\Delta E/kT) \tag{9}$$

where n_+ and n_- are the number of nuclei in the upper and lower energy level, respectively, k is Boltzmann's constant, T is the absolute temperature, and ΔE is the energy separation between the two energy levels. Substituting for ΔE, equation 9 becomes

$$\frac{n_+}{n_-} = \exp(-2\mu B_0/kT) \tag{10}$$

The solution of equation 10 for the excess of nuclei in the lower energy level may be approximated as

$$\frac{n_- - n_+}{n_-} \approx \frac{2\mu B_0}{kT} \tag{11}$$

For protons in a magnetic field of 23,487 gauss at 25°C, the excess population in the lower energy level is only 1.5×10^{-5} nuclei. This slight excess of nuclei in the lower energy state leads to a net absorption of energy. A further result is that the net absorption will depend upon the number of nuclei under consideration, which points out the fact that NMR is relatively insensitive compared with other spectrocopic techniques that operate at much higher frequencies. Since the excess population is directly proportional to B_0, sensitivity enhancement has been one of the major reasons for developing larger magnets during recent years.

Because the energy levels for a collection of identical nuclei are degenerate and therefore equally populated in the absence of a strong magnetic field, the process by which the Boltzmann distribution for the allowed energy levels becomes established is significant. When a sample is placed into a strong magnetic field, a finite amount of time is required for the Boltzmann distribution to reach equilibrium. As the nuclei are not interacting with B_1 at this point and the probability for spontaneous emission is negligible, there must be some process for absorption and emission to occur in order to establish the equilibrium spin distribution.

C. RELAXATION MECHANISMS AND SATURATION

The nuclei are undergoing motions and are interacting with their surroundings (lattice). This interaction with the lattice provides a mechanism for energy transfer between the spin system and the lattice, such that transitions between the energy levels occur. This process is called spin–lattice or longitudinal relaxation and is responsible for establishing the equilibrium-spin distribution. It is a nonradiative, first-order rate process and is characterized by a spin–lattice relaxation time T_1. We shall discuss several mechanisms in Section II.C.1 that can contribute to T_1.

In many cases, it is found that the width of NMR absorption signals is larger than can be accounted for in terms of spin–lattice relaxation alone. It is convenient to define an additional relaxation process, the spin–spin or transverse relaxation time T_2 such that

$$T_1 \geqslant T_2 \tag{12}$$

Spin–spin relaxation is a process in which neighboring nuclei exchange spin orientations by an interaction between their magnetic moments. This process results in no change in the total energy of the system.

With reference to Fig. 1.3, consider the interaction between two $I = 1/2$ nuclei, 1 and 2, which are precessing at the same frequency but with opposite orientations with respect to B_0. Since the rotating component of μ_1 is equal to that of μ_2 in the $x - y$ plane, the rotating component of μ_1 in the $x - y$ plane can be thought of as a second field B_1 acting on μ_2. The resulting interaction can cause μ_2 to flip to the opposite orientation with respect to B_0 while at the same time, μ_2 causes μ_1 to change to the opposite orientation. The result of this process is that two transitions, one absorption and one emission, have occurred with no net change in the total energy of the system. Spin–spin relaxation does not affect the relative population of the energy levels but can affect the line width.

1. Spin–Lattice Relaxation and Saturation

Any process that gives rise to fluctuating local magnetic fields can produce spin–lattice relaxation through the interaction of the local magnetic fields with the spin system. The fluctuating local magnetic fields invariably arise from thermal motions in the sample. The thermal motions of the molecules in the sample are occurring over a wide range of frequencies such that the fluctuating local magnetic fields have frequency components covering a wide range. In many ways, the frequency components of the local magnetic fields behave similar to the frequency of the rotating field B_1 such that frequency components near the precessional frequency can induce transitions among the nuclear spins. The extent to which thermal motions are effective in producing spin–lattice relaxation depends upon the magnitude of the local magnetic fields and the rate at which the fluctuations occur. It is convenient to define a *correlation time*, τ_c, that is a measure of the time required for the local field to acquire a new value. For translational motion, the correlation time may be thought of as the time required for the molecule to move one molecular diameter. In the case of rotational motion, the correlation time is the time required for a molecule to rotate through an angle of one radian.

We shall briefly discuss the following types of processes which give rise to spin–lattice relaxation:

1. Nuclear dipole–dipole interactions
2. Chemical shift anisotropy
3. Scalar coupling
4. Spin rotation
5. Quadrupole interactions
6. Interactions with paramagnetic species

2. Dipole–Dipole Relaxation

Dipole–dipole relaxation results from fluctuating local magnetic fields arising from the interaction of the magnetic moment of another nucleus with the magnetic moment of the nucleus being relaxed. As the molecule tumbles in solution due to Brownian motion, the local field due to another magnetic moment will fluctuate in both magnitude and direction. The interacting magnetic moments may be in the same molecule (intramolecular) or in different molecules (intermolecular). The strength of the interaction depends upon the orientation of the magnetic moments, the strength of the magnetic moments, and the distance separating the magnetic moments. Since other magnetic nuclei are always present in the sample, this is a general relaxation process and is often the dominant relaxation mechanism.

Consider a spin system containing two magnetic nuclei. The intramolecular interaction arises from rotational motion of the molecule. The distance separating the two nuclei remains fixed, as does the magnitude of the magnetic moments, while the orientation of the magnetic moments fluctuates with time. The intermolecular interaction arises from translational motion that is a function of the self-diffusion of the molecules. In general, the correlation times for these two processes will not be equal such that $1/T_1$ is the sum of both terms

$$1/T_1 = 1/T_{1\text{inter}} + 1/T_{1\text{intra}} \tag{13}$$

Since the interaction depends on the motion of the molecules, both relaxation processes are dependent on the viscosity of the medium and the size and shape of the molecules.

Using the theory of Bloembergen, Purcell, and Pound (31), the dipole–dipole relaxation may be expressed as

$$1/T_1 \propto \frac{\tau_c}{1 + 4\pi^2\nu_0^2\tau_c^2} \tag{14}$$

For nonviscous liquids where molecular motion is rapid, τ_c is of the order $10^{-11}-10^{-12}$ such that $1/\tau_c \gg 2\pi\nu_0$ and $1/T_1$ is proportional to τ_c. In this region of correlation time, both T_1 and T_2 are equal. As the liquid becomes more viscous, the correlation time increases and T_1 continues to decrease to the point where $\tau_c(2\pi\nu_0) = 1$. Further increases in the correlation time from this point result in an increase in T_1, and in the limit of slow motion where $1/\tau_c \ll 2\pi\nu_0$, $1/T_1$ is proportional to $1/\tau_c$. In this region however, $1/T_2$ continues to increase linearly with the correlation time such that the line width continues to broaden (31).

The range of T_1 values for the dipole–dipole interaction is from 0.1 to 100 sec. For small organic molecules, T_1 is typically on the order of 10–20 sec and, since molecular motion is rapid, $T_1 = T_2$. However for polymers, the correlation times are such that T_1 is near the minimum and the contribution from incomplete averaging to T_2 controls the relaxation. The viscosity of the solution may be changed by raising the temperature which in turn increases τ_c and results in narrower lines.

Some values of T_1 for small organic molecules are given in Table 1.II. It should be kept in mind that T_1 will not be the same for all nuclei in the sample if they are magnetically different.

TABLE 1.II

Some T_1 Values for Protons In Organic Molecules[a]

Molecule	T_1 (sec)	Temp (C°)
Water	3.6	
Acetic Acid-CH_3	2.4	20
Methyl iodide	3.8	29
Benzene	19.3	25
11% Benzene in CS_2	60	25
Toluene-CH_3	9	25
Toluene-aromatic	16	25

[a] Data taken from reference 185.

3. Chemical-Shift Anisotropy

The magnetic field experienced by a nucleus is determined in part by shielding and deshielding effects of other nuclei in the molecule. The magnitude of the shielding depends on the orientation of the molecule with respect to the applied magnetic field.

On the average, the nucleus experiences an average shielding due to rapid tumbling of the molecule in solution. On a much smaller time scale, however, the shielding may be anisotropic, giving rise to fluctuating local magnetic fields that vary as the molecule rotates in solution. Thus, the local fields provide a relaxation mechanism.

The magnitude of this relaxation mechanism is dependent upon the externally applied magnetic field B_0, since it is related to the chemical shift interaction. In the case of nonviscous liquids where the extreme narrowing limit applies ($1/\tau_c \gg 2\pi\nu_0$), the relaxation due to the chemical shift anisotropy is given by

$$1/T_1 = 2/15\,\gamma^2 B_0^2\,(\sigma_{11} - \sigma_1)^2\,\tau_c \tag{15}$$

The ratio of T_2/T_1 is 6/7.

The range of T_1 values for this relaxation mechanism is of the order of from 10 to 100 sec. This mechanism is thought to be important only at higher magnetic field strengths. At field strengths of the order of 60 k gauss, this mechanism can become comparable with that due to dipole–dipole interaction. An example of relaxation via chemical shift anisotropy has been reported (80). The spin–lattice relaxation time T_1 for $CH_3\,^{13}C\,O_2H$ varies linearly with the square of the frequency over the range of 9–60 MHz.

4. Scalar Coupling

Spin–spin coupling arises from the influence of the magnetic moment of one nucleus on another magnetic nucleus and is due to the difference in the local magnetic field experienced by one nucleus due to the other. If the local magnetic field fluctuates with time, it can provide a relaxation mechanism, scalar coupling. Suppose one nucleus A has a spin $>1/2$. If this nucleus has a short relaxation time compared with that provided by scalar coupling, then the local field experienced by a coupled nucleus B will be an average and, instead of observing the expected multiplet, one observes a single line.

The relaxation due to scalar coupling is given by

$$1/T_1^B = 2\,A/3\,I_s(I_s + 1)\tau_A/1 + (\nu_B - \nu_A)^2\,\tau_A^2 \tag{16}$$

where A is the spin–spin coupling constant in units of radians per sec and τ_A is the relaxation time of nucleus A. An example of this type of relaxation is provided by molecules with protons bonded to nitrogen. Nitrogen-14 has a spin $I = 1$ and is relaxed primarily by a quadrupole relaxation mechanism. As a result, spin–spin coupling is not observed with ^{14}N and one observes only a broadened resonance for protons attached to ^{14}N. In the event that the electron distribution surrounding nitrogen is symmetric, relaxation due to the quadrupole interaction becomes negligible and one can observe spin-coupling between nitrogen and protons (ammonium salts and isonitriles). Similar effects are also observed for protons attached to boron-11 ($I = 3/2$).

Scalar relaxation can also occur when chemical exchange is present. If the exchange rate of nucleus A is much larger than the coupling constant between nuclei A and B, and is larger than $1/T_1$ for both nuclei A and B, then only a single line will be observed, provided the time the nuclei are uncoupled is short compared with the time the nuclei are coupled. An example of this type of relaxation is given by the catalyzed exchange of the hydroxyl proton in alcohols where a single line is observed for the hydroxyl proton. The analysis of the line shape of single lines due to this type of relaxation can provide useful chemical information.

The range of T_1 values due to scalar relaxation is of the order 1–100 sec. This relaxation process can become comparable with or even greater than that provided by the dipole–dipole interaction in cases where the exchange rate is relatively fast.

5. Spin Rotation

When a molecule undergoes rotation, the motion of the electrons in the molecule generates a molecular magnetic moment at the nuclei. This gives fluctuating magnetic fields which provide a relaxation mechanism. The magnetic fields are proportional to the rotational angular momentum which is undergoing changes in both magnitude and direction due to Brownian motion.

The rate of this relaxation process is proportional to the angular momentum correlation time that is the time a molecule spends in any given angular momentum state. This serves to distinguish this relaxation process from the others since τ_c decreases as the temperature increases, whereas the angular momentum correlation time increases with increasing temperature. As a result, T_1 becomes longer as the temperature decreases if this is the dominant relaxation mechanism.

The range of T_1 values for spin rotation relaxation is of the order 10^{-2}–100 sec. For protons, it is not a dominant relaxation mechanism in most organic molecules. However, for small symmetric molecules, it can become a dominant relaxation mechanism for nuclei that have large chemical-shift ranges. Some examples have been found in fluorine-19 relaxation studies (156).

6. Quadrupole Relaxation

The quadrupolar relaxation arises from interactions between the nuclear spins of nuclei with $I > 1/2$ with the electric field gradient at the nucleus. Reorientations of the molecule result in random fluctuations of the components of the quadrupole coupling tensor as a function of time, thereby providing a relaxation mechanism. In the narrowing limit where $\nu_0\tau_c \ll 1$, $1/T_1$ due to this mechanism is proportional to the quadrupole coupling constant $(e^2Qq/\hbar)$ and the correlation time, and is given by

$$1/T_1 = 3/40\,(2I + 3/4I^2)\,(2I - 1)\left(1 + \frac{\eta^2}{3}\right)\left(\frac{e^2Qq}{\hbar}\right)^2 \tau_c \tag{17}$$

where η is the asymmetry parameter (1).

In nonviscous liquids, the molecular correlation time is of the order 10^{-11}–10^{-12} sec and the relaxation time is determined by the quadrupole coupling constant. The quadrupole interaction is the dominant relaxation mechanism for nuclei with $I > 1/2$ and T_1 values range from 10^{-7} to 10^2 sec. In cases where the electron distribution about spin $I > 1/2$ nuclei is symmetrical, the quadrupole coupling constant is zero and relaxation by this mechanism vanishes. Several examples of predominantly quadrupole relaxation may be found in alkyl lithium compounds (106), nitrogen compounds (161), and boron compounds (238). If the quadrupole coupling constant is known from other studies, $1/T_1$ provides a convenient way of determining the molecular correlation time.

7. Interaction with Paramagnetic Species

The presence of paramagnetic species in the sample provides a very effective relaxation mechanism. The magnetic moments of the unpaired electrons on the paramagnetic species provide the fluctuating local magnetic field. Since the electron magnetic moment is much larger than that of the nuclear magnetic moment (by a factor of ~1000), only small concentrations of paramagnetic species are necessary for this relaxation mechanism to become the dominant mechanism. The theory for this process has been considered by Bloembergen, Purcell, and Pound and is given by

$$\frac{1}{T_1} \propto \frac{4\pi^2\gamma^2 nNp\mu^2 eff}{kT} \tag{18}$$

where Np is the number of paramagnetic ions or molecules per cm^3, n is the viscosity of the solution, and μ *eff* is the effective magnetic moment of the paramagnetic species (31).

For an ion to be effective in producing relaxation by this mechanism, the electron spin relaxation time must be long compared with the molecular correlation time. That is to say, the spin of the electron magnetic moment remains fixed while the nuclei are undergoing reorientation. The most effective ions for this type of relaxation are Fe^{3+}, Cr^{3+}, Mn^{2+}, Eu^{+2}, Gd^{+3}, and Cu^{2+}. Less effective ions are Fe^{+2}, Co^{+2}, Ni^{+2}, and the remaining lanthanide ions. Oxygen is one paramagnetic molecule that is fairly effective in this type of relaxation.

In most cases, paramagnetic species produce unwanted line broadening and their presence in NMR samples should be avoided. It is not uncommon for lines to be broadened into the baseline such that only a very broad line is observed if it is detectable at all. However, there are applications where the presence of paramagnetic species can be used to probe the hydration shell and to obtain information on the binding sites of metal ions in biological systems.

D. CHEMICAL SHIFT

Our discussion to this point has focused on the interaction of a magnetic field with either a single nucleus or a collection of chemically equivalent nuclei. When a nucleus possesses a magnetic moment, an NMR absorption is observed for that nucleus when the Larmor equation (equation 3) is satisfied. Either the applied magnetic field or the frequency may be varied to bring about the resonance condition. Normally NMR spectrometers are set up to operate at an essentially constant magnetic field such that a wide range in the frequency (1–100 MHz at 23,487 gauss) is required to observe absorptions from different types of magnetic nuclei. The actual variation in the frequency employed is small (50–10,000 Hz) such that only one type of magnetic nucleus is under consideration during an experiment (i.e., a proton spectrum).

Therefore, a different frequency range is scanned to observe the absorptions from different nuclei.

When an NMR spectrum is obtained on a sample that contains several chemically different types of the same nuclei, say protons, it is found that separate signals are observed for each type of chemically different proton. The separation between the signals for the different protons and a reference compound is called the chemical shift and is perhaps the most important parameter to be derived from an NMR spectrum. Different chemical shifts arise from the fact that the magnetic field experienced by the nucleus depends upon its environment and is not the same as the applied field.

1. Origin of Chemical Shift

The origin of the chemical shift arises from the screening or shielding of the nucleus by the electrons surrounding the nucleus and from electrons in other parts of the molecule. When a nucleus is placed into a magnetic field, the motion of the surrounding electrons induces a secondary magnetic field at the nucleus. The direction of the secondary magnetic field is usually such that it opposes the applied field. Thus, the induced secondary magnetic field screens or shields the nucleus from the applied field, the magnitude of which depends upon the local electron density and on neighboring groups within the molecule. Each different type of proton in the sample will experience a magnetic field that differs from the applied field by a small amount

$$B_{\text{nucleus}} = B_0 (1 - \sigma) \qquad (19)$$

where σ is the screening constant. Substitution of equation 19 into equation 3 gives the resonance condition as

$$\nu = \gamma/2\pi \, B_0 (1 - \sigma) \qquad (20)$$

The screening constant is usually small and, at constant frequency, determines the amount by which the applied field must be increased to satisfy the resonance condition. In practice, it is impossible to determine σ as this would require the measurement of a bare nucleus stripped of its electrons. Instead, a reference compound is employed and all chemical shifts are determined with respect to the absorption of the reference compound.

Tetramethylsilane [TMS, $(CH_3)_4Si$] is employed as the reference compound for ^1H-NMR since it gives a single absorption and appears at a higher applied field than most other proton absorptions. Because the resonance condition may be satisfied by varying either the frequency or the applied field, chemical shifts may be expressed in frequency units (Hz) or field units (gauss). To eliminate the need to specify the magnetic field or the frequency used in measurements of chemical shifts, a method for reporting chemical shifts in nondimensional units is usually employed. For measurements made at constant frequency, the chemical shift δ is defined by

$$\delta = \frac{B_r - B_s}{B_r} \times 10^6 \text{ ppm} \qquad (21)$$

where B_r is the applied field at constant frequency for absorption of the reference, and B_s is the applied field for absorption of a given nucleus in the sample. Alternatively, for measurements made at constant applied field where the frequency is varied, the chemical shift δ is defined by

$$\delta = \frac{\nu_s - \nu_r}{\nu_r} \times 10^6 \text{ ppm} \tag{22}$$

where ν_s is the frequency of absorption for a given nucleus in the sample and ν_r is the frequency of absorption for the reference. Using this convention, most proton chemical shifts will be expressed as positive numbers and approximately 95% of all proton chemical shifts are in the range $\delta = 1 \sim 10$. Protons that are highly shielded will appear at a high applied field or low frequency and will have a small δ value, whereas protons that are less shielded will show absorptions at low field or high frequency and will have high δ values (Table 1.III).

A second convention for reporting chemical shifts was used in the early literature. Using this convention, the chemical shifts in parts per million increases with increasing shielding of the nucleus. The τ scale is defined as

$$\tau = 10.0 - \delta \tag{23}$$

such that values may be readily converted from one scale to the other. Since it has been recommended that the δ scale be adopted as the official scale, further discussions of chemical shifts in this chapter will refer to the δ scale only.

2. Theory of Chemical Shift

Several attempts have been made to account theoretically for the screening constant in terms of the electronic motion in a molecule. Using a second-order perturbation approach, Ramsey has developed a theory that in principle accounts for the shielding in molecules (197). Ramsey considered the screening constant to be the sum of two contributions, a diamagnetic term σ_d and a paramagnetic term σ_p. The diamagnetic term σ_d is positive leading to shifts to higher fields (shielding) and is approximated using the Lamb formula (137) for the shielding of atoms. The paramagnetic term σ_p is negative (deshielding) and results from the fact that the electrons in a molecule are not symmetrically disposed around the nucleus in question.

Application of the Ramsey formulism is difficult for anything except very small molecules (H_2), due to a number of unknown terms that must be approximated in evaluating σ_p. For larger molecules, the two terms become approximately equal leading to their cancellation. Subsequent treatments of the screening constant have used the Ramsey theory as a starting point for further discussion. Instead of trying to calculate the screening constant explicitly, the approach has been to discuss the various factors that can contribute to σ and concentrate on their relative importance and magnitude. Saika and Slichter suggested that an additional term be included to account for contributions from other atoms within the molecule (210). Pople has shown that another term is needed to account for ring currents in aromatic compounds

(182). In our discussions to follow, we shall consider the screening constant to be made up of five contributions

$$\sigma = \sigma_{\text{d}}^{\text{loc}} + \sigma_{\text{p}}^{\text{loc}} + \sigma_{\text{other}} + \sigma_{\text{currents}} + \sigma_{\text{sol}} \tag{24}$$

where $\sigma_{\text{d}}^{\text{loc}}$ is the local diamagnetic term, $\sigma_{\text{p}}^{\text{loc}}$ is the local paramagnetic term, σ_{other} is the contribution from other atoms in the molecule, σ_{current} is the contribution from ring currents, and σ_{sol} is an additional term to account for solvent effects. We shall discuss each of these terms in qualitative fashion as to their relative importance. A more detailed account may be found in the references provided.

a. Local Diamagnetic Screening

Local diamagnetic screening arises from the induced magnetic field due to the circulation of electrons around a nucleus and results in the nucleus being shielded from the applied magnetic field (Fig. 1.6). This contribution for the hydrogen atom may be approximated using the Lamb formula

$$\sigma_{\text{d}}^{\text{loc}} = \frac{4\pi e^2}{3\ mc^2} \int_0^\infty rp\ (r)\ dr \tag{25}$$

where p is the electron density associated with the hydrogen atom at a distance r from the nucleus. Depending upon how the electron density is evaluated, for the hydrogen atom equation 25 leads to

$$\sigma_{\text{d}}^{\text{loc}} = 20\lambda \times 10^{-6} \tag{26}$$

where λ is the effective number of electrons in the hydrogen $1S$ atomic orbital. Equation 26 shows that $\sigma_{\text{d}}^{\text{loc}}$ is of the order of a few parts per million. Since most proton chemical shifts are within this same order of magnitude, it has been suggested that $\sigma_{\text{d}}^{\text{loc}}$ is the primary factor controlling proton chemical shifts.

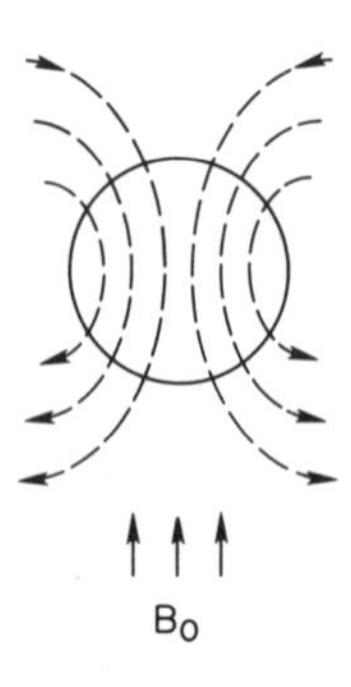

Fig. 1.6. The local diamagnetic screening of an isolated nucleus arising from an induced magnetic field due to the circulation of the electrons around the nucleus.

A second consideration that emerges from equation 26 is that σ_d^{loc} depends on the electron density associated with the atom. Several attempts have been made to correlate proton chemical shifts with substituent electronegativity, assuming that the only influence of the substituent is through inductive and/or resonance effects. For a related series of compounds such as CH_3X, a linear correlation is observed between the proton chemical shifts and the electronegativity of X (223). However, significant variations are observed indicating the importance of other shielding mechanisms.

TABLE 1.III

Some Typical Proton Chemical Shifts[a]

Compound	δ (ppm)
CH_4	.22
Cyclo C_3H_6	.22
CH_3CH_3	.85
$(CH_3)_4C$	.94
HC≡C–H	1.80
$CH_3C(=O)CH_3$	2.04
$C_6H_5\underline{CH_3}$	2.35
$CH_2{=}CH_2$	5.28
C_6H_6	7.27
H_2CO	9.57

[a] Data taken from various sources (74,113,219).

b. Local Paramagnetic Screening

The Lamb formula is not applicable for calculating the screening constant in a polyatomic molecule, since the electron density is not symmetrical about the nuclei. The paramagnetic term makes a negative contribution to the screening constant and arises from the mixing of ground-state wavefunctions with excited-state wavefunctions. Calculation of σ_p^{loc} therefore requires a knowledge of the excited-state wavefunctions. The formula for σ_p^{loc} contains an expression for the difference in energy between the ground state and excited states in the denominator. Usually, a knowledge of the excited-state energies is not available. One approach to overcome this difficulty has been to simplify this term by using an average over the excited states of the molecule, "the average energy approximation" (197). In the case of calculations of the screening in the hydrogen molecule, it is found that σ_p^{loc} contributes approximately 20% to the overall screening (197). For the most part, the contribution from σ_p^{loc} to the screening of protons is considered to be negligible. However, it can become the dominate shielding mechanism for larger nuclei.

c. Screening from Other Atoms in the Molecule

The origin of the effect of other atoms or groups of atoms on the shielding of a nucleus lies in the magnetic anisotropy of these atoms (185). Since the electron density around a proton is relatively low, centers of higher electron density in other parts of the molecule may lead to induced fields that affect the shielding of the proton. This is especially true if the bonding orbitals over which the electrons are distributed are not spherically symmetrical.

To discuss this effect in more detail, consider a model H–A, where A is a center of electron density due to an atom or group of atoms in another part of the molecule, the proton is not necessarily bonded to A. When the sample is placed into a magnetic field, the circulation of the electrons around A generates a magnetic dipole moment at A. The magnitude of the magnetic dipole moment depends on the magnetic susceptibility ψA of A and, for diamagnetic samples will oppose the applied magnetic field. The induced field experienced by H depends upon the orientation of H–A in the magnetic field and is illustrated in Fig. 1.7. Thus, the induced field at A results in deshielding at H if H–A is parallel to the applied magnetic field, and shielding if H–A is perpendicular to the applied magnetic field. Since the molecules are tumbling in solution, the induced field at H is the average of the contributions from the orientations shown in Fig. 1.7 and is given (185) by

$$B_{(\mathrm{H})} = \frac{B_0}{3R^3}(2\psi^{\mathrm{II}} - \psi^{\mathrm{I}} - \psi'^{\mathrm{I}}) \tag{27}$$

where $\Delta B_{(\mathrm{H})}$ is the induced field at H, R is the distance from the proton to A and ψ^{II} and ψ^{I} are the magnetic susceptibilities of A when H–A is orientated parallel and perpendicular to the applied magnetic field, respectively. Except when the electron distribution around A is symmetrical, the induced fields do not cancel due to tumbling. Therefore, either shielding or deshielding is observed at H, depending on the relative magnitudes of ψ^{II} and ψ^{I}.

Various attempts have been made to calculate the magnetic susceptibility of groups of atoms. The most successful approach has been to consider a point magnetic dipole centered at some position in A (183,149). Such extensive approximations are required that the reported magnitude of the susceptibility of various groups often covers a wide

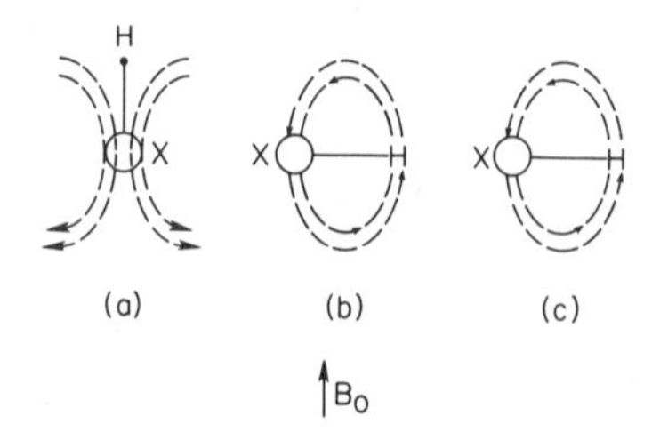

Fig. 1.7. The shielding (a) and deshielding (b and c) of a proton due to an atom or groups of atoms (X) located within a molecule.

range. Nevertheless, the results are useful for predicting the shielding and deshielding effects of these groups.

As pointed out earlier, the local diamagnetic screening is the primary factor controlling proton chemical shifts, as evidenced by the correlation of chemical shifts in methyl derivatives with the electronegativity of the substituent. A similar relationship holds for ethyl derivatives as well. However, the α-proton chemical shift appears at a lower field for an ethyl compound compared with the shift in the corresponding methyl compound. On the basis of inductive effects alone, one might have predicted the opposite result, since a methyl group is known to be slightly electron donating. However, the above trend is completely general in that the chemical shift of a proton appears to lower field with an increasing number of C–C bonds attached to the carbon to which the proton is attached (i.e., CH_4, $\delta = 0.22$; CH_3CH_3, $\delta = 0.85$; $CH_3\underline{CH_2}CH_3$, $\delta = 1.34$) (74).

Examination of the spectrum of cyclohexane at low temperatures reveals two signals separated by 0.47 ppm for the axial and equatorial protons (115). For monosubstituted cyclohexane derivatives at low temperature, where the rate of conversion from one chair conformation to the other is slow, the proton on the carbon α to the substituent exhibits separate resonances for the axial and equatorial conformations. It is now established that, in general, an axial proton is more shielded than an equatorial proton (74). Yet, the origin of the different chemical shifts for axial and equatorial protons in cyclohexanes remains largely unknown. Attempts have been made to rationalize the above shift differences in terms of the anisotropy of the magnetic susceptibility of the C–C and C–H bond (32,163,167). However, calculated values fail to account for the chemical shift difference. While there can be little doubt that there is some long-range shielding effect of C–C and C–H bonds, the origin of this effect remains unknown. About all that can be said at this time is that the theory predicts and experiments show that a nucleus lying along the axis of a single carbon bond will be deshielded, whereas a nucleus lying above or below the bond will be shielded.

The long-range shielding effects of the carbon–carbon double bond are complicated due to the fact that a C=C bond lacks axial symmetry. Hence the induced magnetic field due to the circulation of the π electrons in the double bond depends upon its orientation with respect to the magnetic field. Attempts to calculate this anisotropy using a point dipole approximation (235,184) have led to the following

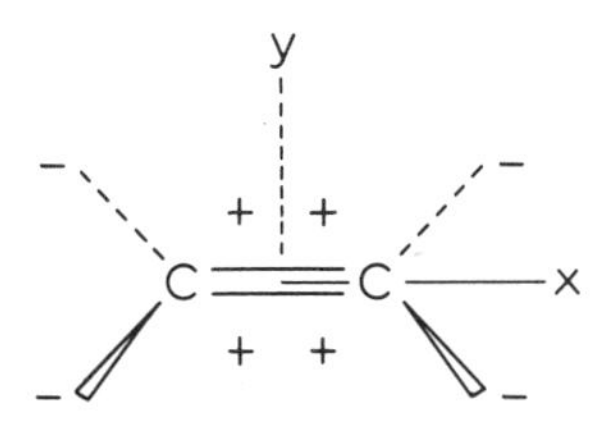

Fig. 1.8. The shielding (+) and deshielding (–) regions due to a carbon–carbon double bond.

conclusions concerning the long-range shielding effect of a C═C bond: (a) protons located in the region above or below the plane of the double bond (*XY* plane) between the two carbons will experience shielding (Fig. 1.8) and (b) protons located in the *XY* plane will be deshielded.

The long-range shielding effect of a C≡C bond is even more dramatic than either a C═C or C–C bond. On the basis of electronegativity effects alone (local diamagnetic shielding), one would expect the chemical shift of acetylenic protons to be further downfield than the shift of olefinic protons. However, the chemical shift of the protons in acetylene is between that of olefinic and aliphatic protons and is only slightly downfield from the chemical shift of ethane. Theory predicts a relatively large anisotropy along the bond axis and a much smaller value perpendicular to this axis as a result of the axial symmetry of the C≡C bond (183). Therefore, if the applied magnetic field is along the axis of the acetylene molecule, a relatively large induced field is generated due to the circulation of the π electrons around the carbon–carbon bond (Fig. 1.9) and a much smaller induced field when the applied field is perpendicular to the axis of the molecule. Therefore, nuclei that lie along the axis of the molecule will experience a shielding effect, whereas nuclei above or below the region of the C≡C bond will experience deshielding (Fig. 1.9). Verification of these predictions has been found experimentally (187,250). Long-range shielding due to a cyano group (C≡N) is predicted to be similar to that of the carbon–carbon triple bond (199,251). However, it has been suggested that the shifts observed in cyano compounds can be rationalized in terms of the electrostatic effect on the diamagnetic anisotropy alone (62).

The chemical shift of formaldehyde (H–C(═O)–H, 9.57 ppm) (219) is considerably downfield from the proton shift of methane (0.22 ppm) (74). While part of this difference is due to the inductive effect of the carbonyl group, the shift difference is too large to be accounted for in terms of electron-withdrawing effects on the local diamagnetic shielding alone. Theoretical calculations of the anisotropy of the carbonyl group predict that a nucleus will experience deshielding if it is located in the plane of the carbonyl group (*XY*) (Fig. 1.10) and shielding if it is located above and below the plane (*XY*) of the bond. Furthermore, shielding is predicted for a small region in the *XY* plane near the carbon end of the carbonyl group (184). Experimental support for these predictions has come from a variety of sources. For example, it is well known that in α,β-unsaturated carbonyl compounds, the *cis*-β-protons are

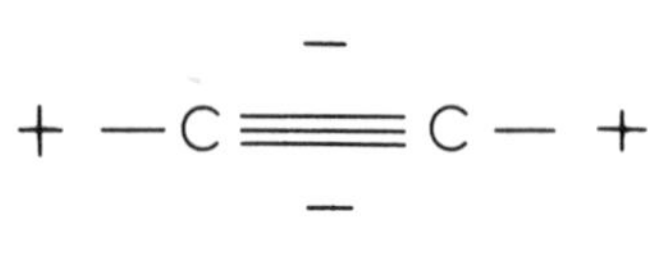

$$\begin{array}{c} - \\ + \;-\mathrm{C}\equiv\mathrm{C}-\; + \\ - \end{array}$$

Fig. 1.9. The shielding (+) and deshielding (−) regions due to a carbon–carbon triple bond.

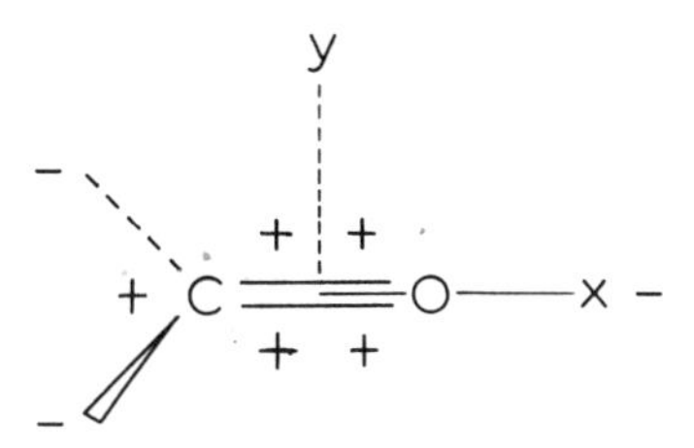

Fig. 1.10. The shielding (+) and deshielding (−) regions due to a carbonyl group.

deshielded with respect to the *trans*-β-protons (114). Additional examples can be found in the spectra of various steriods (26).

d. Ring Currents

The chemical shift of the protons in benzene is deshielded by approximately 2 ppm from those in ethylene, 7.27 versus 5.30 ppm. Since the protons in question are bonded to an sp^2 hybridized carbon in both cases, and the electron density on each carbon is equal, the local diamagnetic screening should be approximately equal in both cases. Therefore, an additional screening contribution is needed to account for the difference in chemical shift.

It is well known that an aromatic molecule possesses excess magnetic susceptibility in the direction perpendicular to the plane of the ring over that parallel to the plane (177). It was suggested that this is due to a ring current resulting from circulation of the π electrons round the orbitals of the ring and induced by the externally applied magnetic field. Pople (182) first suggested that this would lead to a deshielding of aromatic protons relative to ethylene and, using a point dipole approximation, calculated the deshielding to be 1.75 ppm. Refinements of this theory by Waugh and Fessenden (244) and Johnson and Bovey (118) by assuming the ring current to be located above and below the ring lead to a deshielding value of ~2.2 ppm. The results of the above theories show that the secondary magnetic field due to the ring current is opposed to the externally applied field such that protons located inside the region of the ring will experience shielding, whereas protons located outside the region of the ring will be deshielded (Fig. 1.11). Further refinements of the theory by Pople and Untch (187) have suggested that the shielding associated with an annulene ring will

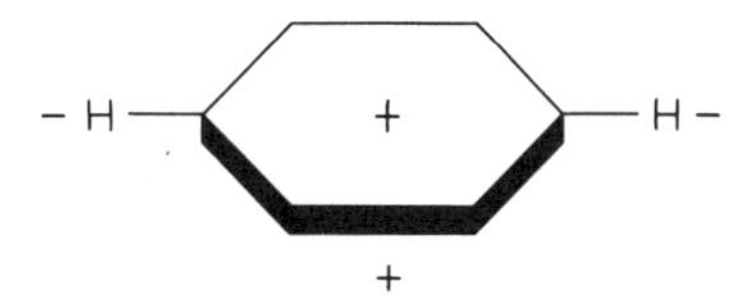

Fig. 1.11. The shielding (+) and deshielding (−) regions due to an aromatic ring.

depend upon the number of π electrons in the ring. Rings containing $4n$ π electrons will exhibit long-range shielding effects (paramagnetic ring current) such that protons located outside the area of the ring will be shielded whereas protons located inside the ring will be deshielded.

Many examples other than benzene exist to support the above conclusions. These include various aromatic hydrocarbons, aromatic heterocycles, annulenes, paracyclophanes, and porphyrins (113). One of the most interesting examples is that provided by the 15,16-dialkyldihydropyrenes (**I**) (160). By varying the size of the alkyl group (R), the distance dependence of the effect of the ring current is illustrated. Furthermore, this ring system may be easily reduced by two electrons to a 16 π-electron dianion illustrating the reversal in the direction of the shielding effects with a paramagnetic ring current (Table 1.IV).

R

R

I

TABLE 1.IV
Chemical Shifts of Some 15,16-Dialkyldihydropyrenes[a,b]

	Alkyl proton shifts			
R	α	β	γ	Exterior proton shifts
Neutral Hydrocarbons				
$\overset{\alpha}{C}H_3$	−4.25			7.95–8.67
$\overset{\alpha}{C}H_2\overset{\beta}{C}H_3$	−3.96	1.86		7.95–8.67
$\overset{\alpha}{C}H_2\overset{\beta}{C}H_2\overset{\gamma}{C}H_3$	−3.95	1.87	0.65	7.95–8.67
Dianions				
$\overset{\alpha}{C}H_3$	21.00	11.70		−(3.19–3.96)
$\overset{\alpha}{C}H_2\overset{\beta}{C}H_3$	21.15	11.70		−(2.50–3.14)
$\overset{\alpha}{C}H_2\overset{\beta}{C}H_2\overset{\gamma}{C}H_3$	21.24	12.59	5.51	−(2.56–3.14)

[a] R. N. Mitchell, C. E. Klopfenstein, and V. Boekelheide (160).
[b] In ppm from TMS.

The question of whether similar ring currents exist in nonaromatic rings has been discussed by a number of workers (41,174). Cyclopropane possesses a diamagnetic susceptibility (136) and it has been suggested that the high-field shift of cyclopropane protons ($\delta = 0.22$) is evidence for a ring current. Theoretical treatments using the above models tend to support this view. Anet and Schenck (14) have considered the possibility of the ring currents in homoaromatic and antiaromatic systems. Using a method based on solvent effects upon chemical shifts, they conclude that a diamagnetic ring current exists in 1,3,5-cycloheptatriene, cyclopentadiene, and norbornadiene, and a paramagnetic ring current in cyclooctatetraene.

e. Solvent Effects

The effect of solvent on the screening of a particular proton may be divided into five contributions: bulk susceptibility of the medium, van der Waals interactions, anisotropy of the susceptibilities of the surrounding molecules, reaction field of the medium, and specific solute–solute interactions (139,205). The need for considering the bulk susceptibility term has been largely eliminated by the use of an internal reference (TMS) and will not be discussed further. However, it must be considered when an external reference is used.

The contribution due to van der Waals interactions is thought to be a result of distortions of the electronic environment of the nucleus by the solvent. This leads to a decrease in the diamagnetic screening of the nucleus and results in a downfield shift. In practice, it has been found that polyhalogenated solvents produce the largest shifts, which can be on the order of 0.5 ppm.

In most cases, the change from a nonpolar solvent to an aromatic solvent such as benzene will result in an upfield shift of the solute protons. It has been suggested that this upfield shift is a result of an average orientation where the solute molecules lie closer to the face of the aromatic ring than to the edge of the aromatic ring. Allowing for an average over all solute–solvent orientations, this results in an upfield shift due to the anisotropy of the aromatic ring. Similar considerations applied to rodlike solvents (CS_2) suggest that the solute lies closer to the axis of the solvent molecule resulting in downfield shifts of the solute protons.

The reaction field contribution is due to the polarization of the surrounding medium by a polar solute that creates an electric field; the reaction field at the solute. Using the Onsager model (179) for the reaction field, Buckingham (38) has developed a theory for the reaction field effect on chemical shifts for a spherical molecule which predicts that the solvent shift should be linear in $(\varepsilon - 1)/(2\varepsilon - n)$ where ε is the dielectric constant of the medium and n is the refractive index of the solute. This relationship has been found to predict correctly the solvent shifts in various systems. However, in other cases it does not hold and the discrepancy is thought to be a result of deviations from the spherical shape. Another theoretical approach has assumed an ellipsoidal shape and arrives at a linear relationship between the solvent shift and $(\varepsilon - 1)/(\varepsilon - \beta)$. Regardless of the model chosen, the practical results are that as the polarity of the solvent is increased, the chemical shifts move downfield, and protons closer to the polar site in the molecule experience larger shifts than those further removed from the polar site in the molecule.

In addition to the solvent shift with aromatic solvents due to anisotropy effects, specific shifts have been observed for polar solutes in aromatic solvents. These shifts are referred to as aromatic solvent-induced shifts (ASIS) and are thought to be due to some type of specific solute–solvent interaction. The results have generally been interpreted in terms of a collision complex brought about by dipole-induced dipole interactions, or some other weak association between the electron donor aromatic solvent and some positively polarized part of the solute molecule. The energy of the interaction is on the order of 1 kcal/mole. For the most part these "complexes" are envisioned as a 1:1 association. The major application of ASIS to date has been to differentiate between various angular methyl groups. Other applications are covered in the two excellent reviews of solvent effects in NMR (139,205). The reader is also referred to the above reviews for a more detailed treatment of general solvent effects.

3. Chemical-Shift Measurements

NMR spectra are normally displayed by using an *x*-*y* recorder in which the amplitude of the signal output from the detector is fed into the *y* axis. The *x* axis input is usually either a function of time or a function of the sweep frequency. The convention is to record the spectrum such that the magnetic field B_0 increases from left to right across the chart. The absolute frequency of a particular absorption signal cannot be conveniently determined. Therefore, the normal convention is to determine relative frequencies of absorptions with respect to some standard reference material.

a. Reference Compounds

The ideal reference compound should not interact with either the solute or solvent and preferably should exhibit a single absorption peak in a region of the spectrum separated from sample absorptions. The standard reference compound for proton spectra in nonaqueous solvents is tetramethylsilane (TMS). Tetramethylsilane, $Si(CH_3)_4$, is a volatile liquid (b.p. 27°C) which facilitates its removal from the sample and the recovery of the solute under investigation. The single peak from the equivalent protons of TMS appears at higher field than most other proton absorptions and thus does not interfere with absorptions from the solute. Normally, a concentration of 1% TMS is sufficient for external locked spectrometers, whereas from 3–5% is required for internal locked spectrometers if the TMS signal is used as the lock signal source.

Since TMS is insoluble in water, it cannot be used as a reference for aqueous solutions. The sodium salt of the compound 2,2-dimethyl-2-silapentane-5-sulphonic acid (DSS), $(CH_3)_3SiCH_2CH_2CH_2SO_3$-Na, has been most widely used as a reference for aqueous solutions. Absorption frequencies are determined relative to the sharp peak given by the equivalent protons of the $(CH_3)_3Si-$ group. The position of this absorption is not affected by the pH of the solution. The use of DSS as a reference has the disadvantage that additional absorptions are present in the spectrum due to the three methylene groups. At low concentrations (1% of DSS, these absorptions will not interfere with the absorptions of the solute. However, if low concentrations of the solute are used, the absorptions due to the methylene peaks may interfere with the solute spectrum. An additionl disadvantage of DSS is the difficulty of separating it from the solute after the spectrum is obtained.

Two methods are employed in NMR spectroscopy for referencing absorptions. The reference compound may either be dissolved directly in the sample under investigation (internal reference) or may be placed in a separate container from that of the sample (external reference). When an internal reference is used, both the sample and reference experience the same magnetic field. This is the most common type of referencing and presents no problems unless the reference interacts with either the solute or solvent, or effects solvent–solute interactions. Interactions of this type would influence the relative positions of the reference and sample absorptions. Usually these interactions are at a minimum when TMS is used as the internal reference.

When an external reference is used, the influence of the reference on the solvent–solute interactions is absent. The usual method is to place the reference in either an especially designed coaxial tube or a capillary tube (a melting point capillary tube) and then place this inside the normal NMR sample tube. The major disadvantage of an external reference is that the sample and reference do not experience the same magnetic field. The magnetization per unit volume induced in the sample (185) depends on the volume magnetic susceptibility K (equation 28).

$$M_0 = K\,B_0 \tag{28}$$

For sufficiently dilute samples, the volume magnetic susceptibility of the sample K_s is essentially that of the solvent. Since the reference is separated from the sample by a glass interface, the volume magnetic susceptibility of the reference K_r will be slightly different. If spherical cells are used, no susceptibility corrections are needed. For the normal cylindrical NMR sample tubes, the magnetic fields at the sample and reference are given by equation 29. Values for K_s and K_r may be obtained from the

$$(B_0)_s = B_0\,(1-2/3\pi K_s) \tag{29}$$

$$(B_0)_R = B_0\,(1-2/3\pi K_R) \tag{30}$$

literature (73) or may be determined by NMR methods (25,70,145). For precise chemical shift measurements using ^{1}H-NMR, the relative frequency between the sample and reference should be corrected to take into account the difference in magnetic field. For other nuclei, where larger frequency ranges are observed for the absorptions, these corrections are not as important.

In many of the earlier NMR studies, different reference compounds such as the solvents themselves were used. It is almost impossible to convert from one reference to another due to the lack of information concerning the concentration of the solution, and the extent of intermolecular association effects. For accurate work, conversion from one internal reference to another is possible only if the absorption frequencies are extrapolated to infinite dilution in the same solvent. Accurate frequencies from the conversion of data obtained with an external reference to an internal reference and vice versa is impossible.

b. FREQUENCY CALIBRATION

The relative positions of the absorptions due to the sample and the reference compound are determined in frequency units. However, the standard for reporting chemical shifts is in the dimensionless units of ppm, which are not dependent upon the operating frequency of the spectrometer. Coupling constants are determined and reported in frequency units (Hertz). Several procedures have been used to calibrate the absorption frequencies in NMR and the method of choice depends upon the complexity of the spectrum, the accuracy desired, and to a certain extent, on the type of spectrometer on which the spectrum is obtained.

For spectrometers without some means of field/frequency stabilization, two methods of calibration are used. The superposition of bands method can only be applied to spectra where the absorptions are well separated and depends upon one's ability to superpose a signal from a sideband of the reference compound on the absorption in question from the sample. The normal procedure is to observe the signal from the sample on an oscilloscope using the fast sweep or on a recorder having a rapid response. The sweep time and filtering are adjusted so that a ringing pattern is observed. A sideband (usually the first sideband) of the reference is generated using an audio-oscillator and its position is adjusted to coincide exactly with the signal from the sample. As the sideband signal is adjusted close to the signal of the sample, distortions in the ringing patterns of the sample signal will be observed. When the two frequencies exactly coincide, the ringing pattern returns to normal. At this point, the frequency of the audio-oscillator, and hence the sample signal, can be read with the aid of a frequency counter. The frequency of the sample signal can usually be determined with an accuracy of $\pm 0.1\%$ using this method, as it does not depend upon the linearity of the sweep. This method is difficult to apply, however, to samples where the signals from the samples are overlapping such that clear ringing patterns are not observed for each peak.

The second method, the audio-sideband technique, uses the interpolation between two sidebands to obtain the frequencies of the sample absorptions with respect to the reference. Sidebands (usually of the reference) of known frequency are introduced by means of an audio-oscillator and the relative positions of the sample signals are determined by interpolation. For routine work, one sideband, placed such that the reference and sideband signals encompass all the sample signals, is usually sufficient. This method suffers from errors due to the nonlinearity of the sweep. For more accurate work, two sidebands of the reference that encompass only a small region of the spectrum are used and this process is repeated to calibrate each region of the spectrum. By taking the average of several such determinations, the frequencies of the sample signals can usually be determined to an accuracy of $\pm 0.5\%$.

For routine spectra on instruments employing some means of field/frequency stabilization, it is possible to record the spectra on precalibrated chart paper. The accuracy of this method is limited by the calibration of the sweep widths and the accuracy of reading the signal position from the chart paper. The sweep width calibration should be checked periodically to insure that it is within the manufacturer's specifications. For more accurate work, sidebands of the reference can be introduced and the interpolation method used. Normally, one would use the smallest sweep width available. With an internal locked spectrometer, the recorder position is directly related to the sweep frequency. By using the smallest sweep width available, the recorder pen can be stopped at the signal of interest and its frequency determined by determining the frequency difference between the control and analytical channels.

With modern Fourier transform NMR spectrometers, a computer routine is available that prints the intensities and frequencies of the peaks in the displayed region of the spectrum. One can position a cursor on any peak, usually the reference standard, and assign a frequency to this peak. The peak printout will then be referenced to the frequency of the peak set using the cursor.

c. Sample Preparation

Several factors, including the state of the sample, the concentration, the solvent, and the presence of impurities, should be considered in the preparation of a sample for NMR analysis. For high-resolution NMR spectra, a liquid compound may be examined as the pure liquid or may be dissolved in a suitable solvent. The latter is usually preferred, especially for viscous liquids, since high viscosity usually leads to signal broadening due to restricted molecular tumbling. To eliminate the dipole–dipole broadening in the solid state, solid samples must be dissolved in a suitable solvent. If the solid has a reasonably low melting point, it may be examined as a pure liquid at high temperatures. The spectra of gases may be obtained, but pressures of several atmospheres may be required in order to obtain reasonable sensitivity. Sample tubes are available which are safe up to about 20 atmospheres pressure (249).

The concentration of the sample should be sufficient to obtain a reasonable signal-to-noise ratio (S/N). For 5-mm o.d. samples tubes, normally from 0.3 to 0.5 ml of volume is required. Depending on the molecular weight of the sample, from 5 to 50 mg of sample is usually sufficient for obtaining the spectrum in a single scan. Microcells and signal-averaging techniques may be used to obtain spectra on smaller samples.

Several factors should be considered in the selection of a suitable solvent. In addition to dissolving the sample, the solvent should not interact chemically with the sample and, preferably, should not contain any hydrogen atoms whose absorptions would interfere with those of the sample. Carbon tetrachloride probably comes closest to being an "ideal" solvent. However, its use is limited by the fact that numerous compounds are not sufficiently soluble in CCl_4. Several relatively inexpensive deuterated solvents of high isotopic purity are commercially available. The most common of these is chloroform-d. However, hydrogen bonding between $CDCl_3$ and the sample may result in substantial shifts from the resonance position in an inert solvent. Solvent shifts are also observed with aromatic solvents. A list of common solvents used in ^{1}H-NMR spectroscopy is given in Table 1.V.

Considerable line broadening may be observed in an NMR spectrum if care is not exercised to insure that paramagnetic or ferromagnetic impurities are eliminated. Any source of unpaired electrons will give rise to line broadening due to a reduction of the relaxation times. Molecular oxygen dissolved in most solvents causes some broadening and should be eliminated by purging the sample with oxygen-free nitrogen or by evacuation using several freeze-pump-thaw cycles.

Ferromagnetic particles, probably coming from a steel spatula or from dust particles in the atmosphere, give rise to considerable line broadening. In addition, solid particles of undissolved sample are another source of line broadening. The ferromagnetic particles may often be removed by using a fairly strong permanent magnet to retain the particle while the solution is withdrawn for transfer to another sample tube. Solid particles may be removed by filtration. Several devices are commercially available for filtering NMR samples (104,249), and it is a good idea to filter each sample routinely.

TABLE 1.V

Common Solvents Used for Proton NMR[a]

Solvent	δ[b]
CCl_4	—
CS_2	—
SO_2	—
Chloroform-d[c]	7.28
Acetone-d_6	2.07
Dimethylsulfoxide-d_6	2.50
p-Dioxane-d_8	3.56
Acetonitrile-d_3	1.96
Methylene Chloride-d_2	5.28
Nitromethane-d_3	4.29
Cyclohexane-d_{12}	1.42
Methanol-d_4[c]	3.34, 4.11
Tetrahydrofuran-d_8	1.79, 3.60
D_2O[b]	4.61
Benzene-d_6	7.24
Pyridine-d_5	7.18, 7.57, 8.57
Toluene-d_8	2.31, 7.10
Acetic acid-d_4[c]	2.06, 11.97
Trifluoroacetic acid[c]	11.34

[a] Data taken from reference 170.

[b] Chemical shift downfield from internal TMS of residual proton signals.

[c] Peak positions are variable depending on hydrogen bonding with the solute and temperature.

E. SPIN-SPIN INTERACTIONS

1. Origin of Spin-Spin Interactions

In 1951, it was reported that the spectra of several liquids contained more lines (muliplets) than could be accounted for on the basis of the number of chemically different nuclei in the sample (98,102,191). This multiplet splitting of resonance lines results from the interaction of a nuclear magnetic moment with the magnetic moments of neighboring nuclei that cause splitting of the energy levels and hence, multiple transitions. The energy of this interaction is expressed as

$$E_{NN} = h J_{NN'} \bar{I}_N \cdot \bar{I}_{N'} \tag{31}$$

where $J_{NN'}$ is the coupling constant between the two nuclei N and N', and $\bar{I}_N \cdot \bar{I}_{N'}$ are the nuclear spin angular moments.

The mechanism proposed for this coupling interaction (198) assumes that the nuclear spins interact through polarization of the spins by the bonding electrons. As an example, consider the interaction between two bonded A and B nuclei of spin $I = 1/2$. If the nuclear spin of A is oriented parallel to the applied magnetic field B_0, the electron near the vicinty of A will tend to orient its spin antiparallel to the nuclear spin due to the pairing of magnetic moments. Since A and B are bonded, the electron in the vicinty of B must have its spin antiparallel to that of A due to the Pauli exclusion principle. The nuclear spin on B will in turn be antiparallel to the electron spin on B with the overall result that the two nuclear spins are oriented in an antiparallel fashion (Fig. 1.12). The opposite orientation with the nuclear spins of A and B being parallel represents a slightly different energy arrangement. Both orientations, parallel and antiparallel, occur to approximately equal extents (Boltzmann distribution) in a bulk sample, since the energy difference between the two orientations is small. When nucleus A undergoes a transition from one orientation with respect to the magnetic field B_0, two transitions occur corresponding to the two different orientations of nucleus B with respect to A. The frequency separation between the two transitions is proportional to the energy of interaction between the nuclear spins.

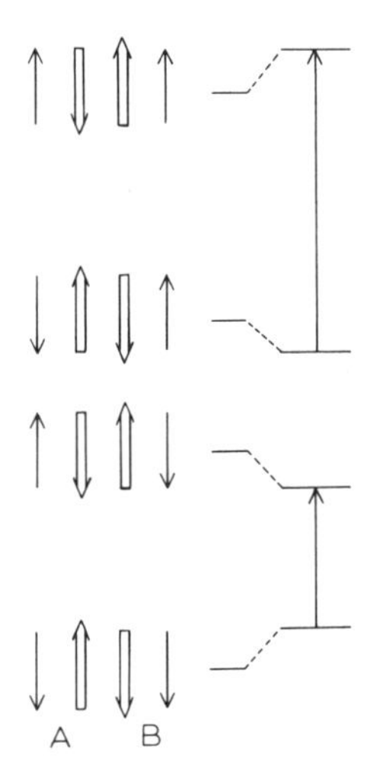

Fig. 1.12. The possible orientations of the magnetic moments (→) and electrons (⇒) of two bonded nuclei A and B, their relative energies, and the two nuclear transitions of nucleus A.

The magnitude of the spin–spin coupling constant J_{AB} is expressed in Hertz (cycles/sec in the older literature) and may be either positive or negative. If the antiparallel orientation of the nuclear spins is the lower energy arrangement, the coupling constant J_{AB} is positive, whereas if the parallel orientation is lower in energy, the coupling constant is negative. The *absolute* sign of a coupling constant cannot be readily determined from a normal high-resolution NMR spectrum. *Relative* signs of coupling constants however may be determined from double-resonance experiments and, in some cases, from the analysis of non-first-order spectra. The absolute sign of the coupling constant between the *ortho* protons in *p*-nitrotoluene has been found to be positive from an experiment in which the molecule was oriented in the magnetic field by a strong electric field (39). A number of relative signs of coupling constants have now been determined with respect to this coupling. Some typical signs of coupling constants are given in Table 1.VI. The relative signs of coupling constants can be of importance in structure elucidation.

TABLE 1.VI

Signs of Some Proton–Proton Coupling Constants[a]

Fragment	Sign[b]
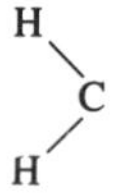 H C H	−
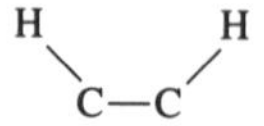 H H C—C	+
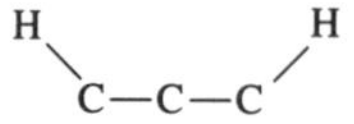 H H C—C—C	−
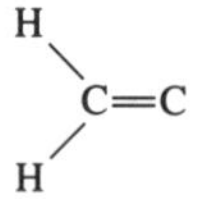 H C=C H	+ or −
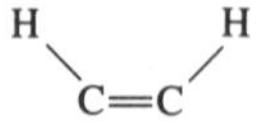 H H C=C	+
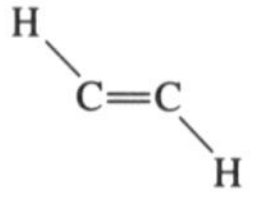 H C=C H	+
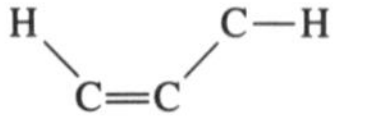 H C—H C=C	−

TABLE 1.VI *(Continued)*

Signs of Some Proton–Proton Coupling Constants[a]

Fragment	Sign[b]
H–C=C–C–H	–
Benzene, H and H on adjacent (ortho) positions	+
Benzene, H and H on meta positions	+
Benzene, H and H on para positions	+

[a] Data taken from reference 183.

[b] Relative signs with respect to J_{C-H} taken as positive.

2. Theory of Spin–Spin Interaction

A theory for the electron-nuclear spin–spin interactions has been proposed by Ramsey (198). The complete Hamiltonian for the motion of electrons in the presence of nuclei which possess magnetic moments is divided into three parts.

$$\mathcal{H}' = \mathcal{H}_1 + \mathcal{H}_2 + \mathcal{H}_3 \quad (32)$$

The first part represents the magnetic shielding of the interactions of the nuclear spins by the electron orbital motion. The second term is included to account for the dipole–dipole interaction between the electron magnetic moments and the nuclear magnetic moments. The final term arises from relativistic effects and is often referred to as the Fermi contact term. This term represents the interaction between electrons in *s* orbitals and nuclear magnetic moments.

Calculation of coupling constants from first principles is complex as it requires a knowledge of the total wavefunctions for the electronic ground and excited states. Several attempts have been made to calculate the coupling constant in the hydrogen molecule with reasonable success. The results suggest that the Fermi contact term accounts for greater than 90% of the spin–spin interaction (22). Several important conclusions concerning coupling constants result from the Fermi contact term. The magnitude of the coupling constant between two directly bonded nuclei depends upon the product of the magnetogyric ratios of the two nuclei $J \alpha \gamma_N \gamma_{N'}$. Thus, if the coupling constant between two nuclei has been determined, the coupling constant between N and an isotope of N' may be calculated provided the magnetogyric ratio for the isotope of N' is known. Secondly, the magnitude of the coupling is directly proportional to the s-electron densities in the bonding orbitals of N and N' used in forming the bond. This has led to correlations of coupling constants with the percent s character in the bond (166).

3. Practical Considerations

In the analysis of NMR spectra (the extraction of the chemical shifts and coupling constants from the spectrum), the chemist is immediately faced with the problem of determining which peaks are due to spin–spin coupling and which are due to the absorptions of uncoupled nuclei. This task is not as difficult as it may appear on first glance. One soon learns the characteristic absorption patterns given by certain groupings of nuclei. From the intensities and number of lines in these patterns, one is able to pick out these multiplets in the spectrum with practice. In more complicated cases where there is considerable overlap of the peaks in a spectrum, one can make use of the fact that the relative intensities of the multiplets are in direct proportion to the number of nuclei giving rise to the peaks.

An important difference between coupling constants and chemical shifts is that coupling constants are independent of the applied magnetic field. When there is some doubt as to whether a particular multiplet is due to spin–spin coupling, or to the overlap of resonances from chemically different nuclei, one can obtain the spectrum at a higher magnetic field. If the multiplet is due to spin–spin coupling, minor intensity changes will be observed in the multiplet but no change in the relative positions of the peaks, whereas if the multiplet is due to different nuclei, the absorptions will be separated more at higher magnetic field. In addition, one can carry out spin-decoupling experiments to determine which multiplets are due to nuclei that are mutually spin-coupled. Obtaining the spectrum in another solvent may also change the spectrum such that characteristic multiplets may be recognized.

Since the spin–spin interaction is transmitted through the bonding electrons, coupling constants are characteristic of the arrangement of the nuclei in a molecule. Coupling between two nuclei may occur even when the two nuclei are separated by several bonds. For aliphatic organic compounds, proton–proton coupling constants are normally observed for protons separated by one, two, three, and sometimes four bonds. Spin–coupling between nuclei separated by a large number of bonds is observable if some of the intervening bonds are π bonds. The magnitude of the coupling constant decreases, in general, as the number of bonds separating the coupled nuclei increases.

Coupling constants are as useful as chemical shifts in analytical applications of

NMR. Since it is usually not possible to calculate a coupling constant from theory, the use of coupling constants relies very heavily upon empirical correlations established through the examination of compounds of known structure. In this respect, it has been found that coupling constants depend upon (a) the geometrical relationship between the two coupled nuclei, (b) the electronegativity of substituents, (c) the nature of the bonds between the coupled nuclei, and (d) solvent effects. Thus, coupling constants provide invaluable information in applications such as structure determination, conformational analysis, and bonding.

F. SIGNAL INTENSITY

1. Theory

One of the major advantages of NMR over other forms of conventional spectroscopy is that the integrated intensity (area) of an NMR absorption signal is directly proportional to the number of nuclei giving rise to the signal. The area under an NMR signal (equation 33) depends on a number of factors including B_0, B_1, T_1, T_2, temperature, and the number and type of nuclei giving the signal.

$$A \propto \frac{NB_1\mu^2}{kT\,(1 + \gamma^2 B_1^2 T_1 T_2)^{1/2}} \tag{33}$$

The dependence of the area on B_0 results from the effect of B_0 on the separation of the energy levels between which the transitions are occurring, and its effect on the relative population of the energy levels. Equation 33 is valid only for relatively low values of the rotating field B_1. The area will increase with an increase in B_1 until the point where $\gamma^2 B_1^2 T_1 T_2$ becomes significant with respect to 1. When this occurs, the area starts to decrease with a further increase in B_1. When the area decreases with an increase in B_1, saturation is said to be occurring. A saturation factor Z_0 is defined as

$$Z_0 = \frac{1}{1 + \gamma^2 B_1^2 T_1 T_2} \tag{34}$$

To help understand this dependence of the area on B_1, consider the NMR process. When B_1 of correct frequency is applied to the sample, transitions occur from the lower to the higher energy level due to absorption of energy, and from the higher to the lower energy level due to induced emission and relaxation. Since the populations of the two energy levels are not equal, there will be a larger number of transitions from the lower to the higher energy level and a net absorption of energy (i.e., an NMR signal). As larger and larger values of B_1 are applied, the population of the two energy levels tends to become more equal, such that a decrease in the net absorption of energy is observed, i.e., a decrease in the area of the signal. The area of the signal depends upon the relaxation times T_1 and T_2, and, because they are usually different for chemically different nuclei in a sample, each peak will saturate at a difference value of B_1. Therefore, care should be exercised in obtaining an integration of an NMR spectrum to insure that B_1 is low enough such that saturation is not occurring.

The peak height of an NMR signal is given by an equation similar to that for the area

$$\text{In} \propto \frac{NB_1\mu^2T_2}{kT\ (1 + \gamma^2B_1^2T_1T_2)} \tag{35}$$

Since equations 33 and 35 are similar, one might question why peak heights cannot be used as an indication of the relative number of nuclei giving rise to the signals, rather than going to the trouble of integrating the signals to obtain the areas. One important difference in equations 33 and 35 is the dependence of the peak height on T_2. Normally, T_2 values are different for chemically different nuclei in the sample. Furthermore, as discussed in Section II.G, the width of an NMR peak is also dependent on T_2 and can be different for chemically different nuclei. Because of this dependence on T_2, peak heights will usually be different for an equal number of chemically different nuclei and cannot be used in quantitative analysis as a measure of the relative number of nuclei giving rise to the signal.

2. Peak Area Measurement

Integration of the NMR resonance line to determine the area beneath the peak is by far the most common and the most practical method of quantitation by NMR. Using modern electronic integrator circuits, integration can be performed with a precision of a few tenths of one per cent. However, in most cases, the reproducibility of the recorder, less than perfect tuning of the instrument, partial saturation of the signal and transient noise results in an overall precision of 1–2%. The precision may be improved by the use of a digital volt meter with decade scaling on the output of the integrator. It is important for the reader to remember here the fundamental difference between precision and accuracy, which are occasionally used incorrectly as interchangable terms. Precision is only a measure of the reproducibility of the observation, in this case the signal integration. There is no relationship between precision and accuracy, as the latter is a measure of the difference between an observed result and a "true" value. Improper standardization or calibration, an error in measurement of the integral value, or some other determinate error will affect accuracy, whereas the precision of the same measurement may be excellent.

There are several precautions that must be observed to obtain both precise and accurate quantitation by peak integration of the NMR signal. First, the instrument must be in good condition. Poor connections of cables, noisy electronic components, and microphonic vibrations will contribute transient signals. The very nature of the integration process will tend to average to zero any noise that is random and has a period much less than the time required for the integration. However, dc transient pulses or low-frequency drift will not be averaged and will cause an error in the integration directly proportional to their magnitude, duration, and frequency of occurrence. Proper precaution must be taken to balance the dc offset of the integrator circuit using a control that is normally accessible on the instrument console. To do this, the spectrum amplitude may be set at its minimum to reduce input to the integrator. The recorder should be set off any resonance to further reduce input. Some spectrometers have an integrator balance "switch" setting to ground the integrator input during the balancing adjustment. The recorder zero may be used to set the pen near the center of the recorder. With the integrator on, adjust the balance control until the integrator drift seen as pen drift is eliminated, or as much so as possible. The integrator output gain should be increased as the adjustment proceeds to increase the sensitivity of the balancing operation. The integrator reset may be operated frequently to keep the pen near the center of the page. This adjustment should be checked before any integration as a dc balance is difficult to stabilize for long time periods. If a persistent leakage of the integrator appears, as indicated by a downward trailing of the integrator output, the amplifier or integrating capacitor should be suspected as failing.

A second important parameter for quality integration is the adjustment of the phase detector. Provided the amplifier balance is properly adjusted, the integral step should show a flat initial baseline as well as a flat plateau after the peak has been passed. It is easy to recognize the effect of incorrect phase detector adjustment as the baseline and plateau of the integral will drift in opposite directions. This is a result of the presence of a component of the dispersion mode signal. In contrast, improper integrator amplifier balance will result in baseline and plateau drift in the same direction. As both may be present simultaneously, it is advised to balance the amplifier as a first step, as this procedure is independent of other spectrometer parameters. When an instrument that has a field/frequency sweep option is used, integration must be performed in the field sweep mode. In the frequency sweep mode, the detector phase

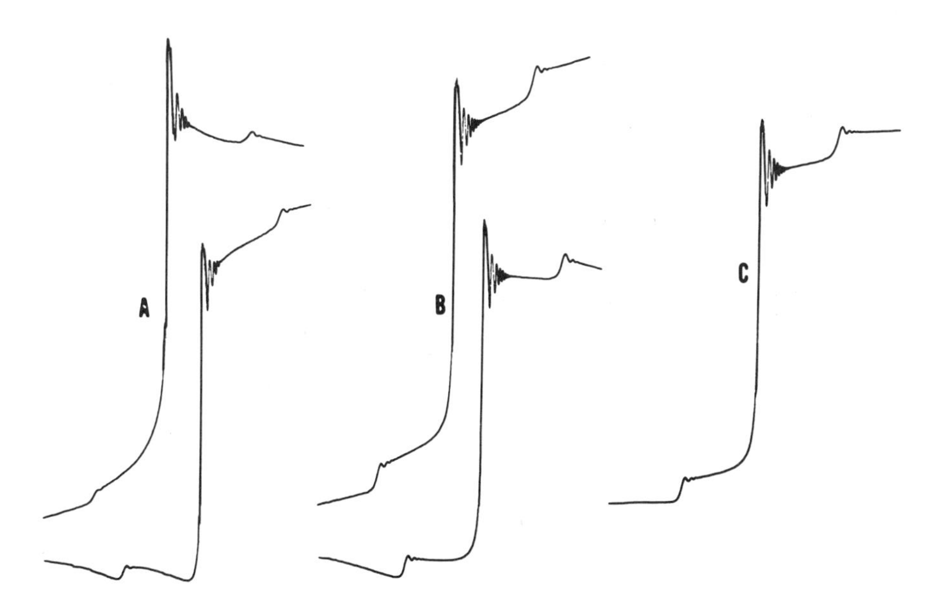

Fig. 1.13. (*a*) Improper settings of the detector phase; (*b*) improper settings of the amplifier balance; (*c*) proper integration showing steps resulting from a sharp singlet and the singlet and the associated spinning sidebands.

is frequency dependent and proper phase adjustment cannot be maintained during the sweep.

When integration is performed, it is best to set the integrator amplitude (gain) at a relatively high value and adjust the spectrum amplitude such that the integral step(s) over the peaks to be integrated extends nearly full scale. This is done for the same reason that a time-averaging device is used to enhance the spectrum. The integrator will tend to average the high-frequency noise. However, if transient dc noise is present at the detector, the integral will be seriously affected; therefore, the output of the spectrum amplitude gain is kept at a minimum.

A parameter that may seriously affect the accuracy (and precision) of integration is the ratio of radio frequency (rf) power (B_1) to sweep rate. There are many opinions concerning the adjustments of these parameters. The major concern when selecting the value of B_1 and sweep rate is the trade-off between good signal-to-noise ratio obtained at the higher B_1 settings and slower sweep rates, and the saturation that will likely occur under these conditions. Several theoretical treatments permit calculations of the values needed to prevent saturation (75,201,248). However, most of these treatments require a knowledge of the values of T_1 and T_2 (or at least their ratio) and a reasonably accurate knowledge of the value of B_1. These parameters can be measured, but doing such would certainly overly complicate the determinations to be performed. The vast literature reporting NMR integrations certainly documents that these tedious procedures are not necessary to obtain good results. However, some empirical scheme is advised to assure accurate results (179). Consideration of only saturation of the signal suggests the best selection of B_1 and sweep time is the lowest possible for each. However, poor signals will invariably result. There are at least three practical schemes to select the settings. First, the sweep rate selected is normally one of the fastest available on commercial spectrometers (~10 Hz/sec). The fast sweep reduces the product of B_1 and sweep time which aids to reduce saturation and decreases the time the integrator must hold its value. The ringing that occurs at the plateau of the integral at these sweep rates does not effect the mean value of the integral. However, one report indicates that integrals with strong ringing tend to be evaluated higher than their true value (214). The value of B_1 may be increased in increments from a low value with an integration taken at each setting. If a single peak is integrated, the value of the integration step height will increase with increasing B_1, reach a maximum, then begin to slowly decrease as saturation begins to occur (179). The value of B_1 just less than that used for the maximum integration is best, however, the value at the maximum may give more precise results as it is less dependent upon B_1 fluctuations. If several resonance peaks in a spectrum are to be integrated, the value of B_1 at which saturation begins may be different for nuclei in different groups of the molecule. In this case, the lowest value of B_1 that will saturate the resonance that has the longest T_1T_2 product can be detected by a change in the ratio of the integrated values. For the most cautious spectroscopist, especially when long T_1 values (such as aromatic protons in degassed samples) are encountered, the area ratios of integrated peaks are most accurately determined using several B_1 settings and extrapolation of the area to zero B_1. In general, a sweep rate of approximately 10 Hz/sec and a B_1 of 0.1–0.2 milligauss is satisfactory for most purposes.

3. Sample Preparation and Procedure for Peak Area Measurements

For purposes of high-resolution spectra required for integration of closely spaced lines, careful selection of the NMR sample tube is required. To obtain the necessary resolution expected of modern instruments, the sample is spun at a rate of about 30 Hz to average the inhomogeneties in the field. In cases where peak separation is not a problem, the integration (not the spectrum) may be as accurately obtained without spinning the sample. There is little evidence that sample spinning is of any direct advantage for integration and a report has appeared that it has no effect. If the sample is spun, spinning side bands resulting from modulations generated by the spinning sample in an inhomogeneous field may be observed in the spectrum unless the spectrometer is very well tuned. These "spinning side bands" should be included in the integral of the peak with which they are associated. Every effort should be made to minimize spinning side-band intensities and therefore the correction made on the area. These side bands will be symmetric about the main peak and removed from it by a frequency shift equal to multiples of the spinning rate of the sample. Their area is derived by modulation of the main peak. Their position may be varied by changing the spinning rate.

G. LINE WIDTHS

Transition probability theory predicts an infinitely sharp absorption line when the frequency of the rf field B_1 exactly equals the frequency corresponding to the energy difference between the two energy levels for the transition in question. In actual practice, however, the lines are broadened by various factors and it appears as if absorption is occurring over a range of frequencies. Therefore, it is convenient to introduce a line shape function $g(\nu)$ which is proportional to the absorption at frequency ν. Experimentally, the absorption usually has a Lorentzian shape and $g(\nu)$ obeys the equation

$$g(\nu) = \frac{a}{b^2 + (\nu - \nu_0)^2} \tag{36}$$

where a and b are constants, ν_0 is the frequency of absorption, and ν is the frequency of the rotating rf field.

Several factors including inhomogeneties in the magnetic field, relaxation, dipole–dipole interaction, and electric quadrupole effects can contribute to the broadening of an NMR absorption peak. Normally, the applied magnetic field B_0 is not constant throughout the entire sample volume such that molecules in different parts of the sample are experiencing slightly different magnetic fields.

$$B = B_0 + \Delta B_0 \tag{37}$$

This results (equation 3) in the absorption occurring over a range of frequencies, $\Delta\nu$, rather than at a discrete frequency, ν_0, so that the observed peak is actually a superposition of absorptions by molecules in different parts of the sample. With the magnets used in commercial NMR spectrometers, the inhomogeneties in the magnetic field limits the observed line width to $\sim$0.1 Hz.

Additional broadening due to relaxation is a result of the finite lifetime of a nucleus in a given energy level. As pointed out previously, thermal motions can induce transitions between energy levels. In general, the lifetime in a particular energy level will be on the order of the spin–lattice relaxation time T_1. The broadening due to T_1 can be estimated using the Heisenberg Uncertainty Principle

$$\Delta E\, \Delta t \approx \hbar \tag{38}$$

Defining the line width at half-height as $\Delta\nu_{1/2}$, substitution of equation 38 into equation 3 and rearrangement gives the relationship

$$\Delta\nu_{1/2} \approx \frac{1}{T_1} \tag{39}$$

Thus with $T_1 = 1$ sec, a line width of about 1 Hz would be expected.

In viscous liquids or solids, interactions between magnetic moments of adjacent nuclei can lead to a greater broadening than predicted by spin–lattice relaxation. Consider two nuclei, 1 and 2, located as a distance r from each other. The field experienced by nucleus 1 due to the presence of nucleus 2 is proportional to μ_2/r^3. Therefore, the field at a given nucleus will depend upon the magnitude of this effect from neighboring nuclei and may either add to or subtract from the field experienced by the nucleus, depending on the orientation of the adjacent magnetic moments with respect to the applied magnetic field B_0. Since all nuclei in a sample will not experience the same dipole interaction, absorption will occur over a range of frequencies leading to a broadened line. This process is similar to the broadening caused by inhomogeneties in B_0, with the exception that the inhomogeneties arise from within the sample itself. This broadening is referred to as that due to spin–spin relaxation. Using the Uncertainty Principle, one can derive a relationship similar to equation 39 between T_2 and $\Delta\nu_{1/2}$ as

$$\Delta\nu_{1/2} \cong \frac{1}{T_2} \tag{40}$$

For a Lorentzian line shape, equation 40 becomes

$$\Delta\nu_{1/2} = \frac{1}{\pi T_2} \tag{41}$$

Normally, with mobile liquid and gaseous samples, the interaction between adjacent magnetic moments averages out due to the rapid tumbling of the molecules, such that spin–spin relaxation is not an important source of broadening. Under these conditions, T_2 and T_1 become approximately equal and the line width can be used to approximate T_1 using equation 41. In any event, the line width provides an estimate on the lower limit of T_1 since $T_2 \leqslant T_1$.

Another source of line broadening exists for nuclei with spin $I > 1/2$. As pointed out earlier, nuclei with $I > 1/2$ have electric quadrupole moments that interact with field gradients resulting from nonsymmetrical electron distribution. Molecular tumbling will lead to fluctuating electric field gradients around the nucleus. These gradients can cause transitions to occur among the nuclear quadrupole energy levels. The effect on the nuclear energy levels is the same as if relaxation were occurring by a magnetic interaction.

III. EXPERIMENTAL AND INSTRUMENTAL ASPECTS OF NMR

Since the first NMR spectra on liquid samples were obtained in 1945 (29,192), the complexity of NMR spectrometers has increased considerably. Fortunately, the difficulty in obtaining NMR spectra has decreased. In this section, we discuss NMR spectrometers in relation to the instrumentation necessary for obtaining a spectrum. In addition, the various factors that influence the appearance of a spectrum and special techniques for obtaining spectra on small samples are also discussed. Finally, the various types of commercially available spectrometers are reviewed briefly.

A. THE SPECTROMETER

Present day commercial NMR spectrometers may be divided into two basic types: (a) the routine spectrometer and (b) the more sophisticated research type spectrometer. Both types of instruments contain complex electronic components and differ primarily in the design and number of components and in the types of experiments they are capable of performing. Regardless of the type of instrument under consideration, or whether it is an older instrument with basically tube-type electronics or a more modern instrument with solid-state electronics, there are several components that are common to all types of high-resolution spectrometers. These include a magnet, a means of magnetic field stabilization, a sample probe, and a radiofrequency oscillator and recorder for display of the spectrum. These components are discussed in more detail below.

1. Magnet

The magnet of an NMR spectrometer is required to produce the condition necessary for the absorption of radiofrequency energy and is designed to produce as homogeneous a magnetic field as possible. One of three types of magnets are currently employed in NMR spectrometers: (a) a permanent magnet; (b) an electromagnet; or (c) a superconducting solenoid. Regardless of the type of magnet used, it should be capable of providing a highly uniform, stable magnetic field with a homogeneity of the order of 3 parts in 10^9. To achieve this homogeneity with permanent and electromagnets, parallel pole caps are used and their design is such as to take into account the relationship between the ratio of the diameter of the pole caps and the gap width, and its effect on the homogeneity.

In addition, electric shim coils (homogeneity coils) are incorporated on the pole faces to allow for additional corrections of the homogeneity (6,7). After the magnet is mechanically shimmed for the best possible homogeneity, the current in the homogeneity coils is varied to improve further the homogeneity in the region of the sample. Controls are usually provided for adjusting a vertical gradient (y axis), a horizontal gradient in the plane parallel to the pole caps (x axis), and a horizontal gradient across the magnet (z axis) (Fig. 1.14). In addition, it is desirable to have the contour of the magnetic field as flat as possible in the region of the sample. Curvature shim coils are provided to bend the magnetic field about the z axis to make corrections in the contour of the field. Additional shim coils to correct the x gradients along the y axis (x-y), the z gradients along the y axis (y-z), and second- and fourth-order corrections are provided on most instruments. The homogeneity is very sensitive to changes in the y gradients and many instruments provide a means for automatically adjusting the y gradient for maximum homogeneity.

The majority of permanent magnets are designed to produce a magnetic field of ~14,092 gauss (60 MHz for protons), although one has appeared recently with a field of 22,100 gauss (90 MHz). Permanent magnets have the advantage that a magnet power supply and a means for cooling the magnet are not required. Furthermore, they usually have excellent long-term field stability. A disadvantage of permanent magnets is, of course, that the magnetic field is fixed and cannot be varied over a wide range. Furthermore, permanent magnets are subject to temperature-dependent variations, so they are usually thermostated to operate at constant temperature. Permanent magnets are also subject to field variations induced by the placement of magnetic materials in the vicinty of the magnet. Nevertheless, the stability and homogeneity of the magnet field from a permanent are entirely satisfactory for use in NMR spectrometers.

The upper limit of the field strength of conventional electromagnets is ~25,000 gauss as a result of homogeneity and stability considerations. A very stable power supply is required to produce the high currents needed. In addition, a means for cooling the magnet and maintaining a constant magnet temperature (±0.1°C) is required. An obvious advantage of an electromagnet over a permanent magnet is the ability to vary the magnetic field strength by varying the current passing through the coils. The stability of a conventional electromagnet is improved by using some type of stabilizing device (190). This usually consists of an additional set of coils on the pole pieces which detect changes in flux across the magnet gap. These coils are connected to some type of galvanometer which detects the current induced in the coils by the changes in flux and applies a correction voltage to another set of coils to

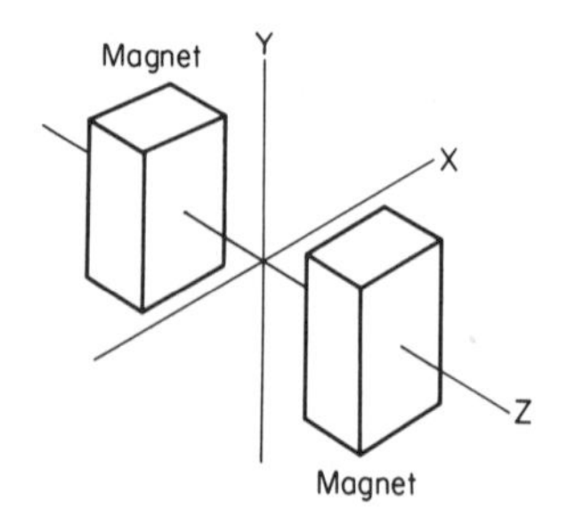

Fig. 1.14. The relative orientations of the x, y, and z axis with respect to the laboratory magnet.

compensate for the change in the magnetic field strength. As an additional precaution to maintain stability and homogeneity of an electromagnet, the room temperature should be thermostated to ±1°C.

The magnetic field strength used in NMR spectrometers has increased considerably in recent years through the use of superconducting solenoids (168). Fields up to ~86,000 gauss (360 MHz for ^{1}H) have been achieved recently. The major problem with superconducting solenoids is obtaining the field homogeneity necessary for NMR applications. Homogeneity coils are used extensively. In addition, the solenoids presently in use operate at liquid helium temperatures. Closed-loop cryostats are used to reduce the helium loss due to evaporation. In spite of the difficulties involved, the use of superconducting solenoids will most likely increase in the future. The advantage in terms of sensitivity and chemical-shift separation far outweigh the difficulties involved in operating a superconducting solenoid.

2. Transmitter

The radio-frequency transmitter supplies the necessary rf power for NMR absorption at a particular magnetic field strength. The range of frequencies covered is ~4–360 MHz for the NMR absorption of all magnetic nuclei with modern spectrometers. The stability requirements for the transmitter are essentially identical to those of the magnetic field. The frequency is usually derived by the multiplication of the frequency of a thermostated, quartz crystal. A means is provided for controlling the B_1 power. The electronic circuits are so highly tuned that separate transmitters are used for different nuclei in many spectrometers. In others, different frequencies are derived by the multiplication of the frequency of a master crystal oscillator or by substitution of different crystals.

There are a growing number of spectrometers utilizing a frequency synthesizer as the frequency source. The frequency of most synthesizers can be controlled to 0.1 Hz or better. These systems have the advantage that several frequencies may be derived from the multiplication, addition, etc., of one frequency source. In this manner, all frequencies are "locked" together such that drifts in the frequencies are controlled by variations in the master frequency source. Provision is made for amplification of the frequency and for controlling the level of the rf power delivered to the probe.

3. Probe

The probe or sample holder is mounted between the pole caps of the transmitter such that its position may be adjusted to place the sample in the region of optimum field homogeneity. In addition, the probe contains an air turbine for spinning the sample, sweep coils, a coil(s) for transmitting the rf signals and detecting the resonance absorption, and a means for controlling the temperature in the region of the sample. The probe body is maintained at constant temperature either by circulating water at constant temperature through coils in the probe, by placing the probe in the thermostated housing surrounding the magnet, from air circulating in the room, or by using a combination of these methods.

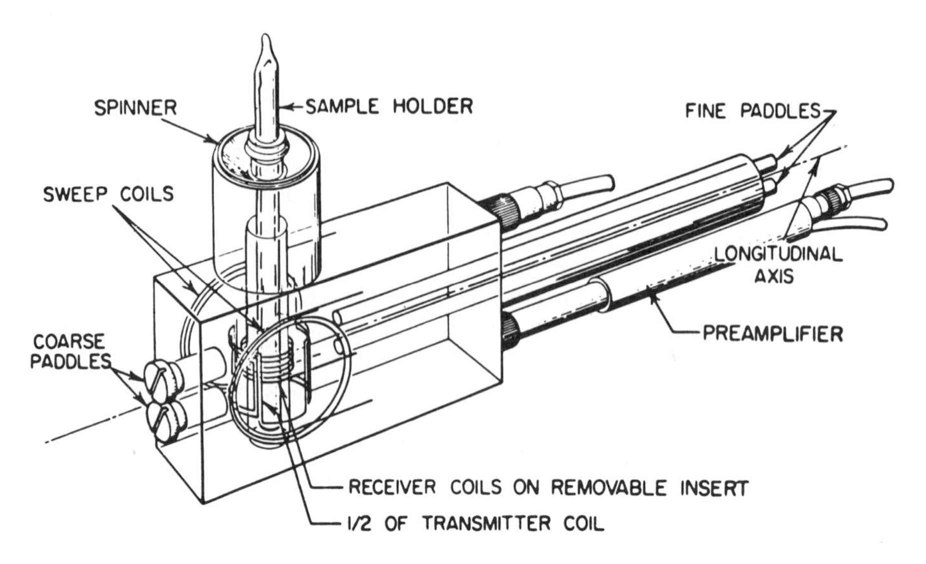

Fig. 1.15. A schematic showing the basic components of an NMR probe. (Courtesy of Varian Associates.)

Molecules in different parts of the sample will experience slightly different magnetic fields due to residual field gradients in the region occupied by the sample. As a result, the absorption will occur over a range of frequencies such that extremely broad lines are observed. The effective homogeneity of the magnetic field is improved by spinning the sample about the y axis. This has the effect of averaging the magnetic field in the x-z plane and reducing the line width due to inhomogeneties in the magnetic field. Spinning rates of 20–40 rev sec^{-1} are normally used. Line widths at half-peak height (resolution) of 0.5 Hz or less are readily obtained, provided the homogeneity coils are properly adjusted. Higher spinning rates can lead to an increase in the noise level due to vibrations as the sample tube rotates. It is essential that uniform sample tubes be used to prevent this unwanted noise. Spinning of the sample also produces unwanted "side bands" (reproductions of the spectral peaks) symmetrically disposed about the spectral peaks at frequencies separated from the spectral peaks by multiples of the spinning frequency. These peaks are easily recognized by their frequency dependence on the spinning rate. Normally they can be effectively removed by proper adjustment of the homogeneity coils and careful selection of samples tubes.

Two basic designs in the detection system of probes are used: (a) the crossed-coil probe or nuclear induction probe (168) and (b) the single-coil probe (192). In the crossed-coil probe, separate transmitter and receiver coils are used. The transmitter coil for producing the B_1 rotating field is wound in two sections such that it surrounds the receiver coil and has its magnetic axis parallel to the x axis. The receiver coil is wound on a glass insert at a right angle to the transmitter coil so that its magnetic axis is also parallel to the x axis. Leakage between the two coils is reduced by placing a Faraday shield between the two coils, as it is impossible in practice to obtain perfect 90° orientation between the two coils. Further reduction in the leakage is achieved by incorporating two paddles (inductors) which are coupled to the transmitter and receiver coils (30). In actual practice, some leakage between the two coils is desirable to serve as a reference for the voltage induced by absorption.

The introduction of some leakage between the transmitter and receiver coils also serves to suppress the dispersion mode signal. When B_0 is located away from its resonance condition value and small amplitudes of B_1 are applied, the magnetization vector M_0 will be along the z axis (Fig. 1.16). As B_0 is brought near the resonance condition, the magnetization vector starts precessing and will move away from the z axis and have components in the x, y plane. At resonance, M_y obtains a maximum value and decreases to zero as B_0 is increased further. A voltage is induced in the receiver coil as a result of the magnetization in the x-y plane. The component M_x is the dispersion signal and the component M_y the absorption signal. If leakage in phase with the x axis is introduced into the receiver coil from the transmitter coil, the dispersion signal is suppressed. In most instruments, however, phase-sensitive detection is employed and the leakage is adjusted to a minimum.

In the single-coil probe, a bridge network is used to detect the NMR absorption (31). A single coil wound on a glass insert, similar to the receiver coil in a crossed-coil probe, is used. The transmitter and a reference signal is balanced by the bridge network and the absorption or dispersion signal is detected as an out of balance emf across the bridge. Most spectrometers using this type of probe utilize a twin-T bridge (192). The emf across the bridge during resonance is a mixture of both the absorption and dispersion mode signals. The dispersion signal can be suppressed by the introduction of a slight imbalance in the bridge tuning network. Adjustments are provided for balancing the phase and amplitude of the bridge network such that only the absorption mode is detected.

The electronic circuits employed in an NMR probe are usually tuned such that they will transmit and receive only frequencies over a very narrow range. Normally, separate probes tuned for one particular frequency are used for the individual nuclei.

The majority of probes have some means for controlling the temperature in the region occupied by the sample such that variable temperature experiments are possible over the range −100°C to 200°C. The range is limited by the expansion and

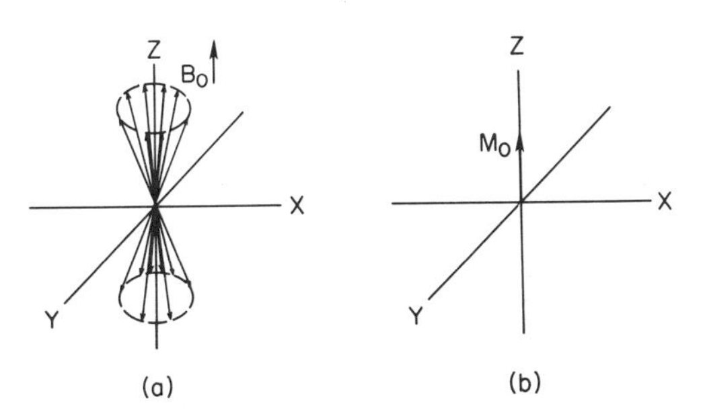

Fig. 1.16. The orientations of the magnetic moments of a collection of identical nuclei when placed into a magnetic field (a) and the resultant magnetization M_0 along the z axis.

contraction characteristics of the glass insert. The insert is usually contained in a vacuum-jacketed dewar to prevent heat loss to the magnet that would result in a loss of resolution and field drifts. Generally, a stream of nitrogen gas at a controlled temperature is passed over the sample. The probe contains a sensing device to control the current to a heating coil located below the sample area. For high-temperature operation, the stream of nitrogen gas is heated to the desired temperature before it passes over the sample. Several different systems have been used for low-temperature operation. In one method, the nitrogen gas is precooled by passing it through a heat exchanger filled with liquid ntirogen and then heating the gas stream in the probe to the desired temperature. A method similar to this uses the boil-off from a liquid nitrogen reservoir as the source of the cold nitrogen gas and the gas is heated in the probe to reach the desired temperature. In another method, the gas stream is split with one part being cooled by passing it through a heat exchanger and then remixing it with the uncooled gas stream. The temperature is controlled by varying the fraction of gas passing through the heat exchanger. In yet another method, gas at high pressure is precooled and then allowed to expand through a Joule-Thompson device. Control of the temperature is achieved by using a sensor in connection with a heating coil. Regardless of the method used, the control of the temperature should be $\pm 1°C$ or better. One problem with accurately maintaining the temperature at the sample is temperature gradients within the sample.

4. Radiofrequency Detection

The signal induced in the receiver coil of the probe due to NMR absorption is extremely weak ($\sim$1 mV), and must be amplified before it can be displayed. The amplification is limited by the noise level resulting from any imbalance in the tuned electric circuits and sample spinning. The signal from the receiver coil is usually fed into a preamplifier mounted directly on the probe to eliminate as much electrical noise as possible. The signal from the preamplifier is further amplified in the receiver and mixed with the output of a local oscillator frequency to produce a beat frequency at an IF frequency, which is further amplified and fed into a phase detector (121). The local oscillator frequency is also mixed with the transmitter frequency to provide a reference signal, which is amplified and fed into the phase detector. The absorption or dispersion signal is selected in the phase detector. The signal is rectified and filtered, and the dc signal is fed into either an oscilloscope or recorder for display.

5. Integrator

Most instruments today employ some type of modulation (109) along with phase-sensitive detection to eliminate baseline drift as a result of improper balance of the circuits. Either field or frequency modulation may be used with identical results. If the output from an audio oscillator is applied to the sweep coils of the probe, the magnetic field at the sample is $B_0 + B_m \cos \omega_m t$, where B_m is the amplitude of the field modulation and ω_m is the angular frequency of the field modulation. If the magnetic field is swept through resonance, signals appear not only at resonance due to B_0 (center band) but also on either side of the main peak at frequencies separated from the main peak by $\pm n\omega_m$, where n is an integer. Usually only the first side band is important. Similar results may be obtained by modulating the transmitter frequency. The frequencies of all the resonances from the receiver coil are mixed with the transmitter frequency. After phase detection, the signals contain both dc and audio components. Any imbalance in the probe will affect only the dc signals. Therefore, one can detect the audio components of the signal, convert to dc, and display the signals. This minimizes the baseline drifts in the spectrum.

The integrator not only contains the components necessary for modulation and phase-sensitive detection, but also for the electronic integration of the area under an absorption signal. As pointed out previously, the area under an absorption signal is directly proportional to the number of nuclei giving rise to the signal and this provides the basis for quantitative applications of NMR. Most spectrometers today are equipped with electronic integration circuits that permit the integration without interference with normal operation of the spectrometer.

6. Sweep Units

As pointed out previously, the NMR absorption condition can be obtained by keeping the magnetic field constant while varying the frequency (frequency sweep) or by holding the frequency constant and varying the magnetic field (field sweep). Most spectrometers provide two types of magnetic field sweep, a slow sweep and a fast, recurrent sweep. Both are linear sweeps. That is to say, the rate of change of the field is constant with time. For the fast sweep, the output from a sawtooth generator is amplified and applied to two small Helmholtz coils usually located on the sides of the probe such that their magnetic axis is in the same direction as the main magnetic field. The sweep time can be varied from seconds to minutes. The absorption signals are usually displayed on an oscilloscope. The return of the magnetic field to its original value for the start of another sweep is so rapid that absorption is not detected on the return. This mode of operation is useful when searching the magnetic field for either the resonance condition or a particular peak. It is also used to monitor the shape of a peak when adjusting the homogeneity coils for the maximum resolution. When the sweep rate is fast with respect to the relaxation times of the nucleus under consideration, the absorption peak will be distorted and ringing will be observed on the trailing edge of the absorption peak. The homogeneity coils are adjusted to produce the maximum ringing.

The slow sweep unit is used to sweep the magnetic field when a recorder presentation of the signals is being used. A dc voltage is applied to the flux stabilizer. The flux stabilizer senses this voltage as an error signal (a field change in the flux of the magnetic field) and supplies a correction voltage to the coils located on the pole faces resulting in a sweep of the magnetic field. The sweep rate may be varied by varying the applied dc voltage. Slow sweep rates may be obtained such that little distortion of the absorption signal occurs and the maximum resolution may be realized. In spectrometers employing some type of field/frequency stabilization, the travel of the recorder arm in the x direction can be synchronized with the slow sweep unit so that the spectra may be recorded on precalibrated charts.

7. Field/Frequency Stabilization

A high degree of stability (1 part in 10^6 or better) of an NMR spectrometer system is achieved by proper control of the magnetic field and the irradiating frequency, as mentioned previously. While this is adequate for most purposes requiring short-term stability, drifts in either the magnetic field or frequency may occur over a longer time period such that broadened signals are observed. Since it is really the ratio of the magnetic field to the frequency that is important for long-term stability, two methods using modulation techniques have been devised so that drifts in the magnetic field are automatically compensated by corresponding changes in the frequency and vice versa. Depending on the spectrometer, the modulation frequency varies from 2 to 5 KHz and, normally, the first sideband is utilized.

One method of field/frequency stabilization referred to as an external lock system, utilizes two separate samples in the probe (8,20). A small, stationary sample tube of water (the control sample) is placed as close as possible to the region occupied by the sample under consideration (analytical sample) so that both samples experience as close to the same magnetic field as possible. A separate set of homogeneity coils are usually provided for the control sample. The center-band frequency is determined by a stable, constant frequency crystal oscillator, while the modulation frequency is varied by a feedback loop such that resonance is maintained at the upper side-band frequency of the sample. The resonance of the control sample is detected in the dispersion mode and the direction of small drifts in the magnetic field is sensed as either a positive or negative error voltage. The error voltage is applied to the feedback loop as a correction so that the control sample is always maintained at resonance. The analytical sample is then field-swept by applying a dc voltage in the normal manner. The stability drift is maintained below 1 part in 10^8/hr with this method. Since the analytical sample and the control sample are in different regions of the magnetic field, a control is provided for correcting the magnetic field in order that precalibrated charts may be used to record the spectra. An external locked system has the advantage that the field/frequency ratio is maintained at all times. A similar arrangement may be used in a frequency sweep mode.

A second method of field/frequency stabilization referred to an internal locked system is also based on sideband modulation and offers certain advantages over an external locked system (84,85,189). The control sample is derived from a sharp peak in the sample and is usually the absorption from the reference material tetramethysi-lane (TMS). Thus, both the analytical and control samples are subjected to the same magnetic field. Separate side-band frequencies and phase-sensitive detectors are used for the control and analytical samples. The control sample is detected as the dispersion mode signal, and error voltages are used in a feedback loop to maintain the magnetic field at resonance. In the frequency-sweep mode of operation, the control sample is maintained at resonance with a fixed-frequency modulation. The analytical sample is swept in a linear manner by varying the modulation frequency of the analytical oscillator. The frequency of the analytical channel is mechanically linked to the recorder arm so that the spectra can be recorded on precalibrated chart paper. In the field-sweep mode of operation, the fixed frequency modulation oscillator is used for the analytical channel while the variable frequency modulation oscillator is used to sweep the control sample linearly. As the control sample modulation is swept, error signals are fed through the feedback loop to sweep the magnetic field through the resonances of the analytical sample. The stability drift of an internal lock is on the order of 1 part in 10^9/hr. With an internal lock system, the lock must be broken each time a sample is changed. This usually presents no problems and the gain in stability and the added advantage of frequency sweep operation for decoupling experiments more than compensates for this inconvenience.

Many commercial spectrometers today offer both external and internal lock capabilities. Furthermore, it is not necessary for the analytical and control channels to detect the same nucleus. Many nuclei have low sensitivity and broad signals, which prevents their use in the control sample. By using a probe that is tuned to two frequencies and two radiofrequency oscillators, it is possible for the analytical channel to detect one nucleus (e.g., ^{13}C) and the control channel to detect another nucleus (e.g., ^{1}H, ^{2}D, or ^{19}F).

B. COMMERCIAL NMR SPECTROMETERS

The developments in NMR instrumentation have focused primarily on three areas: (a) improving the sensitivity and, hence, reducing the sample size required for a spectrum; (b) reducing the difficulties in the day-to-day operation of spectrometers; and (c) reducing the cost of spectrometers. The first commercial NMR spectrometer was introduced in 1953 and operated at 30 MHz for ^{1}H (239). Since that time, improvements in magnet technology have led to instruments operating at 40 MHz in 1955, 60 MHz in 1958, 100 MHz in 1962, 220 MHz in 1966, and 300 MHz in 1971 (239). The first routine type 60-MHz NMR spectrometer was introduced in 1961 (239). Today, it is no more difficult to obtain an NMR spectrum than it is to obtain an infrared or UV spectrum. Bible (27) has reviewed the NMR spectrometers introduced during the period 1963–1969. In this section, we shall discuss the developments in NMR instrumentation since 1970.

1. Routine Proton Spectrometers

Three routine proton spectrometers have been made available since 1970. The EM-300 is a 30-MHz spectrometer using a permanent magnet (239). The low cost (about one-eighth of the cost of the first 30-MHz instrument) and ease of operation of the EM-300 makes it particularly attractive as a teaching instrument. It should also find use in industrial laboratories as an instrument for screening samples and, thus, freeing the research-type instruments for more complex problems. Some disadvantages of the EM-300 are the lack of decoupling and variable temperature accessories and the low frequency.

The EM-360 (239) and R-24A (181) are 60-MHz ^{1}H spectrometers. Both instruments have a sensitivity of ~25:1 (S/N for the methylene resonance of 1% v/v ethylbenzene) and a resolution of 0.6 Hz. Homonuclear decoupling is available but not variable temperature. The simplicity, economy, and ease of operation of these spectrometers make them attractive for a wide variety of routine and, perhaps, research applications.

2. Continuous-Wave Research Spectrometers

The R32 spectrometer introduced in 1972 (181) is a 90-MHz proton spectrometer using a permanent magnet. A resolution of 0.5 Hz and sensitivity of 50:1 are quoted for this instrument. Accessories include variable temperature, field/frequency stabilization, wide sweep ranges, homonuclear decoupling, and other nuclei. Another 90-MHz proton spectrometer using a permanent magnet, the EM-390 (239), has been introduced recently. A resolution of 0.5 Hz and a sensitivity of 50:1 are quoted for the EM-390 (239). Standard accessories include an internal lock with automatic adjustment of the *y*-gradient homogeneity and wide sweep ranges. Optional accessories include spin decoupling, variable temperature, oscilloscope display, and signal averager.

A number of companies other than the original manufacturers now offer Fourier transform accessories for updating the above and older continuous wave spectrometers (68,169). This accessory is almost a prerequisite for running natural abundance carbon-13 spectra and greatly reduces the sample size required for proton spectra. In addition, a 1-mm probe insert is now available for certain instruments (116,239) which further reduces the sample size required. These inserts use melting point capillaries as the sample tube and require $\sim 5\ \mu l$ of sample volume. With this accessory, proton spectra from microgram quantities of sample may be obtained within a matter of minutes using Fourier transform NMR.

3. Fourier Transform Spectrometers

There can be no doubt that Fourier transform (FT)-NMR spectrometers are the instruments of choice for today's NMR work as evidenced by the wide variety of instruments available operating at a proton frequency of 60, 80, 90, and 100 MHz (37,116,181,239). Probe inserts are usually available for 2-, 5-, and 10-mm sample tubes. A multinuclear accessory is available for these instruments covering the range of nuclei from nitrogen to phosphorus.

4. Superconducting Magnet Spectrometers

Advances in superconducting magnet technology have continued to increase the magnetic field strength available on commercial instruments. NMR spectrometers based on superconducting solenoid magnets are available operating at a proton frequency of 150, 200, 250, 270, 300, 360, 400, and 500 MHz (37,116,169,239). Higher fields will most likely be available in the near future. A full line of accessories are available with these instruments, including a range of probe insert sizes and multinuclear capabilities.

5. Spectrometers for Processing Monitoring

Two of the more recent spectrometers for process monitoring are the mini spec PC20 (112a) and the PR-103 (188). Both instruments are pulse NMR spectrometers and are capable of distinguishing signals from protons in different chemical or physical states because of the different spin relaxation times. Either T_1 or T_2 of the sample may be determined. Once the conditions for sample measurements have been determined, several samples per hour may be run, as no preconditioning of the sample is required in most cases. In addition, these instruments are suitable for online process monitoring.

C. EXPERIMENTAL FACTORS

1. Factors Influencing the Resolution and Line Shape

Assuming that the NMR sample has been prepared to the best of one's ability, there are a number of instrumental factors, under operator control, which have a profound effect on the quality of the spectrum obtained. For the majority of samples, the natural line width of the absorption signal is less than the line width due to inhomogeneity in the magnetic field. Thus, the resolution in most cases depends upon the ability of the operator to adjust the current in the homogeneity coils. In addition, the line shape depends upon the rf power level, the phase of the detector, the spinning rate, the sweep rate, and the electronic filtering used before recording the spectrum. The operator should be aware of these factors and make the proper adjustments to obtain the best spectrum possible.

a. HOMOGENEITY OF THE MAGNETIC FIELD

When the spectrometer system is set up in the laboratory, the magnet is adjusted (shimmed) mechanically to obtain the best possible homogeneity of the magnetic field. The remaining gradients are corrected as far as possible by adjusting the current in the homogeneity coils located on the pole faces of the magnet. Before adjusting the homogeneity coils, it is important that the sample be located in the "center" of the magnetic field. Usually, the position of the probe along the z axis is fixed by the probe holder, and controls on the probe holder permit one to adjust the position of the probe along the x and y axes. After obtaining a peak on the oscilloscope using the fast sweep, the position of the probe along the x and y axes is adjusted so that the peak display on the oscilloscope does not vary (assuming no drift in the magnetic field) when the current in the x and y homogeneity coils is varied between the two limits. This adjustment should be made each time the probe is removed from the magnet. Some probe holders are equipped with stops such that, after initial set up, the probe can be returned to the "center" of the magnetic field without the need to readjust the position along the x and y axes each time the probe is changed.

To adjust the current in the homogeneity coils, it is desirable to display the signal from a sample giving a single, narrow, absorption peak. Usually, TMS or chloroform are used for ^{1}H spectra. With the signal displayed on an oscilloscope using the fast sweep, the current in each of the homogeneity coils is adjusted without the sample spinning. One criteria for obtaining the maximum homogeneity is to make the adjustment on each coil for the maximum peak height. If ringing of the peak is observed, the current in each coil can be adjusted for the maximum ringing. Afterwards, spinning of the sample is started and the adjustments of the y and curvature are repeated. Spinning of the sample about the y axis improves the effective homogeneity in the x and z directions. However, spinning does not effectively improve the homogeneity along the y axis or the curvature, both of which are very critical in determining the line width. At this stage, it is usually more convenient to observe the signal on a recorder. The current in the curvature coil is adjusted such that the line shape is symmetrical about the center of absorption. One way to do this is to adjust the curvature such that identical ringing is observed when the field is swept in both directions (Fig. 1.17). The current in the y coil is then adjusted to produce the maximum amount of ringing and/or the maximum peak height.

For normal, day-to-day operation, the homogeneity coils should be adjusted at the beginning of each day. Unless subjected to some external influence, such as a change in the room temperature or a change in the temperature of the magnet cooling water, it will only be necessary to adjust the y and curvature controls when different samples are run throughout the day. For extremely accurate work requiring the maximum resolution, however, it is a good idea to adjust all the homogeneity controls for each sample.

b. Radiofrequency Power Level

The rf power level should be adjusted to avoid saturation effects. Saturation results in a broadening of the absorption signal and will be most noticeable at the center of the absorption signal (Fig. 1.18). Saturation also depends upon the time spent on resonance (the sweep rate). One procedure for determining whether saturation is occurring is to adjust the rf power level to give the desired signal intensity at a particular sweep rate. If the signal intensity does not decrease when the sweep rate is decreased, then saturation is not occurring.

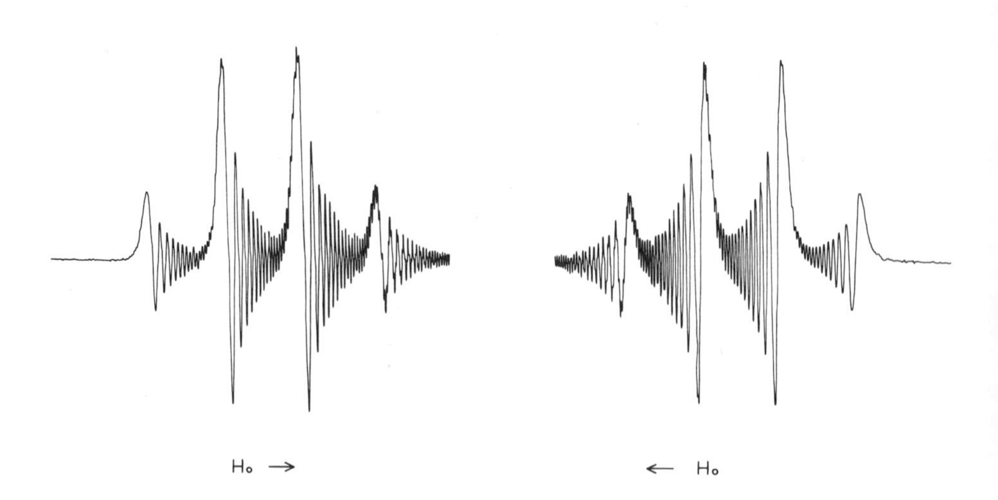

Fig. 1.17. An expansion of the aldehyde quartet in the 60-MHz ^1H spectrum of acetaldehyde.

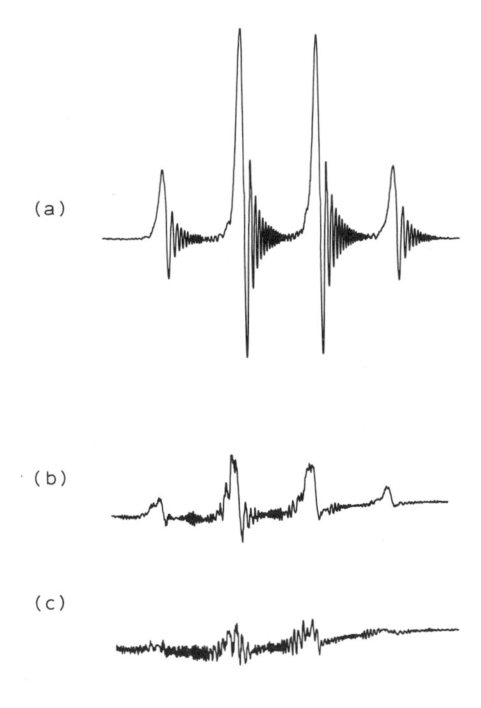

Fig. 1.18. The effect of increasing the rf power level B_1 on the aldehyde quartet of acetaldehyde (*a*) normal spectrum; (*b*) increased B_1; and (*c*) saturation due to a further increase in B_1.

c. Radiofrequency Phase

Incorrect adjustment of the phase-sensitive detector results in signals that are a mixture of the absorption and dispersion modes and will lead to distortion of the absorption signal (Fig. 1.19). The normal procedure for setting the phase is to record the strongest signal in the spectrum at maximum recorder deflection and adjust the phase control so that a flat baseline is obtained. This adjustment should be repeated for each sample, since the phase varies with the magnetic susceptibility of the sample.

d. Sample Spinning Rate

The sample is spun at the rate of 20–40 rev sec^{-1} about the y axis to average the inhomogeneity in the x and z directions. If lower spinning rates are used, the peak may be broadened due to incomplete averaging of the magnetic field. In addition, spinning side bands appear symmetrically disposed about the main peak as a result of modulation effects. The side bands can be effectively removed by careful selection of sample tubes, spinning at a faster rate, and proper adjustment of the homogeneity coils. Very high spinning rates should be avoided because of the problems of vortexing in the sample and the introduction of microphonic noise.

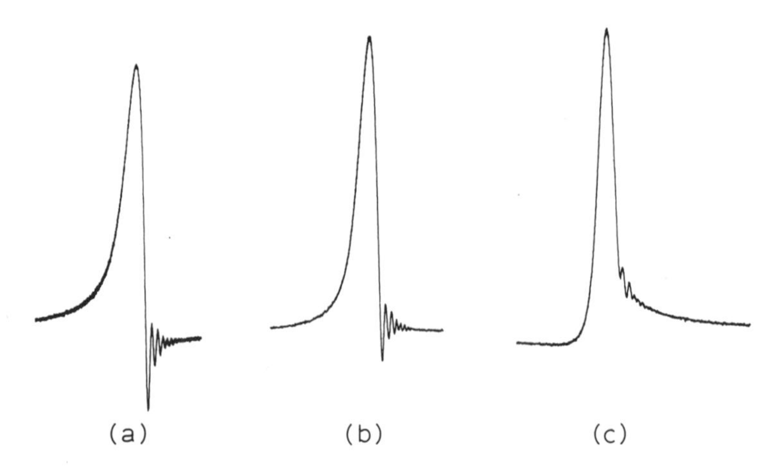

Fig. 1.19. The result of improper phase adjustment (*a* and *c*) on the appearance of an NMR signal and the corrected (*b*) signal.

e. SWEEP RATE

The sweep rate used to record the spectrum can influence the shape of the absorption signal (Fig. 1.20). As mentioned previously, if the field is swept through resonance at a rate faster than the reciprocal of the relaxation times (T_1^{-1} and T_2^{-1}) of the nucleus giving rise to the signal, distortions in the form of ringing or wiggles will appear on the trailing edge of the absorption. The ringing intensity decays exponentially with time. Although ringing is desirable for tuning the homogeneity coils, the distortion can lead to errors when accurate peak frequencies are being determined and one should use slower sweep rates to reduce their effects.

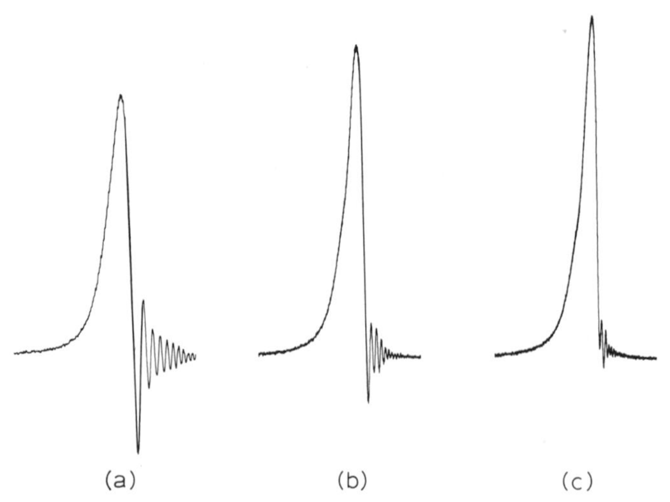

Fig. 1.20. The effect of sweep rate on the appearance of an NMR signal; (*a*) 0.5 Hz/sec; (*b*) 0.2 Hz/sec; and (*c*) 0.1 Hz/sec.

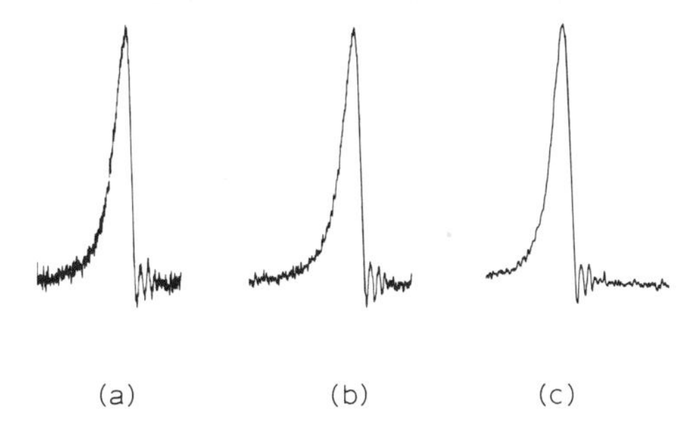

Fig. 1.21. The effect of increasing filtering on the appearance of an NMR signal: (*a*) smallest filter and (*c*) largest filter.

f. SIGNAL FILTERING

Before the signal is displayed after it has come from the phase-sensitive detector, it is filtered to suppress the background noise resulting from short-term fluctuations. Usually time constants on the order of 0.01 – 10 sec are available. The time constant used depends upon the sweep rate used. If longer time constants are used, then slower sweep rates must be employed to eliminate distortions in the signal (Fig. 1.21).

2. Factors Influencing Sensitivity

The major disadvantage with using NMR spectroscopy, compared with other spectroscopic techniques, is the inherent low sensitivity. The signal from the probe that is detected by the receiver is extremely weak and requires a great deal of amplification of the signal before it is finally presented at the recorder. The degree of amplification of the signal is limited by radiofrequency noise background which is amplified along with the signal. The electronic circuits employed in commercial NMR spectrometers are designed to minimize the background noise, and the components are designed and tuned to yield the maximum sensitivity.

a. INSTRUMENTAL FACTORS

For purposes of discussion of the sensitivity of NMR spectrometers, it is convenient to introduce the concept of signal-to-noise ratio (S/N). The standard sample recommended by most manufacturers for determining the S/N is a 1% v/v solution of ethylbenzene in carbon tetrachloride. The spectrum is obtained under optimum conditions and several recordings are made of the methylene quartet (Fig. 1.22). The S/N is then calculated (equation 42) from the average peak height of the tallest peak

$$\text{RMS noise} = \frac{\text{Average P-P noise}}{2.5} \tag{42}$$

$$\text{S/N} = \frac{\text{Average signal amplitude}}{\text{RMS noise}} \tag{43}$$

and the average peak-to-peak baseline noise level. For most 60-MHz spectrometers, a S/N of 5:1 to 15:1 is usually quoted, whereas for most 100-MHz spectrometers, a S/N of 25:1 or greater is common. In actual practice, the S/N specification is usually exceeded by a factor of 2–4, depending on the particular spectrometer.

An analysis of the S/N in terms of the factors that are important in the determination of sensitivity gives the S/N as (87)

$$\mathrm{S/N} = 5.78 \times 10^{-32}\, I(I+1)\,\gamma^{11/4} \frac{1}{T^{3/2}} K \frac{1}{(K\Delta f)^{1/2}} B_0^{7/4} \frac{N_T}{\ell^{1/2}\, D^{1/2}} \tag{44}$$

where K is a factor that depends upon the sweep rate, B_1 field, relaxation times and field homogeneity, F is the noise figure of the receiver, Δf is the filter band width in Hz, ℓ is the receiver coil length in cm, D is the receiver coil diameter in cm, and N_T is the total number of nuclei within the active region of the coil. The factor K includes many of the instrumental factors that are important in determining the line shape. The signal amplitude increases with an increase in B_1 to the point where saturation occurs and then decreases. The maximum B_1 level is dependent on the relaxation times T_1 and T_2 and on the sweep rate used. Better sensitivity is obtained with slower sweep times. However, a longer time is spent on resonance and, hence, the greater the

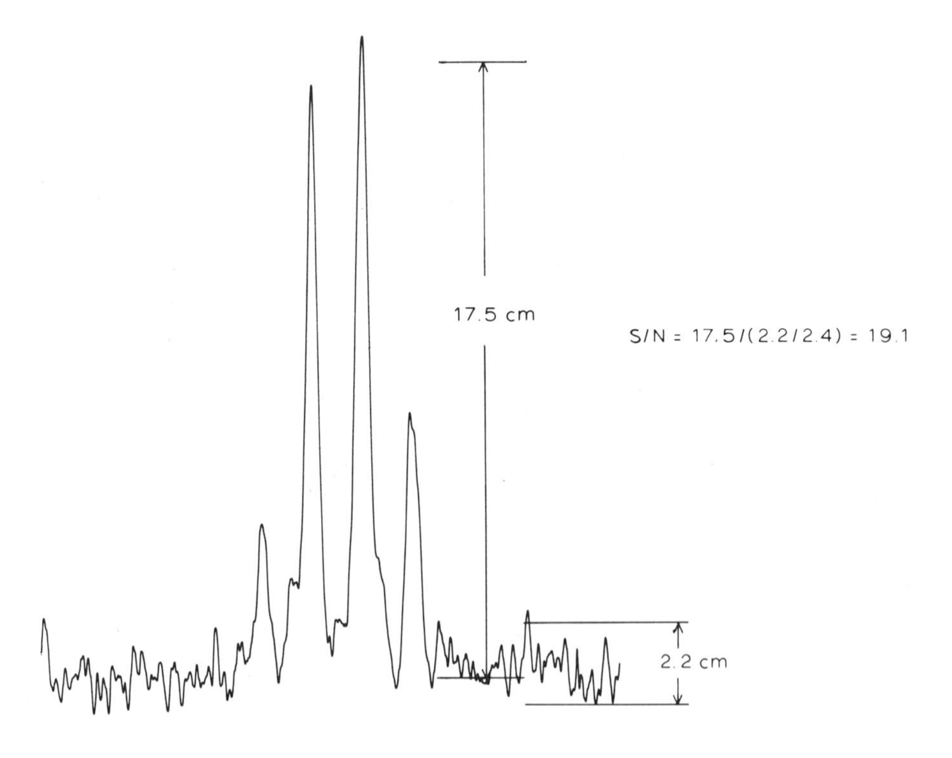

Fig. 1.22. A 60-MHz expansion of the methylene quartet of 1% ethylbenzene and the calculation of the signal-to-noise ratio.

chance for saturation to occur. For dilute samples, a compromise is usually reached between the B_1 level and sweep rate to achieve the best S/N while insuring that saturation is not occurring. Baseline noise may be suppressed by filtering the signal (Δf). The time constant (frequency response) of the filter selected is determined by the sweep rate, since distortions in the line shape of the signal can occur if the time constant is set too long. In practice, the time constant should be greater than the sweep rate in Hz.

The signal intensity is also determined in part by the excess of nuclei (Boltzmann distribution) in the lower energy spin state. Since the excess of nuclei in the lower energy spin state is proportional to the strength of the applied magnetic field B_0 (equation 10), an increase in sensitivity can be obtained by running the spectrum with as large a magnetic field as possible. In theory, the S/N increases with magnetic field B_0 by the factor 1.75. In actual practice, the dependence is somewhat less than this value. For example, a spectrum run at 23.5 kgauss (100 MHz) compared with the spectrum run at 14 kgauss (60 MHz) shows an increase in the S/N of about 1.6. Further increases in the S/N could be realized by running the spectrum on a spectrometer utilizing a high-field superconducting solenoid.

The dependence of the S/N on the temperature (equation 44) is a result of the temperature dependence of the Boltzmann distribution of excess nuclei in the lower energy state. Although an increase in the S/N could be obtained by lowering the temperature, this has not been put into practice due to the added inconvenience of cooling the sample, and to the high melting points of most solvents used in NMR spectroscopy.

As equation 44 shows, the S/N depends not only on the Boltzmann distribution of the nuclei, but also on the total number of nuclei N_T within the region of the coil. This is usually expressed in terms of a filling factor which is the ratio of the sample volume to the coil volume. For maximum S/N, the coil dimensions (equation 44) should be as small as possible. The usual practice is to use thin-walled glass sample tubes with an outside diameter of 5 mm, and have the receiver coil wound on a thin glass insert to maintain a high filling factor.

The most convenient way of increasing the S/N is to increase the concentration of the sample and, hence, increase the number of nuclei in the receiver coil area of the sample. Assuming that a signal can be identified with a S/N of 2:1, a concentration of 0.005–0.02 *M* is needed to detect protons which give a single, sharp peak. If the signal from a proton is split into a multiplet due to spin–spin coupling, higher concentrations are needed. For routine samples, concentrations of ~0.2 *M* will normally give a satisfactory S/N if the molecular weight of the sample does not exceed 300. The volume of sample required depends on the particular design of the probe but is usually somewhere in the neighborhood of 0.3–0.5 ml for a 5-mm o.d. sample tube.

b. Sensitivity Enhancement

In many cases, a sample is either not available in sufficient quantities or is not soluble in suitable NMR solvents to an extent that a satisfactory S/N can be obtained. Methods have been devised to enhance the sensitivity so that usable spectra can be obtained from smaller sample sizes. Some of the more common methods are discussed below. Excellent reviews of these methods are available (75,103,148).

1. Microcells

The S/N of an NMR sample depends upon the number of nuclei coupled to the receiver coil in the probe. An increase in sensitivity by a factor of 2–3 may be obtained by using a microcell which concentrates the sample within the receiver coil area. Two types of microcells are commercially available—spherical and cylindrical (148,249). The disadvantage of both types of microcells is the difficulty in positioning the microcell within the receiver coil. Several types of spherical microcells are available. They may consist of nylon plugs which fit inside a normal 5-mm sample tube and are adjusted to approximate a sphere, or they may be all-glass spherical cells which fit into 5-mm tubes. In addition, all-glass cells are available which contain a fixed, spherical cavity within a 5-mm sample tube.

2. Signal Averaging

One of the limiting aspects of continuous-wave high-resolution NMR is the inherent lack of good lower limits of detection. The minimum detectable quantity of sample that gives a sufficiently good S/N such that accurate quantitative measurements can be made will depend upon the molecular weight of the compound, the number of hydrogen atoms giving rise to a clean, easily integrated signal, the degree of spin–spin splitting, and, of great importance, the instrument used. No one value of detection limit can be given with any certainty, but for a point of discussion the latest commercial proton magnetic resonance spectrometer should be able to provide quantitative data on about 1–10 μg of a compound of a molecular weight of 200. Very often however, the analyst is frustrated by insufficient sample. There are some current aids to this problem and some hope for the future.

There are two fundamental approaches to improve detection limits: increase the signal (viz. microcell) or lower the noise. Assuming that the manufacturer has done all that is possible to reduce electronic noise in the instrument and proper installation has been performed to avoid ground loops and line noise, reduction of the remaining noise may be approached using statistics. If a response consisting of a dc signal, such as an NMR resonance line, is present in conjunction with an ac noise, the S/N may be improved by signal averaging. The signal sums in direct proportion to the number of scans recorded, whereas the noise sums as the square root of the number of scans. Therefore, **n** repeated scans improves the S/N by a factor of $\sqrt{n}$. A signal may be averaged by an analog integration over a period of time. In this regard the NMR integrator is a signal-averaging device. High-frequency noise may be reduced by using a large RC time constant in the integrator. This method is practical for spectroscopic methods involving stationary states with short relaxation times. However, the real danger of saturation of the NMR signal requires a transient signal rather than lengthy averaging of the peak.

Hard-wired, digital data-acquisition equipment such as a computer of average transients (CAT) has become very common as signal enhancement devices for NMR. These devices contain some form of core memory, perhaps in association with a recording device such as magnetic tape or cassettes. The devices perform the function of multichannel (viz. 1024 channels) digital data storage. The NMR spectrum is automatically recorded a preset number of times. Each time, the spectrum amplitude is converted into digital form at equal frequency increments and stored in the corresponding memory address in the CAT. The operator may view the spectrum between scans by displaying the readout on a cathode ray tube. The spectrum may be displayed directly on the NMR spectrometer recorder when desired. The spectrum may be integrated using the spectrometer integrator while recording the spectrum from the CAT.

The advantages of using a computer of average transients to improve the S/N is readily seen because its performance is very close to the theoretical limit. For example, the accumulated spectrum of 100 scans will have very close to 10 times the S/N of a single scan. The few percent deviation from the limit is usually a result of instrumental drift. A decision of the number of scans that will be recorded must be made. The improvement of S/N beyond 400 scans diminishes rapidly and is usually not worthwhile. When the highest resolution is not necessary, it is usually profitable to sweep at a higher rate than is normally used ($\sim$25 Hz/sec) and employ a relatively high B_1 setting. In this way, the signal to noise is improved over a shorter period of total accumulation time than using slower sweep rates and lower B_1 settings. This procedure exercises the frequently encountered trade between resolution and signal-to-noise.

Signal averaging using a CAT is a very time-consuming process that occupies a spectrometer and a rather expensive hardware item for long periods of time. The interruption of a scan procedure by one of the many events that plague instruments can be annoying. The availability of digital computers has evolved other methods of "data smoothing." These procedures may be applied to a single scan, or to spectra that have been averaged using a CAT. The spectrum must be converted to a digital form and entered into a digital computer. This is best done by interfacing an analog-to-digital convertor to the spectrometer and recording spectrum amplitude values at increments of frequency on magnetic tape, paper tape, disk, or directly into computer core memory. Programs may then be written in a high-level language (FORTRAN, BASIC, etc.) to smooth the data. One form of computer averaging is known as boxcar averaging. This technique replaces the set of two or more data points with the mean of those points. For example, five adjacent amplitudes are averaged and the mean replaces the point originally occupied by the third point in the set. The next five points are treated in the same way, and so on. Obviously this leads to serious loss of resolution. A simple refinement of the boxcar method is the "moving window" method. This technique again replaces a set of data points with the mean. However, instead of advancing an entire set, the next set is formed by dropping the first point of the previous set, and adding the point following the previous set. The window is thus moved through the entire data set. This technique does not suffer the severe loss of resolution that the boxcar method does.

These various methods of S/N improvement assume the noise is random in phase

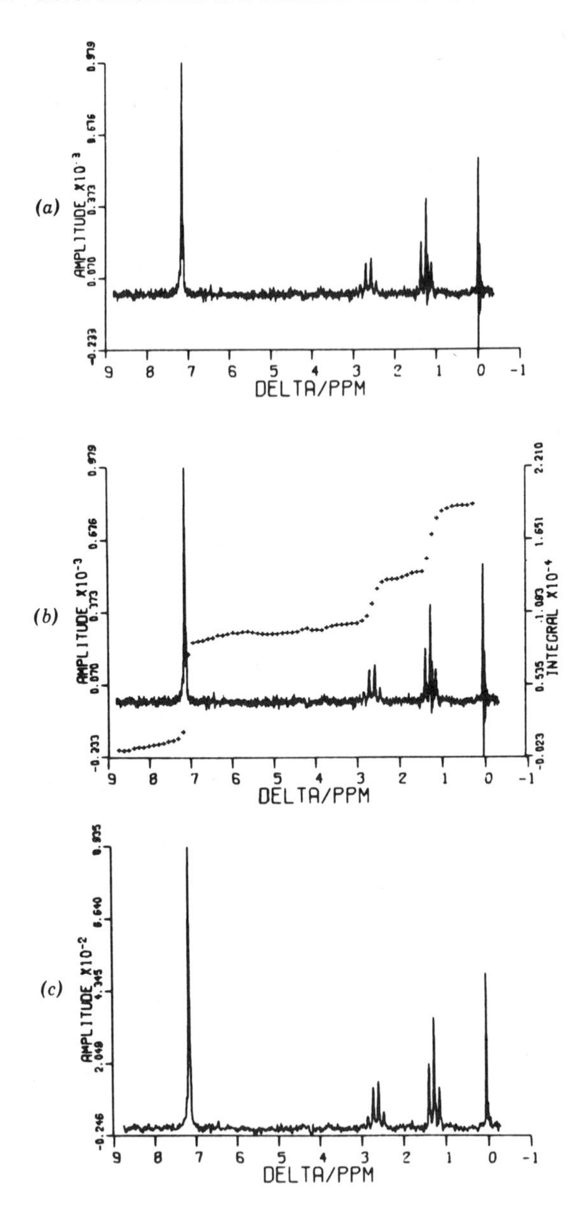

Fig. 1.23. Computer reduced NMR spectra of ethylbenzene: (*a*) raw data, (*b*) with integral trace (++++), and (*c*) after one smoothing.

and amplitude relation to the signal. The smoothing techniques can be of great usefulness when a computer data acquisition system is available. However, they must be used with discretion so that real data are not overly removed and artifacts generated. The programmable computer is much more powerful than the CAT because it may be used not only to acquire the data but also to treat the data, store several spectra as a function of elapsed time, perform digital integration and control the instrument including maintaining tuning (224). Figure 1.23 shows a simple but illustrative example of the acquisition, smoothing, and digital integration using an on line computer.

3. Fourier Transform NMR

With conventional NMR spectrometers such as those described previously, spectra are obtained by sweeping either the frequency or field through the region of interest (the continuous wave or cw-NMR experiment). While this type of equipment can be used with time averaging to improve S/N as discussed above, it suffers from two major drawbacks: (1) considerable time (100 – 1000 sec) must be spent in obtaining each scan of the spectrum to prevent line-shape distortions; and (2) if many scans of the spectrum are required to obtain a suitable S/N, the total time required to complete the experiment can be long (overnight or even over a weekend). Furthermore, much of the total experiment time is wasted as far as obtaining data is concerned because the majority of time spent on any one scan is used in scanning the baseline noise between peaks. Fourier transform (FT) techniques may be used to reduce the time required to achieve a given S/N (76,79).

Most FT-NMR techniques use a method that simultaneously excites all the resonances of a given type of nucleus. The signal induced in the receiver coil is a complex waveform which is a superposition of the resonances of all the nuclei resonating within the frequency range of the exciting device. This complex waveform is sampled as a function of time and is referred to as the time domain spectrum or the free induction decay (FID). The FID is sampled for a few seconds with each response being coherently added and stored in a digital computer. This process of excitation, sampling, and storing is repeated until the desired enhancement in the S/N has been achieved. The frequency spectrum is then obtained by Fourier transformation of the FID and is identical in all respects to the spectrum that would have been obtained using the cw technique with time averaging except that ringing following the peaks is not observed.

The major advantage of FT-NMR over the conventional cw technique for improving the S/N is in the total observation time required to achieve a given S/N. In general, a reduction in the total time by a factor of from 10 to 100 can easily be realized using FT-NMR. As a result, the ^{1}H-NMR spectra of very dilute samples (10^{-3} – 10^{-5} *M*) can be obtained on a routine basis. Furthermore, since a spectrum can be obtained within a few seconds, FT-NMR techniques allow one to examine unstable species with a short lifetime, and to obtain information on chemical and molecular dynamics. Relaxation times for the individual nuclei in a sample may also be determined using FT-NMR (141). For biological samples where the resonance due to H-O-D may be a problem, it is possible in some cases to eliminate this peak entirely by proper choice of the operating procedure (133). An excellent text on FT-NMR is available (79) and a more detailed discussion is presented in Chapter 2.

IV. APPLICATIONS OF HIGH-RESOLUTION ^{1}H-NMR

A. ANALYSIS OF SPECTRA AND DETERMINATION OF STRUCTURE

1. Introduction

The successful use of NMR spectroscopy as an analytical tool depends upon the extraction of the fundamental parameters from the spectrum. The analysis of an NMR spectrum in terms of the chemical shifts and coupling constants is the most important step in the application of NMR spectroscopy. In certain cases, the analysis may amount to the determination of the separation between the peaks in a multiplet and finding the frequency of the center of the multiplet. In other cases, a full mathematical analysis using one of the existing computer programs is required for complete interpretation of the spectrum. Various additional experiments such as spin-decoupling, obtaining the spectrum in a different solvent, obtaining the spectrum in the presence of a shift reagent, and deuterium substitution may be required for the full analysis of a complex spectrum.

Once the parameters have been obtained, they must be assigned to the individual protons of the sample using correlation tables of chemical shifts and coupling constants derived from known compounds. After the assignment has been made, the parameters may be used in conjunction with the correlation tables and empirical relationships to derive additional information about the compound under study.

For unknown compounds, the NMR spectrum alone is not sufficient to establish the structure. One usually has additional information such as a molecular weight, an elemental analysis, a uv spectrum, and an infrared spectrum which, when used in conjunction with the NMR spectrum, allows one to propose a structure consistent with the known data. This structure may then be confirmed by comparison with data on an authentic sample. In this section we shall introduce the nomenclature and discuss the rules for the analysis of NMR spectra. Aids in the analysis of spectra and the use of correlation tables and empirical relationships in the determination of the structure of unknown compounds will be discussed. Finally, we shall briefly discuss the determination and use of relaxation times.

2. Nomenclature

In the discussion of the analysis of NMR spectra, chemists consider only the magnetic nuclei in a compound. The magnetic nuclei in a compound are referred to as the *spin system*. The spin system includes all the interacting magnetic nuclei (those that are spin-coupled with each other) in the compound and it is not necessary for each nucleus to be coupled to every other nucleus in the spin system. In many cases, two or more nuclei may have identical chemical shifts. These nuclei are said to be equivalent. In the analysis of NMR spectra, two types of equivalency are considered, the difference being determined by the coupling constants of the nuclei in question. Nuclei that have identical chemical shifts and identical coupling constants to all other magnetic nuclei in the molecule are said to be *magnetically equivalent*. Nuclei with identical chemical shifts but with different coupling constants to other magnetic nuclei are said to be *chemical shift equivalent*.

Two or more nuclei are chemical shift equivalent if there exists some symmetry element of the molecule such as an axis, center, or plane, that interchanges the nuclei. In addition, nuclei that are interchanged through some exchange process with a rate faster than about once in 10^{-3} sec are also chemical shift equivalent. As an example of chemical shift equivalence, consider *p*-chloronitrobenzene (**II**). This molecule has a plane of symmetry perpendicular to the plane of the ring which bisects the two substituted carbons. As a result, protons 1 and 3 have identical chemical shifts as do protons 2 and 4. However, protons 1 and 3 do not have identical coupling constants to the other magnetic nuclei. The coupling constant between protons 1 and 4 is a *para* coupling whereas the coupling constant between protons 3 and 4 is an *ortho* coupling. These two coupling constants have different values and therefore this molecule contains two sets of chemical-shift-equivalent nuclei.

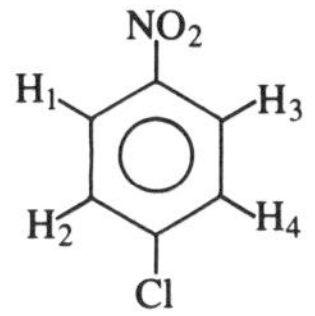

II

As another example, consider the spectrum given by a 1,2-disubstituted ethane, XCH_2CH_2Y. In considering compounds of this type, the authors have found it convenient to consider the compound in terms of the Newman projections (Fig. 1.24). Of the three staggered conformations, only the *anti* conformation possesses a plane of symmetry. However, this is enough to make protons 1 and 2 have identical chemical shifts as do protons 3 and 4. Perhaps this can be seen more easily if we consider the chemical shifts as an average of each of the Newman projections. The chemical shifts of protons 1 and 2 are given by equations 45 and 46 where *p* refers to the populations of the conformations and the subscripts refer to the immediate

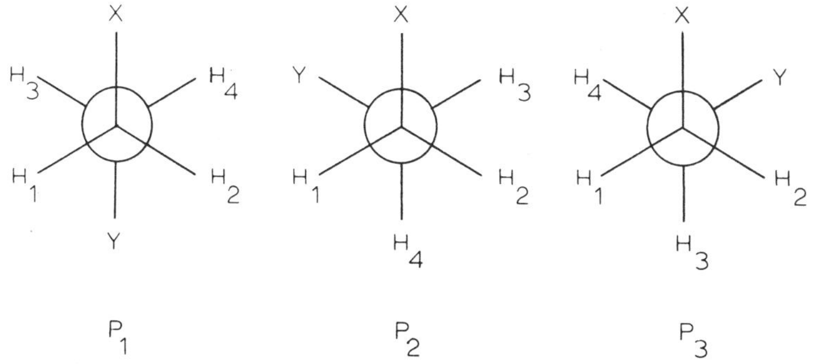

Fig. 1.24. Newman projections for the three most stable rotamers of a 1,2-disubstituted ethane XCH_2CH_2Y.

environment of the protons. Since the population of the two *gauche* conformations are equal

$$\delta H_1 = p_1\,\delta_{Y,H} + p_2\,\delta_{Y,H} + p_3\,\delta_{H,H} \tag{45}$$

$$\delta H_2 = p_1\,\delta_{Y,H} + p_2\,\delta_{H,H} + p_3\,\delta_{Y,H} \tag{46}$$

($p_2 = p_3$), protons 1 and 2 have identical chemical shifts. Similar results are obtained for protons 3 and 4. A similar treatment will show that protons 1 and 2 do not have identical couplings to protons 3 and 4. Considering the coupling constant to be an average of the coupling constants in the three conformations, the coupling constants J_{13} and J_{23} are given by equations 47 and 48 where J_t and J_g refer to *trans* and *gauche*

$$J_{13} = p_1 J_g + p_2 J_t + p_3 J_g \tag{47}$$

$$J_{23} = p_1 J_t + p_2 J_g + p_3 J_g \tag{48}$$

coupling constants, respectively. Since $J_t \neq J_g$, it can readily be seen that $J_{13} \neq J_{23}$.

This same treatment will also show why the two protons in a methylene group adjacent to a chiral center are not equivalent. Consider the Newman projections given in Fig. 1.25 for a compound of the type MCH_2CXYZ. The chemical shifts for protons 1 and 2 are given by equations 49 and 50. Since the populations of the three conformations are not equal, $p_1 \neq p_2 \neq p_3$, the chemical shifts of protons 1 and 2 are not

$$\delta H_1 = p_1\,\delta_{X,Z} + p_2\,\delta_{Y,Z} + p_3\,\delta_{X,Y} \tag{49}$$

$$\delta H_2 = p_1\,\delta_{Y,Z} + p_2\,\delta_{X,Y} + p_3\,\delta_{X,Z} \tag{50}$$

equivalent.

For nuclei to be magnetically equivalent, they must be chemical shift equivalent and have identical coupling constants to all other magnetic nuclei. This usually implies that the bond distances and bond angles with respect to the other nuclei be equal, or that they are equal with fast rotation about a single bond. As an example, consider the compound ethyl bromide, CH_3CH_2Br. If one draws the three staggered

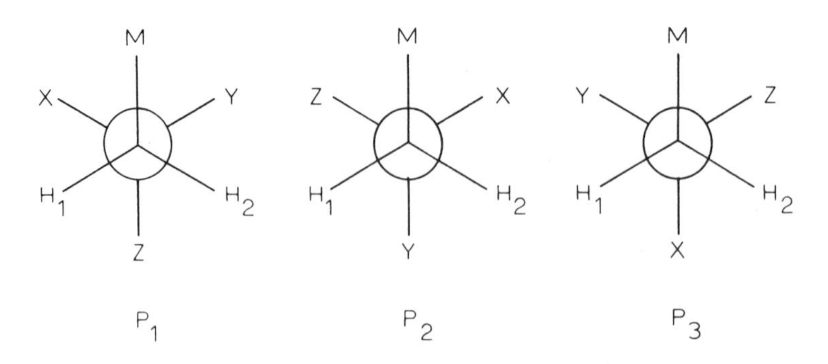

Fig. 1.25. Newman projections for the three most stable rotamers of a 1,1,2-trisubstituted ethane $XZCHCH_2M$.

Newman projections similar to Fig. 1.24 and considers that fast rotation is occurring about the carbon–carbon bond, writing equations for the chemical shifts and coupling constants for the methylene protons will show that the methylene protons have identical chemical shifts and identical coupling constants to each of the three methyl protons. The same is true for the three methyl protons. Therefore, ethyl bromide contains two sets of magnetically equivalent nuclei. Methylene fluoride, CH_2F_2, provides another example. The two protons have identical chemical shifts and identical coupling constants with each of the fluorines. Similarly, the two fluorines have identical chemical shifts and identical coupling constants with each of the protons. The protons and fluorines in methylene fluoride therefore constitute two sets of magnetically equivalent nuclei.

The type of equivalence in a molecule has a large influence upon the number of peaks observed in the spectrum of a spin system and the ease with which the parameters may be obtained from the spectrum. For molecules that contain magnetically equivalent nuclei, the coupling constant between the equivalent nuclei cannot be obtained from the spectrum (i.e., J_{H-H} in methylene fluoride). The spectrum given by a molecule that contains magnetically equivalent nuclei usually contains fewer absorptions and is normally easier to interpret. The spectrum given by a molecule containing chemically equivalent nuclei on the other hand is usually more complex and either a mathematical or computer analysis may be required to obtain the parameters.

A method for designating the magnetic nuclei in a sample in terms of the letters of the alphabet has developed over past years and will be used throughout this text (185). Using this convention, nonequivalent nuclei having chemical shifts that differ by magnitudes comparable to the coupling constants between the nuclei are referred to by the letters A, B, C, D, etc. If additional magnetic nuclei are present in the molecule having chemical shifts separated from the other nuclei by large differences, these nuclei are referred to by letters at the end of the alphabet such as X, Y, Z. Equivalent nuclei are given the same letter of the alphabet.

Nuclei that are magnetically equivalent are given the same letter. For example, methyl iodide, CH_3I, is referred to as an A_3 spin system and methylene fluoride as an A_2X_2 spin system. The number in the subscript refers to the number of nuclei in the equivalent set.

Chemically equivalent nuclei are distinguished from each other by the use of primes as a superscript on the letter. Our example above of *p*-chloronitrobenzene is classified as an AA′BB′ spin system to indicate two sets of equivalent nuclei that are not magnetically equivalent. Additional examples of this notation for spin systems are given in Table 1.VII.

This system of nomenclature is somewhat arbitrary in the choice of letters for the nuclei. For example the notation for CH_3CH_2Br could either be A_3X_2 or A_2X_3. No significance is attached to the relative order in which the nuclei are given. A useful convention might be to start at the high-field end of the spectrum and work towards the low-field end. About the only rule of thumb is that nuclei with different magnetogyric ratios should be distinguished by using letters at opposite ends of the alphabet (for example H-F is an AX spin system).

TABLE 1.VII
Examples of the Nomenclature of Spin Systems

Molecule	Notation
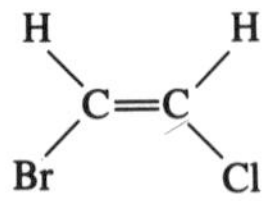 H, H, C=C, Br, Cl	AB
CH_2F_2	A_2X_2
CH_3CH_2Br	A_3B_2 or A_3X_2
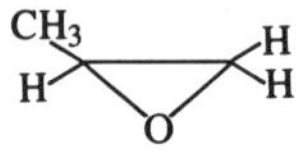 CH_3, H, H, H, O	$ABCX_3$
$BrCH_2CH_2Cl$	AA′BB′
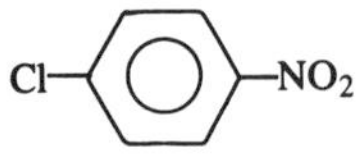 Cl, NO_2	AA′BB′
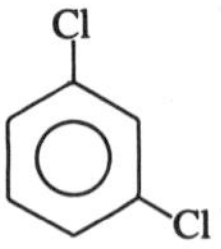 Cl, Cl	AB_2C or AB_2X
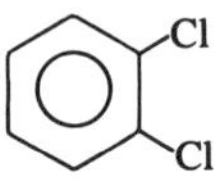 Cl, Cl	AA′BB′
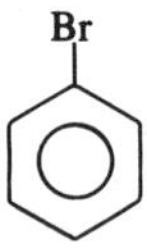 Br	AA′BB′C
$BrCH_2CH_2CH_2Cl$	AA′BB ′CC′

3. Analysis of First-Order Spectra

Under certain conditions, the coupling constants and chemical shifts may be extracted from the spectrum without carrying out a full mathematical analysis. However, it is important that one realize the limitations of the first-order approach to analysis of an NMR spectrum. First-order analysis is applicable only to those spectra

TABLE 1.VIII
Number of Peaks and Peak Intensities for First-Order Multiplets

Number of adjacent nuclei	Multiplet	Intensities
1	Doublet	1:1
2	Triplet	1:2:1
3	Quartet	1:3:3:1
4	Quintet	1:4:6:4:1
5	Sextet	1:5:10:10:5:1
6	Septet	1:6:15:20:15:6:1

that may be classified as arising from magnetically equivalent nuclei or from cases where each nucleus has a different chemical shift. The second condition is that the chemical shift difference between two spin-coupled nuclei must be large compared to the coupling constant. As a general rule, the relationship $J/\delta < 0.05$ must be satisfied for a first-order analysis to be valid. When the above conditions are met, first-order rules relating the number of peaks in the spectrum, the spacing of the peaks in a multiplet, and the intensities of the peaks in a multiplet are valid.

The number of peaks in a given multiplet is related to the number of nuclei coupled to the nuclei giving rise to the multiplet. Consider the case of two groups of magnetic equivalent nuclei A and B coupled to each other. The number of peaks in the multiplet due to A is given by $2nI + 1$, where n is the number of equivalent nuclei B and I is the spin number of B. Similarly, the number of peaks in the multiplet due to B is given by $2nI + 1$ where n is the number of equivalent nuclei A and I is the spin number of A. For those cases where $I = 1/2$ (^{1}H, ^{19}F, ^{31}P), the number of peaks in A is $n + 1$ where n is the number of equivalent nuclei B. The spacing between the peaks in the A multiplet will be equal, and will be equal to the spacing between the peaks in the B multiplet. This spacing is the coupling constant J_{AB}. The chemical shift is the center of the multiplet. For nuclei with $I = 1/2$, the intensities of the peaks in each multiplet are proportional to the coefficients of the binomial expansion $(X + 1)^n$, where n is the number of equivalent nuclei coupled to the nucleus in question (Table 1.VIII).

As the chemical shifts become closer together, noticeable changes occur, first in the relative intensities of the peaks and, as the shifts become even closer, changes also occur in the number of peaks and in the spacing of the peaks. When $0.05 > J/\delta < 0.15$, the spectrum may still be interpreted as a first-order spectrum as far as the number of peaks in the multiplet and the spacing of the peaks is concerned. The intensities of the peaks will no longer be symmetrical about the center of the multiplet. The peaks on the sides between the two multiplets (inner peaks) will show increased intensities while the outer peaks will show decreased intensities. (Fig. 1.26). For this type of spectrum, the chemical shifts will no longer be at the center of the multiplets and the coupling constant may or may not be equal to the spacing between the peaks. One should always carry out a mathematical analysis for spectra of this type to obtain the true chemical shifts and coupling constants. When $J/\delta > 0.15$, the multiplets no longer appear as first-order spectra. The spacing of the

(a)

(b)

(c)

(d)

(e)

Fig. 1.26. An illustration of the dependence of the intensities of an AB spectrum on the chemical shift difference δ_{AB}.

peaks will most likely not be equal to the coupling constant. The intensities of the inner lines increase more as the outer lines decrease in intensity. Furthermore, additional peaks may appear in the spectrum. The complete interpretation of this type of spectrum always requires a full mathematical analysis.

The spectra for spin systems involving nonequivalent nuclei may also be interpreted using first-order rules, so long as the chemical shift differences are large compared to the coupling constants. Consider the case involving three spin 1/2 nuclei AMX, where each nucleus is coupled to the other two. One way to construct the multiplets due to each nucleus is to use a graphical approach and consider the coupling to the other nuclei individually. For nucleus A, start by drawing a single line for the chemical shift of A in the absence of any coupling to the other nuclei (Fig. 1.27). Next, consider the coupling of M with A. Since M is a spin 1/2 nucleus, coupling with A will split the line due to A into a doublet with spacing J_{AM}. Coupling of A with nucleus X will further split each peak of the doublet into doublets with spacing J_{AX}. The result is that the resonance of nucleus A will appear as a quartet with two different spacings, J_{AX}, J_{AM}, each repeated twice (Fig. 1.27). Similarly, the resonance of M and X will also appear as quartets with spacings of J_{AM} and J_{MX} and J_{AX} and J_{MX}, respectively. The coupling constants are assigned on the basis that the repeated spacing (J) will appear in the resonances of both coupled nuclei.

The reverse of the above approach is used to analyze the first-order spectrum of an unknown compound. First, identify the repeated spacings and construct a graph similar to that above in the reverse order.

Before quoting a spacing as a coupling constant, one must ensure that the spectrum is in fact first-order. If the intensities of the peaks are not exactly symmetrical in the multiplets, a full analysis should be carried out. In this case, the spacings may be used as an approximation of the coupling constants in a computer analysis. Finally there

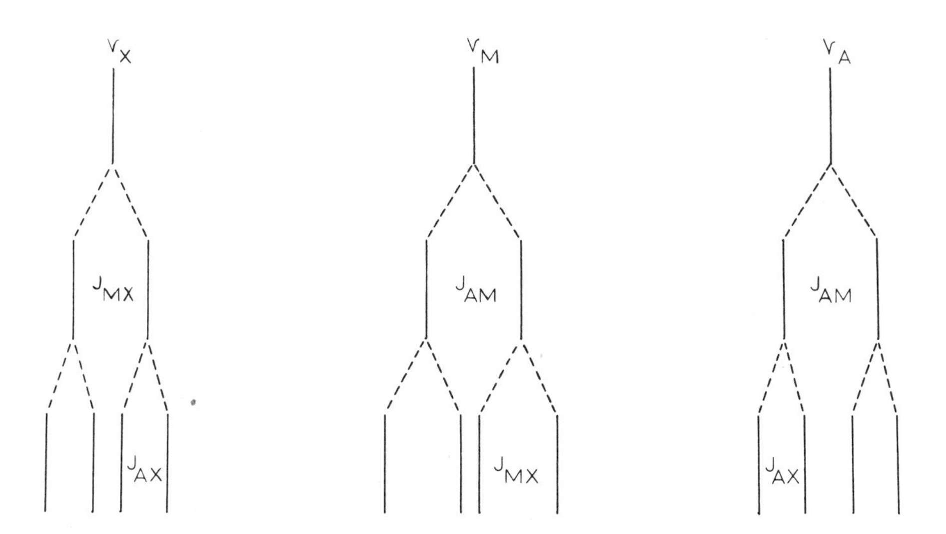

Fig. 1.27. A schematic illustration of an AMX spectrum with the repeated spacings.

are several situations where the spin system contains magnetically nonequivalent nuclei and yet, the spectrum approximates a first-order spectrum (deceptively simple spectra). The analysis of a spectrum of this type may prove difficult unless one can obtain the spectrum under conditions where the simplicity is removed.

4. Analysis of Complex NMR Spectra

The first-order analysis of spectra outlined above is limited in that it applies only to first-order spectra and not to more complex spectra. When the spectrum is not first-order, one must carry out a computer analysis of the spectrum to obtain the chemical shifts and coupling constants. This usually requires that one calculate a theoretical spectrum using one of the existing programs. A first-order approximation may be used to obtain the chemical shifts and coupling constants necessary for calculating the theoretical spectrum. After assigning the transitions in the experimental spectrum to those in the theoretical spectrum, the computer program will calculate a set of chemical shifts and coupling constants that will reproduce the experimental spectrum. If a set of chemical shifts and coupling constants can be found such that when they are used to calculate an NMR spectrum, they reproduce the experimental spectrum in transition frequencies and intensities, the set of parameters are the chemical shifts and coupling constants describing the experimental spectrum. For certain types of spin systems, explicit equations have been derived that permit one to analyze the spectrum without a computer fit. A more detailed account of the analysis of NMR spectra may be found in a number of texts (23,59,73,185,203).

5. Aids in the Analysis of Spectra

As mentioned previously, it may not be possible to completely analyze every spectrum encountered. Spin systems containing equivalent nuclei are particularly difficult because of the fact that not all the allowed transitions are observed in many cases. In other spin systems, deceptively simple spectra may be observed, thereby limiting the parameters that can be obtained from the analysis. On the other hand, with complex spectra it may not be possible to obtain estimates of the parameters to use in the calculation of a trial spectrum for computer analysis. Several methods have been used in the past to aid in the analysis of spectra. Each of these methods is discussed briefly in the following sections.

a. Variations In Magnetic Field

Often, a complex spectrum may be simplified by obtaining the spectrum in a larger magnetic field. As pointed out previously, chemical shifts are field dependent whereas coupling constants are independent of the magnetic field. Therefore, the ratio J/δ may be changed and the spectrum made to approach first-order by obtaining the spectrum in a stronger magnetic field. Commercial spectrometers are available for obtaining proton spectra at 30, 60, 90, 100, 200, 250, 270, 300, 360, 400, and 500 MHz, offering a wide range of possibilities.

The reverse of this process may be used to advantage in certain cases. As a general rule, certain information may not be obtained from a first-order spectrum (i.e., signs of coupling constants and certain coupling constants in cases where the intensities of certain absorptions are such that they are not observed). This information may become available in a more complex spectrum obtained by running the spectrum in a less intense magnetic field. The analysis of a spectrum at two magnetic fields can also be used as a check on the parameters obtained from the spectrum.

b. Isotopic Substitution

Isotopic substitution has been used as an aid in the analysis of NMR spectra in two principal ways: (a) by removing a spin from the system (replacing H with D) and (b) by removing the magnetic equivalence observed in certain cases. Using this technique, it is possible to obtain information from a spectrum that would otherwise prove difficult or impossible to obtain.

One of the most widely used applications of deuterium substitution is that of the identification of labile proton resonances. A typical example is the identification of the –O–H resonance in the spectra of alcohols. After obtaining a normal spectrum, the usual technique is to add a drop of D_2O to the NMR tube and shake for a few minutes to allow D to exchange for H. The spectrum is rerun and the peak due to the hydroxyl proton will not be present in this spectrum. This technique may be used to identify the resonance of any readily exchangeable proton.

A second related area where isotopic substitution has proven valuable in the analysis of spectra is in the simplification of a complex spectrum. Suppose one had a spectrum where it was not possible to assign the resonances to specific protons in the molecule. If one could synthesize the molecule with D at specific positions in the molecule, one could obtain the spectrum of the deuterated material and, from a comparison of the spectra, assign the resonances in the nondeuterated sample. This method can also be used in reverse; that is, if one can assign the spectrum of the nondeuterated material then one can use this technique to determine the site of incorporation of deuterium into a molecule. This has been used in various mechanistic studies.

The removal of resonances from complex spectra is only one aspect of deuterium substitution. The coupling constant between two nuclei is proportional to the product of the magnetogyric ratio of the nuclei. Therefore, substitution of D for a H will affect all coupling constants to the H. The magnetogyric ratio of deuterium, γ_D, is 6.55 times smaller than that of hydrogen. Consequently, substitution of D for H will reduce all coupling constants to the H by a factor of 6.55 ($J_{HD} = J_{HH/6.55}$). Thus, not only are resonances removed from the spectrum, but the remaining resonances are simplified by the reduction in the magnitude of the coupling constants. In many cases, the coupling constants can become negligibly small so that only line broadening is observed in the resonances of the protons that are spin-coupled to deuterium.

The second area where isotopic substitution is used actually results in an increase in the complexity of a spectrum by removing magnetic equivalence. There are several instances where the coupling constant between magnetically equivalent nuclei is needed either for comparison with theoretical predictions or for a base for the discussion of substituent effects. Consider for example the early work on the theory of coupling constant mechanisms where the coupling constant in the hydrogen molecule

H_2 was calculated. The NMR spectrum of H_2 consists of a single line due to the magnetic equivalence of the two hydrogens. This A_2 spin system can be converted into an AX spectrum by substituting one D for H (H–D). The proton spectrum of H–D consists of three equally intense lines due to coupling with deuterium ($I = 1$). From the ratio of J_{HH} to J_{HD} given above, J_{HH} in H_2 can be calculated. Similarly, the substitution of D for H in methane would result in a three-line spectrum from which J_{HH} in methane could be calculated. A variety of coupling constants have been obtained using this technique.

Another method for removing certain magnetic equivalence is to examine the ^{13}C satellite resonances. This does not necessarily require isotope substitution if sufficient sensitivity is available. The ^{13}C isotope ($I = 1/2$) is present in 1.1% natural abundance. In a molecule like $CHCl_3$, 1.1% of the molecules contain ^{13}C which is coupled to the proton. Therefore, the proton spectrum of $^{13}CHCl_3$ gives rise to a doublet that is symmetrically disposed about the single peak due to the $^{13}CHCl_3$ molecules (Fig. 1.28). The ^{13}C satellite spectra can be used to advantage to remove the equivalence in symmetrical molecules. Consider the proton spectrum of benzene which consists of a single line. Approximately 6.6% of the molecules contain a ^{13}C carbon (the probability of having a molecule with two ^{13}C nucleus makes the spin system an A′BB′CC′X spin system. All of the interproton coupling constants can be obtained from the analysis of the ^{13}C satellite spectra on either side of the absorption due to the all ^{13}C molecules. This technique has been utilized to obtain the proton–proton coupling constants in a variety of molecules including ethylene (221), benzene (200), cyclopropane (243), various 1,2-disubstituted ethanes (222), and the three-membered ring heterocycles (164).

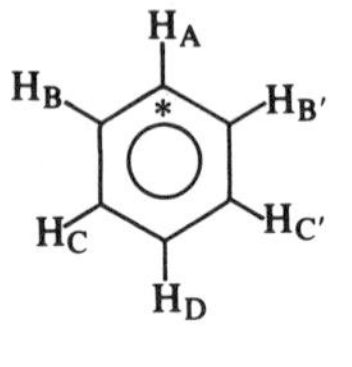

H_A, H_B, $H_{B'}$, H_C, $H_{C'}$, H_D

III

c. DOUBLE IRRADIATION

Double irradiation (resonance) methods are often used in NMR spectroscopy to aid in the interpretation of complex NMR spectra and to extend the range of information that may be obtained from spectra. As the name implies, double-resonance experiments are concerned with the simultaneous application of two radio frequency fields to the sample under consideration. We shall continue to refer to B_1 as the observing field and use B_2 to refer to the second radio frequency field used to perturb the system of nuclei. The notation commonly used for double resonance experiments is $X - \{Y\}$ where X is the nuclear species being observed and Y is the nuclear species being irradiated (21). Two classes of experiments are included under the general topic of double resonance. In homonuclear double resonance, the second radio-frequency field B_2 is applied to the same type of nucleus as that of the observing field B_1, $H - \{H\}$. Heteronuclear double resonance differs in that the field B_2 is applied to a different type of nucleus than that observed by B_1 and it is denoted by $H - \{X\}$.

Several types of experiments are included under homonuclear double resonance and they differ primarily in the strength of the perturbing field B_2. Expressing field strengths in frequency units and using the conversion factor for protons of 1 milligauss equals 4.26 Hz, typical values of B_2 are from 1 to 8 milligauss compared with normal values of 0.1–0.3 for B_1. In principle, homonuclear double-resonance experiments may be carried out by varying the magnetic field B_1, that is, field sweep, or by varying the frequency of B_1, that is, frequency sweep. We shall restrict ourselves here to frequency-sweep experiments since not all types of double-resonance experiments can be carried out using field sweep. Most commercial NMR spectrometers

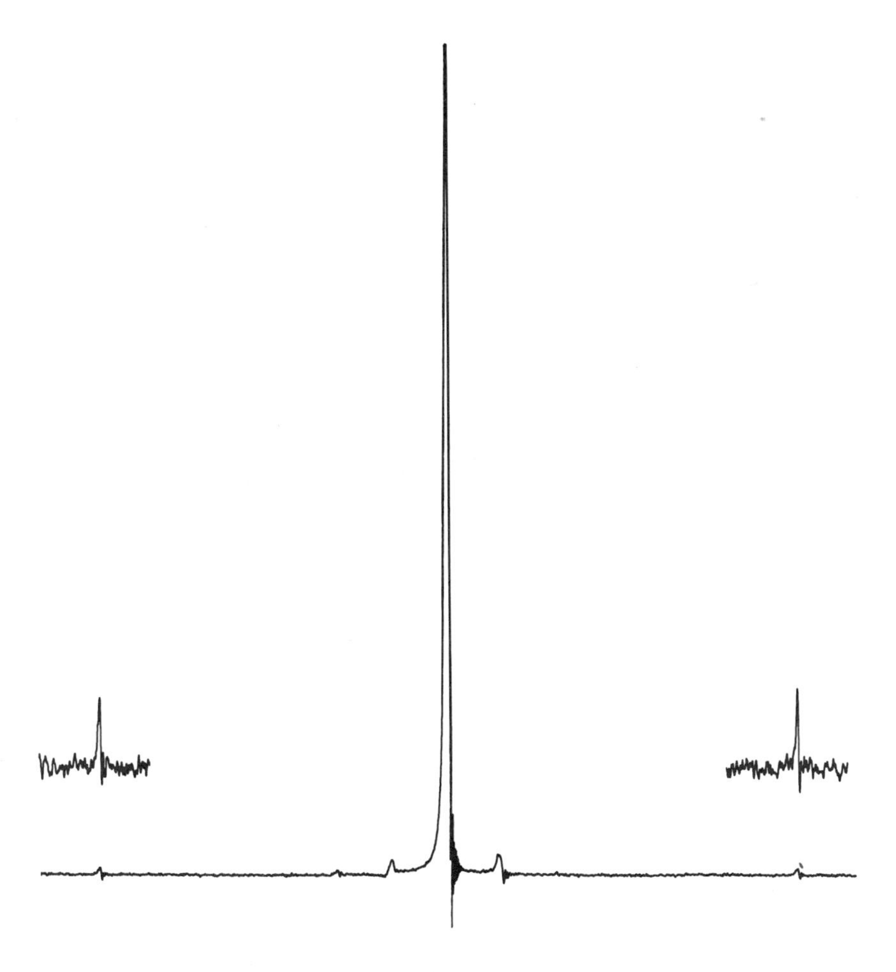

Fig. 1.28. An expansion of the 100-MHz ^{1}H spectrum of $CHCl_3$ showing the ^{13}C satellites.

have, as standard equipment, the accessories necessary for double resonance. Primarily, the basic instrumental requirement is that there be some means provided for field/frequency stabilization. The second radio frequency field B_2 is provided using sideband modulation.

The theory of double-resonance techniques is beyond the scope of this chapter. Excellent reviews of this subject are available (21,112). Among the experiments included under homonuclear double resonance are spin-decoupling, spin-tickling, internuclear double resonance (INDOR), and nuclear Overhauser effect (NOE). Each of these techniques is discussed below as to the experimental details and the type of information obtainable from their use.

During the course of the discussion, we shall refer to lines in the spectrum corresponding to transitions having an energy level in common as connected transitions (9). Connected transitions will be further classified depending on whether the energy of the common level is intermediate between those of the two terminal levels (progressive transition, Fig. 1.29*a*) or whether the energy of the common level falls outside the range of the two other levels (regressive transition, Fig. 1.29*b*). We shall furthermore consider only frequency sweep spectra. In the construction of energy level diagrams, the external magnetic field is considered to lie along the negative Z direction. When considering possible changes in a spectrum caused by the irradiating field B_2, a useful rule of thumb is that B_2 levels that are less than the line widths will cause only population changes, whereas B_2 levels that are equal to the line width or larger will cause changes in the energy levels.

d. Homonuclear Spin Decoupling

Spin decoupling implies that the spectra obtained are equivalent to those that would be observed in the absence of any coupling to the irradiated nuclei. Although this is not generally true, useful information to aid in the interpretation of complex NMR spectra can be obtained from spin-decoupled spectra. The most common

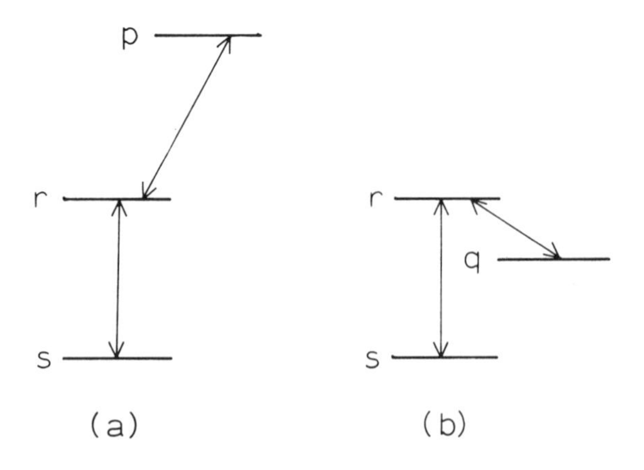

Fig. 1.29. An illustration of (*a*) a progressive and (*b*) regressive transition connecting NMR energy levels.

application is to identify signals from nuclei that are spin-coupled to the nucleus being irradiated. Usually, the multiplet from one nucleus is irradiated with B_2 while observing the effects in the multiplets due to the other nuclei (Fig. 1.30). Another application is the location of a "hidden" resonance in a complex spectrum with overlapping resonances by finding the irradiating frequency ν_2 which causes collapse of splittings arising from coupling to the "hidden" nucleus.

Experimentally, in spin decoupling the irradiating field B_2 is adjusted such that the frequency of B_2, ν_2, coincides with the chemical shift of the nucleus to be irradiated. The level of B_2 is adjusted to a level necessary to bring about spin decoupling, and the spectrum is obtained. Nuclei spin coupled to the nucleus being irradiated appear as if the irradiated nucleus was not present in the spin system, i.e., a reduction in the number of lines in the resonances due to the spin-coupled nuclei is observed. Although it is not strictly correct, the effect of B_2 can be thought of as causing rapid transitions among the possible spin states of the irradiated nucleus such that the remaining nuclei "see" only an average for the irradiated nucleus and not the individual spin states.

True spin decoupling can be observed only when the chemical shift separation between the nuclei in question is much greater than the coupling constant. In the simplest cases in which a single nucleus is coupled to one other nucleus or a group of equivalent nuclei (AX, AX_n), the multiplet signal due to A is reduced to a single peak by irradiation at the resonance frequency, ν_x, of X. To observe complete decoupling, the intensity of B_2 must be sufficiently high such that

$$\frac{\gamma B_2}{2\pi} \gg 2\ |J_{AX}| \tag{51}$$

Estimates of the level of B_2 required for decoupling may be obtained by experiments with one of the test samples (acetaldehyde or ethylbenzene) supplied by the spectrometer manufacturer. At lower levels of B_2 than that required for decoupling, complex spectra are observed. For an AX spin system, these usually amount to spectra in which either four lines are observed for A or a two line pattern is observed in which the splitting is smaller than J_{AX}. Similar effects are also produced if the frequency of B_2 is incorrectly set from the correct value for decoupling, ν_x (112).

Usually, decoupling cannot be obtained when the chemical shift separation is not much larger than the coupling constant between the nuclei. The B_2 level needed becomes comparable to the chemical shift difference such that irradiation of the signals from one nucleus perturbs the other nucleus also. If the chemical shift difference is not infinitely large compared with the coupling constant, the frequency of B_2 for optimum decoupling differs from the resonance frequency of the nucleus being irradiated. The correct frequency is displaced towards the signal being observed by an amount given by (112)

$$\left(\frac{\gamma B_2}{2\pi}\right)^2 \cdot \frac{1}{\nu_A - \nu_X} \tag{52}$$

Caution should be exercised when chemical shifts are determined from decoupled spectra. The levels of B_2 used to produce decoupling are sufficiently high to produce

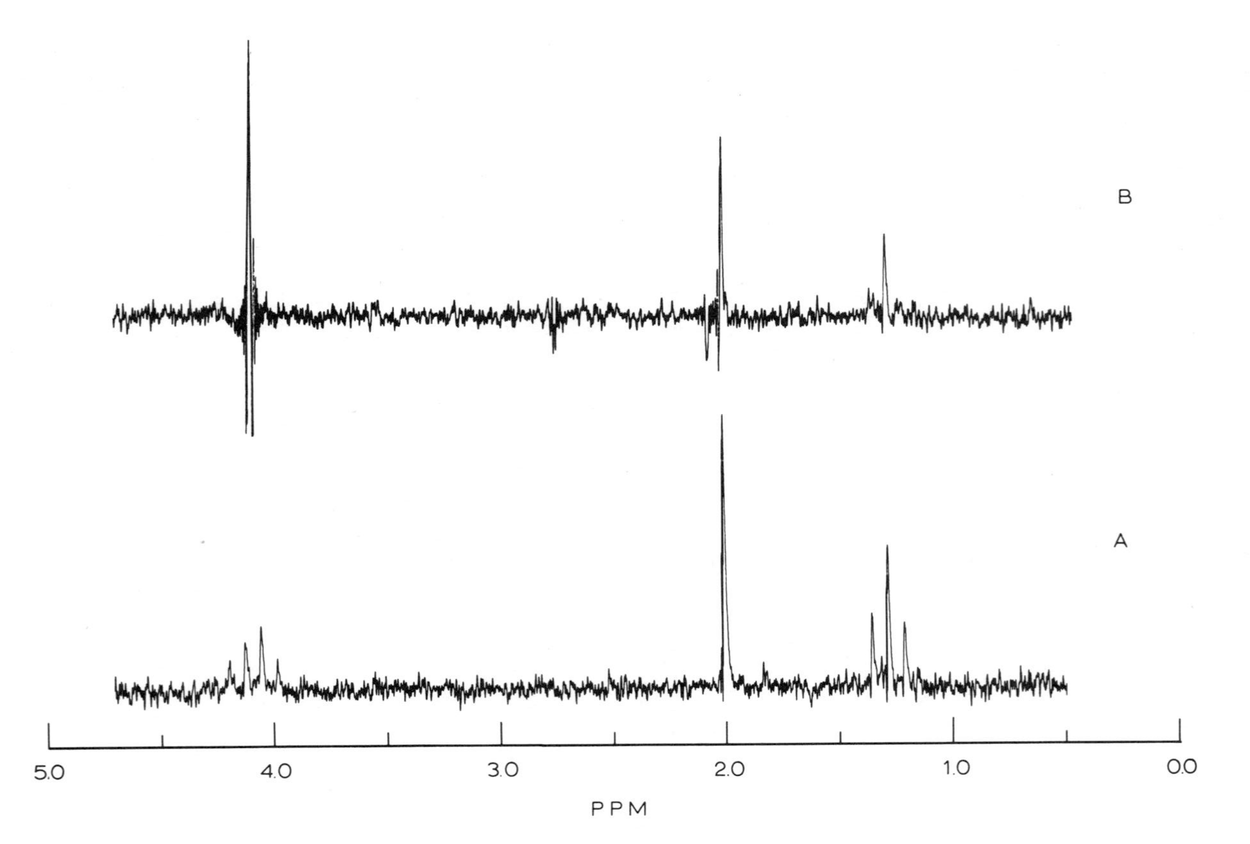

Fig. 1.30. (*a*) The 60-MHz ^{1}H spectrum of ethyl acetate and (*b*) the result of decoupling the methylene protons.

significant changes in the resonance frequencies of the other nuclei. Since the field experienced by the nuclei is a vector sum of B_2 and the polarizing field B_0, all signals are shifted away from the point of irradiation, ν_2, by an amount equal to

$$\left(\frac{\gamma B_2}{2\pi}\right)^2 \cdot \frac{1}{2(\nu_1 - \nu_2)} \tag{53}$$

where ν_1 is the frequency when B_2 is zero. For a B_2 level of 3 milligauss and a separation of 30 Hz between the observing and irradiating frequency, the shift is 2.7 Hz. This is known as the Block-Siegert shift and it affects all signals in the spectrum regardless of whether they are coupled to the nucleus being irradiated or not. An appropriate correction must be made to obtain the true chemical shift value. This shift is greatest for signals near the irradiating signal in a frequency swept spectrum, but in a field-sweep spectrum, all signals experience the same shift.

Finally, one should not use decoupled spectra to obtain integration data for determining relative intensities. The high irradiation level of B_2 can cause saturation of signals near the point of irradiation. The reduction of signal intensity is proportional to the square of the level of B_2 and decreases with increasing separation of the signal from the frequency of B_2. Although the decrease in signal intensity depends upon the relaxation times, it is not uncommon to find a reduction in intensity by 50% for a signal separated from the frequency of B_2 by a B_2 field of 3 milligauss.

e. SPIN-TICKLING

Spin-tickling differs primarily from spin-decoupling in that a weaker irradiating field B_2 is used. Furthermore, only a single nondegenerate line in the spectrum is irradiated, rather than an entire multiplet. If the strength of B_2 is about the same magnitude as that of the width of the line being irradiated (in Hz or milligauss), the result is that transitions that share a common energy level with the irradiated line will be split into doublets (84).

The splitting of the connected transitions is due to the mixing of the unperturbed states in the rotating frame. Irradiation of ν_{rs} (Fig. 1.31) will cause energy level r to be

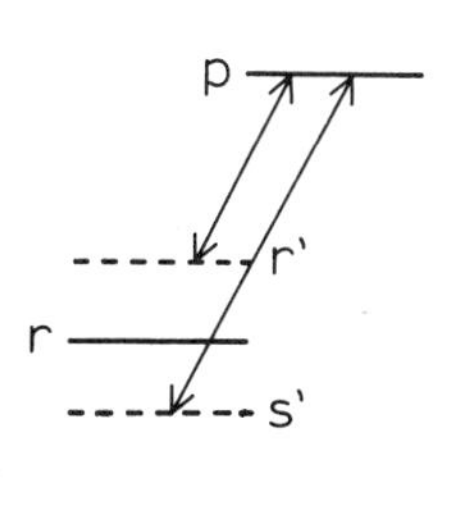

Fig. 1.31. A schematic illustration of the splitting of an energy level during a spin-tickling experiment.

mixed into two new eigenstates r' and s'. The doublet arises from the transitions $r' \rightarrow p$ and $s' \rightarrow p$ in the progressive case or $q \rightarrow r'$ and $q \rightarrow s'$ in the regressive case. For a case such as that in Fig. 1.30 where p and s differ by two quantum numbers, the doublets will be broadened (84) and where s and q have the same spin quantum number, the doublets will be well resolved.

Consider the ABX spectrum given in Fig. 1.32 with $|J_{AB}|, |J_{BX}| > |J_{AX}|$ and $\nu_X > \nu_B > \nu_A$. An energy level diagram for the three-spin system is given in Fig. 1.33 with the correct labeling of the transitions for all positive coupling constants. A spin-state diagram for this case is given in Table 1.IX. If all coupling constants were positive, irradiation of line 12 in the X region with a tickling field should cause lines 7 and 8 in the B region and lines 3 and 4 in the A region of the spectrum to be split into doublets. Saying this another way with reference to the spin-state diagram, irradiating line 12 in the X region should cause those lines in the B region which have the same A spin state as line 12 to be split into doublets and those lines in the A region which have the same B spin state as line 12 to be split into doublets, i.e., lines 3, 4, 7, and 8 should be split into doublets. Clearly from Fig. 1.32 this is not the case. Since lines 1 and 2 in the A region and lines 5 and 6 in the B region are split into doublets, the spin-state diagram should be changed as shown in parentheses in Table 1.IX and the energy level diagram should be relabeled accordingly. Irradiation of line 9 confirms this. Therefore, it is seen from the above results that the coupling constant J_{AB} is opposite in sign from that of J_{AX} and J_{BX}. This method can be used to assign the energy level diagram of larger spin systems. However, the number of experiments increases considerably.

The spin-tickling technique can be used to determine the relative signs of coupling constants and to trace out the energy level diagram for a spin system. Furthermore, this technique can be used to locate transitions which are "hidden" due to overlap with other resonances by varying the frequency of B_2 until the doublet splitting is observed.

f. INDOR

The INDOR (INternuclear DOuble Resonance) (19) technique can be used to obtain information similar to that derived from spin-tickling (131). However, the experimental conditions differ considerably. In INDOR experiments, a smaller B_2 is used compared with that used in tickling experiments. In addition, the intensity of a single line is monitored with the observing field B_1 while B_2 is swept through the spectrum. A change in intensity occurs when B_2 irradiates a line having a common energy level with the line being monitored. A positive signal is produced when the transition observed and that irradiated have a progressive relationship and a negative signal when the transition observed and that irradiated have a regressive relationship. No change in the intensity is observed if there is no coupling between the nuclei or if the two transitions have no common energy level. With reference to the energy level diagram for a three-spin system in Fig. 1.33 if line X_{12} is monitored as B_2 is swept through the A and B portion of the spectrum, one observes positive peaks for A_4 and B_4 and negative peaks for A_3 and B_7 (Fig. 1.34). If the spectrum is not first order, one can construct an energy-level diagram and use the INDOR spectra to assign the transitions.

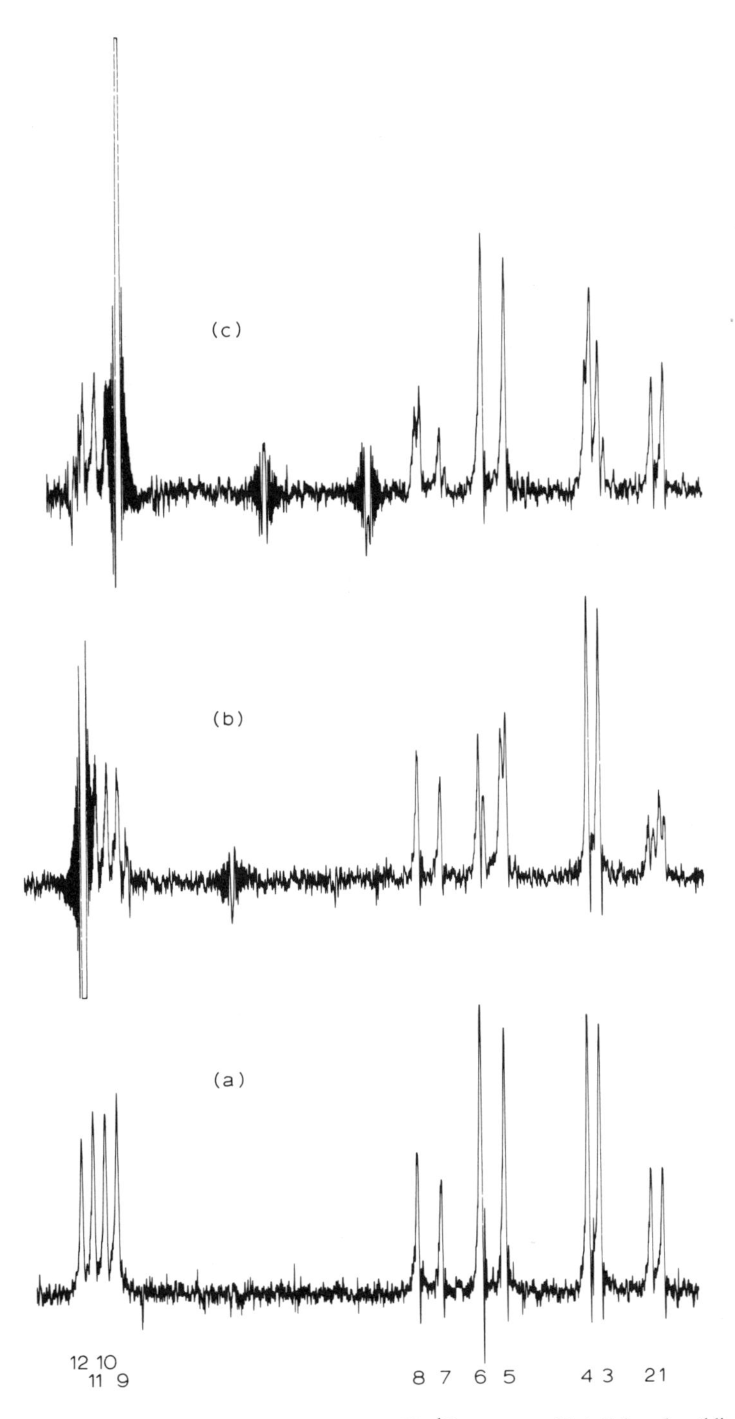

Fig. 1.32. (*a*) An expansion of the ABX part of the 60-MHz ^{1}H spectrum of 1,4-diphenylazetidinone. (*b*) The result of irradiating line 12 with a spin-tickling rf field. (*c*) The result of irradiating line 9 with a spin-tickling rf field.

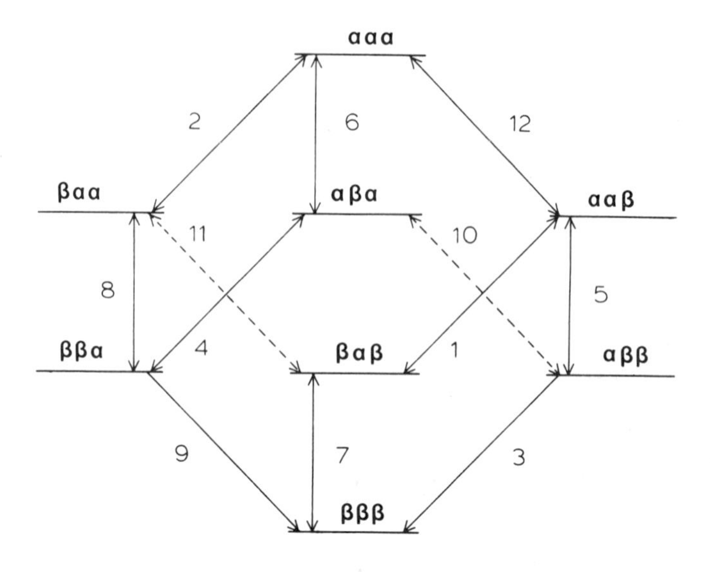

Fig. 1.33. An energy level diagram for a three spin system with the transitions labeled for a negative J_{AB}.

TABLE 1.IX
Spin States Of Neighbors in an ABX System[a]

Neighboring nucleus	X				B				A			
	12	11	10	9	8	7	6	5	4	3	2	1
A	α	β	α	β	α (β	α β	β α	β α)				
B	α	α	β	β					α (β	α β	β α	β α)
X					α	β	α	β	α	β	α	β

[a] The spin states are labeled (α and β) to correspond to those in Fig. 1.33.

The experimental conditions for INDOR determinations are rather rigid in terms of overall system stability. The observing field must be maintained at the center of a line and drifts in B_1 must be less than the line width in order to maintain an even baseline. Furthermore, there should be no short-term variations in the peak *height* due to changes in resolution or uneven spinning. Provision must be made for detecting changes in the intensity of the signal monitored by B_1 with a $Y-t$ recorder. The intensity of B_1 must be sufficiently low in order not to saturate the signal being monitored. Typical values of B_1 are 0.1 to 0.2 Hz. The irradiating field B_2 should be about 0.5 Hz, sufficient to cause population transfer between spin states without perturbing the energy levels.

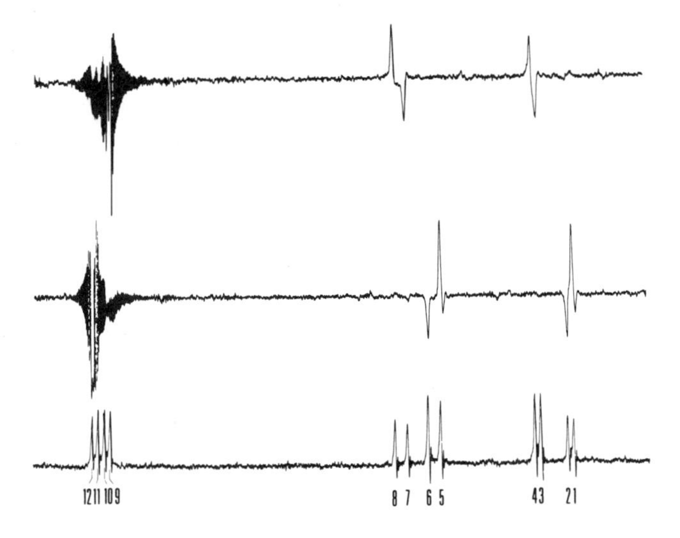

Fig. 1.34. An expansion of the ABX part of the 100-MHz ¹H spectrum of 1,4-diphenylazetidinone and the results of an INDOR experiment with B_1 monitoring lines 9 and 12.

Typical applications of the homonuclear INDOR technique are determination of relative signs of coupling constants and the location and measurement of the multiplicity of hidden signals. In general, if the accessories for INDOR are available, the INDOR method is preferred over spin-tickling for determining relative signs of coupling constants.

g. Intramolecular Nuclear Overhauser Effect

The nuclear Overhauser effect (NOE) (13) differs from the above decoupling techniques in that it does not depend upon two nuclei being spin coupled. NOE experiments are used to provide information about molecular geometries under suitable conditions (18,24,130,171). Experimentally, the method involves the saturation of one signal in the spectrum and observation of changes in the intensities in the other signals. For example, irradiation at the resonance frequency of the 4-methyl protons in 1,2,3,4-tetramethylphenanthrene (**IV**) leads to a 32% enhancement of H_5 and irradiation of the 1-methyl protons results in an 11% increase in the resonance of H_{10} (155).

IV

The magnitude of the intensity changes depends upon internuclear distances between the nuclei concerned. These intensity changes arise from perturbations of the relaxation processes that lead to thermal equilibrium between the spin states. The relaxation of a given nucleus is affected by all surrounding nuclei and depends upon the mean square value of the magnetic field produced by the surrounding nuclei. The relaxation from other nuclei is dominated by short-range interactions and is proportional to r^{-6} where r is the distance separating the nuclei. For protons, an NOE effect is observed for nuclei usually separated by less than about 3.5 Å. The interactions leading to relaxation may be either intermolecular or intramolecular. For dilute solutions in proton-free solvents, only the intramolecular relaxation is significant.

The maximum intensity increase possible, for only dipolar interactions, is 50%. Any other type of relaxation will result in an intensity increase less than 50%. Furthermore, the intensity increase will be reduced if more than one type of nucleus is contributing to the relaxation of the nucleus in question. If a molecule is exchanging rapidly between different conformations, an enhancement may be observed if the internuclear distance is small in one of the conformations and will depend upon the weighted average value of r^{-6}.

Care should be exercised in the preparation of a sample for an NOE experiment. Dissolved oxygen can cause relaxation, thereby reducing the intensity enhancement. Therefore, the sample should be subjected to several freeze-thaw cycles and sealed under vacuum. Solvents with a low concentration of magnetic nuclei such as CS_2 should be used to reduce intermolecular relaxation. Deuterated solvents may be used to reduce the solvent contribution to the relaxation.

For a single line, B_2 levels of the order of 0.1 milligauss are required. Higher values are required to saturate a multiplet. This can lead to problems of saturation of the signal being observed if the irradiated signal is not well separated from the observed signal. Experimentally, it is best to increase the level of B_2 slowly from zero until the maximum signal enhancement is observed. One should always use integration of peak areas for determining the enhancement, as peak height often gives misleading results due to the removal of unresolved small couplings. Furthermore, one should compare the peak area obtained, when B_2 is offset, with a blank region of the spectrum.

h. Heteronuclear Double Resonance

Although the types of experiments are basically the same, there are several important differences between homonuclear and heteronuclear double resonance as far as instrumentation is concerned (21,153). Many of the recent research NMR spectrometers have provisions for heteronuclear double resonance. However, the older instruments must be modified. First, a means must be provided for introducing the second radio frequency to the sample. This has been accomplished by either placing a second transmitter coil in the probe or by double tuning the existing transmitter coil to accept the frequencies of both B_1 and B_2 (153). Second, since the frequency of B_2 usually differs considerably from that of B_1, some means should be provided for locking the frequency of B_2 to the frequency of B_1 in order to compensate for drifts in the magnetic field. The frequency of B_2 can be derived from a frequency synthesizer, from the master crystal oscillator of the spectrometer, or from a stable radio frequency oscillator which is frequency modulated to give sidebands at the proper frequency for double resonance. Third, since proton–other nuclei coupling constants are usually much larger than proton–proton coupling constants, the strength of B_2 is considerably larger in heteronuclear double resonance experiments. The high power levels required usually leads to heating of the probe and provision should be provided for maintaining the probe at constant temperature.

Heteronuclear spin decoupling $H - \{X\}$ has been used to remove the coupling constants to other nuclei in proton spectra. This provides a convenient means not only for simplifying proton spectra for analysis but also for identifying the proton–hetero nuclei coupling constants. If the frequency relationship between B_1 and B_2 is known, heteronuclear nuclear spin decoupling experiments also provide a means for determining the chemical shifts of other nuclei by determining the frequency of B_2 for maximum decoupling. This application is particularly useful for those nuclei that are not very sensitive to NMR detection.

Another major use of heteronuclear spin-decoupling has been to remove the broadening effects due to quadrupole relaxation (21,153). The proton spectra of compounds containing ^{2}D, ^{11}B, and ^{14}N often show line-broadening due to the quadrupolar relaxation of these nuclei. Such broadening complicates the accurate determination of absorption frequencies and may be eliminated by irradiation at the absorption frequencies for these nuclei.

Heteronuclear spin-tickling experiments provide a convenient means for determining the relative signs of proton–other nuclei coupling constants. The techniques are essentially the same as outlined in the previous discussion of homonuclear spin-tickling. The coupling of protons to other magnetic nuclei, present in low natural abundance, is observed in proton spectra as satellite resonances displaced on both sides of the resonance due to species containing other nuclei which are nonmagnetic. The frequencies in the other nuclei spectrum and the relative signs of the coupling constants can be obtained by observing the changes in the satellite resonances while selectively irradiating the resonances due to the other nuclei. Similar techniques have also provided the signs and magnitudes of a variety of coupling constants between two heteronuclei (153). Applications of the INDOR technique can also provide similar information.

One of the more important applications of heteronuclear double resonance has been in $X - \{H\}$ experiments. The NMR of heteronuclei is not as sensitive as that of proton NMR due to the smaller magnetic moments and the lower natural abundance of the heteronuclei. Some sensitivity enhancement is achieved by irradiation over all the proton frequency range using broadband decoupling, which results in the collapse of the multiplet and the concentration of the intensity in a single line. Furthermore, when the protons are directly attached to the heteronuclei, an additional increase in sensitivity may be achieved due to a nuclear Overhauser effect. The broadband proton decoupling technique is now standard practice in ^{13}C-NMR studies where an increase in sensitivity by a factor of ~ 3 may be obtained in certain cases (135).

i. Lanthanide-Shift Reagents

The discovery by Hinckley (111) in 1969 that certain paramagnetic lanthanide β-diketonates can be used to induce stereospecific shifts in the spectra of certain organic compounds has opened up a new area in NMR spectroscopy. Although large shifts induced by paramagnetic lanthanide ions had been observed previously, the ease of application, the small accompanying line-broadening, and the information obtainable from the use of lanthanide-shift reagents (LSR) has greatly expanded previous applications. The initial application of LSR was to reduce the complexity of proton NMR spectra (in many cases the spectra are reduced to first order) (Fig. 1.35), and

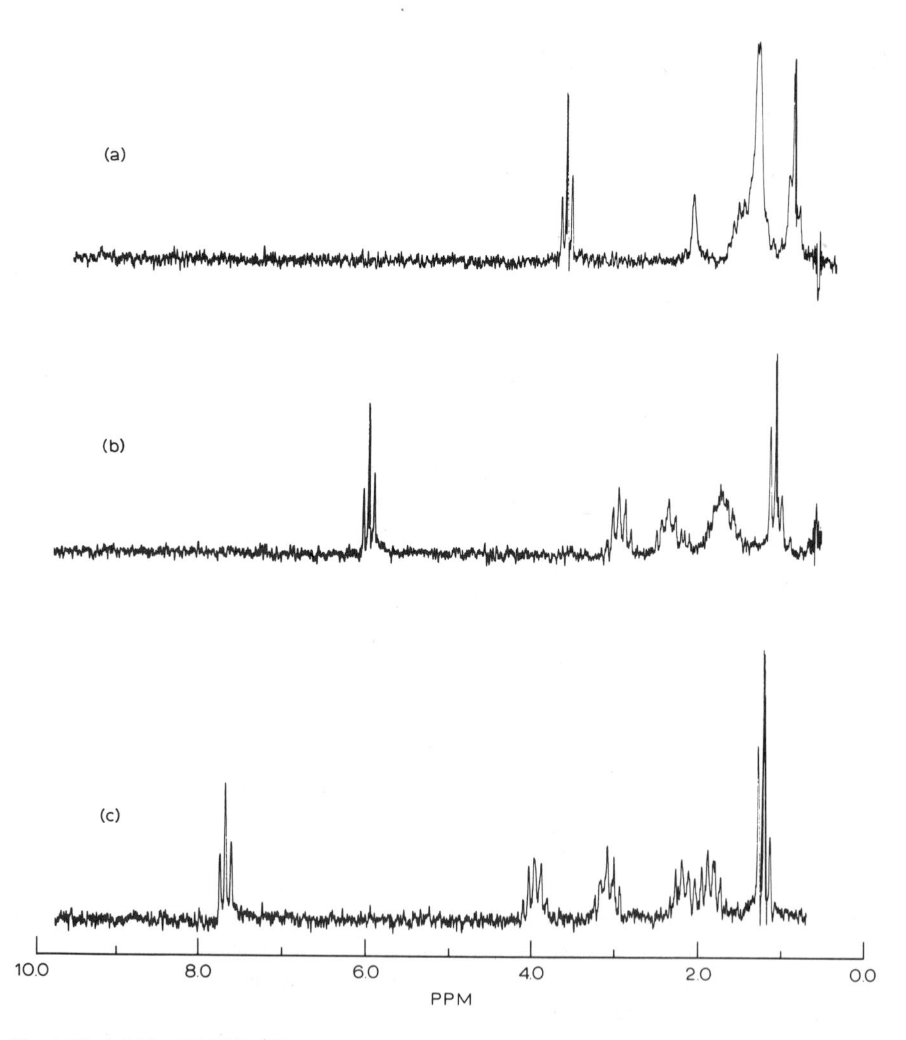

Fig. 1.35. (*a*) The 100-MHz ^1H spectrum of hexanol. (*b*) The spectrum as a result of adding a small amount of $Eu(DPM)_3$ to the sample. (*c*) The spectrum as a result of adding a larger amount of $Eu(DPM)_3$ to the sample.

additional applications have been rapidly forthcoming. Excellent reviews of LSR have appeared (56,157,202).

The usual approach has been to obtain a series of spectra in which increasing increments of a LSR have been added to the sample. A plot of the induced shifts, Δ_{obs}, versus either the concentration of the LSR or [LSR]/[Substrate] is then made. For small relative concentrations of LSR, these plots are usually linear. After assigning the resonances to specific protons and extracting the coupling constants from the "shifted" spectrum, the chemical shifts in the absence of the LSR are found by extrapolating the $\Delta_{obs's}$ back to zero concentration of LSR. It is assumed that the coupling constants obtained from the "shifted" spectrum are identical to the coupling constants in the absence of the LSR. There is some indication that LSRs may affect coupling constants (220). However, using this technique, it is often possible to analyze a complex spectrum as a first-order spectrum.

Another common practice when using LSRs has been to extrapolate the above plots to a 1:1 molar ratio of LSR to substrate and refer to these values of Δ_{obs} at the 1:1 ratio to assess the relative shifting power of various LSRs and the degree of complex formation for various substrates. It now appears that drawing such extrapolations should be avoided since the plots have been found to exhibit curvature at higher LSR/substrate ratios (202).

Several solvents have been used for LSR studies. On the basis of extrapolation of plots similar to that above to a 1:1 molar ratio of substrate to LSR, it appears that carbon tetrachloride is a better solvent (larger induced shifts) than chloroform. Carbon disulfide has also been used. Most of the LSR are hygroscopic and water should be avoided, as it competes with the substrate for the LSR.

The induced shifts observed with LSR depend upon the complexation of the lanthanide ion with a Lewis base site (usually a heteroatom) on the substrate molecule. Although a variety of paramagnetic complexes have been examined for their potential as shifts reagents, the best LSR to date are the *tris*-dipivaloylmethanates (DPM) (**V**) and the *tris*-1,1,1,2,2,3,3-heptafluoro-7, 7-octanedionates (FOD) (**VI**). Their utility as shift reagents is thought to arise from their ability to expand the coordination of the lanthanide by accepting additional ligands.

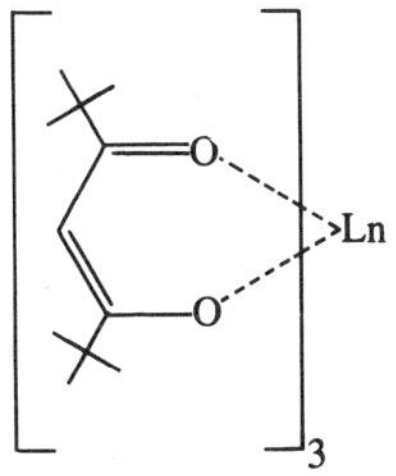

V 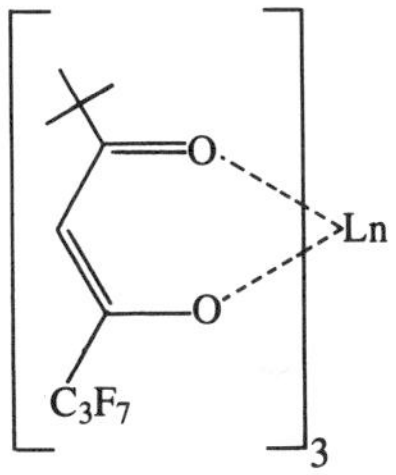 **VI**

Several lanthanide complexes of DPM and FOD have been examined. While large induced shifts have been observed, in many cases the substrate resonances are broadened to the extent that the coupling constants cannot be obtained from the spectra. As a compromise between the magnitude of the induced shift and the accompanying line-broadening, the best lanthanide ions appear to be Eu, Pr, and Yb. These LSRs also exhibit greater solubility in organic solvents normally used in NMR experiments. The induced shifts observed in proton spectra are to lower field with Eu and Yb LSR and to higher field with Pr LSR (202). Both $Eu(NO_3)_3 \cdot 6H_2O$ and $Pr(NO_3)_3 \cdot 6H_2O$ are useful LSRs for aqueous solutions (211). Two mechanisms have been proposed to account for LSR-induced shifts Δ_{obs}; (a) the contact mechanism, Δ_c, and (b) the pseudocontact mechanism, Δ_{pc}. The contact shift mechanism contributes to the induced shift only if there is transfer of unpaired

$$\Delta_{obs} = \Delta_c + \Delta_{pc} \tag{54}$$

spin density from the LSR to the substrate. This transfer of spin density may occur either through direct delocalization of the unpaired spin or by spin polarization. The induced shift due to the contact mechanism is given by (142)

$$\Delta_c = \frac{-2\pi\beta\nu AJ(J+1)gL(gL-1)}{3kt\nu} \tag{55}$$

where β is the Bohr magneton, ν is the nuclear Larmor frequency, A is the scalar coupling constant, J is the electronic spin angular momentum, and gL is the Landé g factor. The contact-induced shift falls off rapidly with distance from the coordination site and, for proton spectra, is thought to be important only for protons in the immediate vicinity of the coordinating site in the substrate molecule. However, for nuclei other than protons, the contact-shift mechanism may be the dominant shift mechanism (25).

The pseudocontact shift arises from a dipolar interaction between the nucleus and the electron spin magnetization of the paramagnetic lanthanide ions and results from the nonaveraging of the anisotropic electronic g tensors in the complexes. Alternatively, the induced shift may be thought of as arising from the magnetic field generated by the unpaired electron spin. If it is assumed that the complex is axially symmetric, the pseudocontact shift is given by (152).

$$\Delta_{pc} = \frac{-\nu B^2 J(J+1)}{9\,kT\mathbf{r}^3}(3\cos^2\theta - 1)(g_z - g_x)(g_z - g_y) \tag{56}$$

where θ is the angle between the distance vector $\mathbf{r}$ joining the lanthanide ion and the nucleus in the complexed substrate and the crystal field axis of the complexed substrate, and g represents the g tensor components. It is further assumed that the crystal field axis of the complex lies along a line joining the lanthanide ion and the coordinating site.

It is thought that the pseudocontact shift mechanism makes the major contribution to LSR-induced shifts of protons. The majority of data have been interpreted in terms of equation 57 where n is the constant

$$\Delta_{pc} = \frac{n(3\cos^2\theta - 1)}{T\mathbf{r}^3} \tag{57}$$

of proportionality. Equation 57 predicts that LSR-induced pseudocontact shifts depend upon the distance separating the lanthanide ion and the nucleus in question, the angle θ (the sign changes at $\theta = 54°$), and the temperature.

The $1/\mathbf{r}^3$ distance dependence of LSR-induced shifts was recognized early and has been adequately demonstrated. The angular dependence of LSR-induced shifts was not considered until it was observed that some resonances are shifted downfield while others are shifted upfield in the same molecule (56,157,202). It now appears that both factors should be considered together. One problem with the angle dependence is the location of the lanthanide ion with respect to the substrate. Computer programs have been written for an iterative procedure whereby the location of the lanthanide ion is varied until a "best fit" is obtained for all shifted nuclei (247). The effects of temperature have not been investigated thoroughly and it is not clear at this time whether the observed effects are due to the temperature *per se* or to the effects of temperature on complex formation (202). It has been observed that the induced shift increases with decreasing temperature in some cases and decreases in others (56,202).

Considerable effort has been devoted to the determination of the stoichiometry of the LSR–substrate complex. Since only one set of resonances are observed for the substrate, there must be fast exchange on the NMR time scale between the substrate molecules. Various equilibria have been considered to derive expressions for the determination of the dissociation constant of the complex and the shift of the

$$\begin{aligned} \text{S} + \text{LSR} &\rightleftharpoons \text{S–LSR 1:1} \\ \text{S–LSR} + \text{S} &\rightleftharpoons \text{S}_2\text{–LSR 1:2} \end{aligned} \tag{58}$$

pure complex. Most of the treatments have been patterned after the studies of charge transfer complex and hydrogen bonding. Caution should be exercised in such studies since interactions of the LSR with impurities or the solvent will affect the results.

Various functional groups have been examined with LSRs. Larger induced shifts are observed with the more basic functional groups. There appears to be a general correlation between the induced shift and the pKa of the substrate (77). Steric factors are also important. Several studies of the ability of various functional groups to coordinate with LSRs have established the following orders of effectiveness: amine > hydroxyl > ketone > aldehyde > ether > ester > nitrile (212), ether > thioether > ketone > ester (105), and phosphoryl > carbonyl > thiocarbonyl > thiophosphoryl (241). Structural factors such as $\alpha - \beta$ unsaturation may change the above order.

Chiral-shift reagents have been prepared and found to be effective in resolving the spectra of enantiomeric mixtures (91,246). Most of these shift reagents use camphor to introduce the optically active center. When a chiral-shift reagent is added to an enantiomeric mixture, it is found that the protons of one enantiomer appear at a lower field than the other enantiomer. These reagents have been used with amines, alcohols, esters, epoxides, and ketones. It appears as if the chiral-shift reagent is forming complexes having different dissociation constants with the enantiomers. There is some evidence that the geometries of the complexes may also be different.

LSRs have found application in a variety of uses other than the analysis of complex spectra. LSR-induced shift data, in conjunction with the $1/\mathbf{r}^3$ and θ dependence, are extremely valuable aids in assigning resonances to specific protons. They also provide information concerning the stereochemistry of the substrate (caution shoud be exercised in conformation studies, since the formation of the complex may change the rotamer populations). Analysis of a mixture of similar compounds is facilitated by the use of LSR.

Recent efforts have been devoted to expanding LSR to weak nucleophiles such as aromatics and olefins. Neither Ag(FOD) nor Yb(FOD) alone are very effective shift reagents for alkenes and aromatics. However, an equal molar mixture of the two produces large downfield shifts (244a,244b). Apparently, the silver-diketonate is bound to the substrate and acts as a bridge between the substrate and LSR. Separation of the signals due to enantiomers of alkenes has been achieved using Ag(FOD)/lanthanide *D*-camphorates, leading to the determination of optical purities (171a). The above combination of reagents should greatly expand the types of functional groups amenable to analysis using shift reagents.

j. Two-Dimensional FT-NMR Spectroscopy

In addition to the advances made in NMR due to the development of modern, high-field, Fourier transform (FT)-NMR spectrometers, another development with great potential impact on the analysis of complex NMR spectra is that of two-dimensional NMR. A normal one-dimensional FT-NMR spectrum is obtained by Fourier transformation of a free induction decay (FID) acquired as a function of an acquisition time t_2. The two-dimensional NMR experiment is divided into three time periods: the preparation, evolution and detection periods. During the preparation period, the nuclear spins are prepared to a specific state, i.e., equilibrium, decoupled, etc. In the evolution period defined by the time t_1, the magnetization due to the spins develops according to some prescribed motion that will influence the detected signal. During the detection period, the signal is acquired for a time t_2 similar to the one-dimensional NMR experiment. This yields a FID signal $S(t_1,t_2)$ as a function of two time variables. A double Fourier transformation then yields the two-dimensional frequency domain spectrum $S(F_1,F_2)$. Many types of two-dimensional spectra (both homonuclear and heteronuclear) are possible depending upon what is done to the magnetization during the preparation and evolution periods. A full account of two-dimensional FT-NMR is beyond the scope of this chapter. Excellent reviews of two-dimensional NMR are available (83a,84a). Here, we shall focus briefly on three types of homonuclear two-dimensional NMR experiments that have proven useful in the analysis of complex ^{1}H-NMR spectra. A general strategy for the analysis of ^{1}H-NMR spectra from complex organic molecules using a variety of one-dimensional and two-dimensional techniques has been presented (103b,c).

The homonuclear two-dimensional J spectroscopy experiment (16a) is a variant of the classical Carr-Purcell spin-echo experiment and uses the following pulse sequence: $90° - t_1/2 - 180° - t_1/2$-acquisition. The experiment is initiated with a nonselective 90° pulse followed by a nonselective refocusing 180° pulse in the middle of the evolution period. The FIDs are collected as a function of t_1 yielding a set $S(t_1,t_2)$ of time-domain spectra. After double Fourier transformation and a 45° tilt of the frequency axes (103a), one obtains a spectrum containing chemical shift information along the F_2 axis and homonuclear coupling constant information along the F_1 axis for spin systems that are weakly coupled. Summation along the F_2 axis yields a spectrum in which each proton appears at its chemical shift as a singlet. Thus, one can obtain a "proton decoupled" proton NMR spectrum. All of the multiplet intensity from each proton is on lines paralled to the F_1 axis. Therefore, one can generate individual cross-sections from each proton yielding the multiplets for the splitting of that particular proton. This method separates the multiplets even though they may overlap extensively in the normal one-dimensional spectrum and permits a first-order analysis of the coupling constant information. Strong coupling is recognized by the appearance of unsymmetrical multiplets at the chemical shift halfway between the shift of the coupled protons (31a). The resolution in the F_1 dimension (coupling constant information) is determined by the reciprocial of twice the increment in t_1, i.e., $2*N*\Delta t_1$ where N is the number of FIDs collected and Δt_1 is the increment in t_1 value between each FID. Therefore, collection of 128 FIDs where Δt_1 is 10 msec will yield a resolution of 0.39 Hz and 256 FIDs would yield a resolution of 0.19 Hz in the F_1 frequency axis. The major disadvantage (if it is a disadvantage considering the information obtained) of this method is the relatively large amount of time (several hours) required to acquire and process the data. It is well to keep in mind that the information obtained may not be available using other methods.

Another useful two-dimensional method is two-dimensional correlated spectroscopy, which manifests connectivities between J-coupled protons. The following pulse sequence is used: 90°-t_1-90°-acquisition (16a). The magnetization components are labeled with their characteristic precession frequencies during the evolution period t_1 and the second 90° pulse causes coherence transfer of magnetization components among all those transitions which belong to the same coupled spin system. The experiment is repeated for a set of equidistant t_1 values normally using an equal number of data points, i.e. 512, in each frequency dimension. After double Fourier transformation, one obtains a square array where the normal one-dimensional spectrum lies along the diagonal. Cross-peaks resulting from proton–proton connectivities appear symmetrically with respect to the diagonal. Normally, a contour plot of a selected region of the two-dimensional spectrum is recorded. This method has the advantage over single-frequency decoupling experiments in that a complete set of J connectivities in a molecule is provided in one experiment and it avoids the problems arising from missetting the decoupling frequency and from limited selectivity of setting the decoupling frequency in crowded spectral regions.

A complete set of NOEs between nearby protons in a molecule may be obtained in one experiment using two-dimensional nuclear Overhauser enhancement spectroscopy (135a). Three nonselective 90° pulses are used in the following pulse sequence: 90°-t_1-90°-t_m-90°-acquisition. As in the above method, the various magentization components are frequency labeled during the evolution period t_1. During the mixing time, t_m, cross-relaxation leads to exchange of magnetization between nearby protons through mutual dipolar interactions. The FIDs are collected for a set of equidistant t_1 values with t_m being fixed. Double Fourier transformation produces the two-dimensional frequency spectrum where one observes peaks on the diagonal disecting the two frequency axes. The diagonal peaks correspond to magnetization components which do not exchange with other components during the mixing time, t_m. Symmetrical off-diagonal peaks appear where there is magnetization transfer due to dipole–dipole cross-relaxation during the mixing time. With short mixing times, i.e., 100 msec, only the shortest proton–proton distances will be observed. With longer mixing times, i.e., 300 msec, longer distances may be observed. By systematically varying t_m, the buildup rates of the NOE may be determined (135b). As above, the spectra are usually presented as a contour plot. In addition to yielding all the NOEs in a single experiment, this method avoids the errors associated with limited selectivity of preirradiation in crowded spectral regions.

It is likely that the area of two-dimensional NMR will continue to develop rapidly in the near future. New pulse sequences to produce specific desired results in spectra are continually being presented and the established methods are being applied to more complex molecules. The authors would not be surprised if in the near future that two-dimensional methods become the methods of choice for the analysis of complex spectra.

B. CORRELATION OF NMR PARAMETERS

When an NMR spectrum is used in addition to other spectral and chemical information, it is often possible to determine uniquely the structure of an unknown compound. In some cases, if a molecular formula is known, the NMR spectrum alone may be sufficient to determine the structure. In more complex cases, the NMR spectrum often provides information that would be difficult or impossible to obtain by other spectral methods.

Basically, there are three types of information available from an NMR spectrum: (a) the chemical shifts; (b) integrated intensity; and (c) coupling constants. Each of these pieces of information provides unique information for use in the determination of structure. The value of the chemical shifts allows one to determine the types of protons present in the sample, i.e, protons bonded to saturated carbons, protons bonded to olefinic carbons, protons bonded to aromatic carbons, protons bonded to carbon in functional groups such as aldehydes, and protons bonded to heteroatoms. In favorable cases, one can further distinguish protons as to specific type such as methyl, methylene, and methine protons. Furthermore, the number of absorptions in the sample allows one to determine the number of chemically different types of protons. The integration provides a determination of the relative number of protons of each chemically different type. If one of the peaks in the spectrum can be assigned to a specific number of protons, a methyl group for example, the total number of protons in the sample can be determined from the relative areas of the remaining peaks. The coupling constants provide information on the geometrical relationship of the protons in the sample. The splitting pattern produced as a result of coupling not only allows one to determine which protons are adjacent in the sample, but also the number of adjacent protons. With experience, one will be able to recognize the splitting patterns in NMR spectra that are characteristic of particular groupings of nuclei or molecular fragments. From a knowledge of the chemical shifts and coupling constants, one can start putting together various fragments of an unknown structure much like one would go about solving a jigsaw puzzle.

The use of chemical shifts and coupling constants in structure determinations relies very heavily on the use of model compounds and the correlations between these parameters and structures derived from the examination of known compounds. Rather than try to cover each possible case that may arise (an impossible task), we shall discuss the most general aspects of chemical shifts and coupling constants in the following sections. There are several excellent compilations of proton NMR data available (34,36,74,113,209,240). These sources should be consulted for additional data on chemical shifts and coupling constants.

1. Chemical Shifts

The examination of the NMR spectra of a wide variety of known compounds has shown that protons bonded to carbon appear in characteristic chemical shift ranges, depending upon the hybridization of the attached carbon and the substitution in the compound. Protons bonded to sp^3 hybridized carbons normally absorb within the range 0.5–5.0 ppm. These proton chemical shifts are influenced primarily by the electronegativity of adjacent substituents (diamagnetic screening) and long-range shielding effects of particular groupings of nuclei, as discussed previously. Of the protons that appear in this range, it is usually easier to distinguish methyl protons from the others, since methyl protons can only appear in a limited number of splitting patterns (singlet, doublet, or triplet) and the signals of a methyl group are more intense than those of methylene and methine protons. In favorable cases where there is no overlap of the absorptions in the spectrum, one should be able to distinguish methylene from methine protons on the basis of the integration.

The chemical shifts of several substituted alkanes which can serve as model compounds are given in Table 1.X. The effect of a single substituent on proton chemical shifts in alkanes depends on the location of the substituent with respect to the proton and the type of proton in question. For protons directly bonded to the carbon to which the substituent is attached H–C–X (α protons), the chemical shifts vary over a range of 3–4 ppm. Protons three bonds removed from the substituent (β protons) vary over a range of ~1 ppm while protons four bonds from the substituent vary over a range of only ~0.3 ppm. Substituent effects on protons further than four bonds removed from the substituent are usually negligible.

From the data in Table 1.X, it can be seen that one can easily distinguish several types of methyl protons. For example, the chemical shift separation between $CH_3–CH_2–$, $CH_3C(=O)–$ and $CH_3O–$ is large enough such that there should be no difficulty in assigning these peaks in an unknown structure. There is however, considerable overlap of the various methyl resonances between 2.0 and 3.0 ppm such that it may be difficult to assign signals uniquely in this region to a particular functional group in the absence of further information.

Several attempts have been made to derive an additivity scheme for predicting the chemical shifts of aliphatic protons in compounds containing a single functional group. One such set of additivity parameters for calculating the chemical shifts of methyl, methylene, and methine protons is given in Table 1.XI. It should be kept in mind that the use of the substituent parameters in Table 1.XI allows one to calculate

TABLE 1.X

Chemical Shifts For Some Alkyl Derivatives With A Single Functional Group[a,b]

X	Methyl	Ethyl		n-Propyl			iso-Propyl	
	CH_3X	CH_3	CH_2X	CH_3	CH_2	CH_2X	CH_3	CHX
H	0.23	0.86	0.86	0.91	1.33	0.91	0.91	1.33
$CH{=}CR_2$	1.73	1.00	2.00					
$-C{\equiv}CR$	1.75	1.15	2.15	0.97	1.50	2.10	1.15	2.59
$-C_6H_5$	2.34	1.21	2.63	0.95	1.65	2.59	1.25	2.89
—CHO	2.18	1.13	2.46	0.98	1.65	2.35	1.13	2.39
—COR	2.10	1.05	2.47	0.93	1.56	2.32	1.08	2.54
$-CO_2H$	2.08	1.16	2.36	1.00	1.68	2.31	1.21	2.56
$-CO_2R$	2.01	1.12	2.28	0.98	1.65	2.22	1.15	2.48
$-CONR_2$	2.05	1.13	2.23	0.99	1.68	2.19	1.18	2.44
—F	4.27	1.24	4.36					
—Cl	3.06	1.33	3.47	1.06	1.81	3.47	1.55	4.14
—Br	2.69	1.66	3.37	1.06	1.89	3.35	1.73	4.21
—I	2.16	1.88	3.16	1.03	1.88	3.16	1.89	4.24
—CN	1.98	1.31	2.35	1.11	1.71	2.29	1.35	2.67
$-NO_2$	4.29	1.58	4.37	1.03	2.01	4.28	1.53	4.44
$-NH_2$	2.47	1.10	2.74	0.93	1.43	2.61	1.03	3.07
—NHCOR	2.71	1.12	3.21	0.96	1.55	3.18	1.13	4.01
—OH	3.39	1.18	3.59	0.93	1.53	3.49	1.16	3.94
—OR	3.24	1.15	3.37	0.93	1.55	3.27	1.08	3.55
—OCOR	3.67	1.21	4.05	0.97	1.56	3.98	1.22	4.94
—SH	2.00	1.31	2.44	1.02	1.57	2.46	1.34	3.16
—SR	2.09	1.25	2.49	0.98	1.59	2.43	1.25	2.93

[a] In ppm.

[b] Data taken from references 74 and 113.

TABLE 1.XI

Substituent Effects on Aliphatic Proton Chemical Shifts[a]

Substituent	Type of hydrogen[b]	Alpha shift	Beta shift
—Cl	CH_3	2.43	0.65
	CH_2	2.30	0.53
	CH	2.55	0.03
—Br	CH_3	1.80	0.83
	CH_2	2.18	0.60
	CH	2.68	0.25
—I	CH_3	1.28	1.23
	CH_2	1.95	0.58
	CH	2.75	0.00
—OH	CH_3	2.50	0.33
	CH_2	2.30	0.13
	CH	2.20	0.00
—OR (sat.)	CH_3	2.43	0.33
	CH_2	2.35	0.15
	CH	2.00	0.00
$—OC_6H_5$, —O—C(=O)R	CH_3	2.88	0.38
—OC(=O)OR	CH_2	2.98	0.43
	CH	3.43	—
—C=C—	CH_3	0.78	—
	CH_2	0.75	0.10
	CH	—	—
—C(=O)—X, X=OH, OR, H, N alkyl, or aryl	CH_3	1.23	0.18
	CH_2	1.05	0.31
	CH	1.05	—
—NRR′	CH_3	1.30	0.13
	CH_2	1.33	0.13
	CH	1.33	—

[a] Taken from reference 225.

[b] These values are to be added to the standard positions: CH_3, δ 0.87; CH_2, δ 1.20; CH, δ 1.55.

approximate chemical shifts only. Normally, the deviations will be within 0.3 ppm. Furthermore, the data refer to dilute solutions (less than 10%) in either $CDCl_3$ or CCl_4. The use of another solvent may lead to significant deviations from the calculated value. The additivity parameters given in Table 1.XI should also prove useful for higher substituted alkanes provided the substituents are located at least six bonds from each other.

An additivity scheme has also been proposed for predicting the chemical shifts of methylene protons where the carbon is attached to two substituents (225). These additivity

$$\delta = 0.28 + \sum \sigma_{\text{eff}} \qquad (59)$$

constants σ_{eff}, known as Shoolery's constants, are given in Table 1.XII. Again, chemical shifts calculated using equation 59 should be regarded as approximate values only. Deviations of up to 0.6 ppm have been observed in some cases. Although, the use of the substituent parameters in Tables 1.XI and 1.XII leads to significant deviations in some instances, they nevertheless have proven useful in assigning protons to specific absorptions in NMR spectra.

The chemical shifts of protons bonded to sp^2 hybridized carbons in olefins are dependent on the remaining substituents attached to the carbons. Data for a large number of compounds have been reported. Examination of this data has led to an additivity scheme that is very useful for predicting the chemical shifts of olefinic protons (67,236). The chemical shift of an olefinic proton can be calculated by taking into account the geometrical relationship of the proton with respect to the substituents by using equation 60

```
H         cis
 \       /
  C==C
 /       \
gem       trans
```

$$\delta = 5.28 + \sum \sigma_{\text{eff}} \qquad (60)$$

TABLE 1.XII

Shoolery's Constants for Calculating the Chemical Shifts of Methylene Protons Bonded to Two Substituents $X—CH_2—Y$[a,b]

X or Y	σ_{eff}	X or Y	σ_{eff}
$—CH_3$	0.47	—CN	1.70
—C=C	1.32	—COR	1.70
—C≡CR	1.44	$—COC_6H_5$	1.84
—I	1.82	$—CONR_2$	1.59
—Br	2.33	—COOR	1.55
—Cl	2.53	—OCOR	3.13
$—C_6H_5$	1.85	—SR	1.64
$—NR_2$	1.57	—SCN	2.30
$—NO_2$	2.46	$—CF_2$	1.21
—OH	2.56	$—CF_3$	1.14
—OR	2.36	$—N_3$	1.97
$—OC_6H_5$	3.23	—NHCOR	2.27

[a] In ppm.
[b] Reference 225.

where 5.28 is the chemical shift of ethylene and σ_{eff} is the additivity parameter for the substituent. Substituent additivity parameters are given in Table 1.XIII. In most cases, chemical shifts calculated using equation 60 are within ± 0.3 ppm of the experimental value. Larger deviations are observed in cases where there is extended conjugation, ring strain, or inhibition of resonance. Chemical shifts for various types of aliphatic and olefinic compounds are given in Table 1.XIV.

The proton chemical shifts of monosubstituted benzenes have been investigated by a number of workers. The chemical shifts of aromatic protons usually lie within the range of 6.0–8.5 ppm (Table 1.XV). A ring current in the aromatic ring is thought to be responsible for the downfield shift of aromatic protons compared to olefinic

TABLE 1.XIII

Substituent Constants for Calculating the Chemical Shift of Olefinic Protons[a]

X	σ_{gem}	σ_{cis}	σ_{trans}
—CH_3	0.44	−0.32	−0.34
—Alkyl	0.44	−0.26	−0.29
—Alkyl-ring	0.71	−0.33	−0.30
—C≡C—	0.50	0.35	0.10
—C=C	0.98	−0.04	−0.21
—C=C (conj)	1.26	0.08	−0.01
—C_6H_5	1.43	0.39	0.06
—Aromatic	1.35	0.37	−0.10
—CH_2O—,	0.67	−0.02	−0.07
—CH_2S—	0.53	−0.15	−0.15
—CH_2Cl, —CH_2Br	0.72	0.12	0.07
—CH_2N	0.66	−0.05	−0.23
—C≡N	0.30	0.75	0.53
—C=O	1.10	1.13	0.81
—C=O (conj)	1.06	1.01	0.95
—CO_2H	1.00	1.35	0.74
—CO_2H (conj)	0.69	0.97	0.39
—CO_2R	0.84	1.15	0.56
—CO_2R (conj)	0.68	1.02	0.33
—CHO	1.03	0.97	1.21
—$CONR_2$	1.37	0.93	0.35
—COCl	1.10	1.41	0.99
—OR (R aliph)	1.18	−1.06	−1.28
—OR (R conj)	1.14	−0.65	−1.05
—Cl	1.05	0.14	0.09
—Br	1.02	0.33	0.53
—NR_2 (R aliph)	0.69	−1.19	−1.31
—NR_2 (R conj)	2.30	−0.73	−0.81
—SR	1.00	−0.24	−0.04
—SO_2—	1.58	1.15	0.95

[a] Data taken from references 67, 236

TABLE 1.XIV

Chemical Shifts of Protons Bonded to Carbon in Some Miscellaneous Compounds[a]

Compound	δ	Compound	δ
	0.22	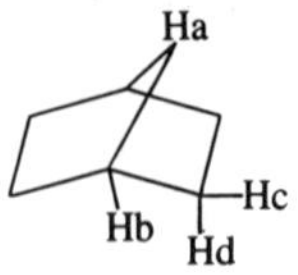 Ha, Hb, Hc, Hd	a 1.21 b 2.20 c 1.49 d 1.18
	1.96	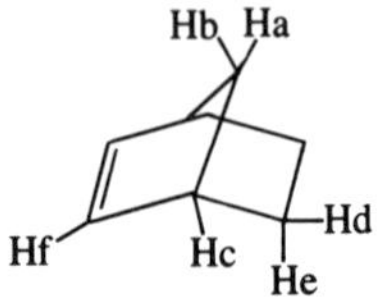 Hb, Ha, Hf, Hc, Hd, He	a 1.32 b 1.07 c 2.83 d 1.57 e 0.94 f 5.95
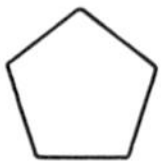	1.51		
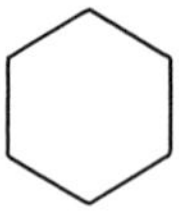	1.44	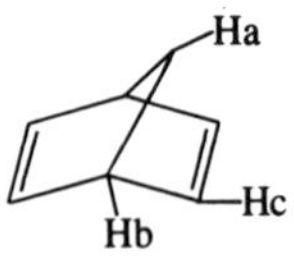 Ha, Hb, Hc	a 1.95 b 3.53 c 6.66
	1.54	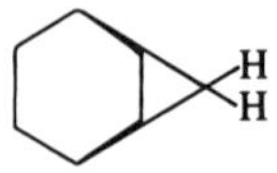 H, H	0.02
$H_2C{=}C{=}CH_2$	4.55	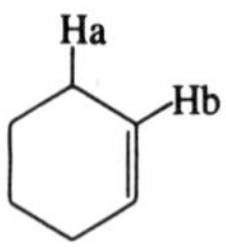 Ha, Hb	a 1.96 b 5.57
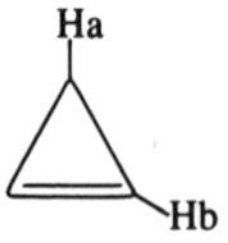 Ha, Hb	a 0.92 b 7.01	 O	1.65

TABLE 1.XIV *(Continued)*

Chemical Shifts of Protons Bonded to Carbon in Some Miscellaneous Compounds[a]

Compound	δ	Compound	δ
Ha, Hb	a 2.57 b 5.97	O, Hb, Ha	a 3.03 b 1.96
Ha, Hb	a 2.28 b 5.60	O, Ha, Hb	a 2.06 b 2.02
CH_2a, Hb	a 5.38 b 0.99	O, H	2.25
H_2a, Hc, Hb	a 4.70 b 2.7 c 1.92	CH_2a, Hb	a 4.55 b 1.5
CH_2a, Hb, Hc	a 4.82 b 2.7 c 1.92	O, Hb, Ha	a 4.73 b 2.72
O	2.54	O, Ha, Hb	a 3.56 b 1.58
O, Ha, Hb	a 3.63 b 1.79	O, O	3.59

TABLE 1.XIV *(Continued)*

Chemical Shifts of Protons Bonded to Carbon in Some Miscellaneous Compounds[a]

Compound	δ	Compound	δ
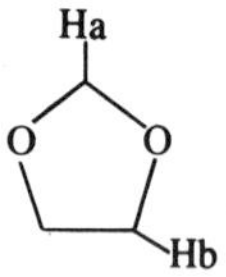 1,3-dioxolane (Ha, Hb)	a 4.77 b 3.77	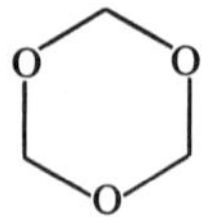 1,3,5-trioxane	5.00
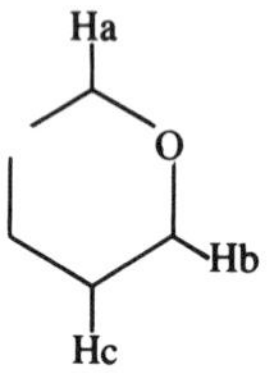 1,3-dioxane (Ha, Hb, Hc)	a 4.82 b 3.80 c 1.68	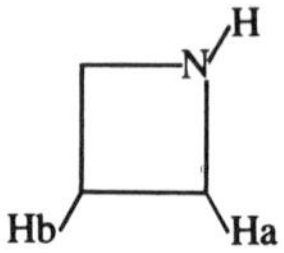 azetidine (Ha, Hb)	a 3.54 b 2.23
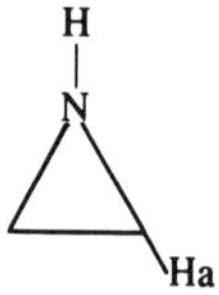 aziridine (Ha)	a 1.48	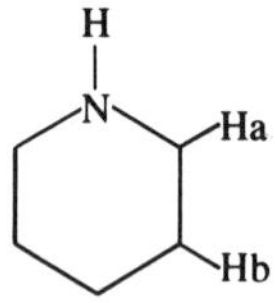 piperidine (Ha, Hb)	a 2.69 b 1.49
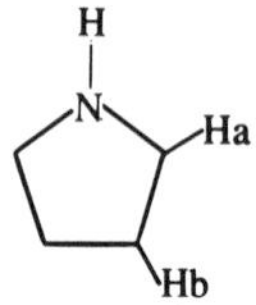 pyrrolidine (Ha, Hb)	a 2.74 b 1.62	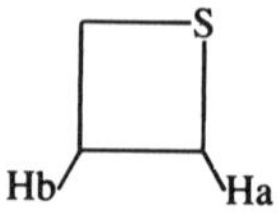 thietane (Ha, Hb)	a 2.82 b 1.93
 thiirane	2.27	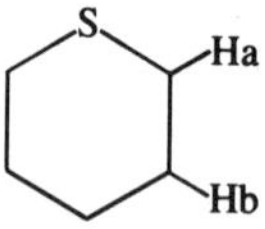 thiane (Ha, Hb)	a 2.57 b 1.6
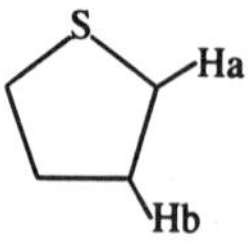 thiolane (Ha, Hb)	a 2.82 b 1.93	 1,3,5-trithiane	4.18

TABLE 1.XIV *(Continued)*

Chemical Shifts of Protons Bonded to Carbon in Some Miscellaneous Compounds[a]

Compound	δ	Compound	δ
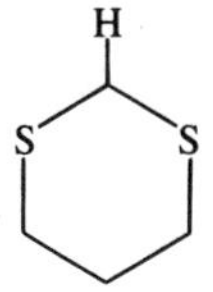	3.69	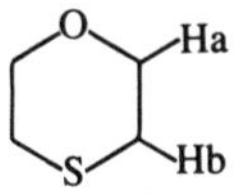	a 3.88 b 2.57
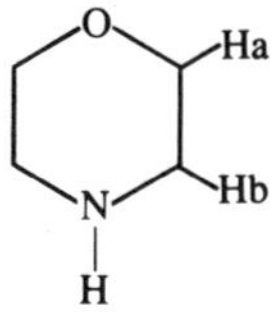	a 3.57 b 2.83	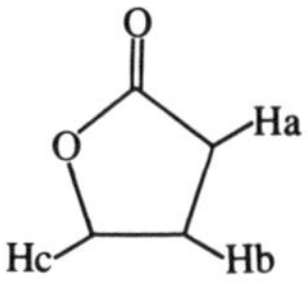	a 2.31 b 2.08 c 4.28
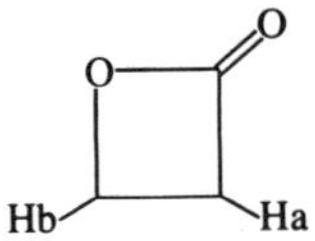	a 3.48 b 4.22	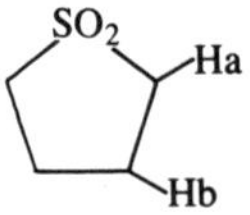	a 2.92 b 2.16
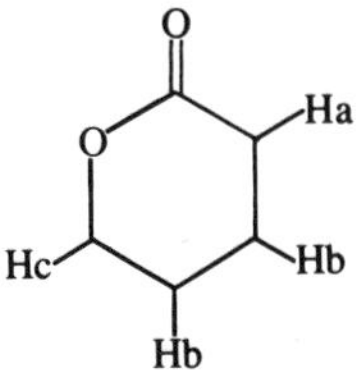	a 2.27 b 1.62 c 4.06	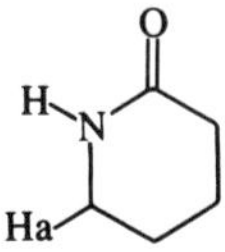	3.17
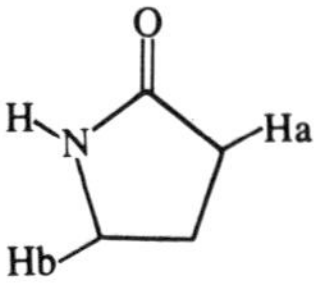	a 2.3 b 3.4		

[a] Data taken from various sources.

protons. Spiesecke and Schneider have considered the effect of substituents on the proton shifts of monosubstituted benzenes (227). They suggested that the *para*-proton shifts are controlled primarily by the resonance effect of the substituent. In support of this, reasonable correlations were found between the *para*-proton shifts and the π-electron density of the attached carbon. Similar correlations were also found with Hammett σ-*para* substituent constants. *Ortho*-proton chemical shifts cannot be explained on the basis of resonance and inductive effects alone. For the halobenzenes, it appears as if there is a major contribution from the diamagentic anisotropy of the substituent. *Meta*-proton chemical shifts do not appear to correlate

TABLE 1.XV

Chemical Shifts for Some Monosubstituted Benzenes[a,b]

X	δ_{ortho}	δ_{meta}	δ_{para}
$—CH_3$	−0.25	−0.17	−0.25
$—C(CH_3)_3$	−0.01	−0.07	−0.13
$—CH{=}CH_2$	0.02	−0.06	−0.13
$—C_6H_5$	0.20	0.03	−0.06
$—C{\equiv}CH$	0.15	−0.05	−0.03
$—C{\equiv}CC_6H_5$	0.19	−0.02	−0.05
$—C{\equiv}N$	0.33	0.17	0.29
—CHO	0.53	0.18	0.27
$—COCH_3$	0.59	0.10	0.18
$—COC_6H_5$	0.43	0.09	0.18
—COCl	0.80	0.20	0.36
$—CONH_2$	0.69	0.17	0.24
—COOH	0.72	0.11	0.22
$—COOCH_3$	0.70	0.07	0.17
$—COOC_6H_5$	0.88	0.15	0.26
—F	−0.30	−0.03	−0.24
—Cl	0.00	−0.07	−0.13
—Br	0.16	−0.12	−0.07
—I	0.37	−0.25	−0.03
$—NH_2$	−0.81	−0.26	−0.65
$—N(CH_3)_3$	−0.69	−0.19	−0.67
$—N^+(CH_3)_3I^-$	0.71	0.39	0.33
$—NO_2$	0.92	0.25	0.38
—OH	−0.46	−0.15	−0.40
$—OCH_3$	−0.50	−0.10	−0.45
$—OC_6H_5$	−0.33	−0.05	−0.29
—Li	0.76	−0.22	−0.30
—MgBr	0.39	−0.20	−0.27
$—SiCl(C_6H_5)_2$	0.31	0.03	−0.14
$—PbCl(C_6H_5)_2$	0.66	0.27	0.10
$—PO(OCH_3)_2$	0.43	0.13	0.21
—SH	−0.06	−0.13	−0.21

[a] Data taken from references 47, 49, 52, 156.

[b] With respect to benzene (7.27 ppm).

with resonance, or with inductive or diamagnetic anisotropic effects of the substituent. It appears as if all three factors are contributing to the *meta*-proton shifts.

Several di- and higher-substituted benzenes have also been examined. From the examination of several *para*-disubstituted benzenes, Diehl (66) has suggested that the effect of substituents on aromatic proton chemical shifts is additive. The effect of a single substituent X on the chemical shifts of a monosubstituted benzene with respect to benzene (Table 1.XV) is denoted as $S_{o;x'}$ $S_{m;x}$ and $S_{p;x}$. The calculation of the proton chemical shifts of a *para*-disubstituted benzene p-C_6H_4XY is given as follows:

$$\delta^{xy}_{o;x} = S_{o;x} + S_{m;y} \quad (61)$$

$$\delta^{xy}_{m;x} = S_{m;x} + S_{o;y} \quad (62)$$

This additivity scheme can also be applied to *meta*-disubstituted benzenes. The respective chemical shifts are

VII

$$\delta_2 = S_{o;x} + S_{o;y} \quad (63)$$

$$\delta_4 = S_{p;x} + S_{o;y} \quad (64)$$

$$\delta_5 = S_{m;x} + S_{m;y} \quad (65)$$

$$\delta_6 = S_{o;x} + S_{p;y} \quad (66)$$

Similar treatment of an *ortho*-disubstituted benzene

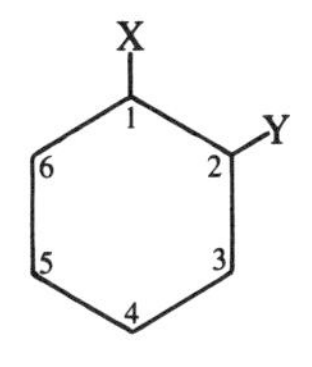

VIII

yields the chemical shifts

$$\delta_3 = S_{m;x} + S_{o;y} \quad (67)$$

$$\delta_4 = S_{p;x} + S_{m;y} \quad (68)$$

$$\delta_5 = S_{m;x} + S_{p;y} \quad (69)$$

$$\delta_6 = S_{o;x} + S_{m;y} \quad (70)$$

For the *para*- and *meta*-disubstituted benzenes, chemical shifts obtained from the above equations agree with experimental values to within ±0.1 ppm in most cases. However, significant deviations are observed for some *ortho*-disubstituted benzenes (60). It has been suggested that the deviations result from steric interactions between the substituents that prevent the substituents from exerting their normal resonance and inductive effects (48). Deviations are also observed when the additivity scheme is applied to 1,2,3-trisubstituted benzenes.

The chemical shifts of a wide variety of substituted heteroaromatic compounds and polycyclic aromatic compounds have been examined. Some representative values are given in Table 1.XVI. In general, the effect of substituents on these ring systems is similar to their effect on benzene chemical shifts. Of special interest are the

TABLE 1.XVI

Representative Chemical Shifts in Aromatic Compounds[a]

Compound	Compound
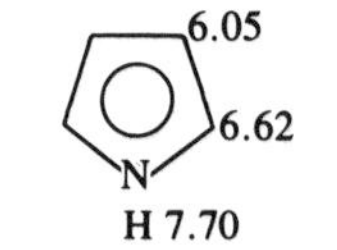 6.05 6.62 N H 7.70	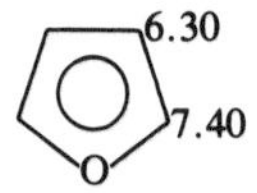 6.30 7.40 O
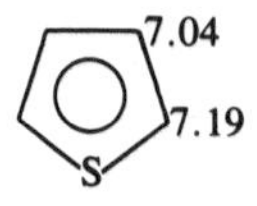 7.04 7.19 S	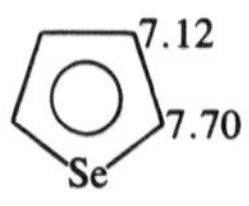 7.12 7.70 Se
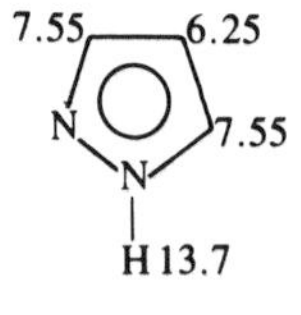 7.55 6.25 N 7.55 N H 13.7	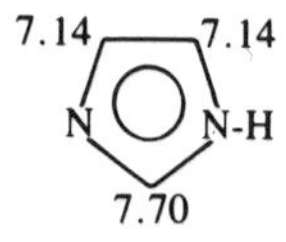 7.14 7.14 N N-H 7.70
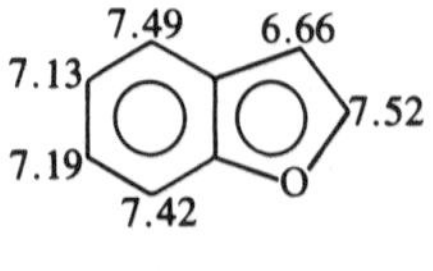 7.49 6.66 7.13 7.52 7.19 O 7.42	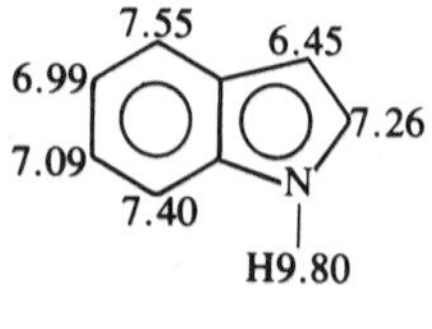 7.55 6.45 6.99 7.26 7.09 N 7.40 H9.80
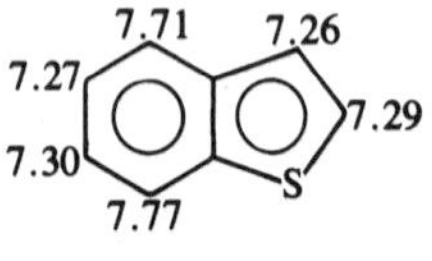 7.71 7.26 7.27 7.29 7.30 S 7.77	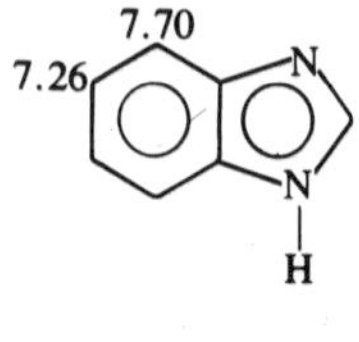 7.70 N 7.26 N H

TABLE 1.XVI *(Continued)*

Representative Chemical Shifts in Aromatic Compounds[a]

Compound	Compound
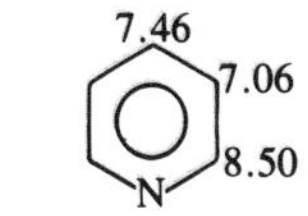 7.46, 7.06, 8.50, N	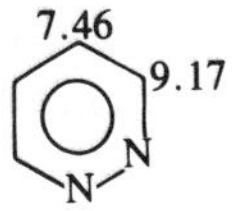 7.46, 9.17, N, N
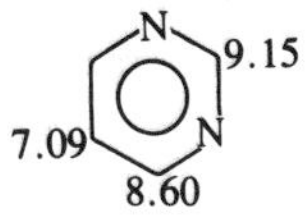 N, 9.15, 7.09, N, 8.60	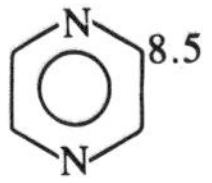 N, 8.5, N
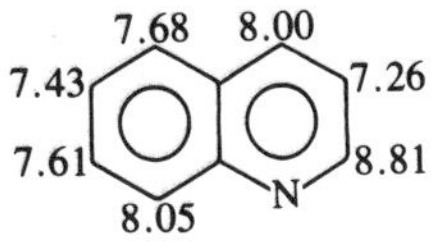 7.68, 8.00, 7.43, 7.26, 7.61, 8.81, 8.05, N	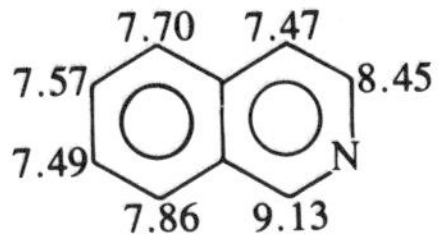 7.70, 7.47, 7.57, 8.45, 7.49, N, 7.86, 9.13
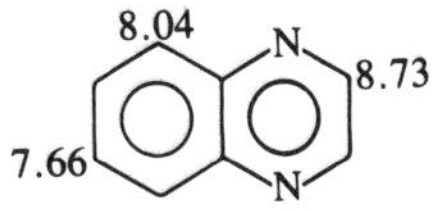 8.04, N, 8.73, 7.66, N	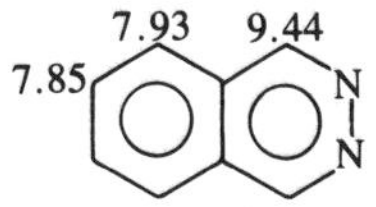 7.93, 9.44, 7.85, N, N
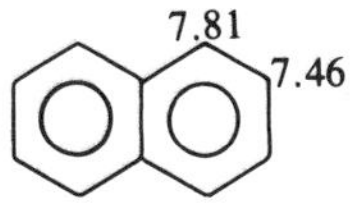 7.81, 7.46	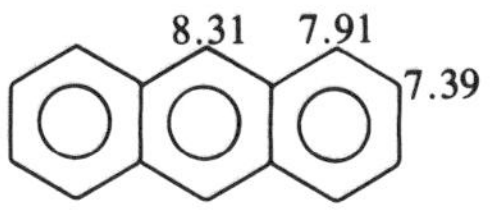 8.31, 7.91, 7.39
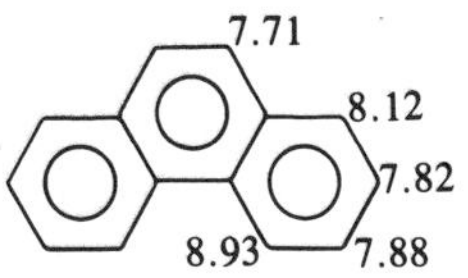 7.71, 8.12, 7.82, 8.93, 7.88	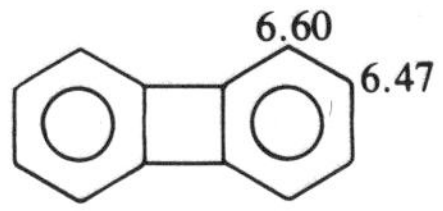 6.60, 6.47
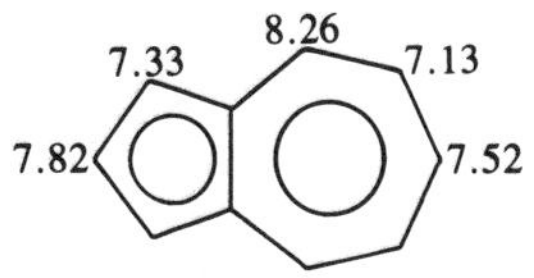 8.26, 7.33, 7.13, 7.82, 7.52	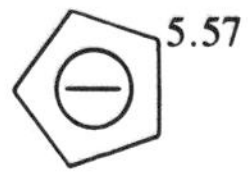 5.57
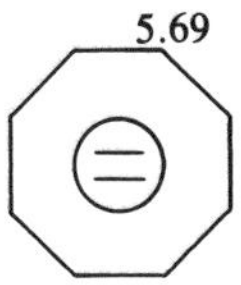 5.69	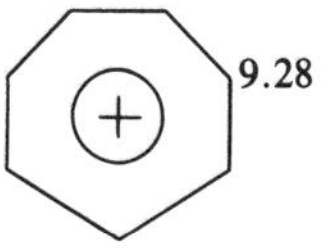 9.28

[a] Data taken from various sources.

chemical shifts of protons which interact sterically (peri-interaction) such as the 1,8-protons on a naphthalene ring system and 1,9-protons on a phenanthrene ring system. These protons are usually shifted downfield in comparison to proton chemical shifts in similar compounds where this interaction is absent. In polycyclic aromatic compounds, the ring current effects of the multiple rings appear to reinforce each other (123). Since this contribution to chemical shifts depends on the distance of the proton in question to the additional rings, protons closer to the attached ring(s) will be shifted downfield with respect to protons further removed from the additional ring(s) (for example the α- and β-proton shifts on naphthalene).

Protons bonded to atoms other than carbon absorb over a wider range than protons bonded to carbon. Protons bonded to oxygen, nitrogen, and sulfur are subject to hydrogen bonding. As a result, their chemical shifts are dependent on the solvent, concentration, and temperature. Furthermore, since these protons are acidic in nature, they readily undergo proton exchange when a trace of acid or water is present in the sample. As a result of this rapid exchange, adjacent protons see only an average for the spin states of these protons and, hence, coupling is usually not observed between protons bonded to carbon and adjacent protons bonded to heteroatoms. If a proton bonded to a heteroatom is suspected in the sample, the peak due to this proton can usually be identified by taking advantage of the exchange property. The usual procedure is to rerun the spectrum after adding a few drops of deuterium oxide to the sample and shaking the solution. The peak due to the proton bonded to the heteroatom will either disappear or reduce in intensity because of exchange of the proton with deuterium. In addition, a new peak due to HOD will appear in the spectrum between 4.5 and 5.0 ppm.

Representative chemical shift ranges for protons bonded to heteroatoms are given in Table 1.XVII. The signal due to the hydroxyl proton of alcohols is usually a sharp singlet due to exchange. If the sample is purified to remove water and acid, it may be possible to slow the exchange and observe coupling with the hydroxyl proton. Furthermore, if two or more hydroxyl groups are present in the compound, it may be possible to observe a separate peak for each group if the exchange is slow. A method for determining the type of alcohol based on slowing the exchange has been presented (55). In highly purified dimethyl sulfoxide, the exchange of hydroxyl protons is slowed such that one observes coupling between the hydroxyl proton and protons on the carbon to which the hydroxyl group is attached. Thus, the hydroxyl proton will appear as a singlet for a tertiary alcohol, a doublet for a secondary alcohol, and a triplet for a primary alcohol.

Since nitrogen-14 has a spin $I = 1$, one should observe coupling between amine protons and the nitrogen. However, there are two factors that usually prevent the coupling from being observed. If the proton on nitrogen is exchanging rapidly, it will "see" an average of the spin states of nitrogen and hence, will appear as if it were not coupled to nitrogen. Second, nitrogen has a quadrupole moment that induces relaxation, thereby decreasing the lifetime of the nitrogen spin states. Thus, a proton attached to nitrogen will see an average for the spin states on nitrogen and will usually appear as a broadened peak. Normally, the N–H proton is exchanging at a rapid rate such that coupling is not observed between N–H and protons attached to the carbon to

TABLE 1.XVII

Chemical Shifts for Protons Bonded to Heteroatoms[a]

Compound type	δ^b
Alcohols	0.5–5.0
Enols	15–19
Phenols	4.0–7.5
Phenols (intramolecular H-bonding)	10.0–12.0
Carboxylic acids	10–13
Oximes	7–10
Primary amines	1.1–1.8
Secondary amines	1.2–2.1
Anilines	3.3–4.0
Amides	5.0–6.5
N-Alkyl amides	6.0–8.2
N-Aryl amides	7.8–9.4
Ammonium salts	7.1–7.7
Thiols	1.0–2.0
Thiophenols	3.0–4.0

[a] Data taken from various sources.

[b] Dependent on concentration, solvent, and temperature.

which nitrogen is attached. This coupling has been observed in cases where the sample was rigorously purified to remove all traces of water. Exchange is also slow in amine salts such that this coupling can be observed. Furthermore, in acid solution, the multiplicity of the signal due to the C–H protons on the carbon to which nitrogen is attached can be used to classify amines in certain cases (10). In addition, the N–H proton may appear as a broadened triplet due to coupling with nitrogen in favorable cases where the quadrupole moment on nitrogen is reduced.

Protons attached to sulfur usually exchange at a slow rate such that coupling is observed between the S–H proton and C–H protons on the carbon to which sulfur is attached. Although S–H protons can be exchanged with D_2O, S–H protons usually do not exchange with hydroxyl, amino, or carboxylic acid protons when these functional groups are present in the same compound.

2. Coupling Constants

The magnitude of the proton-coupling constants provides detailed information on the structure of unknown compounds. The examination of the spectra of a wide variety of known compounds has provided a number of empirical correlations between coupling constants and molecular structure. In most cases, a theoretical explanation of these trends has been presented. After one has carried out the spectral analysis to obtain the coupling constants, the next step is to utilize the coupling constants with the previously established correlations to confirm the presence or absence of particular molecular fragments in the unknown compound. In some cases, the presence or absence of particular molecular fragments may be confirmed by the presence or absence of certain splitting patterns in the spectrum without carrying out the detailed analysis.

Several compilations of coupling constants are available in review articles and texts (33,74,113,230). We shall not attempt to provide a theoretical treatment of coupling constants here. Rather, we will focus on the magnitude of coupling constants, the trends that have been established for coupling constants, and discuss some of the ways in which coupling constants can be used to provide information on the structure of unknown compounds.

In aliphatic compounds, coupling is normally observed between protons separated by two (H–C–H,^2J, geminal coupling) and three bonds (H–C–C–H, ^3J, vicinal coupling). Coupling between protons separated by four or more bonds is usually not observed in aliphatic compounds except in some special molecular arrangements. When one of the intervening bonds is a π bond, a small coupling between protons separated by four bond (H–C=C–C–H, allylic coupling) and five bonds (H–C=C–C–C–H, homoallylic coupling) may be observed. In aromatic compounds, coupling between protons separated by three bonds (*ortho* coupling), four bonds (*meta* coupling), and five bonds (*para* coupling) is normally observed. In addition, coupling may be observed between aromatic protons and protons on the benzylic carbon (benzylic coupling). In cases where the conjugation is extended, such as the cumulenes, coupling between protons separated by as many as nine bonds may be observed.

a. Geminal Coupling Constants

Geminal coupling constants for protons bonded to sp^3-hybridized carbons range in value from +6 to −30 Hz. Some values in representative compounds are given in Table 1.XVIII. In general, electronegative substituents attached to the CH_2 group produce a decrease (more negative value) in the geminal coupling constant. Other factors that influence geminal coupling constants are the H–C–H bond angle and the orientation of lone electron pairs on adjacent heteroatoms. The orientation of π substituents with respect to the H–C–H plane will also influence the magnitude of geminal coupling constants. In most cases, the geminal coupling constants in saturated systems are negative in sign. The exceptions are epoxides, aziridines, and 1,3-dioxalanes where the geminal coupling constant is positive. If a full spectral analysis is carried out and the signs of the coupling constants determined, a large negative coupling constant can usually be safely assigned to a geminal coupling, since the majority of other large coupling constants are positive.

In contrast to saturated systems, the geminal coupling constant for protons attached to an sp^2-hybridized carbon varies over a much larger range (Table 1.XVIII). However, for most ethylene derivatives, the range is smaller than that found in saturated systems. Geminal couplings for protons attached to an sp^2-hybridized carbon are influenced by substituent electronegativity and orientation, and by ring strain for exocyclic methylene groups. A correlation between $^2J_{H-H}$ and the electronegativity of the substituents has been proposed for ethylene derivatives $H_2C{=}CX,Y$

$$^2J_{H-H} = \frac{61.6}{E_x + E_y} = 12.9 \qquad (71)$$

where E_x and E_y are the electronegativities of the substituents. In general, the magnitude of $^2J_{H-H}$ becomes more negative as the electronegativity of the substituents increase.

Geminal coupling constants in both saturated and unsaturated systems may be solvent dependent (226). The general trend is towards a more negative value of J_{gem} as the polarity of the solvent increases, provided the negative end of the dipole moment of the molecule is directed away from the CH_2 group. This solvent-dependence may be used to determine the sign of the coupling constants in favorable cases.

b. Vicinal Coupling Constants

By far, the most informative coupling constant for structural determination is the vicinal coupling constant (H–C–C–H, $^3J_{H-H}$). Representative values of vicinal coupling constants are given in Table 1.XIX. In cases where there is a single bond joining the two carbons, rotation about the carbon–carbon bond is possible. In general, this rotation will be fast such that the observed vicinal coupling constant will be a population-weighted average of the individual conformers. The magnitude of vicinal coupling constants where both carbons are sp^3-hybridized ranges from -0.3 to $+14$ Hz. The major influence on this type of vicinal coupling constants is the dihedral angle between the two C–H bonds (Fig. 1.36). The electronegativity of substituents attached to the two carbons, the orientation of the substituents with respect to the two C–H bonds, bond angles, and bond lengths all influence the magnitude of the vicinal coupling constant. In general, the vicinal coupling decreases as the electronegativity of attached substituents increases.

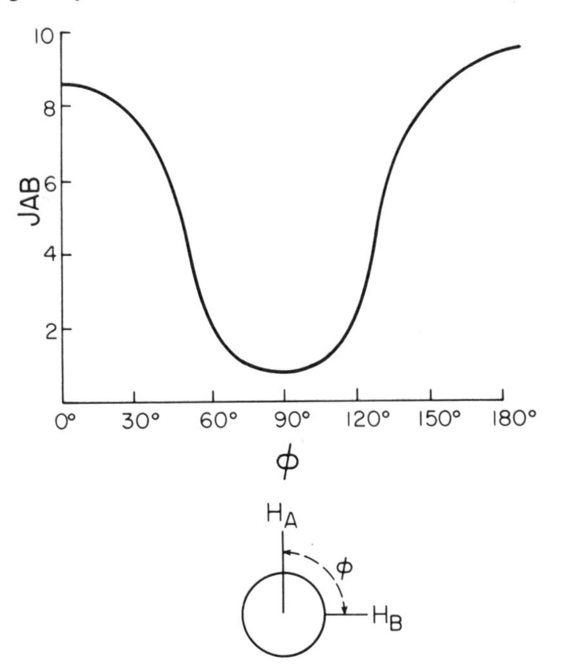

Fig. 1.36. A schematic illustration of the dihedral angle dependence of vicinal coupling constants (the Karplus relationship).

TABLE 1.XVIII
Geminal Proton Coupling Constants[a]

Compound	^2J	Compound	^2J
CH_4	−12.4		−12.0 to −15.0
CH_3X	−9.2 to −16.9		
$CH_2(CN)_2$	−20.4		
$K^+\bar{O}_2CCH_2CHOHCO_2^-K^+$	−15.3		−15.3 to −18.0
	−0.5 to −9.9		
	+4.0 to +6.3		−12.0 to −15.0
	0 to −1.4		−19.0 to −19.5
	0 to +1.5		−11.6 to −15.0
	−5.4		−12.0 to −16.0
	−9.5 to −13.0		−17.0 to −18.9
	−8.0 to −12.0		−8.8 to −10.5

TABLE 1.XVIII *(Continued)*
Geminal Proton Coupling Constants[a]

Compound	^2J	Compound	^2J
		$H_2C{=}O$	+41
(bicyclo[2.2.1]heptene with geminal H,H)	−10.4 to −13.7	$H_2C{=}NR$	+8 to +16.5
		$H_2C{=}CHX$	−3.2 to +7.4
		$H_2C{=}C{=}C<$	−9.0

[a] Data taken from various sources.

On the basis of theoretical calculations, Karplus (124,128) suggested that the relationship between dihedral angle and vicinal coupling constants could be expressed as

$$J = J^\circ \cos^2\phi - C \text{ for } 0^\circ < \phi < 90^\circ \tag{72}$$

$$J = J^{180} \cos^2\phi - C \text{ for } 90^\circ < \phi < 180^\circ \tag{73}$$

where J° (a standard coupling for a dihedral angle of 0°), J^{180} (a standard coupling for a dihedral angle of 180°), and C are constants. For an unsubstituted ethane H–C–C–H fragment), Karplus suggested the values $J^\circ = 8.5$ Hz, $J^{180} = 9.5$ Hz, and $C = -0.3$. A plot of this relationship is given in Fig. 1.36. While the value of C probably remains constant from one compound to another, it is clear that the values of J° and J^{180} are dependent upon the substituents attached the H–C–C–H fragment. Examination of a variety of known compounds has shown that J° and J^{180} may vary from 8 to 16 Hz with J^{180} usually being the larger of the two couplings. This has led to the suggestion that it is probably better to consider the Karplus relationship in terms of a "family of curves" similar to the one shown in Fig. 1.36 (230).

Provided reliable values of J° and J^{180} can be chosen, the application of equations 72 and 73 to derive stereochemical information is usually straightforward. After considering all reasonable conformations and measuring the respective dihedral angles with the aid of models, a series of simultaneous equations can be set up and solved such that they are internally consistent and an estimate of the magnitude of the dihedral angle obtained. With freely rotating systems, an additional unknown is the population of the various conformers. In this case, one can assume reasonable conformers with fixed dihedral angles and set up the Karplus equations to solve for the relative populations. As a typical example, consider the 1,2-disubstituted ethane $X{-}CH_2CH_2{-}Y$. The most likely stable conformers and the corresponding equations are given in Fig. 1.24 and equations 47 and 48. If J_t and J_g are known, or reasonable guesses of their values made, solution of equations 47 and 48 will yield the populations of the conformers.

TABLE 1.XIX

Vicinal Proton Coupling Constants[a]

Compound	^3J	Compound	^3J
CH_3CH_2X	7.0–9.0	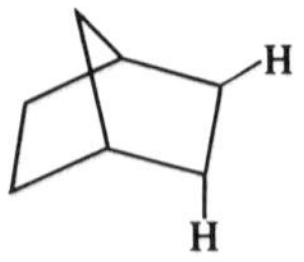 H H	2.5–5.0
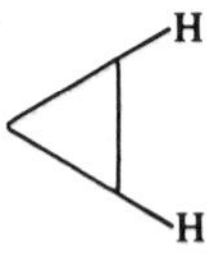 H H	*cis* 2.2–12.5 *trans* 1.4–8.6	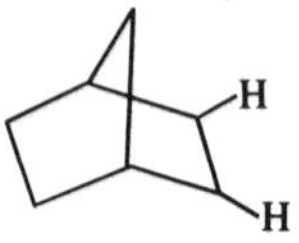 H H	9.0–10.0
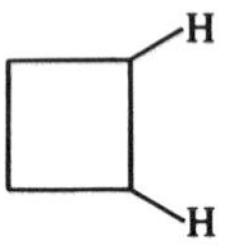 H H	*cis* or 4.0–13.0 *trans*	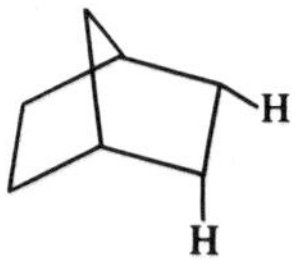 H H	6.0–7.0
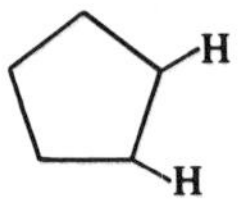 H H	*cis* or 4.0–13.0 *trans*	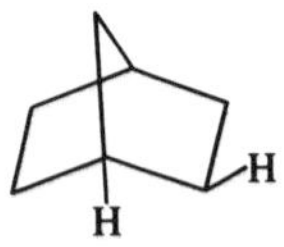 H H	3.0–4.0
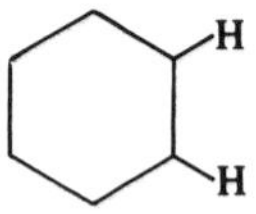 H H	ax–ax 6–14 ax–eq 0–5 eq–eq 0–5	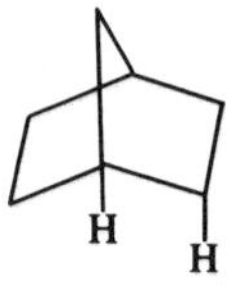 H H	0.0–2.0
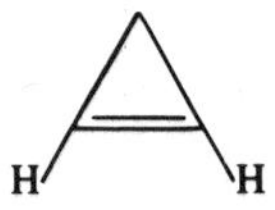 H H	0.5–1.5	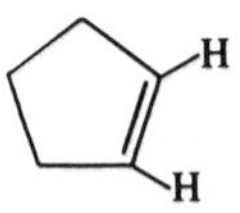 H H	5.1–7.0
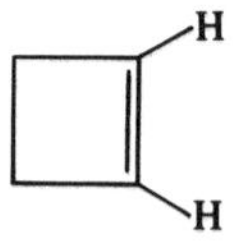 H H	2.0–4.0	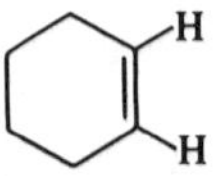 H H	8.8–11.0
$CH_2{=}CH_2$	*cis* 11.5 *trans* 19.0	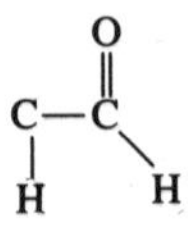 O ‖ C—C H H	1.0–3.0
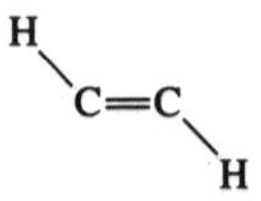 H C=C H	12.0–19.0		

TABLE 1.XIX *(Continued)*
Vicinal Proton Coupling Constants[a]

Compound	^3J	Compound	^3J
$\mathrm{H-C=C-H}$ (both H on the same side)	6.0–12.0	$\mathrm{C=C(H)-C(=O)-H}$	5.0–8.0
		$\mathrm{C=C(H)-C-H}$	4.0–10.0
		$\mathrm{C=C(H)-C(H)=C}$	9.0–11.0

[a] Data taken from various sources.

Vicinal coupling constants in molecular fragments where one or both of the carbons is sp^2 hybridized ($\mathrm{H-\overset{\|}{C}-C-H}$ and $\mathrm{H-\overset{\|}{C}-\overset{\|}{C}-H}$) also appear to follow a Karplus-like dependence on the dihedral angle (230). Rotational averaged values of these couplings are usually in the range 5–8 Hz. For the fragment H–C–C–H, typical values of J_{gauche} and J_{trans} are 1.8–3.7 and 9.6–13.4 Hz, respectively.

The angular dependence of vicinal coupling constants has proven to be of great value in determining the conformations of six-membered rings. Where the chair form of the six-membered ring is the principal form, typical values of the vicinal coupling constants are $J_{\text{ax-ax}}$ 8–13 Hz, $J_{\text{ax-eq}}$ 2–6 Hz, and $J_{\text{eq-eq}}$ 1–5 Hz. Even in cases where the spectrum cannot be or has not been analyzed, the conformation of a particular proton on a cyclohexane ring can be determined provided its signals are separated from other signals. Due to the magnitude of the couplings, the width of the proton multiplet at half-peak height ($\nu_{\frac{1}{2}}$) is characteristic of the conformation of the proton (107). An equatorial proton will have $\nu_{\frac{1}{2}}$ usually smaller than 12 Hz whereas an axial proton will usually have $\nu_{\frac{1}{2}}$ larger than 15 Hz. Not only does this allow the conformation to be determined, but in many cases, the configuration of polysubstituted cyclohexanes may also be determined.

In smaller ring systems, the vicinal coupling constants may or may not provide as useful information concerning the conformation of the ring. As the ring becomes more planar, the dihedral angle between *cis* protons approaches 0° whereas the dihedral angle between *trans* protons decreases from 180° and may approach 90°. As a result, the *cis*-coupling constant increases and the *trans*-coupling constant decreases so that one can no longer assume that the *trans*-coupling will be larger than the *cis* coupling. In three membered rings, J_{cis} is usually larger than J_{trans}. Typical ranges of the coupling are J_{cis} = 2.2–12.5 Hz and J_{trans} = 1.4–8.6 Hz. There is no clear pattern as to whether J_{cis} will be larger or smaller than J_{trans}. It appears that each ring system must be considered separately (230).

TABLE 1.XX

Long-Range Proton Coupling Constants

Compound Type	*J*	Compound Type	*J*
H, C=C, C—, H	−1.0 to −2.0	H, H, —C, C=C, C—	1.0–5.0
H, C=C, C—, H	−0.4 to −1.7	H, —C, C=C, C—, H	1.0–5.0
H, H	+0.5	CHR_2, H	0.6–0.9
O, H, H	−2.1	CHR_2, H	0.2–0.8
$CH_2{=}C{=}CH_2$	−7.37	CHR_2, H	0.1–0.6
$CH_2{=}C{=}C{=}CH_2$	7.01		
H—C=C—C=C—H	0.95		
H, H	1.0–1.4	H, H	1.04
H, H	1.0–1.4	H, H	1.11

TABLE 1.XX *(Continued)*

Long-Range Proton Coupling Constants

Compound Type	J	Compound Type	J
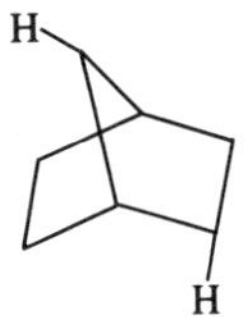	1.7–2.6	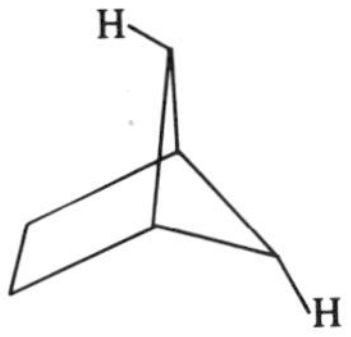	6.7–8.1
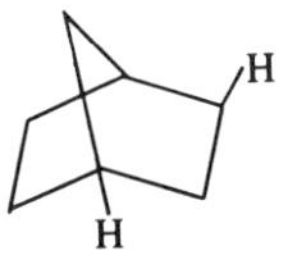	1.0	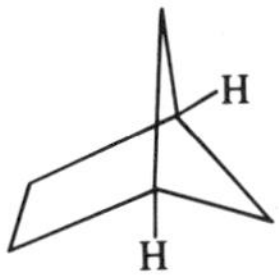	8.0
		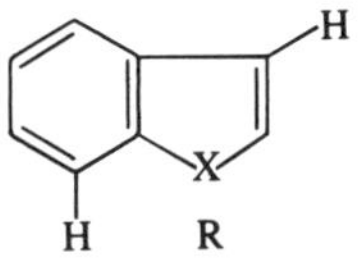	0.6–1.0

[a] Data taken from various sources.

The vicinal coupling constant in the molecular fragment H–C=C–H is just as informative as the previously discussed vicinal couplings in terms of their dependence on structure (Table 1.XIX). A large amount of data has been collected for this type of coupling and it is found that J_{trans} is always larger than J_{cis}. Values for J_{trans} in acyclic olefins are in the range 9.5–19.0 Hz whereas J_{cis} ranges from −2.0 to 11.7 Hz. Both J_{trans} and J_{cis} decrease as the electronegativity of the substituents attached to the H–C=C–H fragment increases. This relationship is approximately linear, so that, if the substituents are known, structural assignments can usually be made when only one isomer is present by comparison of the changes in the coupling constants of the corresponding vinyl derivatives with ethylene. The vicinal coupling in H–C=C–H is also dependent on the ring size in cyclic systems (Table 1.XIX) This dependence on ring size is thought to be due to a dependence of the vicinal coupling on the angle H–C=C. It has also been suggested that this vicinal coupling is dependent on bond order or bond length of the double bond. However, sufficient data on bond lengths are not available to develop trends for use in the solution of structural problems (113).

c. Long-Range Couplings

Coupling constants between protons separated by four or more bonds are referred to as long-range couplings. Usually, in saturated systems, long-range coupling constants are not observed except in some special cases to be discussed later. The most common type of long-range coupling is the allylic coupling (H–C–C=C–H) where both J_{cis} and J_{trans} may be observed. Some representative values are given in Table 1.XX. The allylic coupling is dependent on stereochemistry similar to vicinal coupling constants (230). However, the allylic coupling is considerably smaller and may be either positive or negative in sign (−3 to +2 Hz), depending on the sterochemical arrangement of the protons. This relationship is shown approximately in Fig. 1.37. In most cases, $|J_{trans}| > |J_{cis}|$. However, this relationship is not firmly established to justify its use for structure determination.

Homoallylic long-range couplings (H−C−C═C−C−H) may also be observed (Table 1.XX). These couplings are similar to allylic couplings in both magnitude and stereochemical dependence.

Long-range coupling is also observed in systems with extended conjugation separating the two protons, such as in acetylenes, allenes, and cumulenes (Table 1.XX). The magnitude of this coupling is not attenuated rapidly and coupling between protons separated by as many as nine bonds has been observed. Usually, replacing a proton with a methyl group will not appreciably affect the magnitude of the coupling in conjugated systems.

Long-range coupling is saturated systems is usually observed when the protons are separated by four or five bonds and the protons are located in a planar, zig-zag arrangement. This type of coupling has come to be referred to as coupling-along-a W-path and appears to be independent of either the nature or hybridization of the intervening atoms. Some representative values are given in Table 1.XX. The magnitude of this type of coupling falls off rapidly as the two protons lose coplanarity and is more commonly observed in unsaturated compounds.

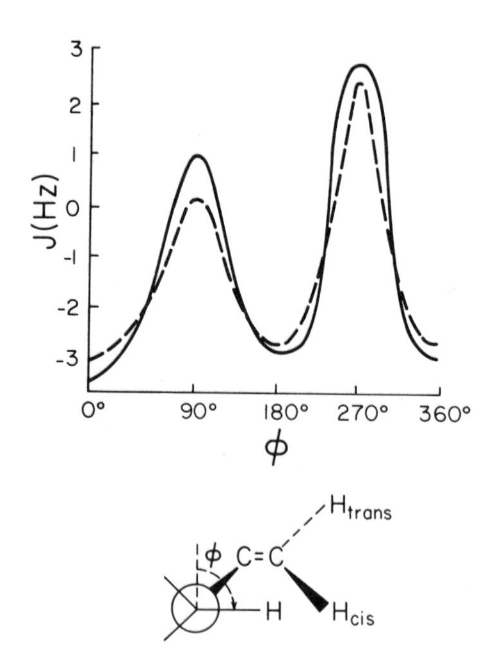

Fig. 1.37. A schematic illustration of the dihedral angle dependence of *cis* (---) and *trans* (——) allylic coupling constants.

d. Coupling in Aromatic and Heteroaromatic Compounds

The proton–proton coupling constants in aromatic and heteroaromatic compounds are of immense value in structural determination. In the case of di- and higher-substituted benzenes, one can easily distinguish between ^3J, ^4J, and ^5J and use these values to determine the substitution pattern. If the heterocyclic system can be identified, the coupling constants can be used to readily determine the substitution pattern. Typical coupling constants in these systems are given in Table 1.XXI. These coupling constants vary regularly with the electronegativity of the substituent and ring size.

From the examination of a wide variety of monosubsituted benzenes, the trends in the coupling constants with substituent electronegativity have been defined (51). The *ortho* coupling, J_{12}, shows a pronounced increase with increasing substituent electronegativity. The *meta* coupling J_{15} shows a similar trend. The *meta* coupling J_{13} and the *para* coupling J_{15} show a decrease as the substituent electronegativity increases. The remaining *ortho* coupling, J_{23}, remains essentially constant with varying substituent electronegativity. It appears that the effect of substituents on the coupling constants of benzenes is primarily an inductive type effect and is rapidly attenuated with increasing distance from the substituent.

An additivity scheme for the coupling constants in disubstituted benzenes has been derived by using the coupling constants obtained from the monosubstituted benzenes. The additivity scheme is given in Fig. 1.38 where J_o = 7.56 Hz, J_m = 1.38 Hz, and J_p = 0.69 Hz are the coupling constants obtained from benzene itself. For *meta*- and *para*- disubstituted benzenes, the agreement between calculated and experimental coupling constants is usually within ±0.1 Hz. Similar agreement is observed for other disubsituted benzenes, except for the case of strongly interacting *ortho*-substituents. Some examples where the additivity scheme fails are *ortho*-di-*t*-butylbenzene, *ortho*-dinitrobenzene (steric interactions between the two substituents, and *ortho*-nitrophenol (hydrogen-bonding between the nitro oxygen and phenolic hydrogen) (48). The breakdown of the additivity scheme for some *ortho*-disubstituted benzenes probably reflects geometrical distortion of the aromatic ring. If the interaction between *ortho*-substituents is taken into account, the additivity scheme should also be applicable to trisubstituted benzenes.

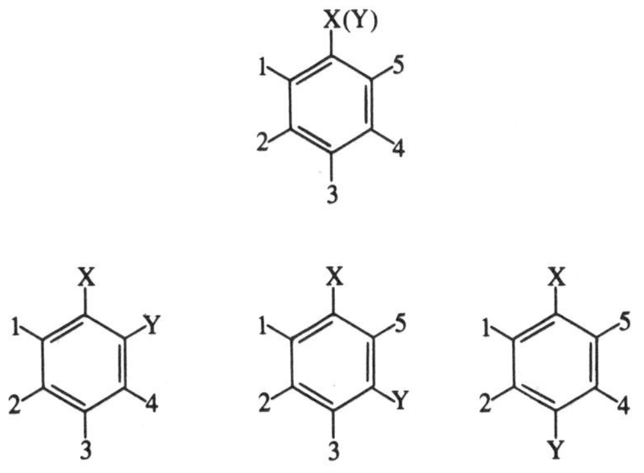

$J_{12} = J^x_{12} + J^y_{23} - J_o$	$J_{12} = J^x_{12} + J^y_{23} - J_o$	$J_{12} = J^x_{12} + J^y_{12} - J_o$
$J_{23} = J^x_{23} + J^y_{23} - J_o$	$J_{23} = J^x_{23} + J^y_{12} - J_o$	
$J_{34} = J^x_{23} + J^y_{12} - J_o$		
$J_{13} = J^x_{13} + J^y_{24} - J_m$	$J_{13} = J^x_{13} + J^y_{13} - J_m$	$J_{15} = J^x_{15} + J^y_{24} - J_m$
$J_{24} = J^x_{24} + J^y_{13} - J_m$	$J_{15} = J^x_{15} + J^y_{13} - J_m$	$J_{24} = J^x_{24} + J^y_{15} - J_m$
	$J_{35} = J^x_{13} + J^y_{15} - J_m$	
$J_{14} = J^x_{14} + J^y_{14} - J_p$	$J_{25} = J^x_{14} + J^y_{14} - J_p$	$J_{14} = J^x_{14} + J^y_{14} - J_p$

Fig. 1.38. The additivity scheme for calculating the coupling constants in disubstituted benzenes.

Substituent effects on the coupling constants in heteroaromatic compounds have been thoroughly investigated and are found to follow similar trends to those found for monosubstituted benzenes. The introduction of a heteroatom (compare substituted benzenes with pyridines in Table 1.XXI) in an aromatic ring results in a decrease in the *ortho*-coupling constant. There is also a decrease in the *ortho*-coupling constant on going from a 6- to a 5-membered ring. This trend is not observed for the *meta*-coupling constants. One of the more interesting trends is the effect of protonation on the coupling constants of heteroaromatic compounds (50). Protonation of a nitrogen heterocycle results in an increase in the *ortho*-coupling constant adjacent to nitrogen (i.e., J_{12} in pyridine) by from 0.5 to 1.5 Hz. This trend appears to be of general nature and may be used to identify the *ortho*-coupling in nitrogen heterocycles. Removal of a proton (i.e., forming the nitranion of pyrrole) results in a corresponding decrease in the *ortho*-coupling constant (61). These trends have been interpreted in terms of substituent effects similar to the effect of substituents on the coupling constants of monosubstituted benzenes (50).

e. PROTON-COUPLING CONSTANTS WITH OTHER NUCLEI

In principle, coupling can occur between any two magnetic nuclei. Fortunately, most organic compounds do not contain magnetic nuclei other than hydrogen. However, when other magnetic nuclei are present, coupling between the protons and other magnetic nuclei is detectable in the proton spectrum. The two most common "other nuclei" appearing in organic compounds are fluorine-19 and phosphorus-31. Both nuclei have spins of $I = 1/2$ and thus couple with the protons in the sample just as if they were an additional proton; the difference being that the coupling constant between hydrogen and another nucleus is usually larger than a proton–proton coupling constant. Some typical values of proton–fluorine and proton–phosphorus coupling constants are given in Table 1.XXII. The magnitude of proton–fluorine coupling constants usually decreases with the number of bonds separating the two nuclei. For proton–phosphorus couplings, the magnitude of the couplings usually follows the trend $^1J > {}^3J > {}^2J$. This trend is also observed for the coupling constants between protons and other magnetic nuclei as well.

TABLE 1.XXI

Coupling Constants in Aromatic and Heteroaromatic Compounds[a]

Compound	J	Compound	J
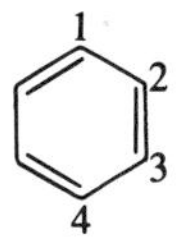	$J_{12} = 6.0-9.5$ $J_{13} = 1.2-3.3$ $J_{14} = 0.0-1.5$	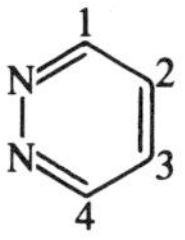	$J_{12} = 5.1$ $J_{23} = 8.0-9.6$ $J_{13} = 1.8$ $J_{14} = 3.5$
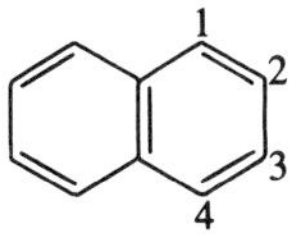	$J_{12} = 8.3-9.1$ $J_{23} = 6.1-6.9$ $J_{13} = 1.2-1.6$ $J_{14} = 0.0-1.0$	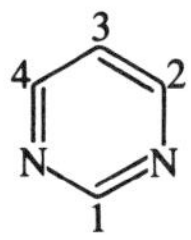	$J_{23} = 4.0-6.0$ $J_{12} = 0.0-1.0$ $J_{24} = 2.5$ $J_{13} = 1.0-2.0$
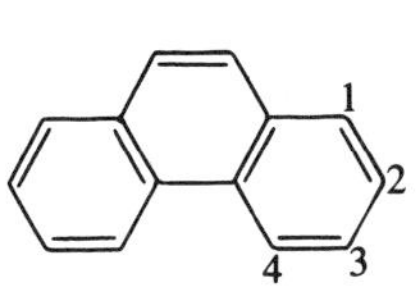	$J_{12} = 8.0-9.0$ $J_{23} = 6.9-7.3$ $J_{34} = 8.0-9.5$ $J_{13} = 0.9-1.6$ $J_{24} = 1.2-1.8$ $J_{14} = 0.3-0.7$	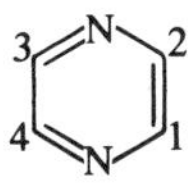	$J_{12} = 1.8-3.0$ $J_{14} = 0.0-0.5$ $J_{13} = 1.3-1.8$
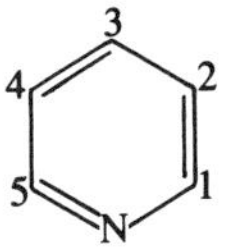	$J_{12} = 4.0-6.0$ $J_{23} = 6.9-9.1$ $J_{13} = 0.0-2.7$ $J_{24} = 0.5-1.8$ $J_{15} = 0.0-0.6$ $J_{14} = 0.0-2.3$	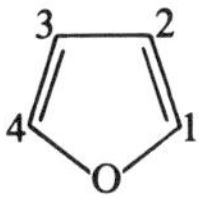	$J_{12} = 1.3-2.0$ $J_{23} = 3.1-3.8$ $J_{13} = 0.4-1.0$ $J_{14} = 1.0-2.0$
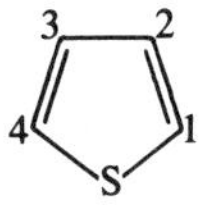	$J_{12} = 4.9-6.2$ $J_{23} = 3.4-5.0$ $J_{13} = 1.2-1.7$ $J_{14} = 3.2-3.7$	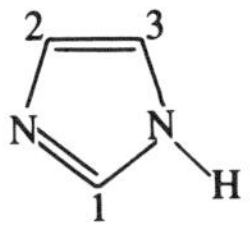	$J_{23} = \sim 1.6$ $J_{12} = 0.8-1.5$
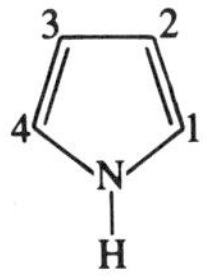	$J_{12} = 2.4-3.1$ $J_{23} = 3.4-3.8$ $J_{13} = 1.3-1.5$ $J_{14} = 1.9-2.2$	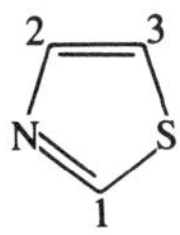	$J_{23} = 3.2$ $J_{12} = <0.5$ $J_{13} = 1.9$
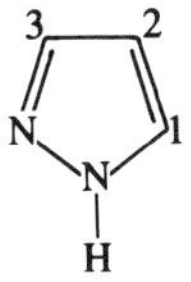	$J_{12} = \sim 1.9$ $J_{23} = \sim 2.0$		

[a] Data taken from various sources.

TABLE 1.XXII

Proton – Fluorine and Proton – Phosphorus Coupling Constants[a]

Compound Type	J	Compound Type	J
—C—C(H)—F	47.5	$H_2C{=}CHF$ (H, H / H, F)	*gem* 84.7 *cis* 20.1 *trans* 52.4
—C(H)—C—F	25.7	$CH_3(H)C{=}C(F)H$	*cis* 2.6 *trans* 2.3
—C—C(F)(H)—F	57.2	C_6H_5F, F / H *ortho*	6.2 – 10.1
—C(H)—C(F)—F	20.8	C_6H_5F, F / H *meta*	6.2 – 8.3
H—C—CF_3	2.0 – 13.0	C_6H_5F, F / H *para*	2.1 – 2.3
H—C—C—CF_3	0.5 – 1.0	$C_6H_5CF_3$, CF_3 / H	0.5 – 1.0
Cyclohexyl fluoride (F, H1, H2)	Fax-H_1 49 Feq-H_1 49 Fax-2ax 43.5 Fax-2eq 3 Feq-2ax <3 Feq-2eq <3	H—C—O—P	0.5 – 12
>PH	180 – 225	H—C—O—P(O)	3.0 – 15
>P(O)H	490 – 710	H—C—C—O—P<	0 – 3.0
>P(S)H	490 – 650		
—P+H	490 – 600		
H—C—P<	0 – 3.0		

TABLE 1.XXII *(Continued)*

Proton - Fluorine and Proton - Phosphorus Coupling Constants[a]

Compound Type	J	Compound Type	J
$H-C-P(O)$	10-15	$H_2C=CH-P$	*gem* 11.7 *cis* 13.6 *trans* 30.2
$H-C-P(S)$	10-15		
$H-C-P^{\mp}$	10-15		
$H-C-C-P$	13.7		
$H-C-C-P(O)$	18		

[a] Data taken from various sources.

C. DETERMINATION OF MOLECULAR STRUCTURE

1. General Procedures

Nuclear magnetic resonance spectroscopy is one of the most widely used tools in the chemical laboratory for the determination of molecular structure. When used in conjunction with infrared spectroscopy and mass spectrometry, few structural problems cannot be solved. The recent commercial availability of NMR spectrometers capable of routine ^{13}C-NMR, as well as other nuclei, has even further enhanced the power of the technique. It is beyond the scope of this chapter to present the details of structure determination using NMR. There are hundreds of examples in the literature in which clever and imaginative use of the magnetic properties of nuclei are employed to solve structural problems. Here, the principles of application of some of these properties will be briefly discussed. The reader will likely find that an in-depth study of the procedures used in the specific area of interest will be of great value. However, a generalized approach to structure elucidation using NMR spectroscopy can be developed. Once a good-quality spectrum is obtained, the usual first task is to extract the chemical shifts and spin-spin coupling constants. These parameters are of prime importance because they provide information on spin or charge densities in the vicinity of the nucleus responsible for the resonance signal, as well as molecular geometries. One major advantage of nuclear magnetic resonance is that useful information may be obtained from these parameters using empirical correlations. This means that an experienced NMR spectroscopist can extract much structural information by a close and careful inspection of the spectrum. At this point, it is extremely useful to have integral curves for the spectral lines. In this way, individual peaks or groups of peaks may be classified in terms of approximate chemical shift values and the number of protons in each group. Occasionally, groups of peaks will be sufficiently well isolated by large chemical shift differences so that characteristic patterns can be recognized such as AB, or a 1:2:1 triplet. In this event, subgroups of the molecule can frequently be identified upon inspection of the spectrum.

Inspection of the spectrum also permits a tentative classification of the spectrum type to be made according to the procedures given in Section IV.A.2. The tentative nature of this classification must be emphasized, as such classification tends to preassume the structure. For example, a spectrum of a five-proton compound such as pyridine may show a pattern of two protons of close chemical shift and a grouping of three other protons. One may give such a pattern a tentative assignment of (AB)(XYZ) simply to indicate the spectral appearance.

At this point a very useful, but often elusive, feature of molecular structure must be considered. Molecular symmetry is very important in interpretation of any type of spectrum. In NMR spectra, a high degree of symmetry in the molecule usually results in nuclei of identical chemical shift at various sites in the molecule. However, these nuclei may have different spin–spin interactions with neighboring nuclei. In the case of pyridine, the spectrum will be of the AA′XX′Y type to account for this. The two protons labeled A and A′ have identical chemical shift values (isochronous) but they are not isogamous (same spin–spin interaction with a given neighboring nucleus).

2. Use of Various Techniques

The above section briefly outlined the general approach to structure determination using high-resolution NMR. As stated, normally the chemical shifts, coupling constants, and relative peak areas are obtained from the initial spectrum. The spectral analysis techniques discussed in Section IV.A. may be used at this point. However, many other chemical and spectroscopic techniques may be employed, and detailed correlation of structure with observed spectroscopic data may provide further information about the sample. Several of these are discussed in Section IV.A.5, however a few more examples are given here to illustrate the methodologies available.

a. Use of Line Width

Occasionally an NMR spectrum of an organic molecule will display one or more resonance lines which have greater line widths than those in the remainder of the spectrum. This observation may be useful in interpretation of the spectrum and elucidation of the structure. Such lines may result from several phenomena. The proton giving rise to the resonance may be bonded to an atom such as nitrogen which has a nuclear quadrupole moment. These nuclei frequently have short relaxation times as a result of interactions between the nuclear quadrupole and molecular dipoles of solvent or solute molecules. The rapid changes in spin orientation result in a broadening of the resonance line of the observed proton nucleus (Section IV.E.2).

The broadening may result from rapid chemical exchange, particularly if a protic solvent is employed. This situation usually occurs when the observed proton is bonded to atoms such as oxygen or nitrogen in carboxylic acids, alcohols, amines, and similar compounds. This case may be identified by the addition of a suitable deuterated compound such as D_2O or CH_3OD such that the labile proton under question is replaced by chemical exchange with a deuterium atom resulting in a decrease in resonance intensity.

Because the protons in the cases described here usually involve extensive hydrogen bonding, the chemical shift is temperature dependent. An increase in temperature will cause substantial chemical shift changes as a result of the temperature dependence of hydrogen bond equilibria. This phenomenon has been used extensively to measure the temperature of NMR probes.

In some cases, the peak width may be determined by intramolecular processes which will show a substantial temperature dependence of width, but little effect on chemical shift. Perhaps the classic example of this phenomenon is the methyl proton resonances of *N*, *N*-dimethylformamide. At room temperature, these protons show two peaks, one for each methyl group, further weakly split by coupling with the formyl proton. However, at 110°C, the peaks are merged into a very broad single peak resulting from more rapid rotation of the C–N bond which is restricted in rotation because of some double-bond character.

Thus, broad peaks in an otherwise normal spectrum should be investigated by isotopic substitution and temperature variation to confirm the source of the enhanced line width. It is also possible to resolve spin-coupling and extract more information by use of strong hydrogen bond acceptors as solvents. For example, dimethylsulfoxide (55) or hexamethylphosphoramide (67) hydrogen bond certain hydroxylic protons such that chemical exchange is sufficiently slowed to permit resolution of the spin-coupling between these protons and neighboring nuclei. As a simple example, primary, secondary, and tertiary alcohols may be identified by the number of peaks in the spectrum of the hydroxylic proton (three, two, and one respectively in the general cases).

b. Use of Vicinal Coupling Constants

Vicinal coupling constants may provide useful structural information. The principles of the method are discussed in Section IV.B.2.b. The Karplus relationships (equations 72, 73) represent the dependence of vicinal coupling constants on the dihedral angle between the spin-coupled protons. Caution must be used in applying the Karplus relationships because the value of the vicinal coupling constants also depend upon the electronegativity of substituents, the C–C–H bond angle and, to a lesser extent, the C–C bond length. Thus, it is good practice to compare vicinal coupling constants of a known compound with those of the compound under question but suspected to have a similar structure. Thus the once tedious task of determining *cis-trans* isomerism about an olefinic double bond is reduced to the measurement of a vicinal coupling constant. In cases of rapid conformational conversions such as substituted cyclohexanes, the observed value of vicinal coupling resulting from the rapid interconversion of axial–axial and axial–equitorial interactions may be used to estimate the fraction of each conformational isomer (11). Using monosubstituted cyclohexanes which have been partially deuterated to give compounds such as VII (or also heptadeutero derivatives) facilitates the studies.

VII

The spectrum of the compound is simplified by double irradiation at the deuterium resonance frequency leaving the ABX ^{1}H-NMR spectrum. From the conformational isomers undergoing rapid interconversion, the observed coupling constant J_{ax} and J_{bx} will be a population weighted average of axial–axial and equatorial–equatorial couplings as given in equations 74 and 75 in which p represents the mole fraction of 6A in the equilibrium.

$$pJ_{aa} + (1 - p)J_{ee} = J_{bx} \tag{74}$$

$$pJ_{ae} + (1 - p)J_{ea} = J_{bx} \tag{75}$$

The values of J_{aa}, J_{ee}, J_{ae}, and J_{ea} must be estimated from the Karplus Relation (Section IV.B.2.b) with due consideration given to substituent electronegativity effects, or by use of rigid model compounds. Once the value of p is obtained, the free energy difference between the two conformational isomers may be obtained using equation 75.

$$\Delta G = -RT \ln \frac{p}{1 - p} \tag{75}$$

These experiments provide for rather easy methods of obtaining this type of information. Rotational isomers may be investigated in a similar manner (Section IV.B.2.b).

c. USE OF ANISOTROPIC SHIELDING

The presence of groups containing double bonds, particularly in compounds of rigid structure, may lead to abnormal chemical shift values for protons many bonds removed from the functional group. This can, however, be a very useful tool in structure elucidation. A classic example is the presence of carbonyl groups in steroids. Consider a structure of the types represented by **VIII** and **IX**:

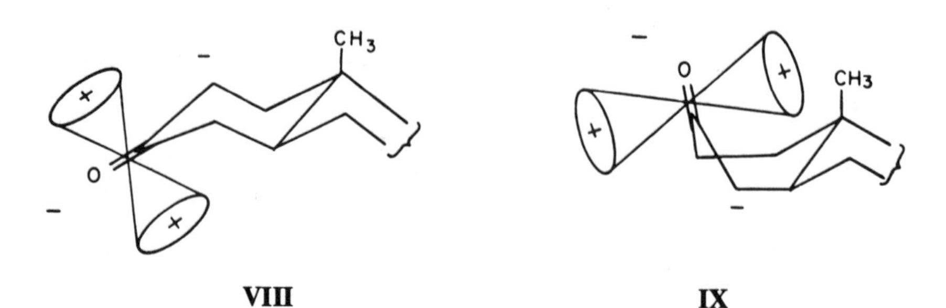

VIII **IX**

Any protons lying in the regions indicated as (plus) + will be shielded by the induced magnetic field generated by the mobile π electrons, whereas any protons lying in the region designated minus (−) will be deshielded. Therefore, the methyl resonance in **IX** is expected to lie at a higher field as is experimentally observed (62). This phenomenon is a valuable tool in the conformational analysis of steroids.

3. Summary

Determinations of molecular structure using an NMR spectrum is an organized deductive process. The organized steps include the acquisition and integration of a spectrum, identification of multiplet patterns, and classification of the patterns according to spectral type. Using chemical shift and spin–spin coupling correlation tables, structural subgroups and spatial arrangements are deduced. These subgroups are assembled into molecular structures by trial and error until one or more structures consistent with the spectrum are obtained. Spin decoupling, nuclear Overhauser effect experiments, solvent changes, isotopic substitution, paramagnetic reagents, or other methods may be employed to confirm the structure and conformation. Each problem is unique, and success depends upon the skill of the spectroscopist. A very useful set of guidelines and extensive correlation tables have been provided by Chamberlain (54). It should also be noted that rarely is NMR used alone in structural elucidation. Commonly other techniques such as mass spectrometry and infrared spectroscopy are also employed. However, very frequently NMR is the first method employed.

D. QUANTITATIVE ANALYSIS

1. Functional Group Determinations

One of the most common problems presented to the practicing analytical chemist is functional group determinations. Therefore, examples of the applications of nuclear magnetic resonance to these problems will be discussed.

a. Determination of Hydroxyl Groups

The simplest and most direct determination of hydroxyl groups is based upon replacement of the hydroxyl hydrogen with deuterium (178). The exchanged hydrogen then appears in the HDO line and may be determined from the area of this peak. In the event of "slow" exchange, in which case the active hydrogen and the HDO show separate resonance lines, a 50-fold excess of deuterium is added to drive the exchange to completion. A qualitative use of this procedure is a common method of identification of active hydrogen resonance lines. If upon the addition of D_2O, a resonance line is seen to decrease in area or disappear, an active hydrogen such as a hydroxyl, amine, carboxylic acid, etc., is suspected. Less labile active hydrogen such as those in esters, methyl ketones, amides, and alkynes may require a catalyst such as pyridine or LiOD. Relative errors using this technique range from near 0% to 5%. However, the method is subject to many interferences and can be used only when pure compounds are to be determined. Because of the strong intermolecular hydrogen bonding of hydroxyl groups, the chemical shifts of their resonance lines are highly temperature and concentration dependent. Therefore, changes, in chemical shift with temperature can confirm their presence (amine hydrogen will behave in a similar way but usually shows broad lines). The chemical shift as a function of concentration may be used to estimate hydroxyl concentration. However, this method is also highly subject to interference from other compounds.

To increase the specificity of hydroxyl group determination, some form of derivatization is commonly used. This approach frequently has the fringe benefit of permitting the determination of whether the alcohols are primary, secondary, or tertiary. Manatt presented a method of utilizing the trifluoroacetic esters of alcohols (54). The derivatization is performed by the addition of trifluoroacetic anhydride to the neat alcohols (or in an inert solvent) using an excess of the anhydride. The reaction can be monitored by scanning the NMR spectrum and observing the disappearance of the hydroxyl resonance lines. The ^{19}F resonance is then taken. Relative to trichlorofluoromethane, the ester ^{19}F resonance will appear at lowest field for primary alcohols, with those for secondary and tertiary alcohols at progressively higher fields. Amines, phenols, and thiols will interfere. The requirement of ^{19}F-NMR is somewhat of a nuisance. The use of dichloroacetic anhydride serves as a similar technique and has the advantage that ^{1}H-NMR may be used to observe the derivative (17).

An exceedingly simple method is to obtain the spectrum of alcohols in dimethylsulfoxide (55). In this solvent, the exchange rate of the hydroxyl proton is sufficiently slow that spin–spin splitting of the hydroxyl proton resonance is observed. The presence of acids or bases may catalyze the exchange and coalesce the splitting pattern. Otherwise, primary alcohols show triplets (except for a quartet for methanol), secondary alcohols show doublets, and tertiary alcohols show singlets for the hydroxyl resonance. In each case the chemical shift is between 5.5 and 4.0 ppm relative to TMS. This method is particularly useful in identification of the source of hydroxyl resonance lines in alkaloids (as quaternary salts), steroids and terpenes. Polyhydroxyl compounds show separate peaks for each hydroxyl group. The chemical shift may be sensitive to isomerism. For example the doublets of *trans*- and *cis*-4-*t*-butylcyclohexyl alcohol are separated by 0.34 ppm. A cautionary note has been issued when using this technique with alcohols which have strong electron withdrawing groups adjacent to the hydroxyl group (237).

Shift reagents are useful in the investigation of alcohols, particularly in the elucidation of structure. Saunders and Williams showed that complex spectra of alcohols such as benzyl alcohol and *n*-hexanol were reduced to simple first-order spectra by the addition of *tris*(pivalomethanato) europium (213). Quantitative applications of this technique have been described (193).

The phenolic hydroxyl group is more easily identified as the chemical shift of the hydroxyl hydrogen appears in the range 8–13 ppm relative to TMS in hexamethylphosphoramide (67). Several precautions must be taken and possible interferences are aldehydes (peak overlap) and acids.

b. DETERMINATION OF CARBONYL GROUPS

Aldehydes are normally easily determined using ^{1}H-NMR. The chemical shift of the formyl hydrogen resonance is in the range 9.6–10.4 ppm relative to TMS. This region of the spectrum is frequently free of other resonance lines and direct integration of the formyl hydrogen resonance may be performed.

Ketones are less easily determined by NMR because there is no direct proton signal. The use of ^{13}C-NMR with proton decoupling would appear to be an obvious choice. However, the chemical shift of the ^{13}C resonance of the carbonyl carbon varies widely and must be identified prior to integration.

Infrared spectra may surpass the importance of NMR in the identification and determination of carbonyl groups although there are ^{1}H-NMR methods which may be useful. The most common method is the derivatization of the functional group, which has been discussed in detail (127). The obvious derivations are hydrazones, and many such derivatization reagents may be used. Methyl ketones are usually easily determined by integration of the methyl proton singlet.

An interesting possibility for ketone mixtures is the technique used by Sudmeier for "counting" peptide moieties (233). By dissolving the carbonly compound in a super strong acid, protonation of the carbonyl oxygen is accomplished. If the proton exchange rate is sufficiently slow, a single resonance line is observed for each protonated carbonyl. Quantitation should be possible because the acidity required to reduce the exchange rate sufficiently is more than adequate to guarantee complete protonation of the carbonyl sites.

c. DETERMINATION OF CARBOXYLIC ACIDS AND RELATED COMPOUNDS

There are several approaches to the determination of carboxylic acids and the choice will be in part determined by the solvent and the presence of other compounds. An obvious and widely used method is the exchange of the active hydrogen for deuterium as already discussed for alcohols in Section IV.D.1.a (178). However, the solvent in which the sample is obtained plays an important role in this choice. In protic solvents, the ^{1}H-NMR resonance lines of the acidic proton may be coalesced with the solvent line and a mixture of acids may show only one merged resonance. In this case, the deuterium substitution technique will be of limited value. In aprotic solvents, however, this method is of significant use.

The use of resonance lines originating from protons on the carbon in the alpha position is important in the determination of carboxylic acids. These proton resonance lines may be directly integrated. However, because there is considerable overlap between the chemical shifts of alpha-CH_2 and alpha-CH protons in acids, considerable caution must be exercised in using this procedure. All of the above procedures become difficult in mixtures more complicated than binary.

There are some possibilities that can be explored when the sample form permits. For example, detailed studies have been conducted on "protonation shifts" of amines, carboxylic acids, and aminocarboxylic acids (143,231,232). The protonation shift is the chemical shift change observed for nonlabile (for example, α-CH_2–)

protons upon protonation of a basic functional group with the addition of acid. The pH (aqueous solution) at which the protonation occurs depends upon the pK_a of the functional group. Therefore, the control of pH can effect the chemical shift of the nonlabile protons and provide some control over overlapping peaks in mixtures of acids or bases with different pK_a values. The magnitude of the shift upon protonation is an aid in assigning the resonance line. A second and perhaps more useful approach to effect chemical shift changes is the esterification of the acids and use of shift reagents to separate overlapping lines (129).

The chemical shift of the merged solvent-acidic proton resonance lines in acid-water mixtures may be used to determine the acid content. This method is applicable to binary acid–water mixtures if the samples are known to be that simple. Formic acid is unique and may be determined in small amounts by integration or peak height measurements of the formyl hydrogen, which is free from most spectral interferences (40).

Salts of carboxylic acids present a problem when NMR determinations are desired as the methods based upon active hydrogen replacement cannot be used unless the sample is first converted to the acid by ion-exchange or some other procedure. The use of an adjacent nonlabile hydrogen is usually the method of choice.

The determination of esters by NMR spectroscopy is normally accomplished by integration or peak height measurement. The complication with esters, however, is that frequently there is spectral overlap between the protons in the acid residue and those from the alcohol residue. Kan has investigated the effect of the polar substituent constant σ^* of the alcohol group upon the chemical shift of the methyl protons (125). The methyl proton resonance of the acetic acid moiety showed shifts of about -1 Hz for higher *n*-alkyl esters to 14 Hz for aromatic esters relative to methyl acetate. It is easily seen that for mixtures of esters, peak overlap can be a significant difficulty. The use of shift reagents for the determination of mixtures of esters should prove valuable.

Amides of carboxylic acids have received a great deal of investigation using NMR spectroscopy. However, much of the attention has been given to the study of the hindered rotation about the carbon–nitrogen bond. Amide–NH proton resonance lines are observed between 5 and 9 ppm relative to TMS. These peaks are somewhat broad as a result of the effect of the nitrogen quadrupole moment on the relaxation time of the –NH proton and $^{14}N-^{1}H$ coupling. However, these protons are easily distinguished from amine protons by chemical shift (0–2 ppm for aliphatic amines and 2–5 ppm for aromatic amines) and the persistence of spin-coupling between the –NH proton and protons on *N*-alkyl groups in monosubstituted amides. Hydrogen bonding of amide hydrogens is strong, therefore solvent, concentration, and temperature may affect the –NH proton chemical shift.

A useful method of quantitation was reported by Leader (140). This method is based upon the formation of an adduct between hexafluoroacetone (HFA) and compounds that have active hydrogen atoms. In addition to amides, the technique may be applied to alcohols, mercaptans, amines, oximes, and other functional groups. Leader tabulated the ^{19}F chemical shifts (Δ) of the adducts measured relative to the HFA-water adduct. There is some correlation of these chemical shift values with the functional group forming the adduct. Some of these are shown in

TABLE 1.XXIII
Chemical Shifts Relative to HFA for Adducts of HFA in Ethyl Acetate

Compound	Δ (ppm)
Methanol	2.68
Ethanol	2.45
1-Pentanol	2.57
Benzyl alcohol	2.80
1,3-Propanediol	2.46
Ethanolamine (OH)	2.38
2-Propanol	1.80
1,4-Cyclohexanediol	1.87, 1.72
tert-Butyl alcohol	1.20
Citric acid	3.08
Phenol	2.73
Ethanethiol	0.95
1-Propanethiol	0.92
2-Propanethiol	1.00
Methylamine	4.11
Benzylamine	3.93
Aniline	3.74
Dimethylamine	−1.85
Benzamide	1.43

Table 1.XXIII. Once identified, the ^{19}F-resonance line for the adduct of interest can be integrated. A fringe benefit is the six fluorine atoms per functional group which amplifies the sensitivity.

d. Determination of Ethers, Epoxides, and Peroxides

This unusual classification of these functional groups is made for a simple reason. One characteristic of a good analytical chemist is that he does not force his problem to be solved by a given technique. In this case, the facts are that functional groups such as the ones in this section are probably better determined by other analytical methods. This is not to say that NMR is useless when identification and/or determination of these functional groups is required. Obviously, the direct application of an NMR spectrum will aid in the elucidation of the structure of compounds bearing these groups. Some direct applications are possible.

For example, vinyl ethers are easily identified in the presence of other vinyl compounds because the ethylenic protons appear at a higher field in the ether than in other vinyl derivations (81).

Organic peroxides have been determined in mixtures using proton resonance lines of protons either alpha or beta relative to the functional group (242). This method was applied to mixtures of peroxides, hydroperoxides, and alcohols.

e. Determination of Olefins

The NMR spectra of olefinic compounds have been studied extensively. The technique is most applicable to the identification and determination of olefins when there is one or more protons on a carbon containing the double bond. Olefinic hydrogen atoms give NMR resonance lines in the range 4–7 ppm from TMS. Only protons on carbon atoms containing highly electronegative substituents or aromatic compounds interfere. When 1, 2-disubstituted compounds are present, the *cis*/*trans* components can be easily determined from the difference in the splitting patterns of the vicinal protons. This is because J_{cis} lies in the range 6–12 Hz whereas J_{trans} shows values between 12 and 19 Hz (Section IV.B.2.b). The group electronegativity of substituents affects the value of these coupling constants. However, because of the minor overlap in the values of J_{cis} and J_{trans}, NMR is one of the fastest methods of elucidating *cis*/*trans* isomers in 1,2-disubstituted olefins.

Frequently, olefinic compounds give complicated NMR spectra because there is great opportunity for magnetic nonequivalence upon substitution about the double bond. The spectrum must be analyzed by the methods given in Section IV.A to obtain accurate chemical shift and coupling constant data. However, many experienced NMR spectroscopists can extract useful and often sufficient information by inspection of these spectra.

Several studies on the determination of specific olefins and the extent of unsaturation of a mixture have been conducted. Stehling and Bartz have published an excellent paper in which chemical shifts, spin–spin coupling constants and spectral patterns were correlated to the structure of 60 mono-olefins (228). These authors used the results to demonstrate that the structure of oligomers of mono-olefins, polymers, of di-olefins and monomer sequence distribution in isobutylene–isoprene copolymers could be determined. This paper is a good starting point for those becoming involved in the use of NMR to investigate olefinic compounds. By careful examination of the spectra as described by Stehling and Bartz, structure olefinic compounds may be determined. However, using the correlations found by these authors, worth-

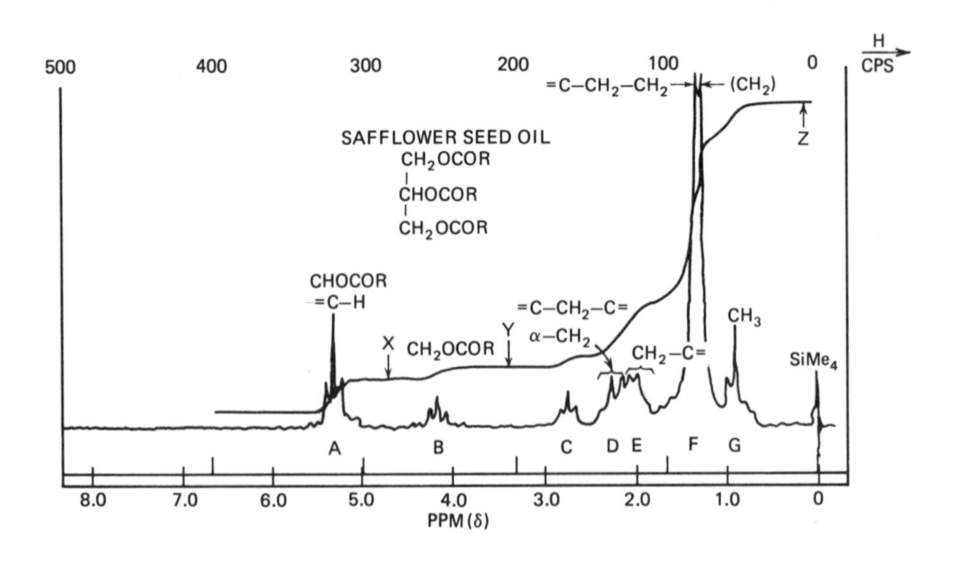

Fig. 1.39. 60-MHz spectrum of safflower seed oil. [*Anal. Chem.*, **34**, 1136 (1962), with permission.]

while results may be obtained by looking at the NMR data of the substituent groups in less detail than total spectrum analysis.

An early paper that effectively demonstrates how NMR can replace an old, tedious, wet chemical procedure was published by Johnson and Shoolery (122). This paper reports a method for the determination of the average molecular weight of fatty acids and triglycerides as well as the unsaturation of these compounds. Figure 1.39 shows an example in the 40 MHz spectrum of safflower seed oil. The integrated area X is proportional to the olefinic proton content, the area given by Y is the area X plus the area resulting from the two methylene groups (4 protons) in each glyceryl moiety. The area at Z is the total area of proton resonance lines in the spectrum. Because it is known that the area given by $Y - X$ results from the four methylenic protons, the area per proton is:

$$\text{Area per proton} = \frac{Y - X}{4} \tag{77}$$

The number of olefinic protons is

$$V = \frac{X - (Y - X)/4}{(Y - X)/4} \tag{78}$$

and the total number of protons is

$$T = \frac{Z}{(Y - X)/4} \tag{79}$$

These relationships prove to be approximate because of overlap of ^{13}C satellite lines, and the authors give corrections for this overlap. The authors show that these data may be used to calculate the average molecular weight with a final equation

$$\text{Mol. wt.} = 120.0 + 7.013T + 6.006V \tag{80}$$

The degree of unsaturation may be expressed as an iodine number using the relationship

$$\text{Iodine number} = \frac{12691V}{\text{Mol. wt.}} \tag{81}$$

Table 1.XXIV gives some comparison of the NMR values and WIJS method results. The technique has been developed as a routine process monitor by interfacing a minicomputer to an A-60D or T-60 spectrometer (224). The computer maintains instrument adjustment, acquires the spectrum and integral, and performs the calculations. Any analyst who has performed a WIJS iodine number will agree that this is progress. The method must be used with discretion when triglycerides which have conjugated double bonds, are determined as illustrated by tung oil in Table 1.XXV. The NMR method is probably more accurate; however the olefinic protons associated with the conjugated double bond are shifted 0.4–1 ppm downfield from those olefinic protons not in a conjugated system.

On occasion, it is desirable to combine chemical reactions with NMR spectroscopy to gain information concerning olefins. For example, the addition of sulfenyl chlorides across double bonds can be performed and monitored in situ (165).

TABLE 1.XXIV
Iodine Number Values of Various Fats

Oil	NMR no.	WIJS no.
Coconut	10.5 ± 1.3	8.0–8.7
Olive	80.8 ± 0.9	83.0–85.3
Peanut	94.5 ± 0.6	95.0–97.2
Soybean	127.1 ± 1.6	125.0–126.1
Sunflower seed	135.0 ± 0.9	136.0–137.7
Safflower seed	141.2 ± 1.0	140.0–143.5
Whale	150.2 ± 1.0	149.0–151.6
Linseed	176.2 ± 1.2	179.0–181.0
Tung	225.2 ± 1.2	146.0–163.5

SOURCE: *Anal. Chem.*, **34**, 1136 (1962), with permission.

Substituents determine whether the addition is Markovnikov or anti-Markovnikov and is useful in structure elucidation. The reaction between iodium nitrate and olefins has been shown as a useful tool to shift resonances downfield because of strong deshielding in the adducts (69). Adducts between mercuric acetate and fatty acid olefinic sites have been shown to permit direct determination of the *cis*/*trans* ratio in these compounds (218). The olefinic proton resonance of the two isomers is separated in the adduct (~0.05 ppm), whereas they overlap in the parent compounds. The choice of solvent is important, and peak height serves as the method of quantitation.

f. Determination of Acetylenic Hydrogen

Acetylenic hydrogen nuclei show strong diamagnetic anisotropy effects resulting in a chemical shift of 2.4–2.7 ppm from TMS. This chemical shift range encompasses the values for methylene and methine protons as well as highly deshielded methyl groups. As a result, the direct identification of acetylenic protons by NMR faces limitations. In addition, the identification of substituted acetylenes is limited by a small acetylenic proton chemical shift range of ~0.3 ppm for a wide variety of substituted acetylenes. There are, however, some useful aids to the identification of acetylenic protons. For example, addition of pyridine to a dilute solution of monosubstituted acetylene in CCl_4 results in a downfield shift of ~1 ppm for the acetylenic proton (132). Acetylenes exhibit long-range coupling (viz. $J_{CH_2\text{-}C\equiv C\text{-}H}$ ~2.9 cps)

TABLE 1.XXV
Average Molecular Weights of Fats from Saponification Values and NMR

Oil	Sap. value	Mol. wt.	NMR mol. wt.
Olive	189.3	887.1	873.7 ± 5.3
Peanut	188.8	891.5	882.3 ± 7.4
Safflower seed	191.5	879.0	874.9 ± 9.3

which is an aid to the identification of acetylenic proton resonance lines. When sufficient concentration or neat samples are available, the ^{13}C satellite may be useful in identifying acetylenic protons. The $J_{^{13}C-H}$ between ^{13}C nuclei bearing the acetylenic proton is ~250 Hz, much larger than the corresponding value for ethylene, benzene, or cyclopropane (~160 Hz).

Once identified, quantitation of the acetylenic proton may be performed by direct integration. The choice of solvent or addition of base such as pyridine may effect a shift to avoid overlap with other lines. The technique of deuterium substitution for active hydrogen using LiOD as a catalyst has been applied to phenylacetylene by Paulson and Cooke (178).

g. Determination of Amines

The use of NMR for the determination of amines would at first appear suspect because the chemical shift of the NH proton resonance lines may appear over a wide range of values depending upon solvent, concentration, presence of acids or bases, temperature, and other factors. In addition, the line shape will vary with similar factors as they determine the rate of chemical exchange of the NH protons. There are fortunately, simple solutions to these problems. The chemical shift may be reproducibly controlled by the addition of a strong acid which protonates the amine forming a salt and reduces the rate of proton exchange to the point that spin–spin coupling between the NH protons and protons on the carbon alpha to the nitrogen is observed. Trifluoroacetic acid is commonly used, as it is a sufficiently strong acid, does not contribute interfering peaks to the proton resonance signal, and is soluble in many organic solvents. The coupling constant between NH protons and the nitrogen is usually approximately 5 Hz. Because of some splitting from the nitrogen nucleus, the N–H resonance lines may not show clear first-order spectra. However, these lines may be integrated for quantitation; remember the protonation contributes a proton to the area. Observation of the resonance from the proton on the alpha carbon is frequently as useful. In trifluoroacetic acid, a primary amine shows a quartet, a secondary amine a triplet, and a tertiary amine a doublet for these lines. This is of particular use for methyl amines.

The use of the hexafluoroacetone (HFA) adducts described by Leader is of importance for amines (140). As with alcohols, thiols, and amides, the adducts between amines and HFA show characteristic fluorine resonance shifts from the free HFA. Many amines may be identified from this shift value. However, the technique suffers, but only slightly with modern instrumentation, from the need for fluorine resonance equipment.

Under the heading of amines it is important to note that this functional group has been widely studied by NMR in terms of the kinetics of proton exchange. Both pulse and continuous wave techniques have been employed. Grunwald has presented a brief but effective review of this topic (96). Caughman has prepared an extensive collection of the results of all N–H proton exchange studies to (1970) (53). Also, considerable attention has been devoted to the study of the inversion at nitrogen centers in amines using NMR methods (43,144,162,216).

Amino acids represent an important class of amines. Many properties of these compounds may be studied by NMR techniques. The quantitative determination of these compounds by NMR can be performed in a straightforward manner, provided the sample is not too complex of a mixture and the compounds are adequately soluble. Again, the use of trifluoroacetic acid as a solvent as a suggested by Bovey and Tiers provides the advantage of reproducible chemical shift values (35). Care must be taken however as some amino acids such as glycine and cysteine have limited solubility in this solvent. The use of NMR as a routine tool for the determination of amino acids may be questioned because there are many other techniques that are suitable, some of which are automated.

E. APPLICATION TO CHEMICAL DYNAMICS

1. Slow Reactions

NMR techniques may be applied to a wide range of studies of chemical kinetics. Similar to equilibrium studies, two subgroups of techniques may also be employed, depending upon the rate of the chemically dynamic process. First, the NMR spectrometer may be used as a quantitative instrument to investigate dynamic processes, such as chemical reactions including isotopic exchange reactions, with half-lives on the order of minutes. In these cases, resonance peaks may be repetitively integrated at known time intervals following initiation of the reaction. The absolute concentration at any time of each chemical species (*viz*, reactant and product) may be calculated from the relative peak areas and mass balance of the nuclei leading to the resonance. The data may then be treated according to conventional methods of rate data analysis. There are obvious applications of this procedure, such as the study of isotopic exchange reactions (^{2}H for ^{1}H), hydrolysis reactions, and so on. In these cases, NMR offers convenience more than other advantages. Using a CAT or computer to acquire rapidly changing spectra permits slightly faster reactions to be investigated. If fast sweep rates (~ 25 Hz/sec) can be tolerated without detrimental loss of resolution, an on-line computer can store repetitive scans of a peak approximately each 5 sec. The peak may then be integrated digitally by the computer. For nuclei with short relaxation times, Fourier transform spectra may be collected each few seconds. These latter methods have the advantage of presenting changes in the entire spectrum during the time of the reaction. In that way, changes at all sites containing resonant nuclei may be monitored.

A variation useful to the study of chemical exchange involving half-lives of a few seconds has been demonstrated (82,83). The technique is based upon the sudden application of a saturating field (B_2) at the resonance frequency of one of two reversibly exchanging nuclei, and the observation of the time dependence of the intensity of the second resonance line. B_2 may be applied using double-resonance techniques. The perturbation of B_2 upon the spin distribution (i.e., the extent of saturation) of the nuclei irradiated will be transferred to the second magnetic environment by chemical exchange. Forsén and Hoffman presented an excellent example of the application of this technique (83). The base-catalyzed exchange of –OH and –CH= protons in the enol tautomer of 2,4-pentanedione was known and proposed to occur though the keto tautomer as an intermediate. An example of the experimental results is shown in Fig. 1.40. The $-OH$ and $-CH_2-$ signals were suddenly saturated and the decay of the $-CH{=}$ proton exchange does occur through the keto intermediate. It should be pointed out that the range of rates which may be studied by this technique is relatively small. Therefore, limited numbers of applications are expected. However, it is a simple method when applicable.

2. Fast Reactions

In the above sections, the application of NMR as a quantitative tool to follow concentration changes during the course of chemical reactions with half lives on the order of minutes was discussed. However, the real significance of the application of NMR to investigations of chemical kinetics lies in the ability to provide information on very fast chemical reactions, even those of a virtual type where reactants and products are chemically identical. Of perhaps even greater importance is the fact that the measurements may be taken while the system is at chemical equilibrium, thus many fast chemical exchange processes may be investigated using conventional cw-NMR spectrometers. There have been many reports of the theory of the effects of chemical exchange on NMR spectra. It is beyond the scope of this work to attempt to develop the theoretical treatment. Several reviews and the references therein will provide an excellent introduction to the theory (13,118). Johnson has pointed out the two major factors that make possible the application of NMR to studies of chemical dynamics (118). First, if chemical dynamic processes lead to magnetic field fluctuations at a nucleus under observation, the precessional frequency of the nucleus will

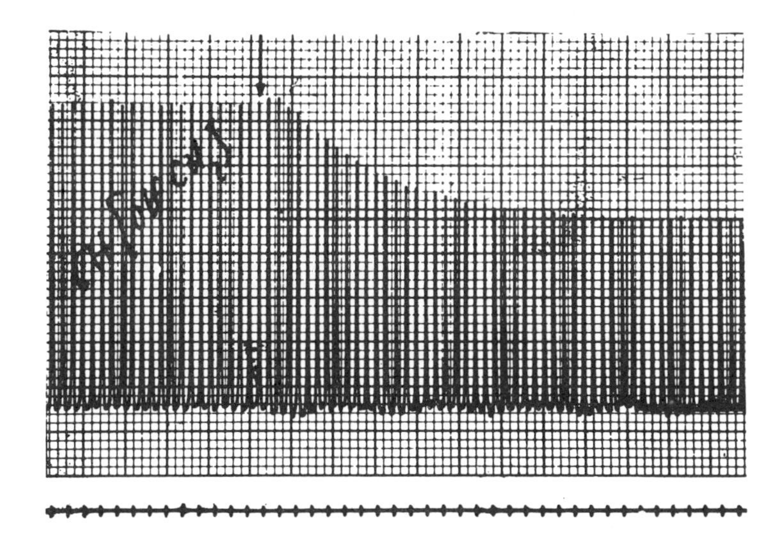

Fig. 1.40. Experimental recording of the decay of signal A ($-CH{=}$) to a new equilibrium value after the saturation of signals B ($-OH$) and C ($-CH_2-$) from a sample of acetylacetone containing a small amount of triethylamine. The markers are spaced at 1-sec intervals and the small arrow indicates the beginning of saturation.

fluctuate accordingly. If the difference in the precessional frequency of the nucleus in two or more environments is comparable in magnitude to the frequency of the fluctuations, there may be an observable effect on the NMR spectrum. Second, the width of NMR resonance lines in the absence of chemically dynamic processes (or other factors such as paramagnetic impurities) is very small, so these effects are readily observed. Even with instrumental contributions line widths are typically only a few tenths of a Hertz. It is important to point out here that within this section labeled "Fast Reactions," relative reaction rates will be termed fast and slow. Within this section, a fast reaction will be one which occurs at a rate larger than the differences in the precessional frequencies (chemical shift) between the nuclei under observation. A slow reaction is the converse.

Before presenting a brief introduction of the theory of the application of NMR to chemical dynamics, a very simple conceptual analogy drawn from the Heisenberg uncertainty principle may be of value. Heisenberg's principle states that the product of the uncertainty in the values of the conjugate variables which describe the energy and position of a particle is approximately equal to Planck's constant. The uncertainty of the energy of a nucleus precessing in a magnetic field may be described by the uncertainty in the precessional frequency, and the uncertainty in its position described by the uncertainty in its residence time in a given magnetic environment.

$$\Delta E \cdot \Delta\tau = \hbar = h/2\pi \quad (82)$$

$$h \cdot \Delta\nu \cdot \Delta\tau = h/2\pi \quad (83)$$

$$\Delta\nu = \frac{1}{2\pi\Delta\tau} \quad (84)$$

The value of $\Delta\nu$ is an approximation of the line width, and $\Delta\tau$ may be expressed as the mean lifetime of the nucleus in a given magnetic environment, τ. Thus, the line width due to chemical exchange is approximated as:

$$\Delta\nu = \frac{1}{2\pi\tau} \quad (85)$$

From this simple approximation, one can see that for a mean time lifetime of 10^{-2} seconds a line width of approximately 16 Hz would be observed that is much greater than line widths obtained in the absence of chemical exchange. Thus, if the nucleus in this case were exchanging between two equally populated sites, an exchange rate ($1/\tau$) of approximately 10^2 sec^{-1} could easily be observed. This simple approach greatly underestimates the applicability of NMR to rate measurements and in diamagnetic systems rates in the range of 10 to 10^5 exchanges per second may be investigated.

Rather than attempt to present a historical and complete development of the theory, an example of an early theory of a two-site exchange case will be given followed by an example of a multisite problem. The purpose of this approach is to show that for simple systems rate data may be obtained with very little mathematical or experimental effort, and that almost amazing insight may be gleaned from complex chemical processes using NMR.

a. Two-Site Exchange

The early theoretical treatment of the effect of chemical exchange on NMR spectra was given by Gutowsky, McCall, and Slichter and has become known as the GMS theory (99). These authors approached the description of chemical dynamics by considering the effect on the Block equations. The Block equations in their original form described the equations of motion of a macroscopic moment of nuclei in the presence of a magnetic field B_0. If the exchange between two sites of different magnetic (and perhaps chemical) environment are to be considered, a Block equation for each magnetically nonequivalent site must be considered. A complex moment defined as

$$G = u + iv \tag{86}$$

where u and v are the respective transverse components of the magnetic moments along and perpendicular to a rotating field B_1. The absorption intensity is proportional to the component v. If there is no exchange between the two magnetically nonequivalent sites A and B, separate Block equations may be written for each site:

$$\frac{dG_A}{dt} + \alpha_A G_A = -i\gamma B_1 B_1 M_{0A} \tag{87}$$

$$\frac{dG_B}{dt} + \alpha_B G_B = i\gamma M_{0A} \tag{88}$$

where
$$\alpha_A = 1/T_{2A} - i(\omega_A - \omega) \tag{89}$$
$$\alpha_B = 1/T_{2B} - i(\omega_B - \omega) \tag{90}$$

and T_{2A} and T_{2B} are the spin–spin relaxation times of nuclei the respective sites. The Block equations must be modified to account for exchange between the two sites. Some assumptions are made. It is assumed that nuclei remain at one site until a sudden jump is made to a nonequivalent site and the nuclear moment during the exchange is neglected. Exchange between equivalent sites will have no effect. The term "interchange" is sometimes used to emphasize that only exchange between magnetically nonequivalent sites are observable. It will be assumed that the mean residence time of a given nucleus on a particular site is constant and expressed as τ_A and τ_B for sites A and B respectively. A statistical correction may be necessary if the two sites are not equally populated;

$$\frac{P_A}{P_B} = \frac{\tau_A}{\tau_B} \tag{91}$$

where P_A and P_B $(= 1 - P_A)$ are the fractional populations of sites A and B. McConnell (150) presented a simplified form of the GMS theory giving a modified form of the Block equation:

$$\frac{dG_A}{dt} + \alpha_A G_A = -i\gamma B_1 M_{0A} + \frac{G_B}{\tau_B} - \frac{G_A}{\tau_A} \tag{92}$$

$$\frac{dG_B}{dt} + \alpha_B G_A = -i\gamma B_1 M_{0B} + \frac{G_A}{\tau_A} - \frac{G_B}{\tau_B} \tag{93}$$

The last two terms on the right hand side represent the modification. Qualitatively, G_B/τ_B in equation 92 represents the increase in magnetization (G_A) resulting from the transfer of spin (nuclei) from site B to site A at a mean rate of τ_B^{-1} sec $^{-1}$. G_A/A represents the decrease in magnetization at site $A(G_A)$ resulting from the transfer of spins from site A to site B. The reverse is the case for equation 93.

Before proceeding with equations, a pause to examine the chemical significance of these terms may be helpful. The Block equations describe the time dependence of the magnetic moment of a nucleus precessing in a magnetic field as a function of several parameters. One parameter which will be seen to be important is $(\omega_0 - \omega)$, the frequency difference between the resonance frequency (ω_0) and the frequency at which the spectrometer is scanning at any instant (ω). Remember, the complex moment G contains a component v which is proportional to the absorption intensity at any frequency of observation (ω). The effect of nuclei exchanging between two nonequivalent sites is to alter the value of v for that nucleus. The way this exchange will affect the spectrum in this two-site case can be seen by solving for v for values of τ_A, τ_B, and ω; that is by calculating a theoretical resonance line. As a specific example consider the exchange of protons between an alcohol and water represented by

$$\mathrm{RO\overline{H}} + \mathrm{HOH} \rightleftharpoons \mathrm{ROH} + \mathrm{\overline{H}OH} \tag{94}$$

where $\overline{\mathrm{H}}$ labels a particular proton. Obviously there are several possible exchange mechanisms (147). If we call the alcohol proton site A and the water proton site B, each interchange between these sites will affect the absorption intensity of the alcohol and water protons in a way described by the solution of equations 92 and 93 for v component.

Two very significant points should be made. First, the exchange process is assumed to be at steady state. This permits the spectrum to be recorded on a solution at equilibrium. Second, if a small rf field (B_1) is used, there will be no change in the total magnetization during a slow sweep of the spectrum (slow passage conditions) and

$$\frac{dG_A}{dt} = \frac{dG_B}{dt} = 0 \tag{95}$$

Solving equations 92 and 93 for G_A and G_B the total complex moment G is

$$G = -i\gamma B_1 M_0 \frac{\tau_A + \tau_B + \tau_A\tau_B(\alpha_A P_A + \alpha_B P_B)}{(1 + \alpha_A\tau_A)(1 + \alpha_B\tau_A) - 1} \tag{96}$$

as was originally obtained by Gutowsky, McCall, and Slichter (99). Of more practical interest to chemists using high resolution NMR spectrometers in the absorption mode is the "imaginary part" of G, that is, the v or absorption made component of G. Gutowsky and Holm (97) found this to be

$$v = \frac{\omega\, \mathrm{Mo}[(1 + \tau T_2^{-1})P + QR]}{P^2 + R^2} \tag{97}$$

$$\text{where} \quad P = \tau\left\{T_2^{-2} - \left[\frac{1}{2}(W_A + W_B) - W\right]^2 + \frac{1}{4}(W_A - W_B)^2\right\} + T_2^{-1} \tag{98}$$

$$Q = \tau\left[\frac{1}{2}(W_A + W_B) - W - \frac{1}{2}(P_A - P_B)(W_A - W_B)\right] \tag{99}$$

$$R = \left[\frac{1}{2}(W_A + W_B) - W\right](1 + 2\tau T_2^{-1}) + \frac{1}{2}(P_A - P_B)(W_A - W_B) \tag{100}$$

Equation 97 may be used directly to calculate the intensity (v) versus frequency (ω) for various values of τ, a simple process using a digital computer. However, the rate of exchange may be considered in three rather arbitary groups of slow, fast, and intermediate. The first two cases permit some simplifying assumptions which are of value. Before considering these, the reader is reminded that equation 97 was specifically derived for an exchange between two nonequivalent sites and is not general. The case of intermediate exchange rates will be considered first and simplifying approximations shown for the limits of fast and slow exchange.

The intermediate exchange rate condition will be considered first as it is the most general in that no approximations may be made in equation 97. Consideration of this condition first will lead to the other categories. Several authors (13,186) have presented solutions to equation 97 using the assumptions

$$P_A = P_B = \frac{1}{2}(\text{equal population of site}) \tag{101}$$

so that

$$\tau_A = \tau_B = 2\tau(\text{note: } \tau \text{ is half the mean lifetime of each site}) \tag{102}$$

and

$$\frac{1}{T_{2A}} = \frac{1}{T_{2B}} \cong 0 \tag{103}$$

The last assumption is that the transverse relaxation times are long; that is the line width in the absence of exchange is small compared to the separation between the lines. Pople, Schneider, and Bernstein show calculated line shapes for this case for illustration (186). However, this assumption would not be made in most real applications. Figure 1.41 shows some line shapes calculated from equation 97 for various parameters as shown on the figure. By comparing experimental spectra of systems undergoing simple two site exchange reactions, the value of τ may be determined. This method has become known as "line shape analysis" although further development to the N site exchange case to be discussed and called "complete line shape analysis" has further important connotations.

As Fig. 1.41 shows, there are extremes of τ values which lie on either end of a central value. First, consider the central case in which $2\pi\tau(\nu_A - \nu_B) = \sqrt{2}$. At this point, the two peaks representing the resonance lines have coalesced into a broad single peak. In practice, this region is the most accurate in determining rate data because the line shape is most sensitive to changes in τ. The extreme values of are defined with respect to $2\pi(\nu_A - \nu_B)$. When $\tau > 10/2\pi(\nu_A - \nu_B)$, the spectrum shows two fully resolved lines and the exchange is "slow." When $\tau \leqslant 0.1/2\pi(\nu_A - \nu_B)$, the spectrum shows a single sharp resonance at a chemical shift equal to $P_A\nu_A + P_B\nu_B$ and the exchange is "fast." The range of τ values that may be determined is then seen to be from approximately $0.1/2\pi(\nu_A - \nu_B)$ to $10/2\pi(\nu_a - \nu_B)$ or roughly a factor of 10^2, with the central value of τ at $\sqrt{2}/2\pi(\nu_a - \nu_B)$. This rule of thumb is very helpful in a prior estimate of whether NMR may be applied to a problem as well as initial adjustments of parameters such as pH or temperature once some information concerning the system has been obtained. The range of rates available is sometimes called the "NMR kinetic window."

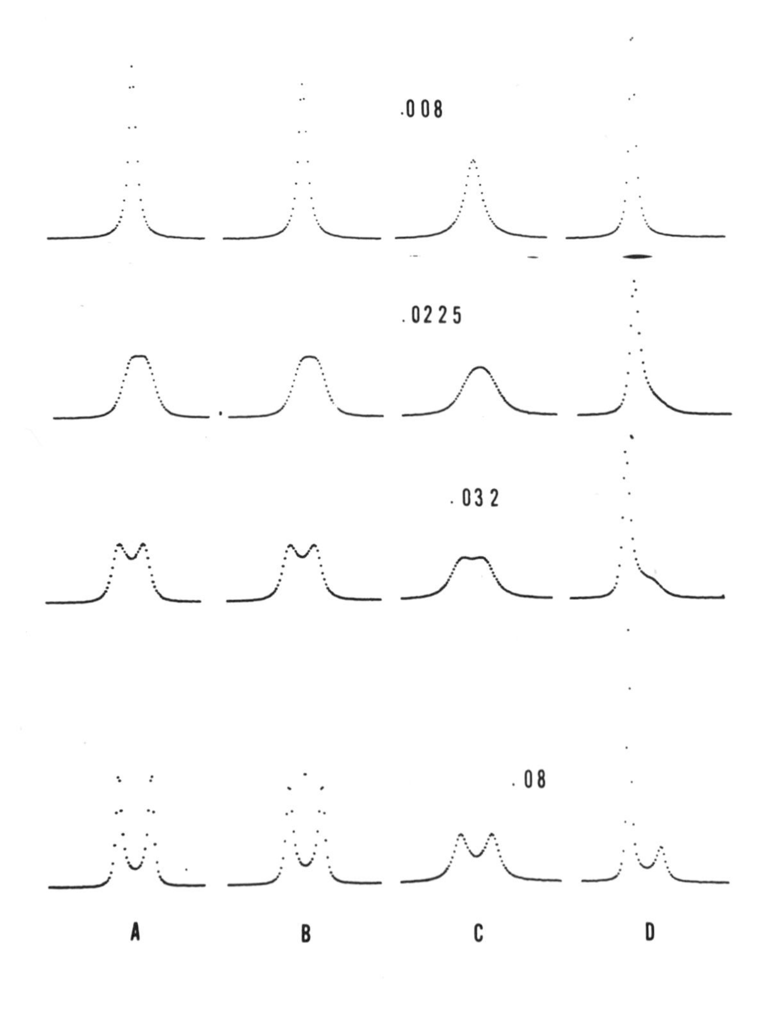

Fig. 1.41. Calculated NMR line shapes for a two-site chemical exchange. The chemical shift difference between the two sites is 10 Hz and the values of τ (seconds) used in the calculations are shown. Column A: T_2 = 15 secs.; P_a = 0.5; column B: T_2 = 1 sec.; P_a = 0.5; column C: T_2 = 0.1 sec.; P_a = 0.5; column D: T_2 = 1 sec.; P_a = 0.75.

In the limit of slow exchange, the line width of half maximum (assuming a Lorentzian line) is given by (204)

$$(\Delta\nu)_{1/2} = \frac{1}{\pi T_2} + \frac{(1 - P_A)}{\tau} \tag{104}$$

which holds for either line by substituting appropriately either P_A or P_B. One can view the term $(1 - P_A)/\tau$ as a quantitative representation of the contribution that random spin exchanges make to the dephasing of nuclear spins. The term $1/T_2$ is a "blank" correction for all other processes leading to line broadening, including magnetic field inhomogeneities and here are assumed equal for both sites and independent of the exchange processes. This is obviously an exceedingly simple method of obtaining rate data, however, caution should be used in extending the method beyond the limits in which the approximations are valid.

A second technique of obtaining τ in the slow exchange limit was introduced by Gutowsky and Holm (97). By setting the derivative of V with respect to $\frac{1}{2}(W_A + W_B) - W$ in equation 97 equal to zero, the expression

$$\frac{1}{\tau} = \frac{1}{\sqrt{2}}(\delta_\omega - \delta_{\omega_e})^{1/2} \tag{105}$$

was obtained where $\delta_\omega = 2\pi(\nu_a - \nu_B)$ is the separation in the limit of no exchange and $\delta\omega_e$ is the corresponding value in the presence of exchange. The apparent peak separation, $\delta\omega_e$, decreases as τ decreases. This is the result of each peak overlapping the other as they become broader. A plot of computer calculated ratios of $\delta\omega_e/\delta_\omega$ versus τ for various values of T_2 can serve as a working curve. This method may be useful in some applications but is of limited general value.

Work done before the availability of computers to almost any laboratory stimulated another technique of obtaining τ values in the slow exchange limit. The intensity ratio method is a further example (204). From equation 97 it can be found that the ratio (R) of the peak intensities to the intensity midway between the peaks of a symmetrical two site exchange ($P_a = P_B$) case is related to τ by

$$\frac{1}{\tau} = \pm\frac{\delta_\omega}{\sqrt{2}}[(R \pm (R^2 - R)^{1/2}]^{-1/2} \tag{106}$$

Again, as in the peak separation method, working curves must be computed. This method suffers the same limitations but gives more accurate values of τ than the peak separation method; probably because the intensity ratios are more precisely measured than peak separations for almost coalesced lines.

As the value of τ decreases (the exchange rate increases), the NMR spectrum of a two-site exchange coalesces into a single line and becomes increasingly narrow. In this region (5)

$$\frac{1}{\tau} = \frac{(\delta_\omega)^2}{4}\left[\frac{1}{T_2'} - \frac{1}{T_2}\right]^{-1} \tag{107}$$

where $\delta_\omega = 2\pi(\nu_a - \nu_B)$ and T_2' is the contribution of the exchange to the transverse relaxation. This equation assumes $P_A = P_B$ and $T_{2A} = T_{2B}$. In principle, equation

107 should permit a larger range of rates to be measured than stated above. However, as T_2' approaches T_2, the line width of a slow passage line can not be used to determine T_2. The use of pulse or rapid passage methods of T_2 measurement can extend this range.

The above example of the two-site exchange theories was presented primarily to provide a conceptual view of the applications of high-resolution (slow passage) NMR techniques and spectrometers to the study of chemical kinetics. The same type of formulation may be used for exchange processes which lead to the coalescence of weakly coupled spin–spin doublets (101); quantum corrections may be required in some cases (95). Although similar closed solutions have been given for three-site (151,175,234), four-site (45), triplets (15), quartets (94), AB patterns (3,126) and a proton–deuterium-coupled (64) systems, this approach has been shown in recent years to be of limited value. The primary reasons for this is that new theoretical approaches to line-shape calculations have developed and the availability of computer hardware and programs permit great flexibility in designing varied solutions to exchange problems. The next section represents a transition of almost 20 years in the theory of NMR line shape calculations.

b. Complete Line-Shape Analysis for Multisite Exchange

The most general treatment of multisite chemical exchange theories is based upon couching the modified Block equations in a "density matrix," which is a format for expressing the relative probabilities of a set of possible exchange paths (3,118,126). An introduction of this concept including specific examples has been given by Johnson and Moreland (120). The use of this approach was developed independently by Kubo (134) and Sack (208). The equation used may be represented as (120):

$$l(\nu)\ \alpha\ Re(\mathbf{P} \cdot \mathbf{A}^{-1} \cdot \mathbf{1}) \tag{108}$$

where $l(\nu)$ is the intensity as a function of frequency, $\mathbf{P}$ is a row vector of the N fractional populations of the nuclear sites, $\mathbf{A}$ is an $N \times N$ complex matrix and $\mathbf{1}$ is a column unit vector. Re represents the "real part of" the result. For an N" site exchange, there would be N equations of the form equation 108 which may be represented as

$$\frac{dG_i}{dt} + \alpha_i G_i = -i\gamma B_1 M_{0_i} - \frac{G_i}{\tau_i} + \sum_{i \neq i} \frac{P_{ij} G_j}{\tau_j} \tag{109}$$

where P_{ij} is the probability of an exchange occurring from site j to site i. The relative equilibrium magnetization M_{o_i} is proportional to the fractional population of that site $P_i\left(\sum_{i\nu l}^{N} P_i = 1\right)$. Retaining the steady state assumption $\left(\frac{dG_i}{dt} = 0\right)$, equation 109 written in the matrix representation would be

$$i\gamma B_1 M_0 \begin{pmatrix} P_A \\ P_B \end{pmatrix} = \begin{pmatrix} -\left(\alpha_A + \frac{1}{\tau_A}\right) & + \frac{1}{\tau_B} \\ + \frac{1}{\tau_A} & -\left(\alpha_B + \frac{1}{\tau_B}\right) \end{pmatrix} \begin{pmatrix} G_A \\ G_B \end{pmatrix} \tag{110}$$

after taking the transpose of both sides:

$$i\gamma B_1 M_0 (P_A P_B) = (G_A G_B) \begin{pmatrix} -\left(\alpha_A + \dfrac{1}{\tau_A}\right) + \dfrac{1}{\tau_A} \\ + \dfrac{1}{\tau_B} - \left(\alpha_B + \dfrac{1}{\tau_B}\right) \end{pmatrix} \tag{111}$$

Because $l(\nu) = lm(\mathbf{G} \cdot \mathbf{l})$, equation (111) is multiplied from the right by $\mathbf{A}^{-1}$ followed by $\mathbf{1}$ to obtain

$$l(\nu) = \gamma B_1 M_0 Re(P_A P_B) \begin{pmatrix} -\left(\alpha_A + \dfrac{1}{\tau_A}\right) + \dfrac{1}{\tau_A} \\ + \dfrac{1}{\tau_B} - \left(\alpha_A + \dfrac{1}{\tau_A}\right) \end{pmatrix} \begin{pmatrix} 1 \\ 1 \end{pmatrix} \tag{112}$$

This method can be readily generalized (119) and a transformation matrix that diagonalizes the $\mathbf{A}$ matrix can be generated (92) which provides for efficient computer simulation of the spectrum. To perform this generalization, $\mathbf{A}$ is defined as

$$\mathbf{A} = \Omega + \mathbf{D} \tag{113}$$

when

$$\Omega_{ij} = -\delta_{ij}\alpha_i \tag{114}$$

$$D_{ij} = P_{ij}/\tau_i (i \neq 1) \tag{115}$$

$$D_{ii} = -\frac{1}{\tau_i} \tag{116}$$

and δ_{ij} is the Kronecker delta. The $\mathbf{A}$ matrix may then be constructed as

$$\mathbf{A} = \begin{pmatrix} -a_1 & o \ldots o \\ o & -a_2 & \vdots \\ o \ldots & & -a_N \end{pmatrix} + \begin{pmatrix} P_{11} & P_{12} \ldots P_{1N} \\ P_{21} & P_{22} & \vdots \\ P_{N1} & \ldots & P_{NN} \end{pmatrix} \times \frac{1}{\tau} \tag{117}$$

where $1/\tau$ represents the total rate of nuclear exchange of all sites. The $\mathbf{D}$ matrix is the probability distribution of exchanges between particular sites. Using the general solution devised by Gordon and McGinnis (92), equation 117 may be efficiently solved by a computer as a function of ν, τ, and P_{ij}. The first problem is to construct the $\mathbf{D}$ matrix by obtaining the probability elements P_{ij} for a particular reaction scheme. Once this matrix is constructed, various values of τ are chosen and the spectrum simulated by calculation of $l(\nu)$ over the range of ν values that encompasses the experimental spectrum. Johnson and Moreland have written a set of rules for obtaining the $\mathbf{D}$ matrix and give examples (120). Upon comparison of the simulated spectra with the experimental spectra over a range of exchange rates (brought about by changes in temperature, concentration, etc.), one decides if they agree. If they do not, a different reaction scheme is chosen and tried. This approach provides a tremendously powerful tool for elucidating the reaction schemes of very fast and complex reactions such as ring rearrangements, hydride shifts, and multischeme exchanges such as proton-exchange reactions. Further specific examples are given by Johnson (118).

c. TRANSIENT METHODS OF RATES STUDIES

The techniques of application of NMR described above were for the most part slow-passage methods. An exception is the transient observation of signal decay in the double resonance method of Forsén and Hoffman discussed in Section IV.E.1 (82,83). The use of pulse techniques were limited in early applications primarily because of instrumental limitations. Commercial instruments were available, but difficult to justify as separate instruments for pulse NMR studies. Several modifications of high-resolution instruments were made (148,90). However, with the advent of commercial spectrometers capable of cw high-resolution and Fourier transform work, various pulsed experiments may be performed more readily. A second consideration in the use of pulsed NMR methods is the limited chemical systems to which it may be applied. This limitation is a direct result of the nature of pulsed experiments.

The principle of the technique is that when a strong rf field B_1 with a frequency equal to the resonance frequency of the nuclear spins is applied, the nuclear magnetization is described by the time-dependent Block equations. The amplitude and duration of the pulse determines the degree to which the magnetic moment vector is rotated about the x axis. For this reason, the pulses are quantitatively referred to as π or $\pi/2$ pulses. For example, after a $\pi/2$ pulse, the magnetization lies along the y axis in the rotating frame and decays with a decay time constant determined by processes that effect T_2, including nuclear exchange. By analysis of the free induction decay of the signal after the pulse, the net T_2' may be determined. As it happens, some clever tricks of a sequence of pulses can be used to minimize field inhomogeneity and diffusion effects. These sequences are those derived by Carr and Purcell (44) and modified by Meiboom and Gill (158). Application of these pulse sequences are followed a series of magnetization bounces or echoes which diminish with time and hence the term "spin-echo" is applied to the methods.

The limitation of spin-echo methods lies in the fact that the strong rf field used for the pulses will have a wide range of frequency components and hence interact with nonexchanging nuclei (of a given isotope) as well as exchanging nuclei. As a result, complicated decay functions are obtained when magnetically nonequivalent nuclei are present. In the extreme, this means that only one type of nucleus (proton for example) may be present in the molecule under study, a serious limitation when organic compounds are investigated. At least one example has been worked out for a case of two non-equivalent protons (2). There are two possibilities to expand the applications. One is to use isotopic substitution for all but the chemical site of interest, for example deuterium for protons. An alternate is to label the exchanging site with deuterium and use an rf pulse at the ^2H resonance frequency. The second, and the one which has lead to the most frequent use of spin-echo methods, is to observe one type of proton which is present in a large excess over all others. For example, protolysis exchange reactions of amines have been studied extensively by observing the spin-echo of the exchanging solvent water protons in solutions so dilute that the nonexchanging amine protons do not contribute significantly to the spin-echo signal (196). A second example is the study of paramagnetic metal binding by macromolecules in water, again in dilute solutions (57,58). With the availability of the Fourier transform instruments, more applications of the pulse methods are expected.

The principle advantage of spin-echo techniques should be specifically pointed out. In general, faster reactions may be investigated using these methods. Reaction rates at which "exchange narrowing" of high resolution lines leads to line widths too narrow to be measured accurately can be studied in many case using spin-echo. Using the techniques described above for investigation of protolysis reactions offers a secondary advantage. To observe high-resolution spectra of the amine directly so that line-shape analysis may be performed requires a concentration sufficiently high as to cause viscosity effect on T_2 and diffusion controlled rate constants, and salt effects on the rate of exchange. These effects are minimized using dilute solutions and observing the spin-echo signal of the solvent (provided the solvent participates in the exchange either as a molecule or atomic exchange).

A comparison of spin-echo and slow-passage techniques for a study of the internal rotation of *N*, *N*-dimethyltrichloroacetamide has been reported (4). The values of the activities energy *Ea* and the pre-exponential factor *A* were found to differ greater than predicted from the standard deviation of the respective methods. Furthermore, a systematic difference of rate was observed. The cause of these differences has yet to be completely explained.

3. Examples of Rate Studies

The application of continuous wave high resolution, wide line, and pulsed NMR methods to the investigation of dynamic chemical processes has grown very rapidly. The variety of chemical systems that has been studied is extensive. A few examples will be given here, not as a review of the subject, but to illustrate the types of problems which may be investigated. The rather straightforward use of NMR as a quantitative tool to follow the course of slow chemical reactions has been discussed in Section IV.E.1 and will not be discussed further. However, it should not be forgotten as an appropriate method in many cases.

a. Intramolecular Processes

Some chemical exchange processes may be brought into a rate range suitable for study by high-resolution NMR by temperature variation. One must keep in mind that here exchange involves an interchange between magnetic environments, and hence intramolecular processes may be investigated as well as intermolecular reactions while the system is at chemical equilibrium. In principle, all that is required is a nucleus that interchanges between two or more magnetic environments at a rate which is suitable to influence the line shape of the NMR spectrum. The magnetic requirements may be met by asymmetry inherent in the molecule to be investigated, or by introducing a probe of the type $-CX_2Y$ into the molecule where X represents the nucleus or group containing the nucleus to be observed.

(1) Rotation about a Chemical Bond

Rotation about a chemical bond may be hindered for a variety of reasons such as steric hindrance or partial double-bond character. Depending upon the degree of restriction to rotation, the rate may be brought within the requirements of NMR by temperature variation. An example is a study of the C(2)–C(3) bond rotation in a series of halogenated methylbutanes (108). Provided the rate of rotation can be slowed sufficiently at low temperature to observe the spectra of each magnetically nonequivalent state (in this case the rotational conformations), the equilibrium population of each state can be estimated. If all three classical rotamers are significantly populated, the problem requires a theoretical treatment as exchange between three nonequivalent sites. In some cases, only two-site treatment is required, as one population may be negligible. In these cases, it is highly advisable to use a computer program such as DNMR (28) which is capable of calculating the complete line shape under various assumed exchange conditions. Because the shape of the NMR spectrum will depend upon the mechanism as well as the rate of exchange, incorrect assumptions may be detected by a lack of agreement between the shape of the experimental and simulated NMR spectra.

In the case of the methylbutanes, the center for magnetic nonequivalence is a core portion of the molecule under study. On the occasions in which no such centers are present, they may be generated by the introduction of diastereotropic probes. Of course, one must consider the possible effects such modifications may have on the process to be investigated. An example is the use of the $-CF_2H$ group to investigate rotational process in triarylmethyl cations (195). Spectra of compounds such as **X**

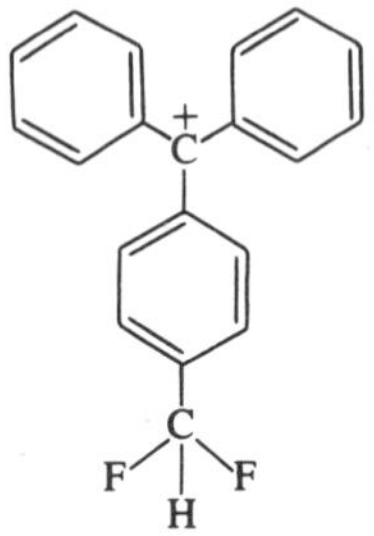

X

show ^{19}F-NMR of the ABX$\rightleftharpoons$BAX type representing interconversions of enantiomers in the propeller forms.

Perhaps the most extensively studied system of internal rotation is those of C–N bonds in amides. The literature is too extensive on this subject to select an outstanding example without offending the authors of other papers. These compound types are of interest in part because they are related to biologically important compounds. Usually, an *N*-methyl or *N*, *N*-dimethyl derivative is employed and the exchange is then treated as a two-site interconversion resulting from rotation about the C–N bond.

(2) Inversions

Two types of inversions will be given as illustrations. Ring inversions of many types have been investigated. The example of the study of the inversion of the cyclohexane **XI** ring, one of the earliest to be investigated using ^{1}H-NMR, is perhaps a classic example of the power of NMR. For each chair–chair interconversion, there is an interchange between the axial (H_1), and equatorial (H_2) protons. As a solution of cyclohexane in CS_2 is cooled,

H_1 H_2

XI

the single NMR resonance of the cyclohexane protons broadens, and below −67°C a doublet is observed. However, the peaks are complicated by spin–spin coupling between the various protons. Yet it is important to realize that this dynamic process may be studied, with no perturbation of the structure. The use of undecadeuterocyclohexane with double irradiation at the deuterium resonance frequency removes the complications resulting from spin–spin interactions (12). The inversion process is then investigated by temperature variation.

A second type of inversion that has been investigated is the inversion of pyramidal atomic configurations. Inversion of nitrogen atom configuration is the most studied example. Again, a diastereotropic probe such as $-CH_2R$ is used. Two types of studies have been performed. The rate of nitrogen inversion may be lowered to the range suitable for NMR by lowering the temperature (42), or by investigating the process in aqueous solution at low pH values (162).

(3) Intramolecular Rearrangements

The exchange between magnetic environments may be the result of some form of rearrangement. Some examples are presented below. In these cases, a theoretical treatment such as that given in Section IV.E.2.b is required.

1,2-Methyl Shift in Heptamethylbenzonium Ion (215) (**XII**)

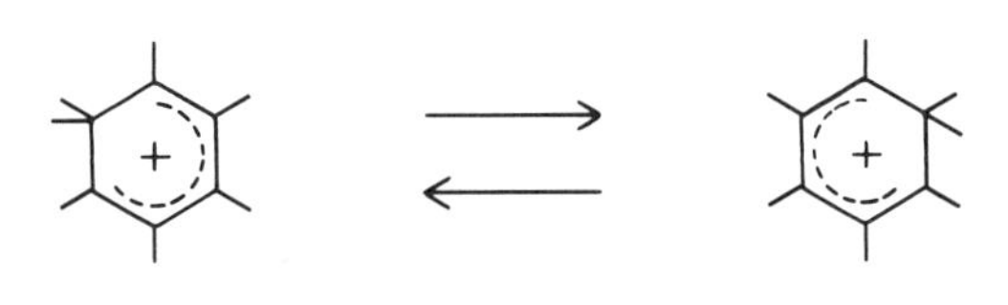

XII

σ-Cyclopentadienyl(triethylphosphine) Copper(I) (245) (**XIII**)

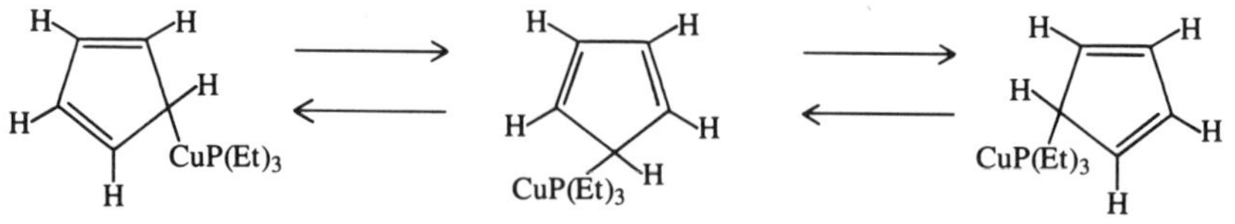

XIII

2,3-Hydride Shift in 2-Norbornyl Cation (217) (**XIV**)

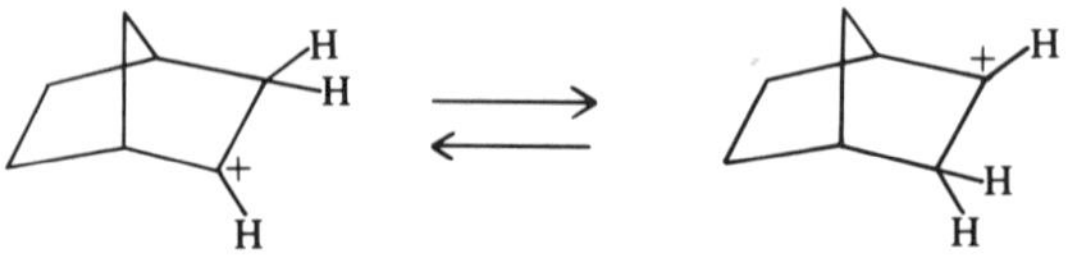

XIV

b. INTERMOLECULAR PROCESSES

For the greater number of cases studied by NMR, intramolecular exchange involves the influence of changes in the magnetic environment (chemical shift) of the observed nuclei upon the NMR spectral line shape. This may also be the case for intermolecular exchange. However, it is also commonly true that spin–spin coupling is influenced by breaking of chemical bonds during intermolecular exchange. In a sense, therefore an additional mechanism for the effect of chemical exchange is available to effect the NMR spectrum. The fundamental concepts of the theoretical treatment are the same as for exhange between magnetic environments. However, the effect of the spin–spin interaction must be taken into account in the mathematical treatment (120). The types of systems that have been studied by ^{1}H-NMR as well as resonance of other nuclei is very extensive and the reader is encouraged to consult one of the several annual reviews of nuclear magnetic resonance to obtain a perspective of the applications reported each year. The availability of multipurpose line-shape simulation programs such as the ones developed by Binisch (28) have made the technique of fast kinetics studied by NMR essentially a routine chemical tool.

(1) Proton Transfer Reactions

Proton transfer reactions were among the earliest fast exchange reactions investigated by ^{1}H-NMR. Grunwald (96) has reviewed some of these. Reactions involving alcohols, amines, amino acids, amides, imidazoles, and other compounds have been studied. Several simultaneous reactions represented schematically in equations 118–121 may be in effect.

$$\mathrm{BH} + \underline{\mathrm{B}} \rightleftharpoons \mathrm{B} + \mathrm{H}\underline{\mathrm{B}} \qquad (118)$$

$$BH + H\underline{B} \rightleftharpoons B + H_2\underline{B}^+ \quad (119)$$

$$BH\ldots(H_2O)_n + \underline{B} \rightleftharpoons B + (H_2O)_n\ldots H\underline{B} \quad (120)$$

$$BH + OH^- \rightleftharpoons B + H_2O \quad (121)$$

Each system must be thoroughly investigated by means of variations in concentrations and pH with close attention paid to salt effects, viscosity of the solution, and any other parameters that may influence the rate of fast reactions.

2. Exchange in Metal Complexes

Early in the development of techniques for the study of fast reactions using ¹H-NMR, Swift and Connick (234) devised an experimentally simple method for the measurement of rates of exchange between ligands bound in a paramagnetic metal complex and free ligands. Their method requires only the measurement of the line width of the free ligand resonance. There is usually a large excess of free ligand present in the solution and the exchange rate may frequently be controlled to some extent by pH adjustment. Examples of the types of systems that have been studied by this approach are metal complexes of ammonia and amino acids (180), the Fe(II) complex of hemin (65), the exchange of methanol and *N*, *N*-dimethylformamide (DMF) from the Mn(II) complex of protoporphyrin IX dimethyl ester (207) and DMF exchange with Ni(II) Schiff-base complexes in which diamagnetic paramagnetic equilibria are involved (206).

In the case of ligand exchange reactions in diamagnetic complexes, more detailed analysis of the line shape is usually involved. However, frequently more information may be obtained. For example, if one considers the metal complexes of multidentate ligands such as ethylenediaminetetraacetic acid (EDTA), it becomes apparent that not only is it possible to investigate the rate of ligand exchange between the free and bound state, but even more detailed studies are possible. Day and Reilley (63) applied ¹H-NMR to the investigation of individual metal ligand bond lifetimes. Before this it had long been accepted that aminocarboxylates formed multidentate chelates. From X-ray data, it was known that in the solid state a given ligand such as EDTA formed different numbers of metal-ligand bonds depending upon protonation of the ligand (71). Mixed aquo- and hydroxy-complexes were also known. However, no direct results had been obtained to determine the lifetime of these metal-ligand bonds.

Consider the predicted effect individual bond lifetime will have on the NMR spectrum of an aminopolycarboxylate chelate. If all metal-ligand bonds have short lifetimes, the spectrum of the nonlabile protons in the molecule will be relatively simple because of internal averaging. This is similar to the fast proton exchange in a partially protonated ligand. As in protonation, the chemical shift of the chelate will be different than the free ligand. An average spectrum would be observed as a result of rapid ligand exchange. On the other hand, if one or more metal-ligand bonds is sufficiently long so that it may be considered permanent by NMR detection, then chemical shifts which are identical in the free ligand may not average in the chelate and a more complex spectrum may result. As an example, two protons on a carbon atom adjacent to an asymmetric nitrogen are expected to be nonequivalent if the nitrogen atom does not undergo inversion. A structure such as

```
     H   R1
     |   |
R—C—N—M——
     |   |
     H   R2
```

where R_1, R_2, and RCH_2- are all different, and the nitrogen-metal bond has a long lifetime meets these requirements. The two protons on the carbon adjacent to the nitrogen are expected to yield the AB-type splitting pattern. The mere presence of an AB pattern in the spectrum confirms qualitatively a long nitrogen-metal bond lifetime. A more detailed examination of these AB patterns may yield quantitative rate data, concerning the bond lifetime.

REFERENCES

1. Abragam, A., *The Principles of Nuclear Magnetism*, Oxford Univ. Press, London, 1961.
2. Alexander, S., *Rev. Sci. Instr.*, **32**, 1966 (1961).
3. Alexander, S., *J. Chem. Phys.*, **37**, 966 (1962).
4. Allerhand, A., and H. S. Gutowsky, *J. Chem. Phys.*, **42**, 1587 (1965).
5. Anbar, M., A. Loewenstein, and S. Meiboom, *J. Am. Chem. Soc.*, **80**, 2630 (1958).
6. Anderson, W. A., in *NMR and EPR Spectroscopy*, Pergamon, New York, 1960, p. 176.
7. Anderson, W. A., *Rev. Sci. Instruments*, **32**, 241 (1961).
8. Anderson, W. A., *Rev. Sci. Instruments*, **33**, 1160 (1962).
9. Anderson, W. A., R. Freeman, and C. A. Reilly, *J. Chem. Phys.*, **39**, 1518 (1963).
10. Anderson, Jr., W. R., and R. M. Silverstein, *Anal. Chem.*, **37**, 1417 (1965).
11. Anet, F. A. L., *J. Am. Chem. Soc.*, **84**, 1053 (1962).
12. Anet, F. A. L., M. Ahmad, and L. D. Hall, *Proc. Chem. Soc.* (London) 145 (1964).
13. Anet, F. A. L., and A. J. R. Bourn, *J. Am. Chem. Soc.*, **87**, 5250 (1965).
14. Anet, F. A. L., and G. E. Schenck, *J. Am. Chem. Soc.*, **93**, 556, 3310 (1971).
15. Arnold, J. T., *Phys. Rev.*, **102**, 136 (1956).
16. Arnold, J. T., S. S. Dharmatti, and M. E. Packard, *J. Chem. Phys.*, **19**, 507 (1951).

16a. Aue, W. P., E. Bartholdi, and R. R. Ernst, *J. Chem. Phys.*, **64**, 2229 (1976).

17. Babiar, J. S., J. R. Barrante, and G. C. Vickers, *Anal. Chem.*, **40**, 610 (1968).
18. Bachers, G. C., and T. Schaefer, *Chem. Rev.*, **71**, 617 (1971).
19. Baker, E. B., *J. Chem. Phys.*, **37**, 911 (1962).
20. Baker, E. B., and L. W. Burd, *Rev. Sci. Instrum.*, **28**, 313 (1957).
21. Baldeschwieler, J. D., and E. W. Randall, *Chem. Rev.*, **63**, 81 (1963).
22. For a review, see Barfield, M., and D. M. Grant, in J. S. Waugh, Ed., *Advances in Magnetic Resonance*, Vol. 1, Academic, New York, 1965, p. 149.
23. Becker, E. D., *High Resolution NMR*, Academic, New York, 1969.
24. Bell, R. A., and J. K. Saunders, in N. L., Allinger and E. L. Eliel, Eds., *Topics in Stereochemistry*, Vol. 7, Interscience, New York, p. 1, 1973.
25. Bernstein, H. J., and K. Frei, *J. Chem. Phys.*, **37**, 1891 (1962).

26. Bhacca, N. S., and D. H. Williams, *Application of NMR Spectroscopy in Organic Chemistry*, Holden-Day, San Francisco, 1965.

27. Bible, Jr., R. H., *Appl. Spectrosc.*, **24**, 326 (1970).

28. Binsch, G., *J. Amer. Chem. Soc.*, **91**, 1304 (1969). (Quantum Chemistry Program Exchange, Indiana University, Programs 140 and 165).

29. Bloch, F., W. W. Hansen, and M. E. Packard, *Phys. Rev.*, **69**, 127 (1946).

30. Block, F., W. W. Hansen, and M. E. Packard, *Phys. Rev.*, **70**, 474 (1946).

31. Bloembergen, N., E. M. Purcell, and R. V. Pound, *Phys. Rev.*, **73**, 679 (1948).

31a. Bodenhausen, G., R. Freeman, R. Niedermeyer, and D. L. Turner, *J. Magnet. Reson.*, **26**, 133 (1977).

32. Bothner-By, A. A., and C. Naar-Colin, *J. Am. Chem. Soc.*, **80**, 1728 (1958).

33. Bothner-By, A. A., in J. S. Waugh, Ed., *Advances in Magnetic Resonance*, Vol. 1, Academic, New York, p. 195 (1965).

34. Bovey, F. A., *NMR Data Tables for Organic Compounds*, Interscience, New York, 1967.

35. Bovey, F. A., and G. V. D. Tiers, *J. Am. Chem. Soc.*, **81**, 2870 (1959).

36. Brugel, W., *NMR Spectra and Chemical Structure*, Academic, New York, 1967.

37. Bruker Scientific, Inc., Elmsford, New York, technical bulletin.

38. Buckingham, A. D., *Can. J. Chem.*, **38**, 300 (1960).

39. Buckingham, A. D., and K. A. McLauchlan, *Proc. Chem. Soc.*, **144**, (1963).

40. Burakevich, J. V., and J. O'Neill, Jr., *Anal. Chim. Acta.*, **54**, 528 (1971).

41. Burke, J. J., and P. C. Lauterbur, *J. Am. Chem. Soc.*, **86**, 1870 (1964).

42. Bushweller, C. H., and W. G. Anderson, *Tetrahedron Lett.*, 129, (1972).

43. Bushweller, C. H., and J. W. O'Neill, *J. Am. Chem. Soc.*, **92**, 2159 (1970).

44. Carr, A. Y., and E. M. Purcell, *Phys. Rev.*, **94**, 630 (1954).

45. Carrington, A., *Mol. Phys.*, **5**, 425 (1962).

46. Carrington, A., and A. D. McLachlan, *Introduction to Magnetic Resonance*, Harper and Row, New York, 1967.

47. Castellano, S., unpublished results.

48. Castellano, S., and R. Kostelnik, *Tetrahedron Lett.*, **5211** (1967).

49. Castellano, S., R. Kostelnik, and C. Sun, *Tetrahedron Lett.*, 4635 (1967).

50. Castellano, S., and R. Kostelnik, *J. Am. Chem. Soc.*, **90**, 141 (1968).

51. Castellano, S., and C. Sun, *J. Am. Chem. Soc.*, **88**, 4741 (1966).

52. Castellano, S., C. Sun, and R. Kostelnik, *Tetrahedron Lett.*, 5205 (1967).

53. Caughman, M. C., *A Review of N-H Proton Exchange Reactions of Nitrogen Compounds*, M. S. Thesis, University of Georgia, Athens, Georgia, 1970.

54. Chamberlain, N. F., *The Practice of NMR Spectroscopy with Spectra-Structure Correlations for Hydrogen-1*, Plenum, New York, 1974.

55. Chapman, O. L., and R. W. King, *J. Am. Chem. Soc.*, **86**, 1256 (1964).

56. Cockerill, A. F., G. L. O. Davies, R. C. Harden, and D. M. Rackham, *Chem. Rev.*, **73**, 553 (1973).

57. Cohn, M., *Biochemistry*, **2**, 623 (1963).

58. Cohn, M., and J. S. Leigh, *Nature*, **193**, 1037 (1962).

59. Cario, P. L., *Structure of High-Resolution NMR Spectra*, Academic, New York, 1967.

60. Cox, R. H., *Spectrochimica Acta.*, **25A**, 1189 (1969).

61. Cox, R. H., unpublished results.

62. Cross, A. D., and I. T. Harrison, *J. Am. Chem. Soc.*, **85**, 3223 (1963).

63. Day, R. J., and C. N. Reilly, *Anal. Chem.*, **36**, 1073 (1964).

64. Day, R. J., and C. N. Reilly, *J. Phys. Chem.*, **73**, 1588 (1967).
65. Degani, H. H., and D. Fiat, *J. Am. Chem. Soc.*, **93**, 4281 (1971).
66. Diehl, P., *Helv. Chim. Acta.*, **44**, 829 (1961).
67. Dietrich, M. W., J. S. Nash, and R. Kelbi, *Anal. Chem.*, **38**, 1479 (1966).
68. Digilab Inc., Cambridge, Mass., technical bulletin.
69. Diner, V. E., and J. W. Lown, *Chem. Commun.*, 333 (1970).
70. Douglass, D. C., and A. Fratiello, *J. Chem. Phys.*, **39**, 3163 (1963).
71. Dwyer, F. P., and D. P. Mellor, Editors, *Chelating Agents and Metal Chelates*, New York, Academic, 1964, Chapter 7.
72. Eastermann, I., and O. Stern, *Z. Phys.*, **85**, 17 (1933).
73. Emsley, J. W., J. Feeney, and L. H. Sutcliffe, *High-Resolution Nuclear Magnetic Resonance Spectroscopy*, Vol. 1, Pergamon, New York, 1965.
74. Emsley, J. W., J. Feeney, and L. H. Sutcliffe, *High-Resolution Nuclear Magnetic Resonance Spectroscopy*, Vol. 2, Pergamon, New York, 1965.
75. Ernst, R. R., in J. S. Waugh, Ed., *Advances in Magnetic Resonance*, Vol. 2, Ch. 1, Academic, New York, 1966.
76. Ernst, R. R., and W. A. Anderson, *Rev. Sci. Instrum.*, **37**, 93 (1966).
77. Ernst, L., and A. Mannschreck, *Tetrahedron Lett.*, 3023 (1971).
78. Evans, Jr., H. B., A. R. Tarpley, and J. H. Goldstein, *J. Phys. Chem.*, **72**, 2552 (1968).
79. Farrar, T. C., and E. D. Becker, *Pulse and Fourier Transform NMR, Introduction to Theory and Methods*, Academic, New York, 1971.
80. Farrar, T. C., S. J. Druck, R. R. Shoup, and E. D. Becker, unpublished manuscript reported in reference 77, p. 60.
81. Feeney, J., A. Ledwith, and L. H. Sutcliffe, *J. Chem. Soc.*, 2021, (1962).
82. Forsén, S., and R. A. Hoffman, *Acta. Chem. Scand.*, **17**, 1787 (1963).
83. Forsén, S., and R. A. Hoffman, *J. Chem. Phys.*, **40**, 1189 (1964).
83a. Freeman, R., *Proc. R. Soc. Lond. A*, **373**, 149 (1980).
84. Freeman, R., and W. A. Anderson, *J. Chem. Phys.*, **37**, 2053 (1963).
84a. Freeman, R., and G. A. Morris, *Bull. Magnet. Reson.*, **1**, 5 (1979).
85. Freeman, R., and D. H. Whiffen, *Mol. Phys.*, **4**, 321 (1961).
86. Frisch, R., and O. Stern, *Z. Phys.*, **85**, 4 (1933).
87. Gabillard, R., and M. Soutif, in P. Grivet, Ed., *La Resonance Paramagnetique Nucleaire*, Centre National de la Recherche Scientifique, Paris, 1955, p. 159.
88. Gerlach, W., and O. Stern, *Ann. Phys. Lpz.*, **74**, 673 (1924).
89. Gil, V. M. S., and J. N. Murrell, *Trans. Farad. Soc.*, **60**, 248 (1964).
90. Ginsburg, A., A. Lipman, and G. Navon, *J. Phys. E.*, **3**, 699 (1970).
91. Goering, H. L., J. N. Eikenberry, and G. S. Koermer, *J. Am. Chem. Soc.*, **93**, 5913 (1971).
92. Gordon, R. G., and R. P. McGinnis, *J. Chem. Phys.*, **49**, 2455 (1968).
93. Gorter, C. J., *Physica*, **3**, 995 (1936).
94. Grunwald, E., A. Loewenstein, and S. Meiboom, *J. Chem. Phys.*, **27**, 630 (1957).
95. Grunwald, E., C. F. Jumper, and S. Meiboom, *J. Am. Chem. Soc.*, **84**, 4664 (1962).
96. Grunwald, E., and E. K. Ralph, *Accounts Chem. Res.*, **4**, 107 (1971).
97. Gutowsky, H. S., and C. H. Holm, *J. Chem. Phys.*, **25**, 1228 (1956).
98. Gutowsky, H. S., and D. W. McCall, *Phys. Rev.*, **82**, 748 (1951).
99. Gutowsky, H. S., and D. W. McCall, and C. P. Slichter, *J. Chem. Phys.*, **21**, 279 (1953).
100. Gutowsky, H. S., L. H. Meyer, and R. E. McClure, *Rev. Sci. Instrum.*, **24**, 644 (1953).

101. Gutowsky, H. S., and A. Saika, *J. Chem. Phys.*, **21**, 1688 (1953).

102. Hahn, E. L., and D. E. Maxwell, *Phys. Rev.*, **84**, 1246 (1951).

103. Hall, G. E., in E. F. Mooney, Ed., *Annual Review of NMR Spectroscopy*, Vol. 1, Academic, New York, 1968, p. 227.

103a. Hall, L. D., G. A. Morris, and S. Sukumar, *Carbohydr. Res.*, **76**, C7 (1979).

103b. Hall, L. D., and J. K. M. Sanders, *J. Am. Chem. Soc.*, **102**, 5703 (1980).

103c. Hall, L. D., and J. K. M. Sanders, *Org. Chem.*, **46**, 1132 (1981).

104. Hamilton Company, Whittier, California.

105. Hart, H., and G. M. Love, *Tetrahedron Lett.*, 625 (1971).

106. Hartwell, G. E., and A. Allerhand, *J. Am. Chem. Soc.*, **93**, 4415 (1971).

107. Hassner, A., and C. Heathcock, *J. Org. Chem.*, **29**, 1350 (1964).

108. Hawkins, B. L., W. Bremser, S. Borcic, and J. D. Roberts, *J. Am. Chem. Soc.*, **93**, 4472 (1971).

109. For a review of modulation, see O. Haworth and R. E. Richards, in J. W. Emsley, J. Feeney, and L. H. Sutcliffe, Eds., *Progress in N. M. R. Spectroscopy*, Vol. 1, Pergamon, New York, 1966, p. 1.

110. Heel, H., and W. Zeil, *Z. Electrochem.*, **64**, 962 (1960).

111. Hinckley, C. C., *J. Am. Chem. Soc.*, **91**, 5160 (1969).

112. Hoffman, R. A., and S. Forsén, in J. W. Emsley, J. Feeney, and L. H. Sutcliffe, Eds., *Progress in N. M. R. Spectroscopy*, Vol. 1, Pergamon, New York, 1966, p. 15.

112a. IBM Instruments Inc., Danbury, Connecticut, technical bulletin.

113. Jackman, L. M., and S. Sternhell, *Applications of Nuclear Magnetic Resonance Spectroscopy in Organic Chemistry*, Pergamon, New York, 1969, p. 94–98.

114. Jackman, L. M., and R. H. Wiley, *J. Chem. Soc.*, 288 (1960).

115. Jensen, F. R., D. S. Noyce, C. H. Sederholm, and A. J. Berlin, *J. Am. Chem. Soc.*, **82**, 1256 (1960).

116. JEOL, U.S.A., Inc., Cranford, New Jersey, technical bulletin.

117. Johnson, C. E., and F. A. Bovey, *J. Chem. Phys.*, **29**, 1012 (1958).

118. Johnson, Jr., C. S., in J. S. Waugh, Ed., *Advances in Magnetic Resonance*, Vol. 1, Academic, New York, 1965, pp. 33–102.

119. Johnson, Jr., C. S., *Am. J. Phys.*, **35**, 929 (1967).

120. Johnson, Jr., C. S., and C. G. Moreland, *J. Chem. Educ.*, **50**, 477 (1973).

121. Johnson, L. F., in *NMR and EPR Spectroscopy*, Pergamon, New York, 1960, p. 60.

122. Johnson, L. F., and J. N. Shoolery, *Anal. Chem.*, **34**, 1136 (1962).

123. Jonathan, N., S. Gordon, and B. P. Dailey, *J. Chem. Phys.*, **36**, 2443 (1962).

124. Karplus, M., *J. Chem. Phys.*, **30**, 11 (1959).

125. Kan, R. O., *J. Am. Chem. Soc.*, **86**, 5180 (1964).

126. Kaplan, J., *J. Chem. Phys.*, **28**, 278 (1958).

127. Karabatsos, G. J., F. M. Vane, R. A. Taller, and N. Hso, *J. Am. Chem. Soc.*, **86**, 3351 (1964).

128. Karplus, M., *J. Am. Chem. Soc.*, **85**, 2870 (1963).

129. Kasler, F., *Quantitative Analysis by NMR Spectroscopy*, Academic, New York, 1973, pg. 127.

130. Kennewell, P. D., *J. Chem. Educ.*, **47**, 278 (1970).

131. For a review, see V. J. Kowalewski, in J. W. Emsley, J. Feeney, and L. H. Sutcliffe, Eds., *Progress in N. M. R. Spectroscopy*, Vol. 5, Pergamon, New York, 1969, p. 1.

132. Kreevoy, M. M., H. B. Charman, and D. R. Vinard, *J. Am. Chem. Soc.*, **83**, 1978 (1961).

133. Krugh, T. R., and W. C. Schaefer, *J. Magnet. Res.*, **19**, 99 (1975).

134. Kubo, R., *Nuovo Cimento, Suppl.*, **6** 1063 (1957).

135. Kuhlmann, K. F., and D. M. Grant, *J. Am. Chem. Soc.*, **90**, 7355 (1968).

135a. Kumar, A., R. R. Ernst, and K. Wuthrich, *Biochem. Biophys. Res. Commun.*, **95**, 1 (1980).

135b. Kumar, A., G. Wagner, R. R. Ernst, and K. Wuthrich, *J. Am. Chem. Soc.*, **103**, 3654 (1981).

136. Lacher, J. R., J. W. Pollock, and J. D. Park, *J. Chem. Phys.*, **20**, 1047 (1952).

137. Lamb, Jr., W. E., *Phys. Rev.*, **60**, 817 (1941).

138. Lasarew, B. G., and L. W. Schubnikow, *Phys. Z. Sowjet*, **11**, 445 (1937).

139. Laszlo, P., in J. W. Emsley, J. Feeney, and L. H. Sutcliffe, Eds., *Progress in Nuclear Magnetic Resonance Spectroscopy*, Vol. 3, Pergamon, New York, 1967, p. 231.

140. Leader, G. R., *Anal. Chem.*, **42**, 16 (1970).

141. Levy, G. C., and I. R. Peat, *J. Magnet Res.*, **18**, 500 (1975).

142. Lewis, W. B., J. A. Jackson, J. F. Lemons, and H. Taube, *J. Chem. Phys.*, **36**, 694 (1962).

143. Leyden, D. E., *Crit. Rev. Anal. Chem.*, **2**, 383 (1971).

144. Leyden, D. E., and R. E. Channell, *J. Phys. Chem.*, **77**, 1562 (1973).

145. Li, N. C., R. L. Scruggs, and E. D. Becker, *J. Am. Chem. Soc.*, **84**, 4650 (1962).

146. Lundin, R. E., R. H. Elsken, R. A. Flath, and R. Teranishi, in E. G. Brame, Jr., Ed., *Appl. Spectroscopy Rev.*, Vol. 1, Marcel Dekker, New York, 1967, p. 131.

147. Luz, Z., D. Gill, and S. Meiboom, *J. Chem. Phys.*, **30**, 1540 (1959).

148. Luz, Z., and S. Meiboom, *J. Chem. Phys.*, **39**, 366 (1963).

149. McConnell, H. M., *J. Chem. Phys.*, **27**, 226 (1957).

150. McConnell, H. M., *J. Chem. Phys.*, **28**, 430 (1958).

151. McConnell, H. M., and S. B. Berger, *J. Chem. Phys.*, **27**, 230 (1957).

152. McConnell, H. M., and R. E. Robertson, *J. Chem. Phys.*, **29**, 1361 (1958).

153. McFarlane, W., in F. C. Nachod and J. J. Zuckerman, Eds., *Determination of Organic Structures by Physical Methods*, Vol. 4, Academic, New York, 1971, p. 139.

154. Manatt, S. L., *J. Am. Chem. Soc.*, **88**, 1323 (1966).

155. Martin, R. H., and J. C. Nouls, *Tetrahedron Lett.*, 2727 (1968).

156. Maryott, A. A., T. C. Farrar, and M. S. Malmberg, *J. Chem. Phys.*, **54**, 64 (1971).

157. Mayo, B. C., *Chem. Soc. Rev.*, **2**, 49 (1973).

158. Meiboom, S., and D. Gill, *Rev. Sci. Instrum.*, **29**, 688 (1958).

159. Memory, J. D., *Quantum Theory of Magnetic Resonance Parameters*, McGraw-Hill, New York, 1968.

160. Mitchell, R. N., C. E. Klopfenstein, and V. Boekelheide, *J. Am. Chem. Soc.*, **91**, 4931 (1969).

161. Moniz, W. B., and H. S. Gutowsky, *J. Chem. Phys.*, **38**, 1155 (1963).

162. Morgan, W. R., and D. E. Leyden, *J. Am. Chem. Soc.*, **92**, 4527 (1970).

163. Moritz, A. G., and N. Sheppard, *Mol. Phys.*, **5**, 361 (1962).

164. Mortimer, F. S., *J. Mol. Spectrosc.*, **5**, 199 (1960).

165. Mueller, W., and P. E. Butler, *J. Am. Chem. Soc.*, **90**, 2075 (1968).

166. Muller, N., and D. E. Pritchard, *J. Chem. Phys.*, **31**, 768 (1959).

167. Musher, J. I., *J. Chem. Phys.*, **35**, 1159 (1961).

168. Naegele, W., in F. C. Nachod and J. J. Zuckerman, Eds., *Determination of Organic Structures by Physical Methods*, Vol. 4, Academic, New York, 1971, p. 1.

169. Nicolet Technology Corporation, Mountain View, California, technical bulletin.

170. *25 N. M. R. Solvents*, The Sadtler Research Laboratories, Inc., Philadelphia, Pennsylvania, 1966.

171. Noggle, J. H., and R. E. Schirmer, *The Nuclear Overhauser Effect, Chemical Applications*, Academic, New York, 1971.

171a. Offermann, C. W., and A. Mannschreck, *Tetrahedron Lett.*, **22**, 3227 (1981).

172. Onsager, L., *J. Am. Chem. Soc.*, **58**, 1486 (1936).

173. Pascual, C., M. Meier, and W. Simon, *Helv. Chim. Acta.*, **49**, 164 (1966).
174. Patel, D. J., M. E. H. Howden, and J. D. Roberts, *J. Am. Chem. Soc.*, **85**, 3218 (1963).
175. Patterson, A., and R. Ettinger, *Z. Elecktrochem.*, **64**, 98 (1960).
176. Pauli, W., *Naturwissenschaften*, **12**, 741 (1924)
177. Pauling, L., *J. Chem. Phys.*, **4**, 673 (1936).
178. Paulsen, P. J., and W. D. Cooke, *Anal. Chem.*, **36**, 721 (1964).
179. Paulsen, P. J., and W. D. Cooke, *Anal. Chem.*, **36**, 1713 (1964).
180. Pearson, R. G., and R. D. Lanier, *J. Am. Chem. Soc.*, **86**, 765 (1964).
181. The Perkin-Elmer Corporation, Norwalk, Connecticut, technical bulletin.
182. Pople, J. A., *J. Chem. Phys.*, **24**, 1111 (1956).
183. Pople, J. A., *Proc. Roy. Soc. (London)*, **A239**, 550 (1957).
184. Pople, J. A., *J. Chem. Phys.*, **37**, 60 (1962).
185. Pople, J. A., W. G. Schneider, and H. J. Bernstein, *High-Resolution Nuclear Magnetic Resonance*, McGraw-Hill, New York, 1959.
186. Pople, J. A., W. G. Schneider, and H. J. Bernstein, *High-Resolution Nuclear Magnetic Resonance*, McGraw-Hill, New York, 1959, p. 222.
187. Pople, J. A., and K. G. Untch, *J. Am. Chem. Soc.*, **88**, 4811 (1966).
188. The Praxis Corporation, San Antonio, Texas, technical bulletin.
189. Primas, H., 5th European Congress on Molecular Spectroscopy, Amsterdam (1961).
190. Primas, H., and H. H. Gunthard, *Rev. Sci. Instrum.*, **28**, 510 (1957).
191. Proctor, W. G., and F. C. Yu, *Phys. Rev.*, **78**, 471 (1950).
192. Purcell, E. M., H. C. Torrey, and R. V. Pound, *Phys. Rev.*, **69**, 37 (1946).
193. Rabenstein, D. L., *Anal. Chem.*, **43**, 1599 (1971).
194. Rabi, I. I., S. Millman, P. Kusch, and J. R. Zacharias, *Phys. Rev.*, **55**, 526 (1939).
195. Rakshys, Jr., J. W., S. V. McKinley, and H. H. Freeman, *J. Am. Chem. Soc.*, **93**, 6522 (1971).
196. Ralph, E. K., and E. Grunwald, *J. Am. Chem. Soc.*, **89**, 2963 (1967).
197. Ramsey, N. F., *Phys. Rev.*, **78**, 699 (1950).
198. Ramsey, N. F., *Phys. Rev.*, **91**, 303 (1953).
199. Reddy, G. S., and J. H. Goldstein, *J. Chem. Phys.*, **39**, 3509 (1963).
200. Read, J. M., R. E. Mayo, and J. H. Goldstein, *J. Mol. Spectrosc.*, **22**, 419 (1967).
201. Reilley, C. A., *Anal. Chem.*, **30**, 839 (1958).
202. Reuben, J., in J. W. Emsley, J. Feeney, and L. H. Sutcliffe, Eds., *Progress in Nuclear Magnetic Resonance Spectroscopy*, Vol. 9, Part 1, Pergamon, New York, 1973.
203. Roberts, J. D., *An Introduction to the Analysis of Spin-Spin Splitting in High-Resolution Nuclear Magnetic Resonance Spectra*, W. A. Benjamin, New York, 1961.
204. Rogers, M. T., and J. C. Woodbrey, *J. Phys. Chem.*, **66**, 540 (1962).
205. Ronayne, J., and D. H. Williams, in E. F. Mooney, Ed., *Annual Report of NMR Spectroscopy*, Vol. 2, Academic, New York, 1969, p. 83.
206. Rusnak, L. L., and R. B. Jordan, *Inorg. Chem.*, **10**, 2686 (1971).
207. Rusnak, L. L., and R. B. Jordan, *Inorg. Chem.*, **11**, 1961 (1972).
208. Sack, R. A., *Mol. Phys.*, **1**, 163 (1958).
209. *Sadtler Standard NMR Spectra*, Sadtler Research Laboratories, Philadelphia, Pennsylvania.
210. Saika, A., and C. P. Slichter, *J. Chem. Phys.*, **22**, 26 (1954).
211. Sanders, J. K. M., S. W. Hanson, and D. H. Williams, *J. Am. Chem. Soc.*, **94**, 5325 (1972).
212. Sanders, J. K. M., and D. H. Williams, *J. Am. Chem. Soc.*, **93**, 641 (1971).
213. Sanders, J. K. M., and D. H. Williams, *Chem. Commun.*, 422 (1970).

214. Sato, T., and Y. Mikami, *Kog. Kag. Zasshi*, **68**, 1401 (1965).
215. Saunders, M., *Magnetic Resonance in Biological Systems*, Pergamon, New York, 1967, p. 85.
216. Saunders, M., and F. Yamada, *J. Am. Chem. Soc.*, **85**, 1882 (1963).
217. Saunders, M., P. von R. Schleyer, and G. A. Olah, *J. Am. Chem. Soc.*, **86**, 5681 (1964).
218. Schaumberg, K., *Lipids*, **5**, 505 (1969).
219. Shapiro, B. L., R. M. Kopchik, and S. J. Ebersole, *J. Chem. Phys.*, **39**, 3154 (1963).
220. Shapiro, B. L., M. D. Johnson, Jr., and R. L. R. Towns, *J. Am. Chem. Soc.*, **94**, 4381 (1972).
221. Sheppard, N., and R. M. Lynden-Bell, *Proc. Roy. Soc.*, **A269**, 385 (1962).
222. Sheppard, N., and J. J. Turner, *Proc. Roy. Soc.*, **A252**, 506 (1959).
223. Shoolery, J. N., and B. P. Dailey, *J. Am. Chem. Soc.*, **77**, 3977 (1955).
224. Shoolery, J. N., and L. H. Smithson, *J. Am. Oil Chem. Soc.*, **47**, 153 (1970).
225. Silverstein, R. M., G. C. Bassler, and T. C. Morrill, *Spectrometric Identification of Organic Compounds*, Wiley, New York, 1974.
226. Smith, S. L., *Top. Cur. Chem.*, **27**, 117 (1972).
227. Spiesecke, H., and W. G. Schneider, *J. Chem. Phys.*, **35**, 731 (1961).
228. Stehling, F. C., and K. W. Bartz, *Anal. Chem.*, **38**, 1467 (1966).
229. Stern, O., *Z. Phys.*, **7**, 249 (1921).
230. Sternhell, S., *Q. Rev.*, **23**, 236 (1969).
231. Sudmeier, J. L., and C. N. Reilley, *Anal. Chem.*, **38**, 1698 (1964).
232. Sudmeier, J. L., and C. N. Reilley, *Anal. Chem.*, **36**, 1707 (1964).
233. Sudmeier, J. L., K. E. Schwartz, and A. J. Senzel, *Inorg. Chem.*, **8**, 2815 (1969).
234. Swift, T. J., and R. E. Connick, *J. Chem. Phys.*, **37**, 307 (1962).
235. Tillieu, J., *Ann. Phys.*, **2**, 471, 631 (1957).
236. Tobey, S. W., *J. Org. Chem.*, **34**, 1281 (1969).
237. Traynham, J. G., and G. A. Knesel, *J. Am. Chem. Soc.*, **87**, 4220 (1965).
238. Tsang, T., and T. C. Farrar, *J. Chem. Phys.*, **50**, 3498 (1969).
239. Varian Associates, Palo Alto, California.
240. *Varian High Resolution NMR Spectra Catalog*, Vols. 1 and 2, Varian Associates, Palo Alto, California, 1963.
241. Ward, T. M., I. L. Allcox, and G. H. Wahl, Jr., *Tetrahedron Lett.*, 4421 (1971).
242. Ward, G. A., and R. D. Main, *Anal. Chem.*, **41**, 538 (1969).
243. Watts, V. S., and H. J. Goldstein, *J. Chem. Phys.*, **46**, 4165 (1967).
244. Waugh, J. S., and R. W. Fessenden, *J. Am. Chem. Soc.*, **79**, 846 (1957).
244a. Wenzel, T. J., T. C. Bettes, J. E. Sadlowski, and R. E. Sievers, *J. Am. Chem. Soc.*, **102**, 5903 (1980).
244b. Wenzel, T. J., and R. E. Sievers, *Anal. Chem.*, **53**, 393 (1981).
245. Whitesides, G. M., and J. S. Fleming, *J. Am. Chem. Soc.*, **89**, 2855 (1967).
246. Whitesides, G. M., and D. W. Lewis, *J. Am. Chem. Soc.*, **92**, 6979 (1970).
247. Willcott, III, M. R., R. E. Lenkinski, and R. E. Davis, *J. Am. Chem. Soc.*, **94**, 1742 (1972).
248. Williams, R. B., *Ann. N. Y. Acad. Sci.*, **70**, 890 (1958).
249. Wilmad Glass Company, Route 40 and Oak Road, Buena, New Jersey.
250. Zeil, W., and H. Buchert, *Z. Phys. Chem.*, **38**, 47 (1963).
251. Zurcher, R. F., in B. Pesce, Ed., *Nuclear Magnetic Resonance in Chemistry*, Academic, New York, 1965, p. 45.

Part 1
Section I

Chapter 2

NUCLEAR MAGNETIC RESONANCE: ^{13}C SPECTRA

BY ANTHONY LOMBARDO AND GEORGE C. LEVY,
Departments of Chemistry, Florida Atlantic University, Boca Raton, Florida, and Syracuse University, Syracuse, New York

Contents

I. INTRODUCTION

Carbon-13 nuclear magnetic resonance (NMR) was first observed in 1957 (118,145). However, the truly explosive growth in ^{13}C-NMR literature which signalled its reaching fruition as a practical spectroscopic technique and research tool did not begin until the early 1970s. A number of current monographs are available (1,29,155,278), along with a continuing series of review volumes (148). These sources replace three original texts (30,156,250).

The long delay between the early origins of ^{13}C-NMR and the full realization of its practical potential was caused by the need for many sophisticated technical developments to bring about its widespread use. These developments were necessitated by formidable difficulties in ^{13}C detection arising from its inherent low sensitivity to the NMR experiment. NMR sensitivity is directly related to the natural isotopic abundance of the observed nucleus and to the third power of its magnetogyric ratio. ^{13}C has a natural abundance of only 1.1% and its magnetogyric ratio is small, about 0.25 that of the proton. These factors combine to make natural abundance ^{13}C-NMR almost 6000 times less sensitive than ^{1}H-NMR. Thus, early experiments, without benefit of sensitivity-enhancement techniques and improved instrumentation, were restricted to highly soluble low-molecular-weight solids, neat liquids, or isotopically enriched compounds. Even under these conditions, signals arising from unique carbon atoms were poorly resolved due to extensive one-bond and longer range ^{13}C-^{1}H spin–spin coupling.

Development of ^{13}C-NMR beyond these initial stages was motivated by recognition of its great potential for the study of organic systems (156,250). With practical problems in detection and spectral resolution alleviated, many advantages over ^{1}H-NMR become apparent. For example, it was known for some time that ^{13}C resonances range over 600 ppm, far surpassing the 20 ppm region that encompasses most proton signals. Furthermore, because the low natural abundance of ^{13}C makes it unlikely that two spins will exist in the same molecule, spin–spin splitting due to homonuclear coupling does not occur in ^{13}C spectra. Application of wideband proton decoupling (69), an example of heteronuclear double resonance mentioned in the preceding chapter, allows poorly resolved ^{13}C-^{1}H multiplets to collapse into sharp singlets. This results in both a sensitivity improvement by concentration of the signals into single lines, and also a large increase in spectral resolution. In fact, due to the broad ^{13}C chemical shift range a sharp singlet is commonly observed in the spectrum for each unique carbon atom, even in complex molecules whose ^{1}H spectra are severely complicated by signal overlap and non-first-order homonuclear spin–spin splittings. Peak assignments are substantially simplified and a sensitive probe into the skeletal framework of organic molecules is realized. Functional group carbon atoms with no directly bound protons (e.g., carbonyls, nitriles, etc.) are observed as are other reaction centers of chemical interest.

Proton noise decoupling (the first form of wideband decoupling) was only the first in a series of technological advances that began to raise ^{13}C-NMR to its present status as a practical research tool enjoying widespread use. Significant sensitivity enhancements were realized by introduction of time-averaging techniques in which the results of several consecutive spectral scans are accumulated in a computer. Since NMR signals add coherently while random baseline noise accumulates only as the square root of the number of scans, the achieved signal-to-noise ratio (S/N) improves with the square root of the number of accumulated scans. Thus, dramatically enhanced spectra became possible, at the cost of extended experimental times.

Subsequently, improvement in sensitivity has been realized by the development of superconducting solenoids operating at increasing magnetic field strengths, e.g., 7 to 12+ Tesla. However, the most important innovation in ^{13}C-NMR, indeed for NMR in general, has been the development of pulse Fourier transform (FT) methods, which make it possible to obtain excellent natural abundance ^{13}C spectra from relatively dilute samples in reasonable periods of time. FT-NMR is uniquely responsible for alleviating those problems in detection that plagued this field for so long, and for making ^{13}C-NMR spectroscopy a powerful research technique and analytical tool.

The nature of ^{13}C FT-NMR and the resultant body of knowledge obtained from it are the topics of this chapter. The evolution of ^{13}C-NMR is graphically demonstrated in Fig. 2.1–4.

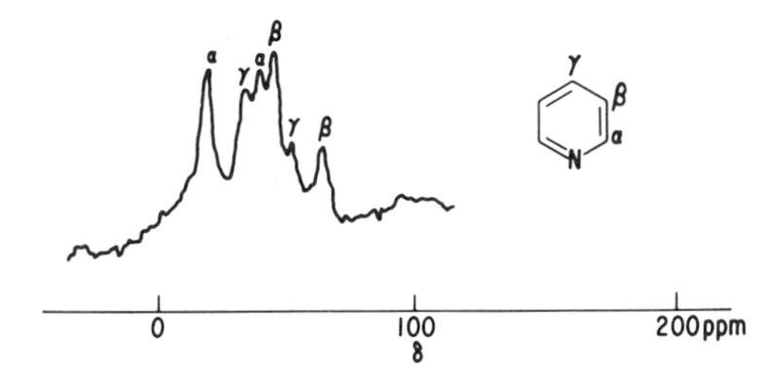

Fig. 2.1. ^{13}C-NMR spectrum of pyridine. Chemical shift scale in parts per million upfield from $^{*}CS_2$. [P. C. Lauterbur, *Ann. N.Y. Acad. Sci.*, **70**, 841 (1958).] (155)

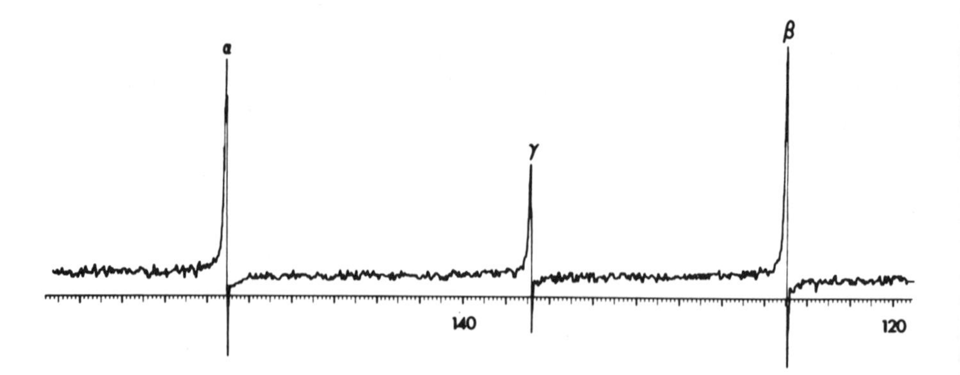

Fig. 2.2. ^{13}C-NMR spectrum of pyridine (^{1}H decoupled). Six 10-Hz/sec, 1000-Hz-wide scans of 2 g of pyridine in approximately 3 ml of total volume. Spectrum accumulated on a Varian C-1024 time averaging computer. Total acquisition time = 600 sec. Chemical shifts in parts per million downfield from TMS (tetramethylsilane) (155).

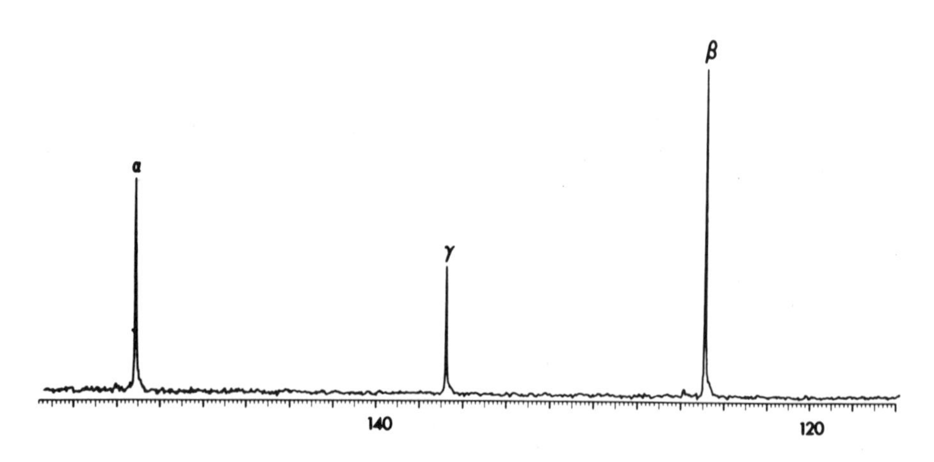

Fig. 2.3. Fourier transform ^{13}C-NMR spectrum of pyridine. Same solution and chemical-shift convention as in Fig. 2.2 but only 6 sec of total acquisition time (155).

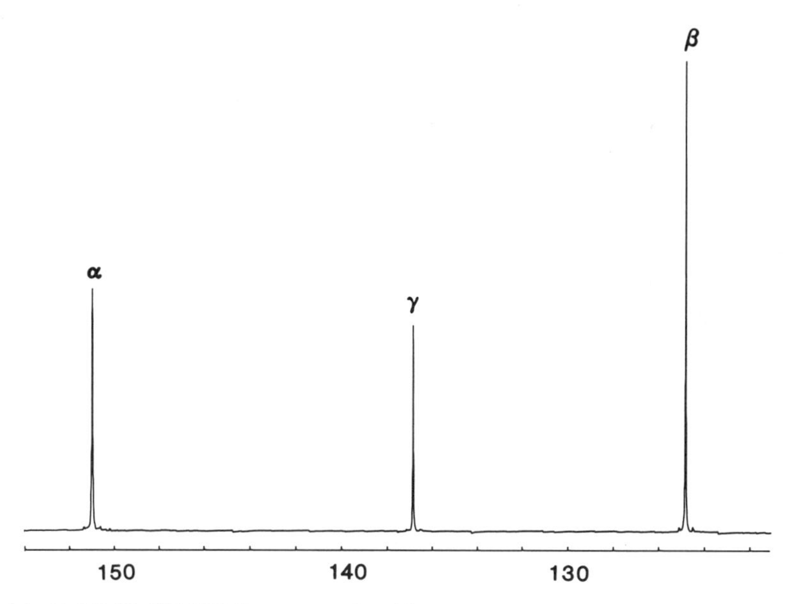

Fig. 2.4. 90.56 MHz ^{13}C FT-NMR spectrum of pyridine; one 1 sec scan (Spectrum courtesy of Linda F. Levy).

II. ^{13}C FOURIER TRANSFORM NMR

A. THE FT-NMR EXPERIMENT

For the first two decades of NMR spectroscopy, the common method of observing magnetically active nuclei was by continuous-wave excitation/detection. This technique, which is discussed in detail in the preceding chapter, utilizes a slow, continuous sweep of the applied magnetic field (field sweep) or of the excitation radio frequency (frequency sweep) giving rise to the term continuous wave, or cw-NMR. At some point in the scan, the field/frequency ratio is exactly that required for resonance of a particular nucleus to occur, generating an NMR signal via absorption or induction from the rf field. There is an inherent inefficiency in this method that arises from the fact that at any given instant, only a single frequency is being monitored. Thus, for ^{13}C-NMR at 4.7 Tesla where the chemical shift range for most molecules is 10 KHz, each 1-Hz-wide resonance line can be observed only 1/10000th of the time. The remaining sweep time is spent observing other peaks or merely tracing baseline.

This inherent inefficiency can be alleviated if the entire band of ^{13}C precession frequencies in a given sample is irradiated and detected simultaneously. This causes essentially instantaneous excitation of all ^{13}C nuclei by absorption of the appropriate frequency components contained in the excitation bandwidth. No time is wasted monitoring baseline between individual resonances. Fortunately, several methods exist for generating a band of excitation frequencies from a single rf signal.

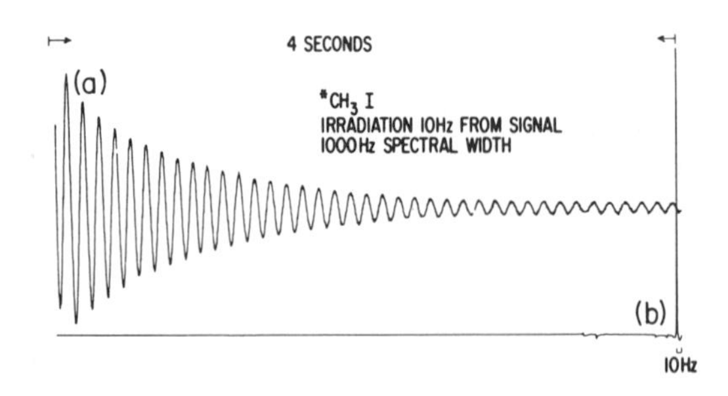

Fig. 2.5. (*a*) Time domain—FID; (*b*) frequency domain—spectrum (*CH_3I, ^1H decoupled) (155).

A pulsed rf signal of frequency ν produces a finite band of rf excitation whose width is inversely proportional to the duration of the pulse, and is centered about ν (236). If the pulse duration is sufficiently short, $\lesssim 25$ μsec, a bandwidth exceeding 10 KHz can be produced.

The response of the sample following excitation, i.e., the rf pulse, is "emission" of all frequencies absorbed by ^{13}C nuclei, as their coherent spin-state populations return to thermal equilibrium values. This process induces in the receiver coil of the spectrometer a signal known as a free induction decay (FID). For absorption of a single NMR resonance frequency, the FID normally has the form of an exponentially decaying sine wave and comprises an intensity versus time (or time domain) spectrum. The frequency of the sine wave is the difference between ν, the nominal radiofrequency of the excitation pulse, and the rf absorption (Larmor) frequency of the observed nucleus. The usual frequency domain (absorption) spectrum of the sample, relative to the carrier, is obtained from the frequency of the FID in the manner shown for the proton-decoupled ^{13}C spectrum (hereafter symbolized by the notation $^{13}C\{^1H\}$) of methyl iodide in Fig. 2.5.

Ordinarily, organic molecules contain several nonequivalent carbon atoms. Excitation by an rf pulse causes absorption of several frequencies and the resultant composite FID is complex. Derivation of a frequency spectrum by direct observation is impossible. Fortunately, the time and frequency domains of a given data set are linked by the mathematical operation of Fourier transformation. Thus, to obtain the NMR frequency spectrum from a complex FID, i.e., to extract all contained frequency components, a Fourier transformation of the time domain signal is performed. Shown in Fig. 2.6 are the FID and its corresponding Fourier transforma-

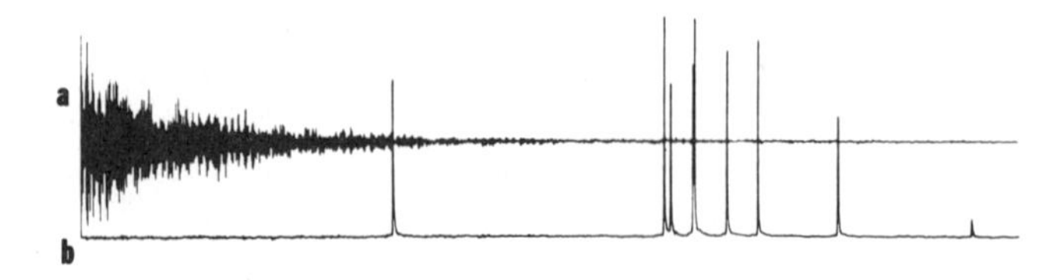

Fig. 2.6. (*a*) Free induction decay of 1-octanol, 300 scans; (*b*) Fourier-transformed spectrum (high-field peak, TMS).

tion obtained from a $^{13}C\{^1H\}$ experiment on 1-octanol. Structural information is contained in the frequency spectrum obtained by Fourier transformation. Other useful information can also be extracted from an FID.

The duration (in seconds) of each individual frequency component defines the resolution (line width in Hertz) of each resonance line. Frequencies that decay rapidly yield broad lines, while those that decay slowly give rise to sharp lines in the frequency domain. The overall shape of an FID is a function of these spectral characteristics and, in turn, is an indication of expected line widths.

It must be emphasized that nothing artificial is done in converting an FID to an NMR frequency spectrum. It should be exactly the same spectrum that would be obtained in a cw field or frequency-sweep experiment. Fourier transformation is required only to extract spectral information from the FID and present it in an easily useable format.

A sensitivity enhancement over cw-NMR arises in a pulse FT-NMR experiment. It is proportional to $(F/\Delta)^{\frac{1}{2}}$ where F is the total chemical shift range and Δ is the line width of the narrowest signal. For ^{13}C NMR the theoretical enhancement is of the order of $(5-25 \text{ kHz}/1 \text{ Hz})^{\frac{1}{2}}$ or 50 – 150.

The enormous sensitivity advantage gained in FT-NMR results from the rapidity with which spectra can be accumulated for time-averaging signal enhancement. In the time required to accumulate 100 cw scans at 250 sec/scan, for example, FT-NMR can easily allow acquisition of $\geqq$25,000 spectra (approximately 1 pulse/sec). Coupled with the inherent advantages of rf pulse excitation described in the preceding paragraph, an overall enhancement of $10^3 - 10^4$ over single cw scans is easily achieved. Stated in another way, *much less time* is required to achieve a specified S/N in FT-NMR than is needed using conventional cw time-averaging methods.

The rapidity with which pulses can be applied to a sample is limited in actual practice by nuclear relaxation times and by spectral parameters utilized in the experiment. These limitations are discussed later. Table 2.I gives a comparison of current ^{13}C capabilities with representative earlier experiments. The numbers are not precise, but are in general agreement with recent results.

B. DETAILS OF NUCLEAR EXCITATION AND RELAXATION IN FT-NMR

It is useful in understanding the details of an FT-NMR experiment to consider the behavior of affected nuclei in terms of magnetization vectors. Like ^1H, ^{13}C is a spin 1/2 nucleus. Thus, in an applied magnetic field, B_0, a Boltzmann distribution of ^{13}C nuclear magnetic moments exists between the two possible spin states, with a slight excess in the lower energy state. This results in a small net magnetization, M_z, which is aligned with B_0 (aligned along the z axis) as shown in Fig. 2.7*a*. Note that M is the resultant magnetization vector of all nuclear magnetic moments precessing about B_0 at the Larmor frequency, ω_0. This representation of magnetic nuclei in an applied field is referred to as the laboratory frame of reference.

The effects of an rf excitation pulse on the nuclei are more easily described by consideration of the spin system in a coordinate frame which is itself rotating at the Larmor frequency. Under this condition, called the rotating frame of reference, the

TABLE 2.I

Natural-Abundance Fourier Transform ^{13}C-NMR Capabilities

Criterion	1970–1972	1975–1977	1980[a]
Typical molar concentration (assumes no molecular symmetry)	0.5–1.0	0.1–1.0	0.05–0.1
Best resolution (organic molecules) (in ppm)	<0.05	<0.02	<0.01
Practical lower limit on molar concentration[b]	0.01	0.002	~0.0005
Resolution in experiments at low concentration[c] (in ppm)	<0.1	<0.1	~0.01
Experimental time required for lower concentration limit[d] (in hours)	12	18	18
Experimental time required to run 0.5–1 *M* solution	~1 min	≲5 sec	<1 sec

[a] Pulsed FT QD, wide-band proton decoupling, heteronuclear field/frequency control, 20-mm sample tube at 50–90 MHz.

[b] To attain S/N ≈12 (signal height over peak-to-peak noise divided by 2.5).

[c] Assumes entire ^{13}C chemical-shift range covered.

[d] To achieve S/N exceeding 25, for practical organic analyses.

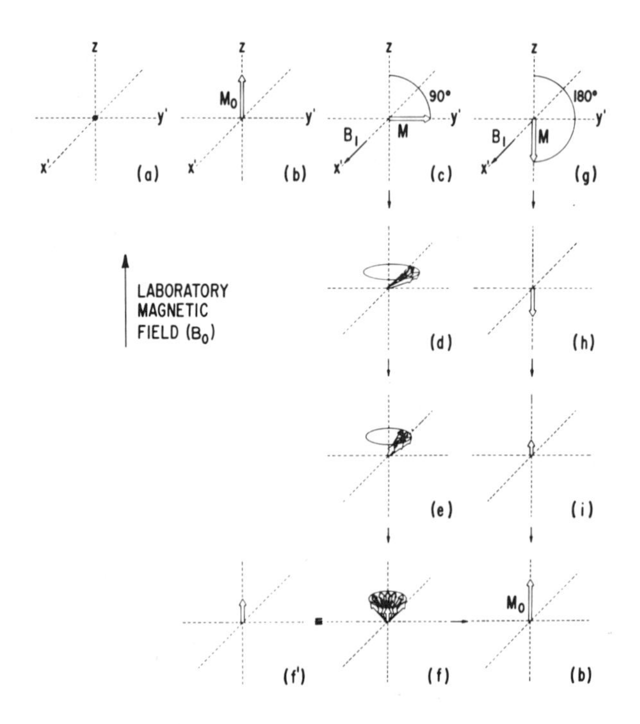

Fig. 2.7. The pulse NMR experiment in the rotating frame. Sequences c, d, e, f, b and g, h, i, b demonstrate relaxation processes (155).

equilibrium nuclear magnetic moments do not appear to precess. Their magnetization vectors, the precessional axis, and the applied static field axis are all coincident. The description of the spin system is thus reduced to a single bulk magnetization vector, M_0, which is stationary and coincident with B_0 (along the z axis which does not change with change of the reference frame).

An rf field, B_1, oscillating at the Larmor frequency is also fixed in the rotating frame. Irradiation of the sample is accomplished by an excitation pulse of B_1 directed along the x' axis and appears as shown in Fig. 2.7b. Applied in this manner, the rf pulse causes M_0 to tip away from the z axis in a clockwise direction toward y'. The angle, α, through which M_0 rotates depends upon the duration of the pulse, τ, according to the equation given below.

$$\alpha = \gamma B_1 \tau \quad (1)$$

By experimentally varying τ (typically between values of 1 and 100 μsec for ^{13}C), any tip angle can be achieved. Thus 30° pulses, 45° pulses, 90° pulses, 180° pulses, etc., are generated as required by the specific needs of the experiment being performed. Since the duration of the pulse, commonly called the pulse width, is so short, the rf power must be high to effect excitation. The angle through which M_0 rotates is called the pulse or flip angle.

Nuclear spin relaxation during excitation is generally precluded by the short pulse widths utilized. Relaxation processes do begin immediately after the pulse is turned off, however. Decay of the $x'y'$ coherence of spin vectors comprising spin–spin relaxation is detected by the receiver coil of the spectrometer, and results in the observed FID. Let us then consider in detail the coupled process of relaxation and signal detection in the context of the rotating frame.

Immediately following an rf pulse of width τ, M_0 is out of alignment with B_0 by the pulse angle α. Two components of the bulk magnetization, M_z and M'_y, thus exist as shown in Fig. 2.7c. Spectrometers are designed to detect signals generated by magnetization in the $x'y'$ plane. Therefore, decay of M'_y as a function of time is directly responsible for the FID. That decay occurs through the spin–lattice and spin–spin relaxation processes.

Spin–lattice relaxation, the loss of energy by excited nuclei to the surrounding liquid or solid "lattice," occurs along the z axis thus causing M_z to increase toward M_0. This process is directly responsible for return of the perturbed nuclear spins to their equilibrium distribution. The process is ordinarily first order with the reciprocal of the rate constant referred to as the spin–lattice relaxation time, T_1. As this proceeds, M'_y must also decrease resulting in decay of the signal. However, M'_y also decreases via a second process, spin–spin relaxation, which occurs in the x-y plane. In this process, nuclei lose phase coherence causing some to precess faster and others slower than the Larmor frequency. In the rotating frame, these spins are no longer aligned with y' but fan out in the $x'y'$ plane. This loss of phase coherence or dephasing also proceeds exponentially, ultimately resulting in loss of signal when the spin vectors become randomly distributed. The time required for this to occur is called the spin–spin relaxation time, T_2.

In practice, T_2 is usually shorter than T_1. Thus, the FID decays to zero before the equilibrium Boltzmann distribution of spin states is restored. After a duration of less than $5T_2$, no useable signal is obtained regardless of the value of T_1.

It is clear from the foregoing discussion that signals of maximum intensity are generated by a 90° pulse while no signal results from an ideal pulse of 180° (no magnetization in the x-y plane). In many ^{13}C studies, pulse angles less than 90° are utilized, sacrificing some signal intensity obtained from each pulse to shorten the period of time required for restoration of M_0. When $T_1 \gg T_2$, the S/N achieved in a specific period of time is ultimately greater using small pulse angles, because a greater number of spectra can be accumulated in that time. In larger organic molecules, ^{13}C T_1 values are short enough to allow rapid data acquisition with 90° excitation pulses. ^{13}C-NMR experiments can require optimization of pulse angles and scan intervals for anticipated spin relaxation in the sample.

C. INSTRUMENTAL REQUIREMENTS FOR FT-NMR

Instrumental requirements for pulsed FT-NMR experiments are discussed in detail elsewhere (236). Only the main features will be covered here.

The basic instrumentation required for FT-NMR consists of a spectrometer coupled to or based on a data acquisition and processing unit. For ^{13}C-NMR of organic systems, spectrometer requirements are similar to those for cw experiments as discussed in the preceding chapter. Some significant modifications are necessary, however.

The pulsed NMR experiment requires a very strong rf signal to bring about complete nuclear excitation in the short time intervals which characterize the usual pulse widths employed. Thus, a high-power rf amplifier is needed. This in turn requires the presence of rf gates to form the pulse and protect the receiver while high-power excitation is in progress.

In a typical ^{13}C-NMR experiment, a large number of FIDs may be accumulated. Consequently, some form of field/frequency stabilization is employed on conventional instruments. Homonuclear lock, often used in cw systems, is inadequate, and proton lock is not feasible since most ^{13}C experiments employ high-power proton decoupling. The use of a heteronuclear lock is the usual system of choice. Those based on deuterium or fluorine are most often utilized. The heteronuclear lock rf channel is also used for making shim adjustments to maximize applied field homogeneity.

^{13}C FT-NMR spectrometers have separate rf channels for use in proton decoupling. As mentioned in the Introduction, extensive spin–spin coupling of carbon nuclei to directly bound and more distant protons results in poorly resolved, low sensitivity multiplets in coupled ^{13}C spectra ($^1J_{CH}$ = 100–200 Hz, (29,155); $^{2-4}J_{CH} \leqq 20$ Hz). Collapse of all ^{13}C-^1H multiplets to sharp singlets is accomplished by essentially simultaneous irradiation of all proton frequencies in the sample. Instrumentally, the frequency of an rf signal from the decoupler channel is set at the center of the proton resonances. It is then modulated in one of a number of ways to produce a finite proton excitation bandwidth eliminating all carbon-proton spin–spin interactions.

Broadband proton decoupling results in more than an order of magnitude increase in the sensitivity of ^{13}C-NMR experiments. This arises not only from the formation of intense sharp singlets, but from an effect concomitant with heteronuclear decoupling called the nuclear Overhauser enhancement (NOE).

The intensity of a line in an NMR spectrum is directly related to the difference in the populations of the spin states involved in the transition causing the line. Broadband irradiation of all proton frequencies causes equalization of the proton spin-state populations. This, in turn, results in perturbation of the ^{13}C spin states from their normal Boltzmann distribution to a distribution in which the lower state gains an increased excess of the nuclei. Excitation will thus give rise to a signal of enhanced intensity.

In a $^{13}C\{^{1}H\}$ experiment, the maximum theoretical NOE is 2.98. This means that if the NOE is fully operative, each ^{13}C signal will have a peak area 2.98 times the total signal area obtained in the absence of proton decoupling. The source of NOE is the dependence of ^{13}C nuclei on protons for dipolar spin–lattice relaxation. In cases where other relaxation mechanisms are operative, the NOE is reduced. Further details of the NOE will be discussed later as they relate to the mechanisms of ^{13}C spin-lattice relaxation.

The level of sophistication of FT-NMR instrumentation far exceeds that of conventional cw systems because of its need for powerful data acquisition and processing equipment. Virtually all modern FT systems have mini-computers integrated into the spectrometer design. In addition to the obvious function of storing and Fourier transforming FIDs, the computer, through operator inputs, typically controls pulse programming, setting of excitation and decoupling rf frequencies, graphics displays, phase corrections, spectrum plotting, and a host of other spectrometer functions. In general then, one of the capabilities the computer provides to FT-NMR experiments is a high degree of automation (159).

High-speed data acquisition, storage, and processing are essential to FT-NMR. The FID signal received is in analog form and thus it must be digitized for storage and Fourier transformation. Digitization is accomplished by an analog-to-digital converter (ADC) which, in compliance with sampling theory, must sample each frequency component at least twice per cycle. In this manner the frequency of each different incoming sine wave is uniquely determined. The FID will contain components as high as the spectral window (sweep width in cw NMR). Thus, for a 10-kHz upper frequency limit, the data acquisition rate must be 20,000 measurements per second with each point converted to digital format for storage in memory and eventual processing.

The configuration of the computer, especially the number of words available and the word length in bits, is a limiting factor in the resolution and overall versatility of an FT system. Modern FT-NMR systems require computer memories of 16K to 40K words or more. A portion of memory is utilized for programming usually leaving 8–23K words for data storage. The mathematical characteristics of Fourier transformation result in frequency domain spectra comprised of one-half the number of data points taken from the FID. Thus, most FT spectra contain 4K to 16K real data points. When the spectral width covers the full ^{13}C chemical shift range, this results in a digital limitation on resolution, at an operating frequency of 50 MHz, of 0.8–3 Hz.

In situations where better resolution is desirable or where available computer memory restricts achievable resolution to poorer values, spectra may be obtained that cover narrower sweep widths. The limiting conditions in such experiments are magnetic field inhomogeneties, natural line widths, and proton decoupling efficiency. However, great care must be taken in using this technique.

The manner in which sweep widths are decreased in FT-NMR is by lowering the sampling rate of the ADC. This limits the highest frequency that can be digitized and stored. However, all precession frequencies within the effective bandwidth of the rf pulse are excited. Thus, if frequencies higher than the maximum defined by the ADC sampling rate exist in the FID, they cannot be accurately represented and digitization assigns to them values lower than the actual frequencies. The result of an incorrect frequency assignment to an incoming signal is "folding" or "aliasing" of that signal into the frequency domain spectrum at an erroneous chemical shift. A signal whose true chemical shift is n Hz higher than the specified spectral width (i.e., SW + n)

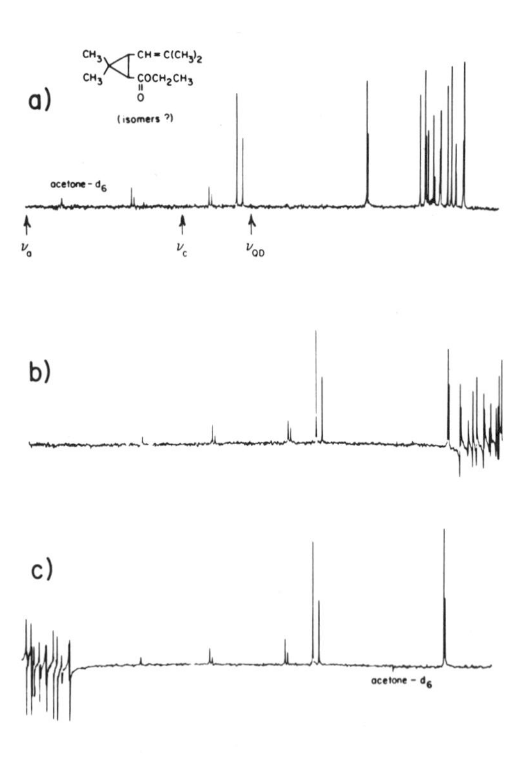

Fig. 2.8. Single-phase and quadrature detection FT-NMR spectra of ethyl chrysanthemumate. Each spectrum obtained at 67.9 MHz, 8 scans, spectral width 15 kHz, ν_a, ν_b, ν_c, indicate placement of rf excitation in the three cases. Spectrum: (*a*) single-phase detection, no lines aliased; (*b*) single-phase spectrum, high-field resonances folded; (*c*) quadrature detection spectrum with high-field resonances folded into the low-field region (irradiation near ν_{QD} would avoid aliasing). Note that spectrum (*c*) demonstrates S/N improvement of $\sqrt{2}$ as predicted for QD FT-NMR. (Spectra, courtesy of T. Gedris and R. Rosanske.) (155)

will appear at (be folded to) a frequency of (SW − *n*) in the absorption spectrum. Aliasing is minimized by placing low-frequency pass filters between the receiver and the ADC. This accomplishes elimination of frequencies much higher than the spectral window, but only limited attenuation of signals just outside the filter setting. Fortunately, folded peaks in phase-corrected FT spectra can usually be identified by a characteristic phase distortion relative to the rest of the spectrum. An example of spectral folding is given in Fig. 2.8*b*.

Another form of aliasing can occur in FT-NMR when single-phase spectrometer receivers are used, since these do not differentiate between frequencies higher or lower than the carrier rf frequency. They only record the frequency *difference* between the carrier and absorption signals, passing +NHz or −NHz as NHz. Thus, spectral folding occurs if the carrier is mistakenly set within the range of sample resonances (i.e., correct and folded peaks are recorded on one arbitrary side of the carrier). In older spectrometers this effect is avoided by setting the carrier at one extreme end of the spectral window. In modern instruments, this problem is alleviated through the use of quadrature phase detection (QD), a technique that allows differentiation of positive and negative frequencies. With QD, the carrier frequency is set near the center of an NkHz spectral window which is determined as ±N/2 kHz (requiring NkHz sampling rate!). Thus, ADC sampling requirements are eased in QD-NMR, especially for superconducting solenoid-based spectrometers. QD schemes also gain a $\sqrt{2}$ signal-to-noise advantage (single-phase detection records signal from only one side of the rf excitation along with noise from both sides). QD-NMR still is subject to down-folding of frequencies higher than half the ADC rate. Folded lines in QD spectra cross the spectral window. Thus a signal folded from +6 kHz around a maximum frequency of ±5 kHz, appears at −4 kHz (see Fig. 2.8*c*.

III. GENERAL CHARACTERISTICS OF ^{13}C-NMR SPECTRA

The purpose of this section is to provide an overview of the general characteristics of ^{13}C-NMR spectra; detailed coverage of those characteristics, especially chemical shifts, is reserved for Section IV. Some techniques which are useful in spectral interpretation are also introduced in discussing the general aspects of spin–spin coupling to ^{13}C.

A. CHEMICAL SHIFTS

One fundamental advantage of ^{13}C-NMR over ^{1}H-NMR is the much broader range of ^{13}C chemical shifts. While most ^{1}H resonances fall within a spectral width of 10 ppm, all known ^{13}C chemical shifts in diamagnetic molecules cover a range of about 600 ppm. In most cases, however, complete ^{13}C spectra occur over a width slightly surpassing 200 ppm, still a significantly increased dispersion over the shielding range for protons.

As in ^{1}H-NMR, ^{13}C chemical shifts are referenced to tetramethylsilane (TMS), with positive values indicating shifts to higher frequencies (lower fields). TMS was adopted as a universal reporting standard for ^{13}C-NMR in the 1970s. Previously, carbon shifts were referenced to an external capillary of ^{13}C-enriched CS_2 ($*CS_2$). Positive shifts reported relative to the $*CS_2$ external standard are opposite in direction from shieldings reported relative to TMS. In practice, solvent resonances are often used as secondary standards with chemical shifts adjusted to the TMS scale. Table 2.II lists a number of solvent resonances for use in reporting ^{13}C chemical shifts. Older $*CS_2$-based shifts can be converted to the TMS scale using the appropriate value in Table 2.II. All shieldings given in this chapter are referenced to TMS. The limits of the normal ^{13}C chemical-shift range are marked by carbonyl carbons, which are the most deshielded, and hydrocarbon methyl groups, which can appear 10–20 ppm from TMS.

Various factors which affect ^{1}H chemical shifts also affect the chemical shifts of ^{13}C. Thus, carbon hybridization, substituent electronegativities (inductive effects), and neighbor group anisotropy all contribute. However, some variance in their individual importance relative to ^{1}H shieldings exist. For example, the effect of hybridization on ^{13}C shifts is directly analogous to its effect on protons, i.e., sp^3 carbons are most shielded and sp^2 carbons most deshielded, with sp carbons intermediate. Electronegative substituents deshield carbon as they do protons.

These trends are substantially magnified in the case of ^{13}C, however. Resonances of differently hybridized carbons are separated by approximately 50 ppm, and a single fluorine substituent can cause a downfield shift in the signal of the substituted carbon of 80 ppm. On the other hand, neighboring group anisotropy, which causes dramatic effects in the context of the proton chemical shift range, is relatively unimportant in ^{13}C-NMR, generally causing shifts of a few ppm or less (29,155).

The sensitivity of ^{13}C-NMR to hybridization and inductive effects has, in the past, led to the misconception that ^{13}C shifts depend entirely on local electron densities. In fact, that notion is erroneous. Factors exist that make major contributions to ^{13}C resonance frequencies through various mechanisms that have no direct bearing on local charge densities.

One example is the β effect, which has not been satisfactorily accounted for by theory. In alkanes (and for that matter in all other molecules) carbons are universally deshielded by β substituents. This is true even for electronically neutral alkyl substituents. Typically β-substituent effects are on the order of 10 ppm, exceeding in some cases the effect of α-substitution. Direct substitution by heavy halogens (iodine and in some cases, bromine) results in another unusual effect. These substituents shield the carbon to which they are bonded, sometimes quite markedly. In fact, the effect increases with multiple halogen substitution (see Table 2.IX). These effects can be illustrated with a series of 1-halopentanes.

Substitution of F, Cl, or Br on C-1 of pentane causes an appreciable deshielding of that atom as expected. This deshielding parallels the electronegativity of the substituent. However, substitution with iodine, results in shielding of C-1 by 7.4 ppm (1). On the other hand C-2, the atom β to the substituent, is deshielded by about 10 ppm, the same extent to which the other halogens affect its resonance position.

Remote substituents can also influence ^{13}C chemical shifts through a process that is linked to stereochemical considerations (1,29,65,155,278). The magnitude and, sometimes, the direction of the induced "steric" shift is dependent upon the specific conformational relationship existing between the substituent and the affected atom and also, for heteroatomic systems, on electronic considerations. For example, γ-gauche alkyl substituents can give rise to upfield shifts in the range of 2–6 ppm, with the larger values corresponding to rigidly fixed substituents. These shifts can be a valuable aid in analyzing the spectra of complex organic systems. Specific peaks can be assigned on the basis of expected configurations. Conversely, observation of induced shifts, by comparison with model compounds, can be of help in determining molecular geometry. Specific examples of this effect and its use will be discussed later.

Additional chemical shift factors unique to ^{13}C-NMR which are not completely understood arise from electric field effects (15), mesomeric (resonance) effects, and isotope effects. Because a complex set of factors can be influential in determining ^{13}C resonance frequencies, theoretical calculations of carbon shifts have only achieved limited success.

Experimentally, it has been found that the effects of substituents on ^{13}C resonances are often additive and thus allow successful predictions of chemical shifts by empirical methods. Using tables of derived substituent parameters of alkanes and other compounds, shifts can be calculated to within 0.5–2 ppm of the experimentally obtained values. This excellent agreement corresponds to predicting proton shifts to $\lesssim$0.1 ppm when the total relative shift ranges of the two nuclei are taken into consideration. ^{13}C shielding calculations of this type have proven very useful in peak assignments for a wide variety of systems. Examples will be discussed in further detail in Section III. Some general classes of organic compounds and their ^{13}C chemical-shift ranges are shown in Fig. 2.9.

TABLE 2.II

TMS-Based ^{13}C Chemical Shifts for Common Standards and Solvents

Compound, solvent	Chemical shift[a]	
	Protio compound	Deuterio compound
Toluene (methyl carbon)	21.3	20.4
Cyclohexane	27.5	26.1
Acetone (methyl carbon)	30.4	29.2
Dimethylsulfoxide	40.5	39.6
Methanol	49.9	49.0
Methylene chloride	54.0	53.6
Dioxane	67.4	66.5
Chloroform	77.2	76.9
Carbon tetrachloride	96.0	
Benzene	128.5	128.0
Acetic acid (*CO)	178.3	
CS_2	192.8	
CS_2 capillary	193.7	

[a] In parts per million from internal TMS; ±0.05 ppm at 38°C.

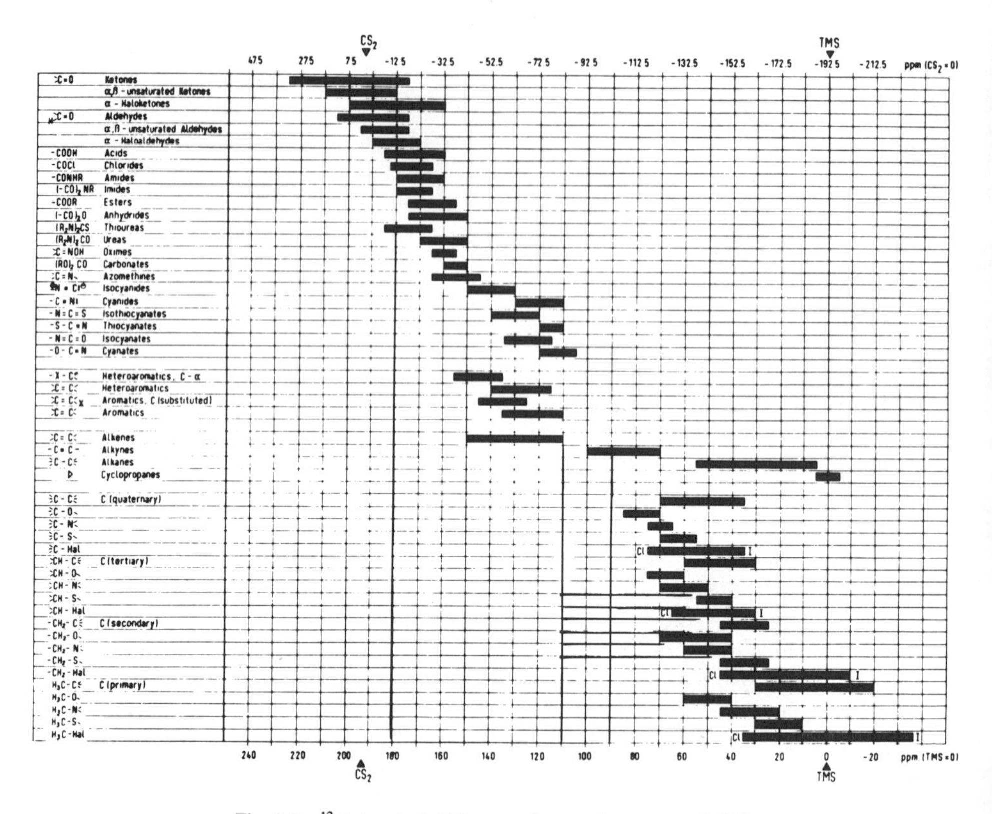

Fig. 2.9. ^{13}C chemical shift ranges in organic compounds (29).

B. SPIN-SPIN COUPLING

Three types of spin–spin interactions are possible in ^{13}C-NMR: coupling between two ^{13}C nuclei, coupling between a ^{13}C nucleus and a proton, and coupling between a carbon and some other magnetically active nucleus in the compound. In natural abundance ^{13}C-NMR experiments, $^{13}C-^{13}C$ coupling is not observed because of the low probability of two ^{13}C nuclei being present in the same molecule. One-bond homonuclear ^{13}C coupling constants, $^{1}J_{CC}$, can be obtained, however, from spectra of isotopically enriched materials or under high-sensitivity conditions from natural abundance samples. $^{1}J_{CC}$ values are dependent upon hybridization, bond angles, and the electronegativities of directly bound substituents (1,29,155,278).

Observation of $^{13}C-^{1}H$ spin–spin coupling is precluded when wideband proton decoupling is employed for resolution and sensitivity enhancement in ^{13}C-NMR. Frequently, however, the elimination of $^{13}C-^{1}H$ coupling information, while giving rise to spectra of greatly improved appearance, causes difficulty in assigning peaks. Fortunately, it is possible, with minimal losses in resolution and sensitivity, to recover coupling information when needed. One of the techniques used is off-resonance decoupling.

1. Decoupling Methods

In a single-frequency off-resonance decoupling experiment, the ^{1}H decoupling frequency is kept at high power, but its center is set 500–1000 Hz outside the proton region to be irradiated and the wideband generator is turned off. In the resultant spectra, all one-bond $^{13}C-^{1}H$ coupling patterns appear with the observed couplings appreciably reduced in magnitude relative to actual $^{1}J_{CH}$ values. A two-fold benefit results: spectral resolution is not adversely affected by the appearance of multiplets due to their reduced bandwidths; the existence of multiplets allows assignment of nonprotonated carbons, CH, CH_2, and CH_3, as singlets, doublets, triplets, and quartets, respectively (although CH_2 groups often give rise to nonideal band shapes).

The magnitudes of the observed residual couplings in an off-resonance decoupled spectrum are a function of the actual $^{1}J_{CH}$ values, decoupling power, and decoupler offset. Overlapping signals can often be resolved if necessary for interpretation, by adjustment of the decoupler offset. Importantly, good sensitivity is preserved since NOE remains operative and long-range $^{13}C-^{1}H$ coupling is absent.

Another method that can be used to reacquire $^{13}C-^{1}H$ coupling information for use in spectral assignments is single-frequency proton decoupling. It is closely analogous to proton homonuclear decoupling, as described in the preceding chapter. In this technique, a specific proton resonance, determined from the ^{1}H NMR spectrum of the compound, is selectively irradiated at low rf power. In the ^{13}C NMR spectrum, the signal for the carbon bound to that proton (or equivalent protons) will collapse to a singlet, uniquely assigning that signal to a particular carbon atom. All other proton-bearing carbons retain a residual $^{13}C-^{1}H$ coupling. Carbon signals in the sample can also be assigned by taking advantage of this relationship between ^{1}H and ^{13}C lines through automated, successive single-frequency decoupling experiments. Ideally, one might limit this procedure by choosing only those proton signals that are well resolved enough to yield discrete resonance frequencies for irradiation. However, as shown in Fig. 2.10, sets of single-frequency decoupled spectra obtained by proton irradiation at (arbitrary) 0.5 ppm intervals can be of great value in assigning peaks for compounds as complex as strychnine.

A rigorous explanation of spin–spin coupling based on theoretical considerations is complex (140). However, one-bond $^{13}C-^{1}H$ and $^{13}C-^{13}C$ couplings in organic molecules can be estimated by semiempirical means. For example, $^{1}J_{CH}$ values, which normally range from 120–350 Hz, have been directly related to the degree of *s* character of the carbon hybrid orbital and to the electronegativities of other substituents on the carbon atom. Thus $^{1}J_{CH}$ values involving sp^3 carbons are lowest ($\sim$125 Hz) while $^{1}J_{CH}$ for *sp* carbons are highest ($\sim$250 Hz). The simple formula, $J_{CH} = 5 \times \%s$, approximates the couplings observed in hydrocarbons. Substitution of electronegative atoms on carbon increases $^{1}J_{CH}$ significantly. Bond strain and C–H internuclear distance also appear to contribute to $^{1}J_{CH}$.

One-bond C–C coupling constants are also proportional to the *s* character of the *two* nuclei involved.

Two-bond C–H and three-bond C–H couplings exhibit strong dependence on substituents and hybridization in a manner similar to J_{CH}. Typical $^{2}J_{CH}$ values range

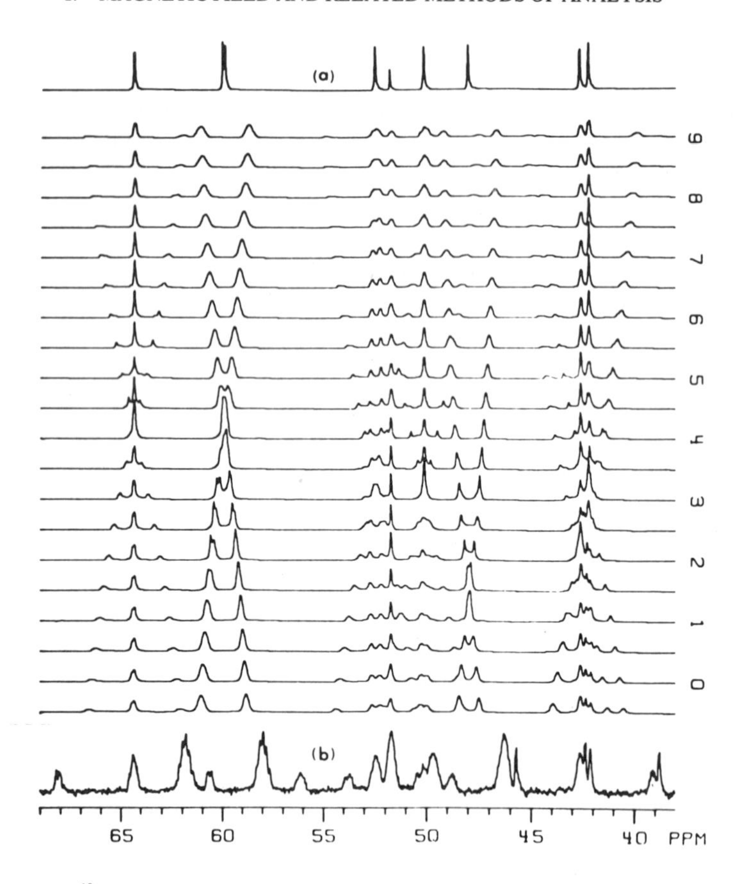

Fig. 2.10. Partial ^{13}C spectra for strychnine (39–69 ppm): (*a*) fully proton decoupled; (*b*) proton coupled; others, sequentially single-frequency decoupled, stepping through the proton spectrum at 0.5-ppm intervals (L. F. Johnson, in G. C. Levy, Ed., *Topics in Carbon-13 NMR Spectroscopy*, Vol. 3, Wiley-Interscience, New York, 1979.) (155)

from -6 to $+10$ Hz while $^3J_{CH}$ ranges from $+3$ to $+15$ Hz. Vicinal couplings show a dependence on dihedral angle.

Spin–spin couplings of carbon to ^{31}P and ^{19}F are frequently observed. Coupling to nuclei such as ^2H, ^{14}N, ^{17}O, and $^{35/37}Cl$ are not usually seen since these species possess electric quadrupole moments. Coupling to ^{15}N or ^{27}Si is generally precluded due to their low natural abundances.

It is evident from the foregoing discussion that spectral peak assignment is enormously facilitated when coupling information is available. Ideally, coupled spectra would be the most useful variety for structural analyses. In practice, even if resolution is acceptable, the severe sensitivity reduction that results from peak multiplicity and loss of NOE places unrealistic time demands on achieving the S/N required for this application. It is possible, however, to generate proton-coupled ^{13}C spectra with retention of NOE and the signal enhancement it affords. The technique utilized is gated decoupling (83).

The presence or absence of NOE and/or proton decoupling depends entirely upon the *off/on* status of the decoupler during two distinct phases of the pulsed FT-NMR experiment. Decoupling is obtained when proton irradiation is in effect during acquisition of the FID and is absent if the decoupler is off while data are being acquired. The status of the decoupler during an individual acquisition has no effect on NOE observed for that acquisition. Appearance of NOE results from proton irradia-

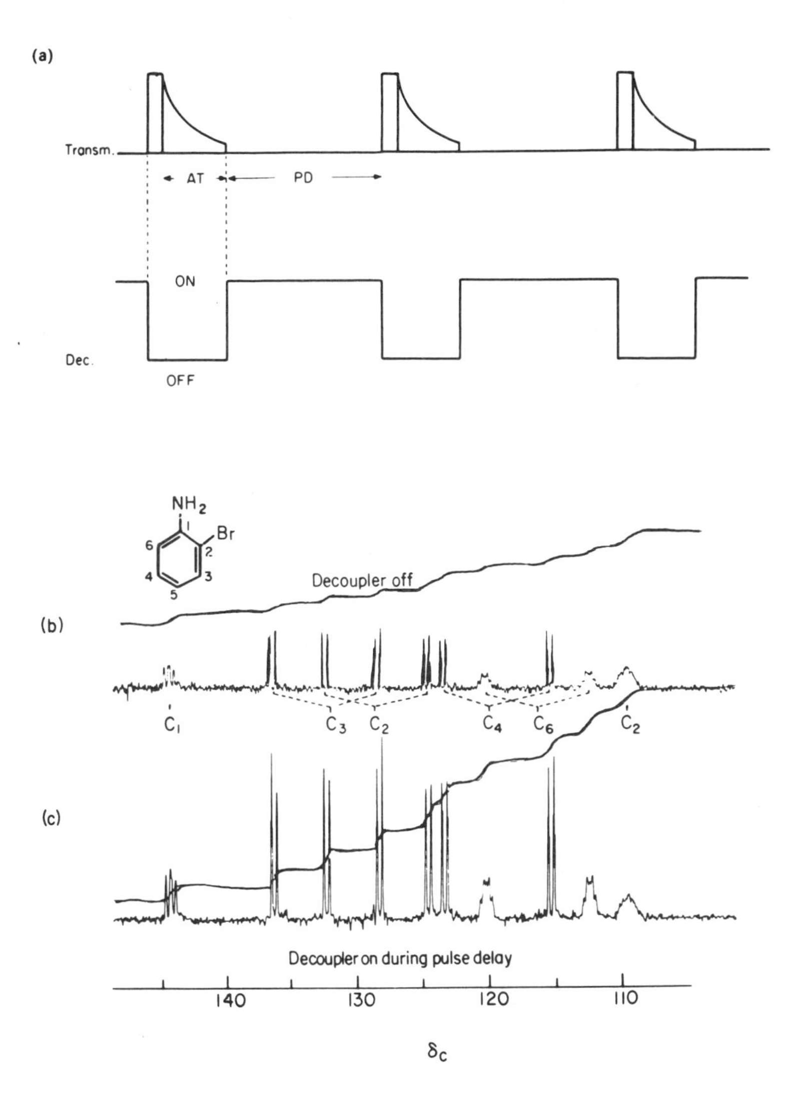

Fig. 2.11. (*a*) Timing diagram in a pulse-modulated (gated) decoupling experiment which permits the retention of heteronuclear spin–spin coupling and nuclear Overhauser effect. The decoupler is on during a pulse delay (PD), but off during the data acquisition period (AT); (*b*) 20-MHz single-resonance spectrum of 2-bromoaniline in deuteroacetone; (*c*) same conditions as in (*b*) but with the decoupler gated in the manner described under (*a*). (278)

tion being present prior to the excitation pulse. It is in this time period that the population of ^{13}C nuclei in the lower spin state is augmented, at a rate defined by the ^{13}C spin–lattice relaxation process. Thus, in the proton-coupled ^{13}C gated decoupling experiment, the decoupler is switched on for a period of time, τ, sufficient to allow the perturbed spin-state populations responsible for NOE to be achieved. It is then switched off and the excitation pulse is applied immediately. Since the short duration of the pulse precludes any relaxation during its application, the perturbed ^{13}C spin system undergoes excitation and NOE is ultimately observed in the resulting spectrum. However, the decoupler is left off during acquisition of the FID, so normal couplings are observed. The overall result is a proton-coupled ^{13}C spectrum with NOE. This is illustrated in Fig. 2.11*bc* for the compound, 2-bromoaniline (278). Figure 2.11*a* gives a schematic representation of the control sequence of this gated decoupling experiment. This relatively complex sequence, which is repeated many times in the course of acquiring a large number of transients, is entirely controlled by the computer.

2. Determination of NOE

It is of interest to note that inversion of the control sequence just described affords quantitative determination of NOE, itself an important experimental parameter. A fully decoupled ^{13}C spectrum with NOE is obtained when proton irradiation is in effect both preceding and during data acquisition. If the decoupler is kept *off* between data acquisitions but turned on during acquisition (reversing the sequence of Fig. 2.11*a*), the resulting spectrum will be decoupled but will have no NOE, provided that the intervals between data acquisition periods meet certain requirements (generally the interval should exceed 10–15 times the longest T_1 for any carbon of interest). Comparison of the signal intensities (peak areas) obtained in gated and continuous wideband decoupled experiments yields the degree of enhancement achieved in the latter.

It is possible to determine NOE values extremely accurately using the DNOE pulse sequence. In this experiment, the decoupler is gated *on* τ seconds *before* the ^{13}C excitation pulse is applied. A set of spectra obtained with τ values ranging from 0 to several times T_1, gives the time evolution of the NOE and thus allows calculation of both NOE and T_1. Use of regression analysis techniques can allow NOEs to be determined to 1–5% accuracy in favorable cases (141).

C. SPIN-RELAXATION PARAMETERS

1. Spin-Lattice Relaxation Processes

There are several mechanisms that can lead to spin-lattice relaxation for ^{13}C nuclei:

1. Dipole–Dipole Relaxation. Spin–lattice relaxation for ^{13}C nuclei can arise from fluctuating fields as a result of dipole–dipole interactions with neighboring magnetic nuclei (or with unpaired electrons).

2. Spin–Rotation Relaxation. Small molecules and freely rotating CH_3 groups

can be effectively relaxed by a mechanism involving quantum-rotational states of the molecule or the group. In these cases spin–rotation relaxation often competes with dipole–dipole relaxation for protonated carbons, whereas the spin–rotation interaction dominates the relaxation of nonprotonated carbons.

3. Chemical-Shift Anisotropy. Significant anisotropy (directionality) in the shielding tensor of a nucleus can give rise to fluctuating magnetic fields when the molecule tumbles in solution (relative to the fixed laboratory magnetic field).

4. Scalar Relaxation. A ^{13}C nucleus that is spin–spin (or scalar) coupled to a nucleus X that is undergoing rapid spin–lattice relaxation may itself be relaxed because of the fluctuating scalar interaction between the two nuclei. This mechanism is normally encountered when the X nucleus has a spin I, which is $>1/2$ (and X is relaxed by a mechanism known as quadrupolar relaxation), but scalar relaxation of ^{13}C nuclei may also occur when $X = {}^{1}H$. *Scalar relaxation of ^{13}C nuclei is generally confined to spin–spin relaxation.* (The spin–spin relaxation time, T_2, defines the resonance line widths but does not involve saturation of energy-level populations). In a few cases, scalar *spin–lattice* relaxation of ^{13}C nuclei does occur. When X is undergoing (quadrupolar) spin–lattice relaxation very rapidly and when the Larmor (resonance) frequency of X is close to the Larmor frequency for ^{13}C, scalar spin–lattice relaxation of ^{13}C nuclei may be competitive with other mechanisms.

In cases where scalar relaxation of ^{13}C is limited to spin–spin relaxation, line broadening is often evident in the NMR spectra. For example, ^{13}C nuclei attached to ^{14}N nuclei frequently appear as broad resonances.

While any of the above relaxation mechanisms may dominate ^{13}C relaxation in individual cases (149), carbons in most moderate-to-large-sized organic molecules relax by $^{13}C-{}^{1}H$ dipolar relaxation and these ^{13}C nuclei exhibit essentially full nuclear Overhauser enhancements. However, experiments at higher magnetic fields may show increased contributions from chemical shift anisotropy relaxation and hence reduced NOEs. However, polymers and highly associated smaller molecules violate the extreme spectral narrowing region (where molecular tumbling is rapid relative to the inverse of the spectrometer resonance frequency). In this case, despite dominant dipolar relaxation, theory dictates that observed nuclear Overhauser enhancements will be reduced (at the limit of slow motions, to a maximum value near 0.1).

2. Dipole–Dipole Relaxation

Dipole–dipole relaxation arises from local magnetic fields associated with magnetic nuclei (nuclei with spin $I \neq 0$). If two neighboring magnetic nuclei are placed in an external magnetic field B_0, each nucleus sees a total magnetic field comprised of B_0 and a contribution from the local magnetic field of the other nucleus. The strength and the direction of this localized interaction depends on the magnetic moments and the internuclear separation of the two nuclei and their internal orientation *relative* to B_0. When molecular motions are rapid, as they are in a liquid, the relative orientation of the two nuclei with respect to B_0 is constantly changing. The rapid internal reorientation of the two nuclei gives rise to rapid fluctuations of the localized (and thus of the total) magnetic field as seen by each nucleus; these fluctuations effect relaxation of the nuclei.

Relaxation of ^{13}C nuclei in organic molecules usually results from dipole–dipole interactions with protons. This relaxation follows the general equation 2

$$\frac{1}{T_1^{DD}} = \sum_{i=1}^{n} \frac{\gamma_C^2 \gamma_H^2 \hbar^2 \tau_C^{\text{eff}}}{r_{CHi}^6} \tag{2}$$

where γ_C and γ_H are the magnetogyric ratios for ^{13}C and ^1H nuclei, r_{CHi} is the through-space distance between the ^{13}C nucleus and proton i, and τ_C^{eff} is the effective reorientational correlation time. Dipole–dipole interactions are summed for all relevant protons i.

The r_{CHi}^{-6} distance dependence for the ^{13}C dipolar relaxation rate means that *intermolecular* contributions may usually be ignored. In fact, for protonated carbons, only *bonded* hydrogens make major dipolar contributions to T_1^{DD}. In the case of protonated carbons the dipolar relaxation rate is then simplified, following equation 3:

$$\frac{1}{T_1^{DD}} = N_H \gamma_C^2 \gamma_H^2 \hbar^2 r_{CH}^{-6} \tau_C^{\text{eff}} \tag{3}$$

when N_H is the number of directly attached hydrogens (each one has a magnetic field), and r_{CH} is now represented by the C–H bond distance, assumed to be constant at 1.09×10^{-8} cm.

The average frequency (or, more accurately, the frequency distribution) of a localized fluctuating magnetic field determines its effectiveness in causing spin relaxation. Frequencies of motions for molecules in solution range from very slow to very fast. In the case of very small molecules, most of the individual molecules will be "rotating" at rates of $>10^{12}$ revolutions per second while a few will be "rotating" much more slowly. Actually, the movements of a molecule in solution are very complex; characteristic rotation rates cannot be dissected easily from overall tumbling motions. Fortunately, it is generally not necessary to consider separation of the various molecular motions. It is usually sufficient to describe an effective correlation time, τ^{eff}, which represents the *average* time for a molecule to rotate through 1 radian (2π rad/sec = 1 revolution/sec or 1 Hz). Dipole–dipole relaxation is best effected by rotational motions with frequencies comparable with the resonance frequency, which for ^{13}C nuclei is 10^7–10^8 Hz or 10^8–10^9 rad/sec. A correlation time τ^{eff}, close to the reciprocal of the resonance frequency in rad/sec leads to most efficient dipolar spin–lattice relaxation of a ^{13}C nucleus. A graphical representation of the effect of τ^{eff} on T_1 values is shown in Fig. 2.12. This figure also shows what happens to the spin–spin relaxation time T_2 as τ^{eff} changes.

The correlation time for most molecules falls on the left-hand side of Fig. 2.12. Typical τ^{eff} values for small molecules would be 10^{-12}–10^{-13} sec; for reasonably large organic molecules τ_c may be as long as 10^{-10} sec.

Large organic molecules (e.g., MW > 300) tumble relatively slowly in solution because of their size (inertial effects) and the extensive reordering of solvent

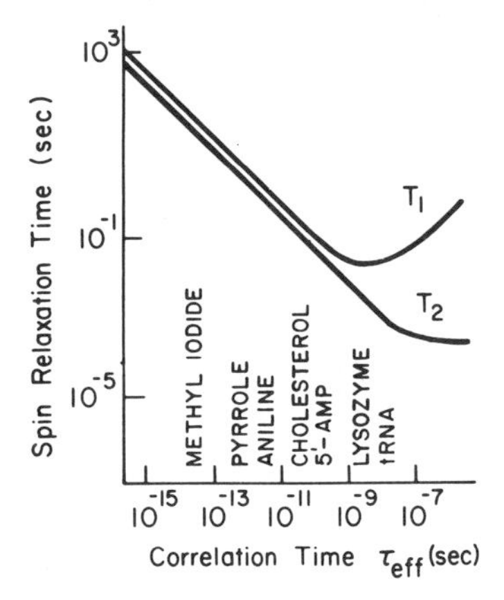

Fig. 2.12. Relationship between T_1 (and T_2) and molecular tumbling as represented by τ_{eff} (assumes dipolar relaxation) (155).

molecules that are swept aside during molecular rotation (this second effect also depends on molecular symmetry).

3. Separation and Identification of Relaxation Contributions

For most applications, to obtain meaningful information from ^{13}C spin–relaxation data, it is *absolutely necessary* to know the mechanism(s) responsible for the observed T_1 values. When more than one mechanism is contributing, a semiquantitative, or, better, a quantitative assessment of at least one mechanistic contribution is necessary. In fact, this is usually a straightforward procedure. The observed ^{13}C relaxation *rate* is dissected according to equation 4

$$R_1^{\text{obs}} = \frac{1}{T_1^{\text{obs}}} = \frac{1}{T_1^{\text{DD}}} + \frac{1}{T_1^{\text{SR}}} + \frac{1}{T_1^{\text{CSA}}} + \frac{1}{T_1^{\text{SC}}} + \frac{1}{T_1^{\text{e}}} \tag{4}$$

In fact, several other separations such as *intramolecular* and *intermolecular* ^{13}C–^{1}H dipolar terms, may be effected, depending on the degree of identification possible.

The identification of mechanistic contributions to T_1 requires a number of experiments. Assuming that extreme spectral narrowing condition applies, T_1^{DD} (^{13}C–^{1}H) is identified from the observed NOEF in ^{13}C{^{1}H} experiments:

$$T_1^{\text{DD}} = T_1^{\text{obs}} \cdot \frac{1.98}{(\text{NOEF})_{\text{obs}}} \tag{5}$$

Spin–rotation relaxation is identified by its temperature dependence, T_1^{SR} shortening with increasing temperature, whereas T_1^{DD} and T_1^{CSA} are easily quantified from measurements at two or more magnetic fields. Scalar and/or electron–nuclear spin–lattice relaxation processes are not identified from any one characteristic, but from a combination of structural effects and temperature and frequency dependencies. Table 2.III lists ^{13}C spin–lattice relaxation data for representative organic molecules, along with comments about mechanistic contributions.

TABLE 2.III

^{13}C Relaxation Times in Small and Intermediate-Sized Molecules (155)

Compound	Carbon	T_1^{obs} (Sec)	NOEF	T_1^{DD}	Other Mechanisms, Comments
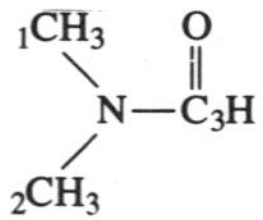 $_1CH_3$, $_2CH_3$, N—C_3H, O	1	18.1	1.3	29	T_1^{SR}; electrostatic ordering, internal CH_3 motion
	2	11.1	1.7	13	
	3	20.2	1.4	29	
		29	1.6	35	T_1^{SR}: high degree of symmetry
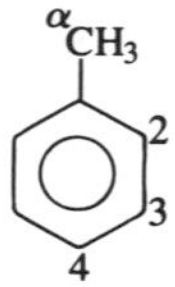 $^{\alpha}CH_3$, 2, 3, 4	1	89	0.56		T_1^{SR} for C-1 and C-α; high degree of symmetry; rapid internal rotation of CH_3 gives short T_1^{SR}
	2	24	1.6		
	3	24	1.7		
	4	17	1.6		
	α	16	0.61		
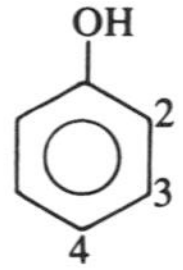 OH, 2, 3, 4 (in CCl_4)	1	18.4	$\gtrsim$1.8		Fully dipolar; H bonding; preferred rotation with shorter T_1 for C-4; dilution gives longer T_1
	2	2.8	2.0		
	3	2.8	2.0		
	4	1.9	2.0		
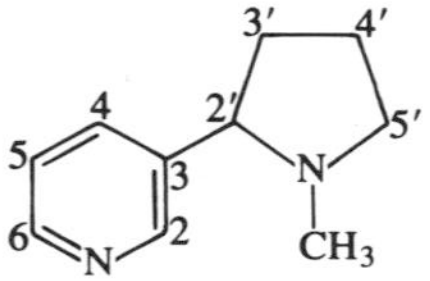 2, 3, 4, 5, 6, N, 2′, 3′, 4′, 5′, N, CH_3	2	5.0			Internal rotations of pyridine ring and N-CH_3 indicated
	3	39.5			
	4	4			
	5	4.5			
	6	3.0			
	2′	4.5			
	3′	2.0			
	4′	2.0			

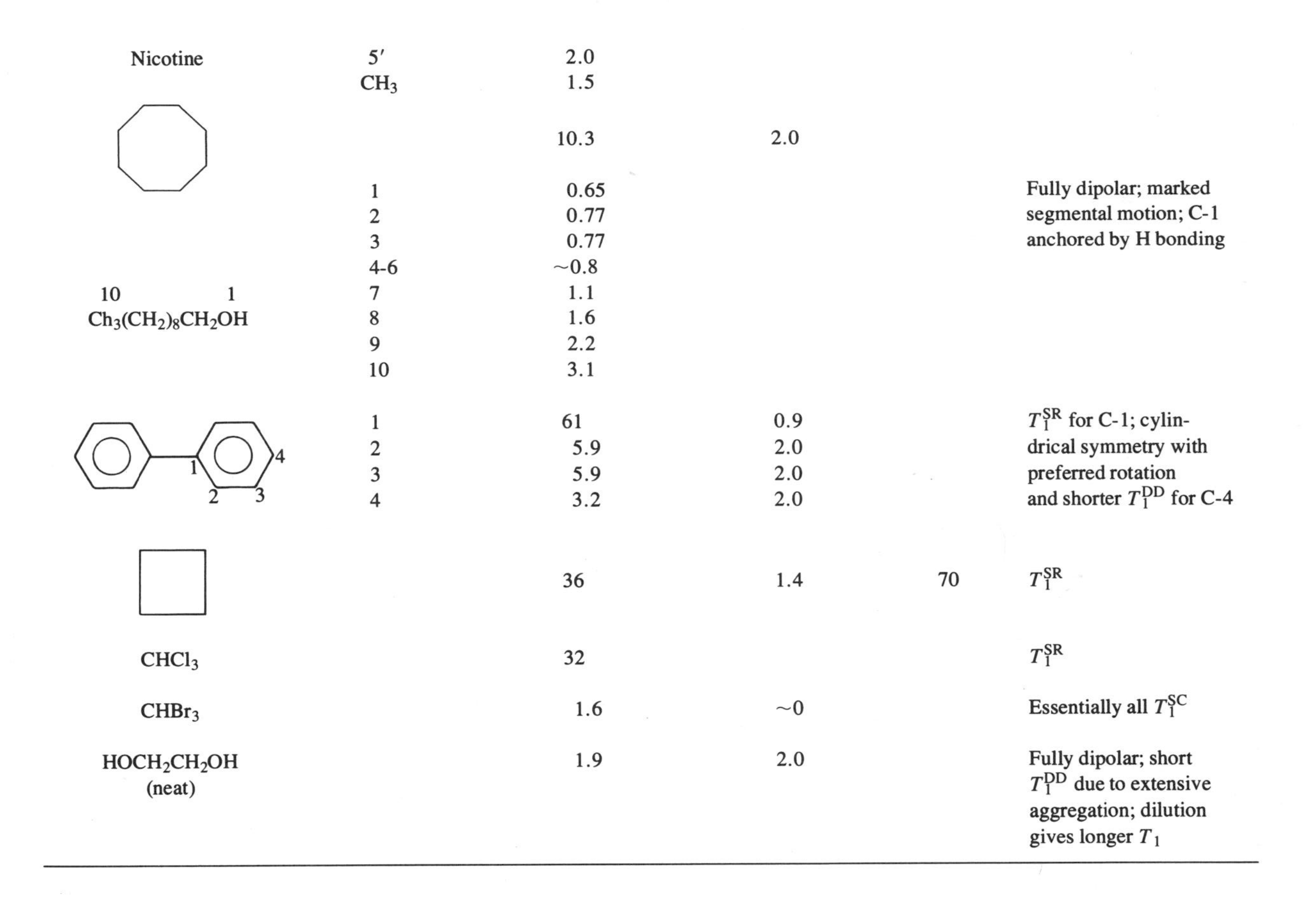

Nicotine	5′	2.0			
	CH_3	1.5			
(cyclooctane structure)		10.3	2.0		
10 1 $Ch_3(CH_2)_8CH_2OH$	1	0.65			Fully dipolar; marked segmental motion; C-1 anchored by H bonding
	2	0.77			
	3	0.77			
	4-6	~0.8			
	7	1.1			
	8	1.6			
	9	2.2			
	10	3.1			
(biphenyl structure, positions 1–4)	1	61	0.9		T_1^{SR} for C-1; cylin-drical symmetry with preferred rotation and shorter T_1^{DD} for C-4
	2	5.9	2.0		
	3	5.9	2.0		
	4	3.2	2.0		
(cyclobutane structure)		36	1.4	70	T_1^{SR}
$CHCl_3$		32			T_1^{SR}
$CHBr_3$		1.6	~0		Essentially all T_1^{SC}
$HOCH_2CH_2OH$ (neat)		1.9	2.0		Fully dipolar; short T_1^{DD} due to extensive aggregation; dilution gives longer T_1

4. Determination of ^{13}C Spin-Lattice Relaxation Times

Measurements of T_1 are designed to monitor M_z as a function of a time parameter, to describe its approach to M_0.

Pulse FT methods for T_1 measurements include inversion-recovery (271), fast-inversion-recovery (37), progressive saturation (84), saturation-recovery (174), and dynamic NOE/gated-decoupled (85) pulse sequences. Generally, the slower sequences (inversion recovery and dynamic NOE) are less useful for natural-abundance ^{13}C T_1 measurements. The fast-inversion-recovery Fourier transform (FIRFT) and saturation-recovery Fourier transform (SRFT) pulse sequences can be recommended for routine use, but the progressive-saturation Fourier transform (PSFT) sequence gives large errors with misset 90° pulses and should be avoided except under exceptional conditions. [A recent paper (17) concludes that the FIRFT sequence generally has the highest efficiency.]

Several papers discuss experimental methods for T_1 measurements, treating sources of errors and outlining requirements for accurate determinations (7,110,158,283). Other papers discuss statistical methods for efficient analysis of the experimental data (111,142,146,223).

Relaxation studies have several applications (recent reviews: 277,285). Measurements of T_1 in small molecules often yield information about very rapid internal group rotations ($>10^{13}$ sec^{-1}) and the nature of molecular symmetry and steric and bonding interactions; whereas T_1 measurements in large molecules can be used to facilitate spectral assignments in very complex ^{13}C-NMR spectra and also to gain insight into molecular configuration. As in small molecules, these T_1 measurements may be used to detect fast internal motions, such as CH_3 group rotations. It is also possible to learn a great deal about the geometry of short-lived "charge-transfer" and hydrogen-bonded molecular complexes.

D. INTEGRATION OF ^{13}C-NMR SPECTRA

Accurate peak areas are routinely obtained in ^1H-NMR and comprise an extremely valuable tool in spectral interpretation. Unfortunately, some difficulties exist in performing analogous measurements in ^{13}C-NMR. Under typical experimental conditions, no direct correlation is found between integrated peak areas and the number of carbon nuclei forming each peak. The loss of this important correlation arises from two principal sources: (1) variant spin-lattice relaxation times among different carbon atoms, which can result in differential saturation effects under rapid pulsing conditions, and (2) variable NOE among nonequivalent nuclei in the sample. For these reasons, peak areas are not routinely measured in ^{13}C spectra. However, meaningful integration data can be obtained from experiments designed for that purpose.

For each unique carbon resonance to yield maximum signal intensity, all nuclei should be fully relaxed to their thermal equilibrium populations prior to each excitation pulse. However, in many samples different types of carbon atoms often have very different T_1 values. The range of values can be from less than 0.1 sec up to ca. 100 sec, although generally T_1 values vary by less than a factor of 10–25 in an individual molecule. In experiments that employ rapid pulse repetition rates, those nuclei with the longer spin–lattice relaxation times will not be fully relaxed between scans; partial saturation from previous pulses causes their signals to be reduced in intensity. Therefore, the accumulated signals do not reflect the actual number of atoms being irradiated. To alleviate this effect, an experiment designed to produce quantitative intensity data requires a delay between pulses sufficient to allow all relevant carbon nuclei to be sequentially excited from their thermal equilibrium states. A pulse delay of 4 to 5 times the longest T_1 value must be used to comply with this requirement.

Insertion of an appropriate delay into the pulse sequence of a ^{13}C FT-NMR experiment still does not guarantee that *all* signals will accurately reflect their numbers of associated nuclei. Difficulties can still arise due to the second effect mentioned above, unequal NOEs among nonequivalent nuclei in the sample.

The NOE for any broadband, heteronuclear decoupled FT-NMR experiment can be defined as I_D/I_0, the integrated intensity ratio of decoupled (I_D) to undecoupled (I_0) resonances of the observed nucleus. For ^{13}C {^{1}H} experiments its theoretical maximum value is 2.98. This corresponds to a nuclear Overhauser enhancement factor (NOEF), defined as the NOE-1, of 1.98. Maximum NOE is observed in proton-decoupled ^{13}C-NMR experiments only when spin–lattice relaxation of the ^{13}C nucleus is exclusively dipolar (with protons). If other relaxation pathways are utilized by some nuclei (e.g., chemical-shift anisotropy), their observed NOEs are reduced accordingly and reflect only that fraction of the overall relaxation process which is dipolar.

Larger organic molecules (mol. wt. 250–400) are particularly favorable when ^{13}C peak integrations are required. ^{13}C T_1 values are short ($\lesssim$0.1 sec for CH$_x$ carbons; $\approx$1 sec for nonprotonated carbons) and NOE values are either full or equivalent (except for nonprotonated sp^2 or sp carbons, when measurements are made above 2 Tesla). In these cases extended pulse (scan) intervals are not necessary and use of specialized techniques is not required.

For those samples that are not characterized by short, dipolar ^{13}C T_1 values, paramagnetic relaxation agents and gated decoupling can be combined to circumvent the effects of partial saturation and variable NOE. Use of paramagnetic species known as relaxation reagents can partially alleviate difficulties caused by long relaxation times and variable NOE effects. Table 2.IV lists several commonly used paramagnetic relaxation agents. Among the nonshifting compounds, Cr(acac)$_3$ and Cr(dpm)$_3$ are most suitable for dissolution in organic solvents. If a sufficient quantity of the reagent can be introduced into the sample, highly efficient ^{13}C relaxation via interaction with the electron spin becomes dominant. This has the combined effect of shortening T_1 values and suppressing all NOEs, since ^{13}C–^{1}H dipolar relaxation is circumvented.

However, it should be noted that two potential problems exist. First, carbons undergoing efficient ^{13}C–^{1}H dipolar relaxation may show considerable residual NOE even in the presence of an essentially saturated solution of relaxation agent. Second, all of these agents except Cr(dpm)$_3$ have the ability to associate with specific

TABLE 2.IV

Characteristics of Paramagnetic Reagents in ^{13}C Spectroscopy

Compound Type	Commercial	Typical concentration (M)	Interacting Functional groups
Lanthanide-shift reagents (Lu, Pr, Yb chelates, etc.)	Yes	10^{-1}	These reagents cause shifts; C=O, ether, amine, alcohol, etc.
Tris[acetylacetonatochromium(III)], $Cr(acac)_3$	Yes	10^{-2} to 10^{-1}	Outer-sphere coordination; ligands hydrogen-bond to all acidic hydrogens, OH, NH_2, etc.; electrostatic interactions with polar groups
Tris[dipivaloylmethanatochromium(III)], $Cr(dpm)_3$	No[a]	10^{-2} to 10^{-1}	Inert as tested.[a]
Tris[dipivaloylmethanatogadolinium(III)], $Gd(dpm)_3$	Yes	10^{-4} to 10^{-3}	*Inner-sphere* coordination to basic nitrogens, etc.

[a] See reference 151.

molecular sites, thereby producing gradations in T_1 augmentation; carbons closer to the reagent have shorter T_1 values (152).

If the paramagnetic relaxation agent does not effect quantitative suppression of ^{13}C NOEs, gated decoupling may be efficiently employed for this task (237). This combination of techniques is particularly attractive since scan intervals can be kept short in the presence of the paramagnetic agent. The only requirement remains in establishing the data acquisition duty cycle, and/or the lower limit for the scan interval.

Finally, instrumental factors exist that can cause difficulties in ^{13}C peak area measurements. To generate accurate integration data, uniform excitation over the entire spectral width must be achieved. A pulse amplifier powerful enough to accomplish this is, therefore, required. For ^{13}C-NMR at 50 MHz, an excitation field of 20 kHz is adequate to effectively cover the normal 200 ppm carbon chemical shift range. At higher magnetic-field strengths, sweep widths become necessarily larger requiring still more powerful pulse amplification for uniform excitation. The advent of quadrature detection has significantly alleviated many potential problems in this area since it allows the rf excitation frequency to be placed at the center of the spectral window.

A second instrumental factor of potential difficulty relates to limitations in the computer storage capacity. Individual peaks may be defined by as few as two to three data points in a digitized spectrum. This can result in an erroneous representation of such peaks, sometimes assigning to them an artificially reduced intensity. It is important to be aware of this possibility. In practice, resonances defined by a few points usually yield satisfactory if not quantitative values since integrations are not as sensitive to limits on digitization as are peak heights. For highest accuracy it is necessary to have better peak representation, with a minimum five or more data points defining even resolved single lines.

IV. DETAILED ANALYSIS OF ^{13}C SHIFTS AND COUPLINGS

A. ^{13}C CHEMICAL SHIFTS

1. Hydrocarbons

The sensitivity of ^{13}C-NMR as a probe of organic molecular structure is dramatically exemplified by the spectra obtained from saturated hydrocarbons. Proton spectra of these systems are characterized by broad absorption bands caused by inevitable overlapping resonances and extensive non-first-order and virtual spin-spin coupling. Direct ^{1}H spectral interpretation is all but impossible, and distinguishing isomers is very difficult. Proton-decoupled ^{13}C spectra, on the other hand, yield a sharp singlet for each unique carbon and chemical shifts are profoundly affected by directly bound and nearby carbon atoms. Direct interpretation and, thereby, differentiation of isomers becomes relatively straight-forward. The superiority of ^{13}C-NMR over ^{1}H-NMR in this application is demonstrated in Fig. 2.13 where the ^{1}H and ^{13}C spectra of 3-methylheptane are given.

Table 2.V shows ^{13}C shieldings for a series of alkanes. The sensitivity of paraffin ^{13}C chemical shifts to α, β, and (to a lesser extent) γ substituents is demonstrated by

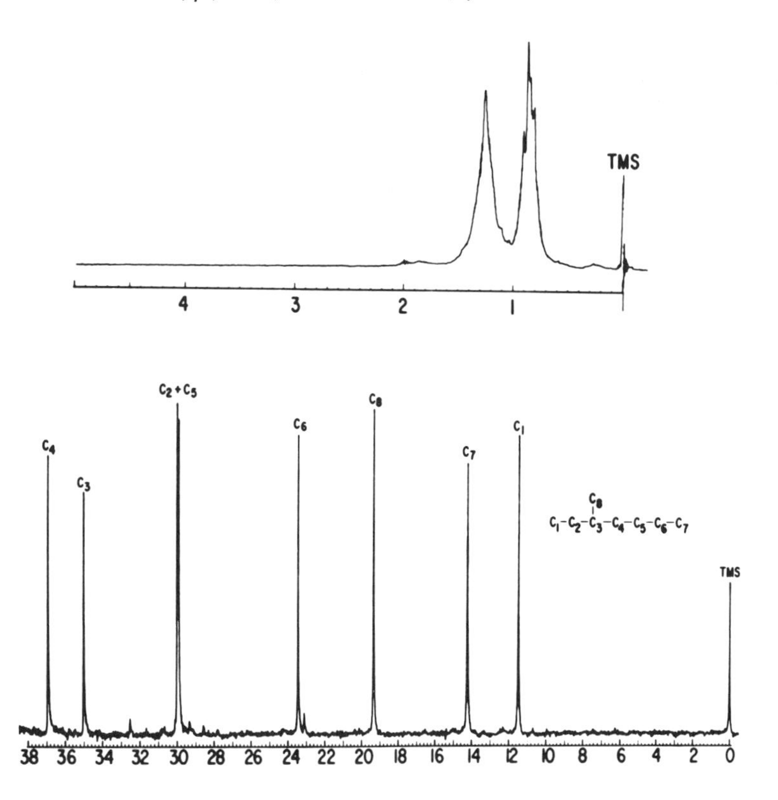

Fig. 2.13. Proton NMR spectrum (*above*) and ^{13}C-NMR spectrum of 3-methylheptane (155).

the essentially constant shieldings observed for C-4 to C-6 in homologs higher than *n*-octane. Grant and Paul were the first to characterize these trends through empirical additive substituent parameters which included changes resulting from chain branching (96). Later, Lindemann and Adams extended this approach, adding parameters to account for steric interactions at branching junctions (161). Those relationships make it possible to predict quite accurate ^{13}C shieldings in acyclic alkanes (1,155,278).

Table 2.VI gives ^{13}C shieldings for a number of cycloalkanes. Prediction of ^{13}C shieldings in these systems is best effected by utilizing the shieldings of the parent molecule. Thus, a set of functional group and skeletal substituent parameters may be derived for each type of molecular framework (278). Through the work of Grant and co-workers, chemical shifts in methyl-substituted cyclohexanes can be calculated with success using parameters based on geometric and conformational factors (48,204). So-called "steric" shifts are accounted for. The efficacy of this approach is confirmed by the fact that the initial spectral assignment of cholesterol was made based largely on these substituent effects.

Relative to alkanes, sp^2 carbons (alkene and aromatic) are considerably deshielded; the usual range of their chemical shifts being 100–145 ppm. Generally, alkene ^{13}C shieldings occur 100 ± 20 ppm downfield of the corresponding resonance in a saturated system. For example, the C-2 chemical shift in 2,3-dimethyl-2-butene is 122.8 ppm, while the value for C-2 in 2,3-dimethylbutane is 34.0 ppm (155). Within the olefins, terminal vinyl carbons are more shielded than those which are alkyl substituted. As noted earlier, alkyne shieldings appear between those of sp^3 and sp^2 carbons. Reflecting this is the 82.0 ppm C-3 chemical shift in 3-hexyne (155). As

TABLE 2.V

^{13}C Chemical Shifts (in ppm) for Representative Linear and Branched Alkanes (2)

Compound	C-1	C-2	C-3	C-4	C-5
Methane	−2.3				
Ethane	5.7				
Propane	15.4	15.9			
n-Butane	13.0	24.8			
n-Hexane	13.7	22.7	31.8		
n-Octane	13.9	22.9	32.2	29.5	
n-Nonane	13.9	22.9	32.3	29.7	30.0
n-Decane	13.9	22.8	32.2	29.7	30.1
2-Methylpropane	24.1	25.0			
2-Methylbutane	21.8	29.7	31.6	11.3	
2-Methylpentane	22.3	27.6	41.6	30.5	13.9

TABLE 2.VI

^{13}C Chemical Shifts (in ppm) of Cycloalkanes

Cyclopropane	−2.9	Cycloheptane	28.5
Cyclobutane	23.3	Cyclooctane	26.9
Cyclopentane	26.5	Cyclononane	26.1
Cyclohexane	27.3	Cyclodecane	25.3

a substituent, a triple-bond shields an attached carbon by 5–15 ppm relative to the corresponding alkene.

Alkene shieldings can be calculated by empirical methods analogous to those described for alkanes (219,258). The substituent parameters and correction terms utilized were developed by Roberts and co-workers (58). Alkyne shifts can be similarly predicted and the effects of a triple bond on alkyl groups can be accounted for empirically (59,116).

Of course, ^{13}C spectra of aromatic carbons are of great interest to the organic chemist. Not only do they provide yet another important means of characterizing this important class of compounds, but relationships between aromatic ^{13}C shieldings and the electronic nature of delocalized rings comprise an experimental probe into the efficacy of theoretical treatments of benzenes and benzenelike molecules. Theoretical treatment of aromatic ^{13}C chemical shifts has been discussed extensively (57,192,246).

^{13}C-NMR of benzene derivatives will be covered in detail in the next section where substituted hydrocarbons of all types are discussed. It is sufficient to state at this time that the carbons of benzene itself resonate at 128.5 ppm, and substitution into the ring creates a range of ^{13}C shieldings from 110–170 ppm. In polynuclear aromatic systems, the quaternary carbons generally appear at lowest field. Nonalternant aromatic hydrocarbons (127) display a broader ^{13}C shielding range (14 to 22 ppm) than do alternant systems such as naphthalene (8 ppm), phenanthrene (9.2 ppm), and pyrene (6.4 ppm) (166,254).

Polycyclic aromatic hydrocarbons have received considerable attention because their π electron systems are amenable to theoretical descriptions. Thus, a number of papers exist that deal with correlations between theoretical parameters and ^{13}C chemical shifts (72,107,193). It has been shown that carbon shifts are not solely related to π electron density, but that both π and σ charge density variations contribute to shieldings in alternant and nonalternant hydrocarbons (127).

2. Substituted Hydrocarbons

The use of additive empirical substituent parameters can be extended to substituted alkanes. Table 2.VII summarizes approximate shielding changes expected at C-1 (bonded to substituent), C-2, and C-3 upon replacement of an alkane methyl group by a polar substituent. Note that the observed trends are not linear with respect to substituent electronegativities. A sample calculation on 2-pentanol demonstrates the usefulness of these parameters (155). The poorest agreement between the calculated and observed values is for C-3; the error in calculation is only +0.7 ppm relative to an observed shift of 41.9 ppm.

Empirical substituent parameters may also be derived for the direct replacement of H by some group or heteroatom, X. However, these must contain a constant contribution from α and β shift effects caused by alkyl groups as shown in Table 2.VII. For example, the ^{13}C shift at C-2 of 2-fluoronorbornane can be estimated by adding 60 ppm (from Table 2.VII) and 8 ppm (the extent to which a methyl group deshields an α-carbon) to 30.1 ppm [the chemical shift of C-2 in norbornane (101)]

TABLE 2.VII
Approximate Changes in ^{13}C Chemical Shifts on Replacement of a Methyl Group by a Polar Substituent

Substituent	C-1	C-2	C-3
−OR	+48	−2	−2
−OH	+40	+1	−3
−OCOR	+43	−3	−1
$-NO_2$	+54	−6	
$-NH_2$	+20	+2	−1
$-NR_2$	+33	−3	
$-NH_3^+$	+17	−2	−3
−I	−15	+1	+1
−Br	+12	+2	−1
−Cl	+23	+2	−1
−F	+60	−1	−2
−COX	+15	−5	0
−COOR	+10	−1	−1
−COOH	+12	−3	−1
−CN	−2	−1	−1
−SH	+2	+2	−2

yielding 98.1 ppm. This compares satisfactorily with the observed shift (95.9 ppm) (101), but is not quite as accurate as some of the predictions previously noted.

Substituent parameters derived for substituted alicyclic systems are qualitatively similar, but different in absolute value from those used for acyclic molecules. Furthermore, they must be derived separately for each type of molecular framework for highest accuracy.

Extensive work has been done on alkanes substituted with common functional groups. In the cases of alcohols (62,217) and amines (61,222) additivity parameters have been derived. Contributions from substituent conformational effects on ^{13}C resonance positions were evaluated and have proven to be an invaluable aid in assigning various alcohol (154,253) and amine (26,27,65,66,67,266,267) structures. Protonation of amines shifts the resonance positions of both α and β carbons, but the shifts may reflect either shielding or deshielding of those nuclei. Protonation shifts may be calculated, but the empirical equation utilized requires 15 parameters (222). ^{13}C shieldings in carboxylic acids and carboxylate ions have been reported and substituent parameters derived (211,266). Essentially C-α and C-β are deshielding by introduction of this group, while C-γ is shielded. Surprisingly, ionization of an aliphatic acid results in a deshielding of the carboxylic carbon atom. Table 2.VIII summarizes the ^{13}C chemical shifts of some representative alcohols, amines, and carboxylic acids.

^{13}C chemical shifts in alkyl halides are interesting because they display solvent and concentration dependencies, and are subject to the previously mentioned "heavy-atom effect" (123). The more extensively fluorinated and chlorinated compounds exhibit solvent effects that may reflect hydrogen-bonding interactions (160). An

TABLE 2.VIII

^{13}C Chemical Shifts (in ppm) for Selected Alcohols, Amines, and Carboxylic Acids (49–58)

Compound	C-1	C-2	C-3	C-4
Methanol	49.3			
Ethanol	57.3	17.9		
1-Propanol	63.9	26.1	10.3	
1-Butanol	61.7	35.3	19.4	13.9
2-Propanol	25.4	63.7		
2-Butanol	22.9	69.0	32.3	10.2
3-Pentanol	10.1	30.0	74.1	
Methylamine	28.3			
Ethylamine	36.9	19.0		
1-Aminopropane	44.58	27.4	11.54	
2-Aminopropane	26.45	42.96		
1-Aminobutane	42.33	36.75	20.47	14.16
2-Aminobutane	23.97	48.79	33.42	10.81
Diethylamine	44.43	15.72		
Di-*n*-Propylamine	52.34	23.94	11.98	
Trimethylamine	47.56			
Triethylamine	46.92	12.60		
Tri-*n*-propylamine	56.76	21.24	12.01	
Cyclohexylamine	50.4	36.7	25.7	25.1
Acetic acid	176.8	21.1		
Propionic acid	179.9	27.8	9.0	
Butyric acid	178.9	36.3	18.5	13.4
Acetate ion	181.0	24.0		
Propionate ion	184.4	31.3	10.8	
Butyrate ion	183.4	40.2	19.9	13.9

The alcohol and acid shifts were corrected to TMS from CS_2 by subtracting the CS_2-relative shifts from 192.8.

TABLE 2.IX

^{13}C Chemical Shifts (in ppm) of Halomethanes (55)

Fluoromethane	Chloromethane	Bromomethane	Iodomethane
65.7	23.9	9.0	−21.7
Difluoromethane	Dichloromethane	Dibromomethane	Diiodomethane
104.4	52.9	20.4	−55.1
Trifluoromethane	Chloroform	Bromoform	Iodoform
113.5	76.5	11.1	−140.9
Carbon tetrafluoride	Carbon tetrachloride	Carbon tetrabromide	Carbon tetraiodide
117.5	95.4	−29.7	−293.5

interesting observation is that effects of multiple chlorine substitution appear to be additive (~25 ppm per chlorine), while increased bromine and iodine substitution causes accelerated shielding. Representative alkyl halide ^{13}C chemical shifts are given in Table 2.IX.

Heteroatoms in saturated rings appear to influence ^{13}C shieldings not only through a simple substituent effect, but also by introducing conformational factors (65,66). Representative saturated heterocycle ^{13}C data are given in Table 2.X (68,194,207).

^{13}C chemical shifts of olefinic carbons are appreciably affected by monosubstitution (113,165,184). Electronegative groups, as might be expected, deshield C-1, the directly bonded carbon atom, except in the cases of I and Br where shielding via the heavy-halogen effect is observed. Substitution by oxygen (e.g., methoxyl) causes the greatest deshielding of C-1 but has the added interesting effect of drastically shielding the β-carbon atoms, C-2. For example, the olefinic ^{13}C chemical shifts in methylvinyl ether are 153.2 ppm for C-1 and 84.1 ppm for C-2, compared with the ^{13}C shift of ethylene (123.5 ppm).

Geometric isomers of disubstituted ethylenes are difficult to distinguish by ^{13}C-NMR. Some substituents cause shielding of the carbons in the *cis* isomer relative to the *trans*, while others have the opposite effect (183). Differentiation of alkene groups from aromatic ones is best done by proton NMR. However, vinylic ^{13}C resonances can be assigned in the presence of aromatic carbons by single frequency proton decoupling experiments since the ^{1}H resonance frequencies to be irradiated can be precisely determined (132,143,215).

^{13}C shieldings for substituted acetylenes are available in the literature (45,247,281).

As mentioned earlier, the relationship between aromatic ^{13}C shieldings and the charge distributions in substituted aromatic systems has been a topic of extensive investigations. Theoretical approaches have limited applicability, for it is neither

TABLE 2.X

^{13}C Chemical Shifts (in ppm) in Saturated Heterocycles (61,62)

Three-membered ring	Four-membered ring	Five-membered ring	Six-membered ring
N–H: 18.2	N–H: 45.3, 19.3	N–H: 47.1, 25.7	N–H: 47.5, 27.2, 25.5
O: 40.8	O: 72.8, 23.1	O: 67.9, 25.8	O: 68.0, 26.6, 23.6
S: 18.7	S: 26.0, 28.0	S: 31.7, 31.2	S: 29.3, 28.2, 26.9

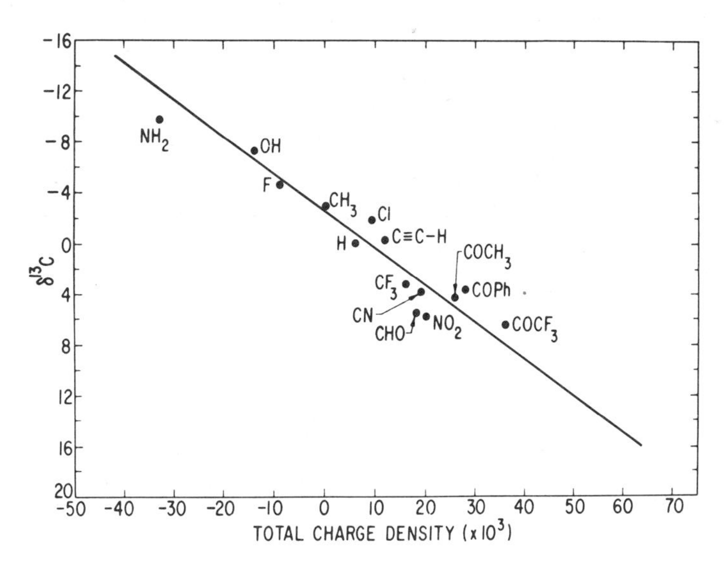

Fig. 2.14. Plot of total charge density (CNDO/2) versus chemical shift (referenced to benzene) for the *para* position of substituted benzenes (155).

anticipated nor observed that a direct correlation between aromatic ^{13}C chemical shifts and calculated local electron densities exists (72,166,193). Other factors such as charge polarization, variation in bond order, and average excitation energy also influence the ^{13}C shieldings. When some of these other factors show little variation over a series of related compounds, or change in a compensatory manner, then a correlation with electron density may become apparent. An example of such a relationship is the special case of benzene carbons *para* to a substituent. Their chemical shifts have been shown to correlate with CNDO/2 charge densities (Fig. 2.14) (192).

Substituent constants like those of Hammett have also been used in attempts to correlate aromatic ^{13}C shieldings and the electronic nature of aromatic systems (192,246). More recently, dual-substituent parameter (DSP) treatments have been employed (95). The most successful relationships have been derived for *para*-carbon atoms in monosubstituted benzenes (32,33,256). Poorer correlations with *meta*- and *ortho*-^{13}C shifts are explained as resulting from combinations of steric, inductive, and field effects (2,49,119).

An interesting aspect of aromatic ^{13}C-NMR is the existence of solvent effects on the chemical shifts of ring carbons in substituted molecules. These effects derive from the varying degrees of interaction that occur between specific substituents and the various solvents utilized to determine spectra. It is reasonable to expect that the manner in which a substituent influences aromatic ^{13}C shieldings should show some dependence on its own external associations. In most cases where solvent effects are observed, the *para* carbons appear to be most sensitive to the perturbations caused by varying interactions.

Solvent effects have been used to derive solvent-dependent substituent parameters

TABLE 2.XI

Substituent Effects on ^{13}C Chemical Shifts in Monosubstituted Benzenes (155)

Substituent	C-1	*ortho*	*meta*	*para*
Br[a]	−5.5	+3.4	+1.7	−1.6
CF_3[a]	+2.6	−3.3	−0.3	+3.2
CH_3[a]	+8.9	+0.7	−0.1	−2.9
CN[a]	−15.4	+3.6	+0.6	+3.9
C≡C—H[a]	−6.1	+3.8	+0.4	−0.2
$COCF_3$[a]	−5.6	+1.8	+0.7	+6.7
$COCH_3$[a]	+9.1	+0.1	0.0	+4.2
COCl[a]	+4.6	+2.4	0.0	+6.2
CHO[a]	+8.6	+1.3	+0.6	+5.5
COOH[a]	+2.1	+1.5	0.0	+5.1
Cl[a]	+6.2	+0.4	+1.3	−1.9
F[a]	+34.8	−12.9	+1.4	−4.5
H[a]	0.0	—	—	—
NH_2[a]	+18.0	−13.3	+0.9	−9.8
NH_3^+[a,g]	+0.1	−5.8	+2.2	+2.2
NO_2[a]	+20.0	−4.8	+0.9	+5.8
OCH_3[a]	+31.4	−14.4	+1.0	−7.7
OH[a]	+26.9	−12.7	+1.4	−7.3
Ph[a]	+13.1	−1.1	+0.4	−1.2
SO_2NH_2[b]	+15.3	−2.9	+0.4	+3.3
O^-[c]	+39.6	−8.2	+1.9	−13.6
OPh[c]	+29.2	−9.4	+1.6	−5.1
NMe_2[c]	+22.6	−15.6	+1.0	−11.5
CH_2OH[c]	+12.3	−1.4	−1.4	−1.4
I[e]	−32.0	+10.2	+2.9	+1.0
$SiMe_3$[c]	+13.4	+4.4	−1.1	−1.1
$CH{=}CH_2$[c]	+9.5	−2.0	+0.2	−0.5
CO_2Me[d]	+1.3	−0.5	−0.5	+3.5
COCl[c]	+5.8	+2.6	+1.2	+7.4
CHO[c]	+9.0	+1.2	+1.2	+6.0
COEt[d]	+7.6	−1.5	−1.5	+2.4
Li[f]	−43.2	−12.7	+2.4	+3.1

[a] Parts per million relative to internal benzene standard; positive shifts deshielding. Solute concentration, 10% in CCl_4.

[b] Parts per million relative to benzene. Data obtained relative to internal TMS and converted using δ_C = 128.5 for benzene. Solute concentration ~10% in $CDCl_3$.

[c] Neat liquids examined unless noted otherwise. Data converted using δ_C = 128.7 for benzene.

[d] Original data obtained relative to CS_2. Conversion δ_C = 192.8 was used for CS_2.

[e] See footnote *e*. Conversion δ_C = 193.7 was used for CS_2.

[f] Solvent unspecified.

[g] CF_3CO_2H solvent.

that can be utilized to estimate the relative reactivity of aromatic systems with strongly solvated functional groups (192).

Protonation of substituents also affects ring carbon shieldings. With aniline for example, substantial chemical shift changes are observed in its spectrum as obtained in methanesulfonic acid (NH_2 protonated) when compared with its spectrum in carbon tetrachloride. The *ortho* and *para* carbons are deshielded considerably and the *meta* to a lesser extent (8.1, 11.4, and 1.0 ppm, respectively) (192). Conversely, the substituent-bound ring carbon (*ipso* carbon) is shielded by −17.6 ppm.

It is clear that aromatic ^{13}C solvent-induced chemical shift changes can be used to probe solvent–solute interactions in systems of this kind. They can sometimes also be used for the practical purpose of resolving fortuitously coincident carbon resonances (149).

Despite the susceptibility of aromatic ^{13}C chemical shifts to the influences just described, substituent effects are remarkably additive. Table 2.XI summarizes the ^{13}C shieldings for a number of substituted benzenes (relative to benzene carbons at 128.5 ppm) (57,164,192,213,242,246). Chemical shifts for polysubstituted aromatics are predictable on the basis of the additivity of the chemical shifts of the corresponding monosubstituted benzenes.

Heterocyclic aromatic systems have been studied extensively, many of those studies again focusing on the question of experimental chemical shift correlations with theoretical parameters. Extensive ^{13}C chemical shift data of representative heteroaromatics including fused rings and some which contain more than one heteroatom, are available in the literature (42, 63, 64, 73, 74, 75, 81, 87, 89, 91, 93, 99,100,114,209,210,224,274,287).

3. Functional Groups

a. Groups Containing Carbonyl Carbon Atoms

Table 2.XII provides a representative list of chemical shifts for the carbonyl carbon atom as it occurs in important functional groups of organic compounds. The data given serve as illustrative examples of the observations that follow in discussing specific functional groups and factors which affect the shielding of their C=O carbon atoms.

(1) Aldehydes and Ketones

The most deshielded of all carbonyl carbons occur in saturated ketones. Substitution of alkyl groups into the α position causes further deshielding in a roughly cumulative manner (120). This effect is also observed in cyclic ketones.

Ring size variations in cyclic ketones produce a varied effect, with cyclopentanones being deshielded relative to cyclohexanones and cyclobutanones (23,101,252,279). Shieldings in carbonyls adjacent to saturated rings are also affected by ring size (176). An adjacent (conjugated) aromatic ring shields the carbonyl carbon by 8–10 ppm (55). *Meta* and *para* substituents have little further effect, but *ortho* groups can deshield the carbonyl by 2–8 ppm (55).

TABLE 2.XII

Carbonyl Chemical Shifts (in ppm) in Various Functional Groups

Compound	δ (ppm)
Dialkyl ketones (91)	
$CH_3-C(=O)-CH_3$	205.2
$CH_3CH_2-C(=O)-CH_3$	207.7
$CH_3CH_2-C(=O)-CH_2CH_3$	210.1
$(CH_3)_3C-C(=O)-C(CH_3)_3$	216.5
Cyclic ketones (92,93)	
	208.2
	213.9
	208.8
	216.2
	215.3
Haloketones (96–98)	
$ClCH_2-C(=O)-CH_3$	200.7
$ClCH_2-C(=O)-CH_2Cl$	194.9

Compound	δ (ppm)
***α,β*-Unsaturated ketones (93,96,98,99)**	
$CH_2=CH-C(=O)-CH_3$	198.1
	196.5
	208.1
	197.1
	155.1
Aromatic ketones (95)	
$Ph-C(=O)-CH_3$	196.9
$Ph-C(=O)-Ph$	194.8
Quinones (100,101)	
	187.0
	180.2

TABLE 2.XII *(Continued)*

Carbonyl Chemical Shifts (in ppm) in Various Functional Groups

Compound	δ (ppm)
$Cl_3C-C(=O)-CH_3$	186.3
$F_3C-C(=O)-CH_3$	187.7
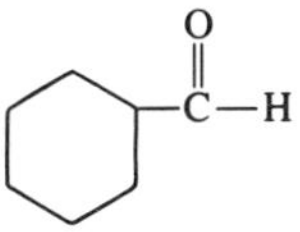 $C_6H_{11}-C(=O)-H$ (cyclohexyl)	202.7
$CCl_3-C(=O)-H$	175.9
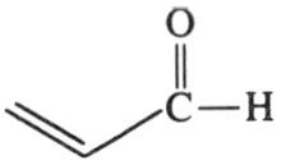 $CH_2=CH-C(=O)-H$	193.3
$Ph-C(=O)-H$	190.7
Carboxylic acids (96,102)	
$H-C(=O)-OH$	166.7
$CH_3-C(=O)-OH$	177.3
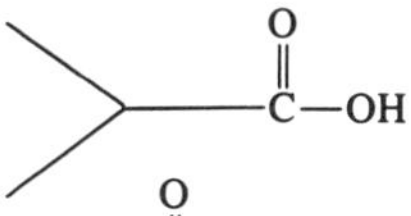 $(CH_3)_2CH-C(=O)-OH$	184.8
$CH_2Cl-C(=O)-OH$	174.7
$CCl_3-C(=O)-OH$	168.0
$CF_3-C(=O)-OH$	166.0
$CH_2=CH-C(=O)-OH$	173.2
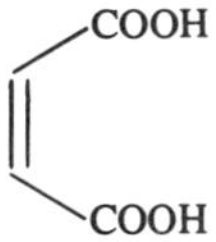 $HOOC-CH=CH-COOH$ (cis)	167.1
Aldehydes (96–98)	
$CH_3-C(=O)-H$	200.5
Carboxylic esters (93,96,102,103)	
$H-C(=O)-OCH_3$	160.7
$CH_3-C(=O)-OCH_2CH_3$	170.3
$CH_2=CH-C(=O)-OCH_3$	164.5
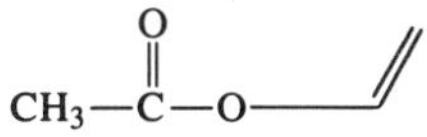 $CH_3-C(=O)-O-CH=CH_2$	167.7
$Ph-C(=O)-OCH_3$	167.0
$Ph-C(=O)-OCH_2CH_3$	164.9
$CH_3-C(=O)-C(=O)-OCH_2CH_3$	161.6
four-membered lactone (ring C=O, O)	171.2
five-membered lactone (ring C=O, O)	178.0
six-membered lactone (ring C=O, O)	175.2

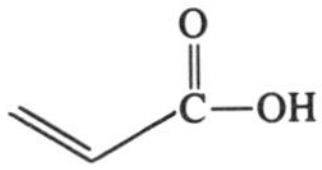

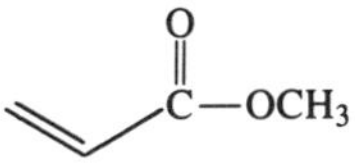

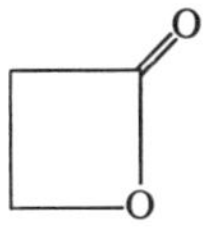

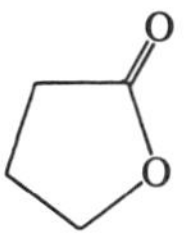

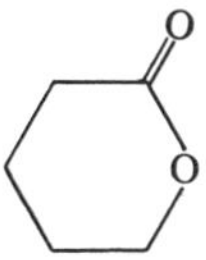

TABLE 2.XII *(Continued)*

Carbonyl Chemical Shifts (in ppm) in Various Functional Groups

Compound	δ (ppm)	Compound	δ (ppm)
		Acid anlydrides (104)	
		$CH_3—C(=O)—O—C(=O)—CH_3$	167.3
$HOOC—CH=CH—COOH$ (trans)	166.6		
		$CH_3CH_2—C(=O)—O—C(=O)—CH_2CH_3$	170.8
$Ph—C(=O)—OH$	173.5	Amides (107,108,109)	
		$H—C(=O)—NH_2$	165.5
cyclic $CH_2CH_2—C(=O)—O—C(=O)$ (five-membered ring)	173.1	$H—C(=O)—N(CH_3)H$ (CH₃ trans to O)	163.4
cyclic $CH_2CH_2CH_2—C(=O)—O—C(=O)$ (six-membered ring)	168.2	$H—C(=O)—N(H)CH_3$ (H trans to O)	166.7
$CF_3—C(=O)—O—C(=O)—CF_3$	151.5	$H—C(=O)—N(CH_3)_2$	162.7
		$CH_3—C(=O)—NH_2$	172.7
cyclic $CH=CH—C(=O)—O—C(=O)$ (five-membered ring)	165.9		
		$CH_3CH_2—C(=O)—NH_2$	177.2
cyclic $ClC=CCl—C(=O)—O—C(=O)$ (five-membered ring)	158.6		
		$CH_3—C(=O)—N(CH_3)H$	171.6
		$H—C(=O)—N(CH_3)Ph$	162.1
benzo-fused cyclic $C_6H_4—C(=O)—O—C(=O)$	163.6		
		$Ph—C(=O)—N(CH_3)_2$	170.8
$Ph—C(=O)—O—C(=O)—Ph$	162.8		

TABLE 2.XII *(Continued)*

Carbonyl Chemical Shifts (in ppm) in Various Functional Groups

Compound	δ (ppm)
Acid halides (96,105,106)	
$CH_3-C(=O)-Cl$	170.5
$CH_3-C(=O)-Br$	166.5
$CH_3-C(=O)-I$	159.8
$CH_3CH_2-C(=O)-Cl$	174.5
$Ph-C(=O)-Cl$	168.7
$ClCH_2-C(=O)-Cl$	169.7
$Cl_2CH-C(=O)-Cl$	166.2
$Cl_3C-C(=O)-Cl$	163.5
$CH_3-C(=O)-NHPh$	168.3
$CH_3-C(=O)-NH-C_6H_4-NO_2$	169.2
Imides (110)	
$CH_3-C(=O)-NH-C(=O)-CH_3$	171.8
Succinimide (NH)	179.1
N-Phenylsuccinimide (N Ph)	177.9
Maleimide (NH)	172.5
N-Phenylmaleimide (N Ph)	170.0
Phthalimide (NH)	169.7
N-Phenylphthalimide (N Ph)	167.7

Ketone carbonyl resonances are appreciably shielded by both α-halogen substitution (162,175,191,251) and α, β-unsaturation (23,50,101,175,191,251). The carbonyl carbon of acetone, for example, is shielded by approximately 5 ppm per α-chloro substituent. Conjugation of one double bond with a keto group shields the carbonyl carbon by about 10 ppm (cyclopentanone and 2,3-cyclopentenone are exceptions). Of special note is the highly shielded carbonyl of cyclopropenone. There exists a substantial contribution to its structure by the dipolar canonical form which imparts considerable single-bond character to the carbonyl bond (50).

Carbonyl ^{13}C chemical shifts, especially those of ketones, are sensitive to solvents, particularly solvents capable of hydrogen bonding to the carbonyl oxygen. This sensitivity makes it difficult to compare shieldings from different studies. It is not uncommon for differences of more than 1 ppm to be reported for the same compound under similar conditions.

Quinone carbonyl resonances occur in the range 170–190 ppm with the *ortho* isomer usually being more shielded than the *para* (8,19,20,117,179). The chemical shifts of aldehydic carbons generally parallel those of corresponding ketones but are shielded by 5–10 ppm (175,191,251). Aldehyde carbonyl resonances are usually more intense than those of ketones at the same concentration, due to shorter relaxation times which result from the directly bound proton, and in small-sized molecules, due to higher NOEs.

(2) Carboxylic Acids and Their Derivatives

Carboxylic acid carbonyl resonances occur between 165 and 185 ppm (44,251). Chemical shift trends caused by α-substitution parallel those of ketones; α-alkyl groups deshield cumulatively, α-chlorines shield cumulatively, and α, β-unsaturation also results in carbonyl shielding. Interestingly, carboxylate ions are deshielded by up to 5 ppm (107).

Carbonyl shifts in esters are affected by α-substituents and by the nature of the group in the alcohol residue. The former effect is usually more pronounced than the latter (23,41,44,101,202,251). The overall range of shifts is 160–180 ppm.

In both acid anhydrides and acid halides, carbonyl carbons are shielded relative to that of the corresponding carboxyl group. The shielding in anhydrides is ~ -10 ppm (134,190), while in acid chlorides a -4 to -8 ppm upfield shift is observed. Acid bromide and iodide carbonyls are further shifted by -5 and -8 ppm, respectively (39,94,97,251).

(3) Amides and Imides

Amide carbonyl resonances occur in the range 160–180 ppm (28,34,157,234). *N*-monoalkylation results in a conformational effect on the carbonyl chemical shift as shown in Table 2.XII for *N*-methylformamide (157). An extensive compilation of *N,N*-dialkylamide shieldings, with *N*-alkyl groups ranging from 1 to 12 carbons, has been published (88). Benzamide carbonyl shifts exist at 169 to 170 ppm and are independent of *para* substitution (128,147).

Imide carbonyl shieldings fall in the same range as those of amides. Deshielding results from incorporation of the functional group into a five-membered ring. Both aliphatic and aromatic *N*-substituents cause carbonyl shielding as does introduction of unsaturation (43,190).

b. OTHER FUNCTIONAL GROUPS

Table 2.XIII gives a representative list of chemical shifts for carbons contained in noncarbonyl functional groups. Specific shifts or shift ranges are presented for the typical carbonyl derivatives: ald- and ketoximes (189,190); imines, phenylhydrazones, and semicarbazones (189); nitriles (101,124,169,180,191); diazo compounds and ketenes (3,78,199); and carbodiimides (6). Data on selected thio- and selenocarbonyl carbons are also shown (14, 40, 76, 79, 98, 124, 131, 169, 180, 197, 205, 206, 212,240,248,269).

It is of interest to note that in all applicable cases, α-alkyl substitution again induces deshielding of the derivatized carbon, albeit to different degrees in different functional groups. A chemical-shift dependence on geometric isomerism is also observed where such structural variations exist (35,82,189). Finally, terminal carbons in diazo compounds and ketenes are unusually shielded relative to normal alkenes (3,78,199).

TABLE 2.XIII

Chemical Shifts (in ppm) of Functional Carbon Atoms in NonCarbonyl Functional Groups

Oxime carbons (107a,111)			
HO–N=C(CH_3)(CH_3)	154.3	Cyclohexanone oxime (N–OH on ring C=N)	159.4
(C_6H_5)(H)C=N–OH (OH on the H side)	149.6	HO–N=C(C_6H_5)(H) (HO on the C_6H_5 side)	146.4
HO–N=C(CH_3)(CH_2CH_3) (HO on the CH_3 side)	159.2	(CH_3)(CH_2CH_3)C=N–OH (OH on the CH_2CH_3 side)	158.7
Imines (111b)	168 – 175 (aliphatic) 157 – 163 (aromatic)		
Semicarbazones (111b)	158 – 160 (aliphatic)		
Phenylhydrazones (111b)	145 – 149 (aliphatic) 145 – 146 (aromatic)		

TABLE 2.XIII (Cont.)
Chemical Shifts (in ppm) of Functional Carbon Atoms in NonCarbonyl Functional Groups

Nitrile carbons (93a,98a,112a,113a)		PhCN	118.7
CH_3CN	117.7	$CH_2{=}CH{-}CN$	117.5
$(CH_3)_3C\text{-}CN$	125.1		
Cyclohexyl–CN	121.5	$N(CH_2CN)_3$	115.1
Diazo compound and ketene carbon atoms (114-116)			
CH_2N_2	23.3	$CH_2{=}C{=}O$	194.6 (2.7 α-carbon)
Ph_2CN_2	62.3	$Ph_2C{=}C{=}O$	201.4 (48.6 α-carbon)
Carbodiimide carbon atoms (117)			
$(CH_3)_2CHN{=}C{=}NCH(CH_3)_2$	140.2		
Thio- and selenocarbonyl carbon atoms (112,113,118–121)			
$CH_3{-}C({=}S){-}NH_2$	207.2	$Ph{-}C({=}Se){-}NHCH_3$	204.6
$CH_3{-}C({=}S){-}N(CH_3)_2$	199.6	$CH_3{-}C({=}Se){-}N(CH_3)_2$	202.0
$CH_3{-}C({=}S){-}NHPh$	200.4		
$Ph{-}C({=}S){-}NHCH_3$	199.7		

B. ^{13}C SPIN–SPIN COUPLING

Using the modern instrumentation and appropriate gated pulse sequences described in earlier sections, directly bound and long range C–H couplings can now be routinely measured for most organic systems in reasonable experimental times. Direct C–C couplings are more difficult to determine, but a substantial body of data has been compiled. The following discussion covers representative examples of C–H and C–C coupling constants, factors that affect them, and trends in their magnitudes induced by those factors.

1. Aliphatic Hydrocarbons and Their Substituted Derivatives

Table 2.XIV gives a representative listing of $^1J_{CH}$ values for a series of substituted and unsubstituted alkanes. Three parameters can be seen to contribute to $^1J_{CH}$ in these systems: carbon hybridization which sets the average value at ~125 Hz in simple molecules; angular distortion which increases the coupling constant (233,282); substitution by electronegative groups which also causes increases in $^1J_{CH}$ (126,172).

Long-range C–H coupling constants ($^2J_{CH}$ and $^3J_{CH}$) found in alkanes and repre-

TABLE 2.XIV
$^1J_{CH}$ (in Hertz) in Alkanes and Substituted Alkanes (119,121)

Compound	$^1J_{CH}$	Compound	$^1J_{CH}$
H-CH_3	125.0	H-$CH_2C{\equiv}CH$	132.0
H-CH_2CH_3	124.9	H-CH_2NH_2	133.0
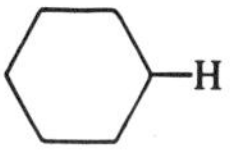 —H	123.0	H-CH_2CN	136.1
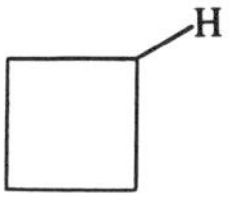 H	134.0	H-CH_2OCH_3	140.0
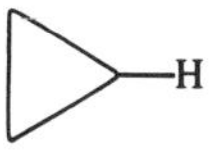 —H	161.0	H-CH_2OH	141.0
H-CH_2COPh	125.7	H-CH_2OPH	143.0
H-CH_2CHO	127.0	H-CH_2NO_2	146.7
H-CH_2COOH	130.0	H-CH_2F	147.9

sentative derivatives are shown in Table 2.XV (13,77,172,273,282,286). A Karplus-type relationship between $^3J_{CH}$ and dihedral angle exists (108).

Directly bonded and long-range C–C coupling constants typical of saturated systems are shown in Table 2.XVI (12,77,125,195,249,272,282).

One-bond and long-range C–H coupling constants found in a wide variety of olefinic compounds are provided in Table 2.XVII (11,36,46,103,126,171,172,183). Note that the same factors which affect $^1J_{CH}$ in alkanes do so in alkenes as well. Thus, simple sp^2 carbons have $^1J_{CH}$ values ~30 Hz greater than sp^3 carbons and bond-angle distortion and electronegative substituents cause $^1J_{CH}$ to increase. The large value for cyclopropene reflects the substantial *sp* character expected for its vinylic carbons (102).

Values of $^2J_{CH}$ are small and vary irregularly (178). $^3J_{CH}$ values have a geometric dependence and are generally larger for *trans*- than for *cis*-coupled carbon atoms (108). As in saturated systems, a Karplus-type relationship for $^3J_{CH}$ in olefins has been derived (236). The correlation with coupling constant values has been surveyed (11,36,46,133,139,270).

$^1J_{CC}$ in ethylene is 67.6 Hz (272), indicating the expected dependence on *s* character in the two bonding carbon orbitals. Substitution by electronegative groups increases $^1J_{CC}$. Alkyl substitution has the same effect as illustrated by $^1J_{C_3C_4}$ in 4-*n*-propyl-3-heptene with a value of 73.1 Hz. The limited olefinic character of the C_2–C_3 bond in butadiene is evidenced by $^1J_{C_2C_3}$ being 53.7 Hz (16). Two-bond C–C

TABLE 2.XV
Long-Range $^{13}C-^{1}H$ Coupling Constants (in Hertz) in Alkanes and Substituted Alkanes (123,125,126,127)

Compound	Range	$^{n}J_{CH}$
CH_3CH_3	^2J	−4.5
$CH_3CH_2\mathbf{CH_3}$	^2J	−4.3
$CH_3\mathbf{CH_2}CH_3$	^2J	1.9
$ClCH_2CH_2Cl$	^2J	−3.4
$CH_3\mathbf{CH}CH_3$ (with OH on CH)	^2J	0.7
$CH_3CH\mathbf{CH_3}$ (with X on CH)	^2J	−4.0 to −4.7
CH_3CHCH_3 (with X on CH)	^3J	4.6 to 5.8

TABLE 2.XVI
J_{CC} Values (in Hertz) in Alkanes (123,126a,129,130)

Compound	J Type	Value
H_3C-CH_3	^1J	34.6
H_3C-CH_2OH	^1J	37.7
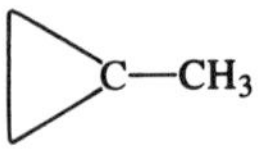 cyclopropyl C—CH_3	^1J	44.0
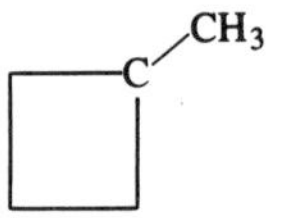 cyclobutyl C—CH_3	^1J	36.1
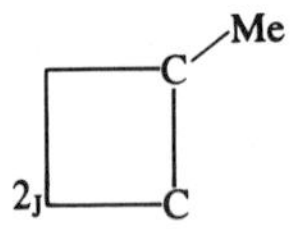 cyclobutyl C—Me; C; ^2J	^1J ^2J	29.1 8.1

couplings in alkenes are quite low and of limited usefulness. Interestingly, $^3J_{CC}$ has been measured in butadiene and has a value of 9.05 Hz (16).

2. Aromatic Hydrocarbons and Their Substituted Derivatives

One-bond and long-range C–H coupling constants have been precisely determined for benzene and many of its substituted derivatives. In benzene itself the values

TABLE 2.XVII

$^{13}C-{}^{1}H$ Coupling Constants (in Hertz) for Alkenes and Substituted Alkenes (121,127,128,129,130)

Compound	$^{1}J_{CH}$	$^{2}J_{CH}$	Compound	$^{1}J_{CH}$	$^{2}J_{CH}$
$H_2C=CH_2$	156.4	−2.4	$H\alpha(Cl)C=CH\beta H\gamma$	198	β 7.5 α −7.9
$H_2C=CHF$	159.1		$H\alpha(Cl)C=CH\beta H\gamma$	162	α 6.9
$HIC=CHI$ (H, H cis; I, I cis)	187.9	11.0	$HClC=CClH$ (H, H trans)		0.8
$HIC=CIH$ (H, H trans)	194.2	−1.4	$HClC=CHCl$ (H, H cis)		16.0
$H_2C=CHF$	200.2		cyclopropene	228.2	
			cyclobutene	170.0	

are $^{1}J_{CH}$ = 157.65 Hz, $^{2}J_{CH}$ = 1.16 Hz, $^{3}J_{CH}$ = 7.63 Hz, and $^{4}J_{CH}$ = −1.22 Hz (104). Values selected from a comprehensive study on a number of derivatives are given in Table 2.XVIII (71). In these systems, $^{1}J_{CH}$ and $^{3}J_{CH}$ values are comparable in magnitude with couplings in olefinic counterparts and $^{3}J_{CH}$ always exceeds $^{2}J_{CH}$ for a given compound. $^{1}J_{CH}$ has an apparent dependence on ring size, with the general trend being that coupling decreases with larger rings (71).

Incorporation of a heteroatom such as nitrogen into a ring tends to increase J_{CH} values. For example $^{1}J_{C2H2}$ in pyridine is 177.4 Hz, $^{2}J_{C2H3}$ is 3.1 Hz, but $^{3}J_{C2H4}$ is 6.8 Hz, a little smaller than the corresponding value in benzene (105). Extensive C–H coupling data for some five-membered heterocycles have been published (275).

Spin–spin coupling between bound carbons within several classes of aromatic rings have been determined from both natural abundance spectra and spectra obtained from singly and doubly enriched materials. Extensive compilations of coupling constants exist in the literature (106,272). Generally, $^{1}J_{CC}$ values are in the range of 57–60 Hz, values which are increased by electronegative substituents (21,109). Long-range ring carbon coupling constants are less than 10 Hz with $^{3}J_{CC}$ values being greater than $^{2}J_{CC}$ (106).

TABLE 2.XVIII

$^{13}C - ^{1}H$ Coupling Constants (in Hertz) in Monosubstituted Benzenes (137)

X	$^{2}J_{12}$	$^{3}J_{13}$	$^{4}J_{14}$	$^{1}J_{22}$	$^{3}J_{26}$	$^{3}J_{35}$
F	−4.89	10.95	−1.73	162.55	4.11	9.02
NO_2	−3.57	9.67	−1.75	168.12	4.45	8.18
OCH_3	−2.79	9.22	−1.51	158.52	4.80	8.73
CHO	0.29	7.19	−1.26	160.95	6.25	7.58
CH_3	0.54	7.61	−1.40	155.89	6.59	7.91
$Si(CH_3)_3$	4.19	6.34	−1.10	156.14	8.63	7.26

One-bond coupling constants between aromatic and directly bound substituent carbons are somewhat larger than in aliphatic derivatives. Long-range values are less than 4 Hz; $^{3}J_{CC}$ is usually greater than $^{2}J_{CC}$ (177,272).

3. Functional Groups Containing Carbonyl Carbon Atoms

Table 2.XIX provides a representative list of one- and two-bond coupling constants for a series of compounds containing a formyl carbon atom, i.e., a carbonyl carbon with a directly attached hydrogen. $^{1}J_{CH}$ values display a strong dependence on the

TABLE 2.XIX

Direct and Two-Bond $^{13}C - ^{1}H$ Coupling Constants (in Hertz) involving Formyl Hydrogen Atoms (145,147)

Compound	$^{1}J_{CH}$	Compound	$^{1}J_{CH}$	$^{2}J_{C\alpha H}$
Ph—C(=O)—H	173.7	CH_3—C(=O)—H	172.4	26.6
H—C(=O)—OH	222.0	$ClCH_2$—C(=O)—H		32.5
H—C(=O)—$O^{-}_{(aq)}$	194.8	Cl_2CH—C(=O)—H	198.0	35.8
H—C(=O)—$N(CH_3)_2$	191.2	Cl_3C—C(=O)—H	207.0	46.3
H—C(=O)—OCH_3	226.2	$CH_3CH_2CH_2$—C(=O)—H	170.3	24.8
H—C(=O)—F	267.2	CH_3CH_2CHCl—C(=O)—H	183.9	26.5
H—C(=O)—H	172.4	CH_3CH_2CHBr—C(=O)—H	183.3	27.6
		CH_3CH_2CHI—C(=O)—H	183.2	28.9

TABLE 2.XX
^{13}C Coupling Constants (in Hertz) for $C_AH_3C_BOX$ Compounds

Substituent X	$^1J_{C_AC_B}$	$^2J_{C_BH_A}$
H	39.4	6.6
CH_3	40.1	5.9
I	46.5	7.5
$N(CH_3)_2$	52.2	5.95
Br	54.1	7.5
Cl	56.1	7.45
OH	56.7	6.70
OC_2H_5	58.8	6.79

nature of the other group bound to the carbonyl carbon (126,172). They have been correlated with C–H infrared stretching frequencies as well as with σ_I and σ^* substituent constants (218). It is of interest that the $^2J_{C\alpha H}$ values shown for aldehydes, coupling between the formyl hydrogen and carbon α to the carbonyl, are quite substantial. In carbonyls of a wide variety, $^2J_{CH}$, coupling between the carbonyl carbon and hydrogens on adjacent carbons is on the order of 6–7 Hz.

$^1J_{CC}$ varies widely with the specific type of carbonyl compound (97). Examples of these couplings are given in Table 2.XX.

V. TECHNIQUES

A. METHODS USED IN ASSIGNING SPECTRA

1. Decoupling Techniques and Two-Dimensional FT-NMR

The use of a variety of decoupling methods to obtain various forms of spin–spin splitting information to aid in assigning ^{13}C peaks has already been discussed. A recently introduced technique that utilizes variations of gated decoupling, shows great promise in further facilitating the analyses of complex spectra through the acquisition of coupling information. It is called two-dimensional FT-NMR (2DFT).

Forms of 2DFT afford spectra that are presented as a function of two independent frequency parameters (e.g., ^{13}C observation frequency versus ^1H decoupler frequency; ^{13}C chemical shifts versus ^1H chemical shifts, etc.) (24,86,259). The spectra result from Fourier transformation of data collected as a function of two independent time variables that are contained in an experimental time frame comprising three distinct periods (24,86,259). The specific two-dimensional information content obtained is determined by the experimental conditions imposed on the spins during the three periods that define the experiment.

In a particularly useful application of 2DFT called δ-sorting, $^{13}C-{^1H}$ multiplets are sorted by ^{13}C chemical shifts on the second frequency axis (188). If the $^{13}C\{^1H\}$

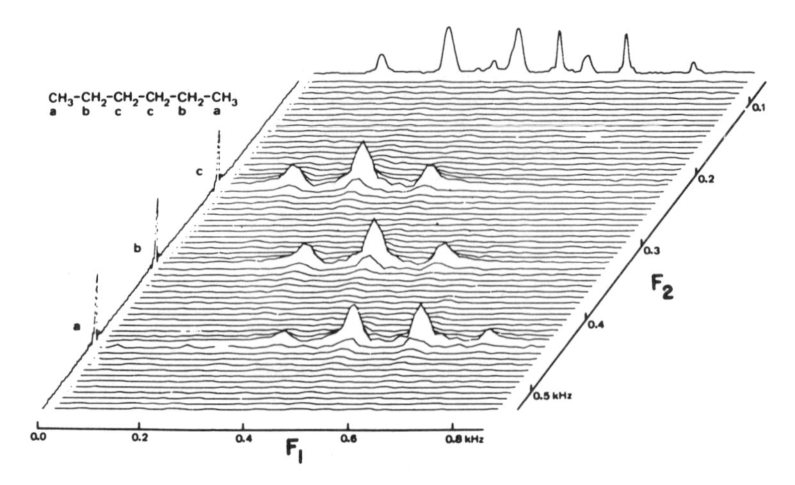

Fig. 2.15. ^{13}C-NMR δ-sorting experiment, performed before 1975 (188).

spectrum is completely resolved (which is usually the case), the δ-sorted spectrum affords fully coupled multiplets free of overlap in the "second dimension". Furthermore, the conventional (overlapped) coupled spectrum can be viewed as a projection of the individual multiplets on one frequency axis, while the ^{13}C {^{1}H} spectrum appears as a projection on the other. This application of 2DFT is illustrated in Fig. 2.15. It is evident that peak assignments are greatly facilitated by the information content in such a spectrum.

Although 2DFT ^{13}C-NMR is not yet routine, it has already demonstrated an enormous potential for assigning the spectra of complex compounds. The wide variety of experiments performed and spectral information obtained also attest to its versatility (24,86,187,259). Problems that include lower effective sensitivity, difficult phase correction, and the general need for updated software are being worked on in several laboratories.

2. Deuterium Labeling

Resonances of deuterated carbons are readily assignable. This fact derives from three important effects of deuterium substitution on ^{13}C-NMR signals: (1) deuterated carbons exhibit no NOE; (2) resonance lines are split by residual ^{13}C–^{2}H coupling; (3) resonance lines can be broadened by the effect of the deuterium quadrupole on ^{13}C spin–spin relaxation times. These factors combine to substantially reduce the signal intensities of ^{2}H-labeled carbons relative to protonated ones. Thus, comparison of spectra of specifically labeled compounds with the corresponding spectra of the nonlabeled molecules renders the assignment of resonances straightforward.

3. Lanthanide-Shift Reagents

Because ^{13}C resonances are usually free of overlap, lanthanide chemical-shift reagents (LSRs) have not been as widely utilized in ^{13}C studies as in ^{1}H-NMR where the vastly increased ^{1}H chemical shift dispersion induced by LSRs can be of enormous help in assigning peaks that overlap in their absence (241). Nevertheless, LSRs do increase the range of ^{13}C chemical shifts as well, and this has been taken advantage of to aid peak assignments in complex carbon spectra (5,31,60,92,244,255,280). Assignments are usually based on comparisons of experimental lanthanide-induced shifts (LIS) with calculated pseudocontact shifts expected from coordination of a specific site in the molecule to an LSR. Geometric factors are of great importance, and conformational information is often a by-product of these studies.

B. NEW PULSE SEQUENCES AND EXCITATION METHODS

A number of specialized pulse sequences and alternative excitation modes presently exist for FT-NMR spectroscopy. One technique, stochastic resonance, utilizes random modulation of a low-power rf signal to effect excitation rather than a discrete high-power rf pulse (70,130). Linked with a Hadamard transform prior to FT, this excitation mode produces the usual frequency-domain NMR spectrum (70,130,288). One advantage of stochastic resonance derives from the fact that broad ^{13}C spectral ranges may be measured without the need for high-power amplifiers and high-power probe technology. Nevertheless, no other advantages over single-pulse FT-NMR really accrue and stochastic excitation has not gained widespread use.

A modification of stochastic (random) modulation of the rf field produces an excitation scheme that has considerable advantages. A Fourier synthesis of the excitation frequency spectrum defining a particular application is computed. The function generated is used to modulate a sequence of short, low-level rf pulses. In this manner selected or tailored excitation can be effected (115). This is particularly useful since it allows the effective excitation window to be chosen to exclude one or more regions (solvent peaks, perhaps), or to apply different degrees of excitation to different regions (saturation of selected resonances).

Another technique developed to allow selective excitation of one region of an NMR spectrum is rapid-scan or correlation FT-NMR. This scheme does not use rf pulses at all. Rather, the response from a rapid spectral scan is correlated with the response obtained under identical conditions for a single sharp line (e.g., TMS). Fourier transformation then yields the normal absorption spectrum (badly distorted lines caused by the rapid scanning are corrected by the correlation scheme). Since resonance conditions are met sequentially, and the width of the scan can be specified, suppression of peaks just outside of the scan range by a factor of several hundred can be realized.

Appreciable sensitivity enhancement has been realized in coupled spectra through the combined use of gated decoupling and specialized pulse sequences. In one application known as *selective population transfer*, spectral peak assignments are simplified and $^{13}C-X$ coupling constants (sign and magnitude) can be precisely determined (121,122,245). It has been shown that cross-polarization methods originally developed for sensitivity gains in ^{13}C-NMR of solids can be applied to increase ^{13}C sensitivity in liquids as well. One method utilizes J (scalar) coupling to effect cross-polarization (22). Another technique called INEPT (insensitive nuclei enhanced by polarization transfer) circumvents the stringent requirements of J cross-polarization (186).

C. SPECIAL APPLICATIONS OF ^{13}C-NMR

1. Synthetic Polymers

^{13}C NMR has been used with great success to investigate structure and conformation in synthetic polymers in both the liquid and solid phases (22,185,186,225,229). The broad dispersion of ^{13}C chemical shifts and a low tendency for dipolar line broadening of ^{13}C signals are particularly advantageous for this application.

Through examination by ^{13}C-NMR, polymers have been characterized as to their tacticity (conformational sequences) and block lengths (when present) in copolymers. Head-to-head and tail-to-tail monomer additions can be observed and differentiated in vinylic polymers (214). Unusual chain-branching patterns at levels below one branch per 1000 main-chain carbons have been characterized (9), providing insight into the actual polymerization process.

Synthetic polymeric systems are amenable to quantitative analysis by ^{13}C-NMR as discussed in the next section. Accurate end-group analyses are relatively straightforward. Finally, polymer ^{13}C relaxation parameters can be obtained. These data give insight into the complex molecular motions of large molecules in solution with relatively rapid group segmental motions superimposed on slow reorientation of the chains themselves (285).

2. Biopolymers, Biosynthesis, and Metabolic Pathways

Studies of biopolymers by ^{13}C-NMR have increased rapidly and to the extent that excellent reviews of this research are now available (29,52,138,155,262). Observation of biopolymer ^{13}C resonances at natural isotopic abundance cannot yet be considered routine due to the limited concentrations generally available (ca. $\lesssim 10^{-3}$ M) and the often restricted availability of samples. However, instrumentation and techniques now in use allow practical application of ^{13}C-NMR to peptides (18,51,53,54,137,261), proteins (4,56,198,264), lipids (231,235), glycans (203), nucleic acids (25,135,136,216,230), and a number of the larger nonpolymeric biomolecules (155).

Many of the difficulties associated with the use of ^{13}C-NMR to study biopolymers can be alleviated by ^{13}C enrichment. Numerous approaches encompassing this technique have been taken, from specific site enrichment via synthetic means to blanket enrichment through biosynthesis by organisms grown in appropriate nutrient mediums. Successful ^{13}C labeling enhances sensitivity to the extent that work on dilute solutions becomes practical. However, problems still exist with broad resonance lines (an unavoidable relaxation phenomenon) and the overall resolution of large numbers of peaks. Despite these difficulties, extensive application of ^{13}C-NMR to investigation of biomolecules of all sizes and complexity is expected to continue at a brisk pace.

A particularly exciting corollary field to the study of individual biomolecules, is the use of ^{13}C-NMR in elucidating biosynthetic pathways. Many advantages over the classic technique of ^{14}C labeling, among them the absence of any need for molecular degradation, exist. Several excellent reviews of this imaginative research have appeared (144,181,182,243,257).

The applicability of ^{13}C-NMR to studies on metabolic pathways has also been demonstrated. Glycolysis in anaerobic *Escherichia coli* cells has been observed and followed in great detail with positive identification of all end products from C-1 ^{13}C enriched glucose (263). In another study the time course of porphobilinogen metabolism by a strain of live cells in a glucose-rich medium was determined (232).

3. ^{13}C-NMR of Solids

It is now possible to obtain high-resolution ^{13}C spectra from a wide variety of solid materials (229). This has been realized through the development of three techniques; dipolar decoupling, magic-angle spinning, and cross polarization. When used together, line broadening normally associated with motionally restricted samples is removed, and inefficient ^{13}C spin-lattice relaxation is circumvented.

Line broadening in ^{13}C NMR spectra of solids arises jointly from static dipolar interactions, almost exclusively with protons in organic compounds, and from anisotropy in ^{13}C-shielding tensors. In isotropic liquids, these effects are averaged to zero due to rapid reorientation of the molecules. In solids, $^{13}C-{}^{1}H$ static dipolar line broadening is removed by dipolar decoupling. Line broadening caused by chemical-shift anisotropy is eliminated (or greatly attenuated) by spinning the sample at an angle θ with respect to the static magnetic field, thus reducing the term which appears in equations dealing with dipolar and chemical shift anisotropy interactions, $(3\cos^2\theta - 1)$, to zero. This angle, 54.7°, is called the *magic angle*.

An additional difficulty exists in ^{13}C-NMR of solids even when high resolution is achieved by dipolar decoupling and magic angle spinning. Because ^{13}C T_1 values in solids can be quite long, it is necessary that a delay of several T_1 be inserted in the pulse sequence. Thus, experimental times required for acceptable sensitivity can be inconvenient if not prohibitive. To alleviate this problem, a technique which allows spectral accumulations at rates greater than those permitted by the T_1 values must be utilized. Cross-polarization (CP) is such a method. In CP, polarization of the ^{13}C nuclei is achieved using the greater equilibrium polarization of the protons. Further advantage arises from the fact that the proton spins recover more rapidly than the carbon spins, actually in a time closer to T_2 (ca. 100 μsec) rather than a longer T_1 (112,208).

Utilizing the three techniques just discussed results in high-resolution ^{13}C-NMR experiments of solids known as CPMAS (for *c*ross-*p*olarization *m*agic-*a*ngle *s*pinning) experiments. Examples of the success of CPMAS studies are spectra of frozen liquids (90), including the first high-resolution ^{13}C spectra of carbocations in the solid state (163), and detection of individual carbon resonances in solid proteins (201). A particularly excellent CPMAS solid-state spectrum is shown in Fig. 2.16 (196).

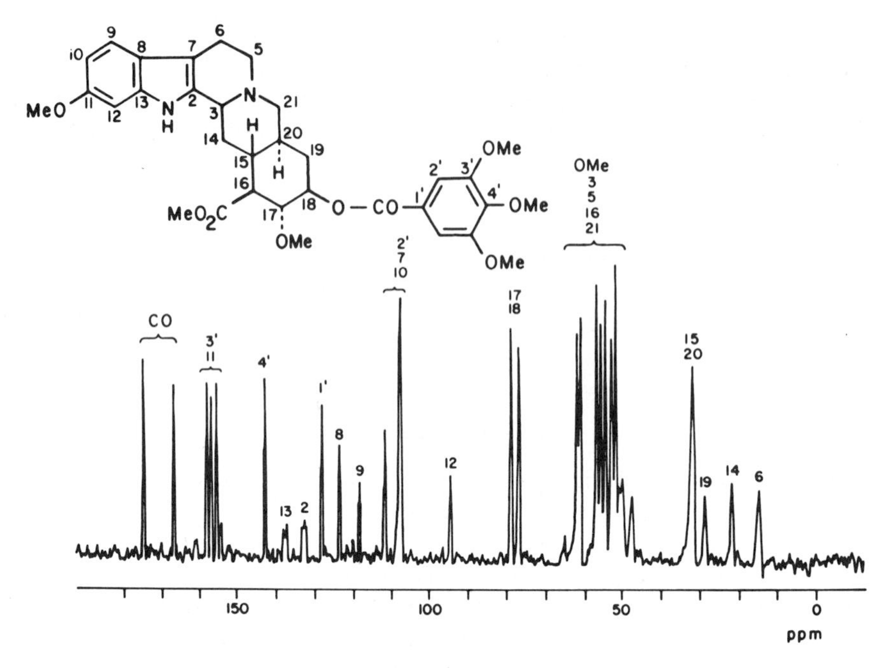

Fig. 2.16. 75.46-MHz carbon-13 solid-state CPMAS spectrum of reserpine obtained on a Bruker CSP-300. Side band removal was accomplished by a block-multiplication technique (300 scans plus 364 scans). (Spectrum courtesy of Dr. D. Müller, Bruker Analytische Messtechnik (GmbH.) (155).

D. QUANTITATIVE ANALYSIS BY ^{13}C NMR

Previous sections have discussed the effects of spin-relaxation and other factors on peak area measurements in ^{13}C-NMR. This section presents several types of quantitative ^{13}C-NMR applications, with emphasis on experimental techniques that can be used to optimize these studies.

1. Analysis of Simple Mixtures of Organic Compounds

Combined use of paramagnetic relaxation agents and gated-proton-decoupling facilitates routine analysis of mixtures of simple organic molecules (note that relaxation agents may not give desirable results) (150). Shoolery demonstrated the advantages of this approach graphically with spectra of the compound acenapthene (4 of 7 spectra are given in Figs. 2.17–20) (237).

Of course, if paramagnetic relaxation agents cannot be added to a sample, then alternative procedures must be used. In this case, it becomes necessary to characterize ^{13}C spin–lattice relaxation for the sample of interest (or for a closely related sample). In samples where ^{13}C relaxation is favorable (short T_1 values, full NOEs: larger organic molecules) FT-NMR pulse conditions are characterized by short pulse intervals and continuous wideband proton decoupling. In samples where ^{13}C relaxation is not favorable, pulse intervals as long as 300–500 sec may be required, along

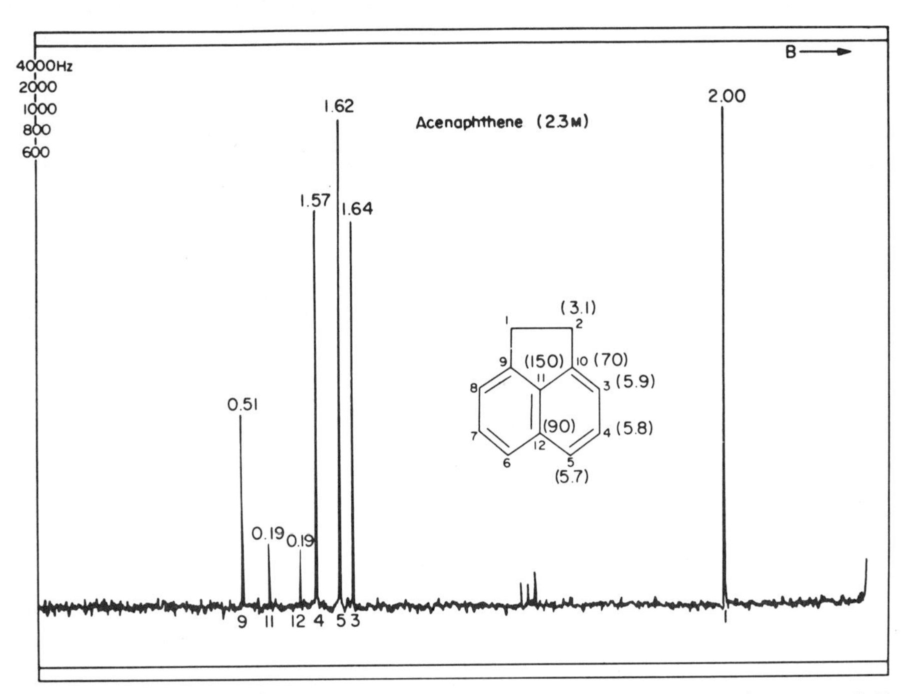

Fig. 2.17. Acenaphthene ^{13}C spectrum run under normal conditions showing assignments and T_1 relaxation times (237).

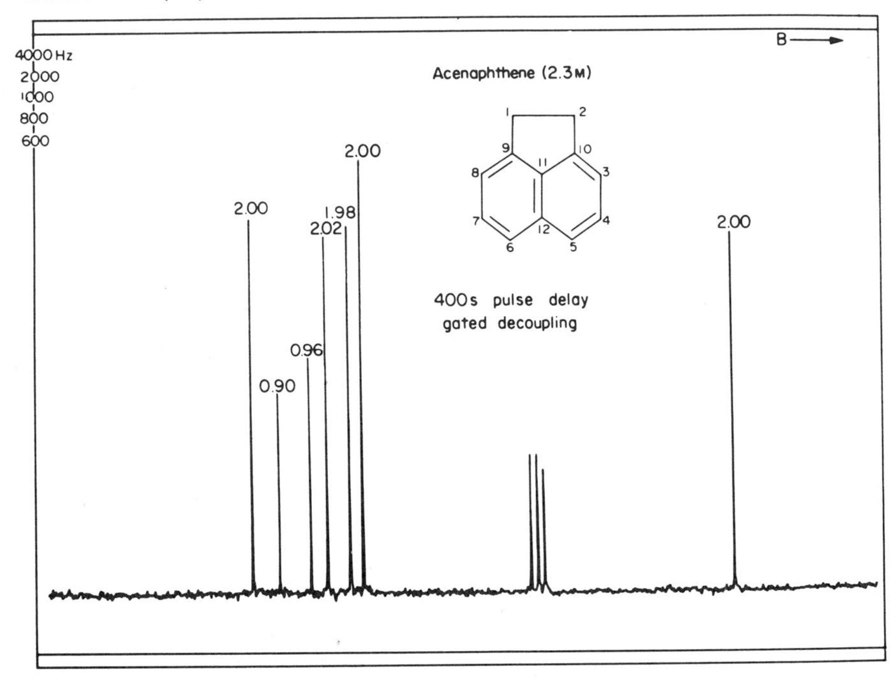

Fig. 2.18. Acenaphthene ^{13}C spectrum with 400 s pulse delay and decoupler gated "off" during delay period (237).

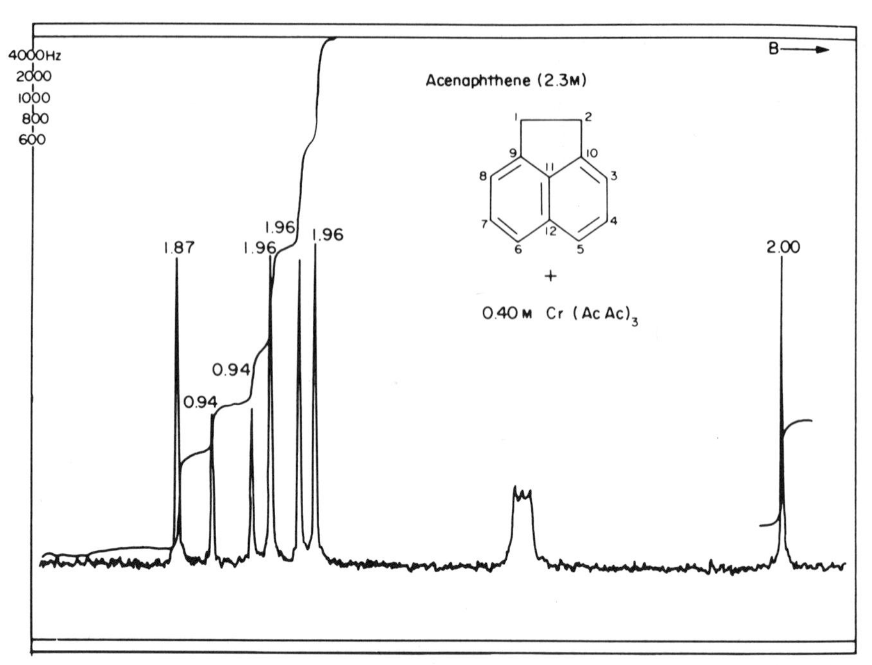

Fig. 2.19. Acenaphthene ^{13}C spectrum with 0.4 M $Cr(AcAc)_3$, run with 1-s pulse repetition rate, continuous noise decoupling, and 45° ^{13}C pulse flip angles (237).

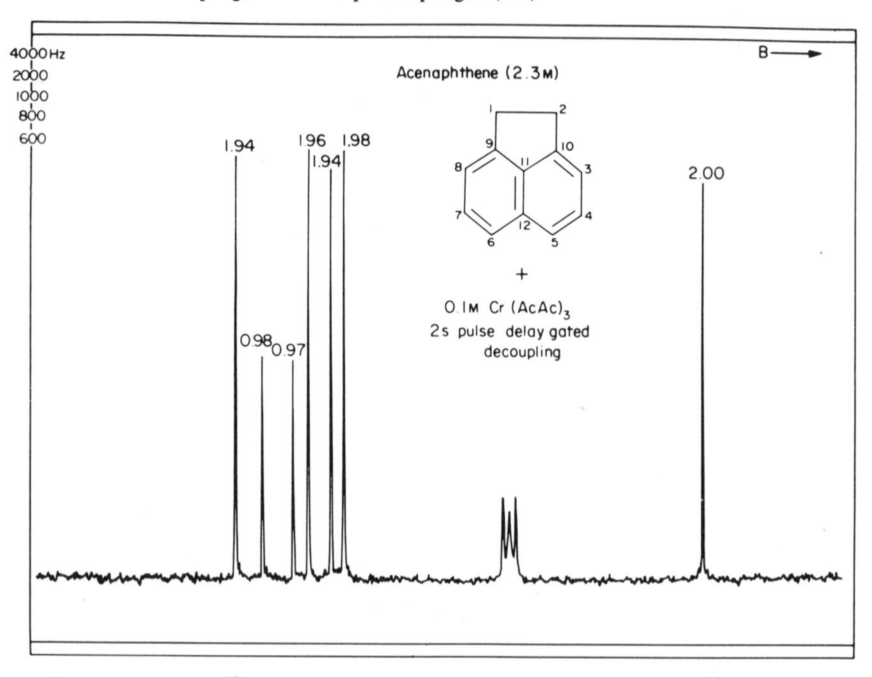

Fig. 2.20. Acenaphthene ^{13}C spectrum with 0.1 M $Cr(AcAc)_3$ run with 2 s pulse delay and decoupler gated "off" during delay period (237).

with gated decoupling to suppress variable NOEs. This can be very inefficient. Some improvement in efficiency can be effected without use of standard relaxation agents. If the sample can be oxygenated to 1 atm O_2 for ca. 30 sec., then T_1 values will not exceed 20–30 sec; and the gated decoupling experiment will become more efficient.

Analysis of mixtures of isomers can generally be effected in a straightforward manner. Usually, ^{13}C spin-relaxation is similar for closely related compounds, and thus rapid pulsing conditions may be used. It is also possible to utilize only resonances of protonated carbons, ignoring peak areas representing the slower and possibly nondipolar relaxing quaternary carbons. Examples have been reported (38,153,173,200,221,260).

2. Analysis of Complex Mixtures: Fuels

^{13}C-NMR analyses of fossil fuels have become of increasing interest. Techniques have been developed to obtain "high resolution" ^{13}C spectra from samples as intractable as raw coal (168,265) and coal-tar pitches (80). These studies utilize specialized techniques such as cross-polarization and magic-angle spinning to reduce spectral line widths and enhance sensitivity. The cross-polarization process, however, prevents straightforward quantitative analysis of these materials since $^{13}C-^{1}H$ distances in these heterogeneous solids can exceed practical requirements for effective cross-polarization. In this case, carbons that are isolated from proton spins will yield attenuated signals, in the limit making no contribution to the spectra!

Other fuel materials examined include oil shales (167,268) and crude oils (228,238).

3. Analysis of Lipids and Other Biomolecules

Several reports have indicated quantitative analysis of lipids, including unsaturated lipid ratios (220,226,227,237,239).

The quantitative analysis of the amino acid L-hydroxyproline in meat protein has been reported (129). An internal standard was chosen such that the chemical shift of one of the reference peaks was in close proximity to a resonance line of L-hydroxyproline to minimize effects from pulse excitation or detection which can be frequency separation-dependent. Both T_1 and NOE variations were checked for these two resonances, such that no errors could be attributed to differences in these values. Using standard conditions of 29 min per sample, standard deviation in the amino acid content of about 0.15% was observed. The authors concluded that ^{13}C NMR did in fact provide a much faster and more flexible analysis than standard amino acid analyses.

4. Quantitative Analysis of Polymers

Quantitative analyses of synthetic polymers includes evaluation of molecular weight and chain termination branching, copolymer composition, and polymerization characterization. Examples include analysis of low- (9,10,47) and high-density (214) polyethylenes, other vinyl polymers (214) and silicone rubber copolymers (284).

^{13}C-NMR can be used to quantify local stereochemical placements, with pentad and heptad structures sometimes available (214). Polymer end-group analyses are straightforward, at least to levels $\lesssim 1\%$; with modern instrumentation, end-group or chain-defect analyses can *sometimes* be performed to 0.01%.

It is, of course, necessary to account for ^{13}C spin-relaxation characteristics (T_1 and NOE) to perform meaningful quantitative analyses. Early reports indicated that stereochemical placements did not result in variation of ^{13}C T_1 values but recent studies have shown that large T_1 differentials may be observed for different placements.

Low-density polyethylenes (LDPE) have been examined to determine the various types of side chain present. In one respect LDPEs are quite favorable for quantitative studies, since main- and side-chain carbons all show full Overhauser enhancements. On the other hand, early ^{13}C analyses of these polymers did not anticipate large gradations in ^{13}C T_1 values. High-sensitivity ^{13}C spectra of high-density (linear) polyethylenes show that modern, high-dynamic-range FT-NMR spectrometers allow determination of chain branches and end groups to the level of 0.01% in these systems.

^{13}C-NMR quantitative analysis of solid polymer samples presents additional difficulties. These arise from both the basic relaxation-spin physics in the solids and from limitations imposed by solids NMR experiments. CPMAS experiments, for example, circumvent long ^{13}C T_1 values due to immobility of a carbon site, but these experiments do not respond equally to ^{13}C spins isolated from protons or undergoing internal motions.

VI. SUMMARY AND FUTURE PROSPECTS

The 1970s witnessed development of ^{13}C NMR spectroscopy as a powerful tool for analysis of complex organic structures. Sensitivity improvements in NMR instrumentation during this decade have extended practical natural abundance ^{13}C-NMR to levels that were unthinkable, even in the early 1970s. Quantitative analysis by ^{13}C-NMR can now be performed on diverse samples, including synthetic polymers, complex fossil fuel mixtures, whole meats and vegetables, and viable whole seeds.

Prospects for the future include powerful new applications in ^{13}C-NMR imaging, sometimes coupled with direct metabolic analysis, versatile solid-state ^{13}C-NMR analysis of organic solids, and ultra-high-resolution ^{13}C-NMR spectral analysis of complex organic molecules of mw 300–1000 or more.

Improvements anticipated for methods and instrumentation include computer software advances to make *practical* and *efficient* two-dimensional FT ^{13}C-NMR studies, to allow accurate quantitative analysis of small spectral lines in the presence of ultra-large peaks, and to allow automated multiple ^{13}C NMR spectral analyses on one sample (already possible!) or on a large set of samples (with automatic sample changing, shimming and lock, and heuristic analyses).

Sensitivities of future spectrometers will improve further in an evolutionary manner, from use of higher magnetic fields and from closer optimization of probe designs. In this way the limits for a 1990 spectrometer operating at 14–19 Tesla might be two to five times higher than currently possible at 11.7 Tesla, the highest magnetic field currently in use for ^{13}C-NMR. Of course, use of the highest magnetic fields presents major difficulties, particularly for proton decoupling; current technology greatly restricts sample size, limiting sensitivity advantage to the case of limited availability materials. It is possible that a new breakthrough in instrumental technology will produce more marked increase in ^{13}C-NMR sensitivity by 1990, but the nature of that breakthrough is elusive.

REFERENCES

1. Abraham, R.J., and P. Loftus, *Proton and Carbon-13 NMR Spectroscopy*, Heyden, London, 1978.
2. Adcock, W., B.D. Gupta, and W. Kitching, *J. Org. Chem.*, **41**, 1498 (1976).
3. Albright, T.A., and W.J. Freeman, *Org. Magn. Resonance*, **9**, 75 (1977).
4. Allerhand, A., R.F. Childers, and E. Oldfield, *Ann. N.Y. Acad. Sci.*, **222**, 764 (1973).
5. Ammon, H.L., P.H. Mazzocchi, and E.J. Colicelli, *Org. Magn. Resonance*, **11**, 1 (1978).
6. Anet, F.A.L., and I. Yavari, *Org. Magn. Resonance*, **3**, 327 (1976).
7. Armitage, I.M., H. Huber, D.W. Live, W. Pearson, and J.D. Roberts, *J. Magn. Resonance*, **15**, 142 (1974).
8. Arnone, A., G. Fronza, R. Mondelli, and J. St. Pyrek, *J. Magn. Resonance*, **28**, 69 (1977).
9. Axelson, D.E., G.C. Levy, and L. Mandelkern, *Macromolecules*, **12**, 41 (1979).
10. Axelson, D.E., L. Mandelkern, and G.C. Levy, *Macromolecules*, **10**, 557 (1977).
11. Bachmann, K., and W. vonPhilipsborn, *Org. Magn. Resonance*, **8**, 648 (1976).
12. Barfield, M., S.A. Conn, J.L. Marshall, and D.E. Miller, *J. Am. Chem. Soc.*, **98**, 6253 (1976).
13. Barfield, M., J.C. Marshall, E.D. Canada, and M.R. Willcott, III, *J. Am. Chem. Soc.*, **100**, 7075 (1978).
14. Bartels-Keith, J.R., M.T. Burgess, and J.M. Stevenson, *J. Org. Chem.*, **42**, 3725 (1977).
15. Batchelor, J.G., *J. Am. Chem. Soc.*, **97**, 3410 (1975).
16. Becher, G., W. Lüttke, and G. Schrumpf, *Angew. Chem.*, **12**, 339 (1973).
17. Becker, E.D., J.A. Ferreti, R.K. Gupta, and G.H. Weiss, *J. Magn. Resonance*, **37**, 381 (1980).
18. Belich, H.E., J.D. Cutnell, and J.A. Glasel, *Biochemistry*, **15**, 2455 (1976).
19. Berger, S., and A. Rieker, *Chem. Ber.*, **109**, 3252 (1976).
20. Berger, S., and A. Rieker, *Tetrahedron*, **28**, 3123 (1972).
21. Berger, S., and K.P. Zeller, *Org. Magn. Resonance*, **11**, 303 (1978).
22. Betrand, R.D., W.B. Moniz, and A.N. Garroway, *J. Am. Chem. Soc.*, **100**, 5227 (1978).
23. Bicker, R., H. Kessler, A. Steigel, and G. Zimmerman, *Chem. Ber.*, **111**, 3215 (1978).
24. Bodenhausen, G., R. Freeman, R. Niedermeyer, and D.L. Turner, *J. Magn. Resonance*, **26**, 133 (1977).
25. Bolton, P.H., and T.L. James *Biochemistry*, **19**, 1388 (1980).
26. Booth, H., and D.V. Griffiths, *J. Chem. Soc.*, *Perkin Transact. II*, **1975** 111.
27. Booth, H., D.V. Griffiths, and M.L. Jozefowic, *J. Chem. Soc.*, *Perkin Transact. II*, **1976** 752.
28. Bose, A.K., and P.R. Srinivasan, *Org. Magn. Resonance*, **12**, 34 (1979).
29. Breitmaier, E., and W. Voelter, *^{13}C NMR Spectroscopy*, Second Ed., Verlag Chemie, Weinheim, 1978.
30. Breitmaier, E., and W. Voelter, *^{13}C NMR Spectroscopy*, Verlag Chemie, Weinheim, 1974.

31. Briggs, J., F.A. Hart, G.P. Moss, and E.W. Randall, *Chem. Commun.*, **1971**, 364.
32. Bromilow, J., and R.T.C. Brownlee, *J. Org. Chem.*, **44**, 1261 (1979).
33. Bromilow, J., R.T.C. Brownlee, D.J. Craik, V.O. Lopez, and R.W. Taft, *J. Org. Chem.*, **44**, 4766 (1979).
34. Buchi, R., and E. Pretsch, *Helv. Chim. Acta*, **58**, 1573 (1975).
35. Bunnell, C.A., and P.L. Fuchs, *J. Org. Chem.*, **42**, 2614 (1977).
36. Butler, R.S., J.M. Kead, Jr., and J.H. Goldstein, *J. Molec. Spectrosc.*, **35**, 83 (1979).
37. Canet, D., G.C. Levy, and I.R. Peat, *J. Magn. Resonance*, **18**, 199 (1975).
38. Carman, C.J., and C.E. Wilkes, *Macromolecules*, **7**, 40 (1974).
39. Carroll, F.I., G.N. Mitchell, J.T. Blackwell, A. Sobti, and R. Meck, *J. Org. Chem.*, **39**, 3890 (1974).
40. Carroll, F.I., A. Philip, and C.G. Moreland, *J. Med. Chem.*, **19**, 521 (1976).
41. Christl, M., H.J. Reich, and J.D. Roberts, *J. Am. Chem. Soc.*, **93**, 3453 (1971).
42. Clark, P.D., D.F. Ewing, and R.M. Scrowton, *Org. Magn. Resonance*, **8**, 252 (1976).
43. Combrisson, J., J.P. Lautié, and M. Olomuczki, *Bull. Soc. Chim. Fr.*, **1975**, 2769.
44. Couperus, P.A., A.D.H. Clague, and J.P.C.M. vanDongen, *Org. Magn. Resonance*, **11**, 590 (1978).
45. Crain, Jr., W.O., W.C. Wildman, and J.D. Roberts, *J. Am. Chem. Soc.*, **93**, 990 (1971).
46. Crecely, K.M., R.W. Crecely, and J.H. Goldstein, *J. Molec. Spectrosc.*, **37**, 252 (1971).
47. Cudby, M.E.A., and A. Bunn, *Polymer*, **17**, 345 (1976).
48. Dalling, D.K., and D.M. Grant, *J. Am. Chem. Soc.*, **94**, 5318 (1972).
49. Dawson, D.A., and W.F. Reynolds, *Can. J. Chem.*, **53**, 373 (1975).
50. Dehmlow, E.V., R. Zeisberg, and S.S. Dehmlow, *Org. Magn. Resonance*, **7**, 418 (1975).
51. Deslauriers, R., E. Ralston, and R.L. Somorjai, *J. Molec. Biol.*, **113**, 697 (1977).
52. Deslauriers, R., and I.C.P. Smith, *Topics in Carbon-13 NMR Spectroscopy*, Vol. 2, G.C. Levy, Ed., Wiley-Interscience, New York, 1976, Chap. 1.
53. Deslauriers, R., I.C.P. Smith, G.C. Levy, R. Orlowski, and R. Walter, *J. Am. Chem. Soc.*, **100**, 3912 (1978).
54. Deslauriers, R., and R. Somorjai, *J. Am. Chem. Soc.*, **98**, 1931 (1976).
55. Dhami, K.S., and J.B. Stothers, *Can. J. Chem.*, **43**, 479 (1965).
56. Dill, K., and A. Allerhand, *J. Am. Chem. Soc.*, **99**, 4508 (1977).
57. Ditchfield, R., and P.D. Ellis, *Topics in Carbon-13 NMR Spectroscopy*, Vol. 1., G.C. Levy, Ed., Wiley-Interscience, New York, 1974, Chap. 1 and references cited therein.
58. Dorman, D.E., M. Jautelat, and J.D. Roberts, *J. Org. Chem.*, **36**, 2757 (1971).
59. Dorman, D.E., M. Jautelat, and J.D. Roberts, *J. Org. Chem.*, **38**, 1026 (1973).
60. Duggan, J.C., W.H. Urry, and J. Schaefer, *Tetrahedron Lett.*, **1971**, 4197.
61. Eggert, H., and C. Djerassi, *J. Am. Chem. Soc.*, **95**, 3710 (1973).
62. Ejchart, A., *Org. Magn. Resonance*, **9**, 351 (1977).
63. Elguero, J., A. Fruchier, and M. delCarmen Pardo, *Can. J. Chem.*, **54**, 1329 (1976).
64. Elguero, J., C. Marzin, and J.D. Roberts, *J. Org. Chem.*, **39**, 357 (1974).
65. Eliel, E.L., K.H. Pietrusiewicz, *Topics in Carbon-13 NMR Spectroscopy*, Vol. 3, G.C. Levy, Ed., Wiley-Interscience, New York, 1979, Chap. 3.
66. Eliel, E.L., V.S. Rao, and K.M. Pietrusiewicz, *Org. Magn. Resonance*, **12**, 461 (1979).
67. Eliel, E.L., and F.W. Vierhapper, *J. Org. Chem.*, **41**, 199 (1976).
68. Ellis, G., and R.C. Jones, *J. Chem. Soc.*, *Perkin Transact. II*, **1972** 437.
69. Ernst, R.R., *J. Chem. Phys.*, **45**, 3845 (1966).

70. Ernst, R.R., *J. Magn. Resonance*, **3**, 10 (1970).
71. Ernst, L., V. Wray, V.A. Chertokov, and N.M. Sergeyev, *J. Magn. Resonance*, **25**, 123 (1977).
72. Ewing, D.F., *Org. Magn. Resonance*, **12**, 499 (1979).
73. Faure, R., A. Assaf, E.J. Vincent, and J.P. Aune, *J. Chim. Phys.*, **75**, 727 (1978).
74. Faure, R., J. Elguero, E.J. Vincent, and R. Lazaro, *Org. Magn. Resonance*, **11**, 617 (1978).
75. Faure, R., J.P. Galy, E.J. Vincent, and J. Elguero, *Can. J. Chem.*, **56**, 46 (1978).
76. Filleux-Blanchard, M.L., *Org. Magn. Resonance*, **9**, 125 (1977).
77. Finkelmeier, H., and W. Lüttke, *J. Am. Chem. Soc.*, **100**, 626 (1978).
78. Firl, J., and W. Runge, *Angew. Chem. Internat. Ed. Eng.*, **12**, 668 (1973).
79. Firl, J., W. Runge, W. Hartman, and H. Utikal, *Chem. Lett.*, **41** (1975).
80. Fischer, P., J.W. Stadelhofer, and M. Zander, *Fuel*, **57**, 345 (1978).
81. Florea, S., W. Kimpenhaus, and V. Farcasan, *Org. Magn. Resonance*, **9**, 133 (1977).
82. Fraser, R.R., K.L. Dhawan, and K. Taymaz, *Org. Magn. Resonance*, **11**, 269 (1978).
83. Freeman, R., *J. Chem. Phys.*, **53**, 457 (1970).
84. Freeman, R., and H.D.W. Hill, *J. Chem. Phys.*, **54**, 3367 (1971).
85. Freeman, R., H.D.W. Hill, and R. Kaptein, *J. Magn. Resonance*, **7**, 82 (1972).
86. Freeman, R., and G. Morris, *Bull. Magn. Res.*, **1**, 5 (1979).
87. Fringuelli, F., S. Gronowitz, A.B. Hörnfeldt, I. Johnson, and A. Taticchi, *Acta Chem. Scand.*, **28B**, 125, 175 (1974).
88. Fritz, H., P. Hug, H. Sauter, T. Winkler, and E. Logemann, *Org. Magn. Resonance*, **9**, 108 (1977).
89. Fruchier, A., E. Alcade, and J. Elguero, *Org. Magn. Resonance*, **9**, 235 (1977).
90. Fyfe, C.A., J.R. Lyerla, and C.S. Yannoni, *J. Am. Chem. Soc.*, **100**, 5635 (1978).
91. Gainer, J., G.A. Howarth, W. Hoyle, and S.M. Roberts, *Org. Magn. Resonance*, **8**, 226 (1976).
92. Gansow, O.A., M.R. Willcott, and R.E. Lenkinski, *J. Am. Chem. Soc.*, **93**, 4295 (1971).
93. Garreau, M., G.J. Martin, M.C. Martin, M. Morel, and C. Paulmier, *Org. Magn. Resonance*, **6**, 648 (1974).
94. Gasic, M.J., Z. Djarmati, and S.W. Pelletier, *J. Org. Chem.*, **41**, 1219 (1976).
95. Godfrey, M., *J. Chem. Soc., Perkin Transact. II*, **1977**, 769.
96. Grant, D.M., and E.G. Paul, *J. Am. Chem. Soc.*, **86**, 2984 (1964).
97. Gray, G.A., P.D. Ellis, D.D. Traficante, and G.E. Maciel, *J. Magn. Resonance*, **1**, 41 (1969).
98. Gray, G.A., G.E. Maciel, and P.D. Ellis, *J. Magn. Resonance*, **1**, 407 (1969).
99. Gronowitz, S., I. Johnson, and A. Bugge, *Acta Chem. Scand.*, **30B**, 417 (1976).
100. Gronowitz, S., I. Johnson, and A.B. Hörnfeldt, *Chemica Scripta*, **7**, 76, 211 (1975).
101. Grutzner, J.B., M. Jautelat, J.B. Dence, R.A. Smith, and J.D. Roberts, *J. Am. Chem. Soc.*, **92**, 7107 (1970).
102. Günther, H., and T. Keller, *Chem. Ber.*, **106**, 1863 (1973).
103. Günther, H., and H. Seel, *Org. Magn. Resonance*, **8**, 299 (1976).
104. Günther, H., H. Seel, and M.E. Günther, *Org. Magn. Resonance*, **11**, 97 (1978).
105. Günther, H., H. Seel, and H. Schmickler, *J. Magn. Resonance*, **28**, 145 (1977).
106. Hansen, P.E., *Org. Magn. Resonance*, **11**, 215 (1978).
107. Hansen, P.E., *Org. Magn. Resonance*, **12**, 109 (1979).
108. Hansen, P.E., J. Feeney, and G.C.K. Roberts, *J. Magn. Resonance*, **17**, 249 (1975) and references cited therein.
109. Hansen, P.E., O.K. Poulsen, and A. Berg, *Org. Magn. Resonance*, **7**, 475 (1975).
110. Hanssum, H., W. Maurer, and H. Rüterjans, *J. Magn. Resonance*, **31**, 231 (1978).
111. Hanssum, H., and H. Rüterjans, *J. Magn. Resonance*, **39**, 65 (1980).

112. Hartmann, S.R., and E.L. Hahn, *Phys. Rev.*, **128**, 2042 (1962).

113. Hatada, K., K. Nagata, and H. Yuki, *Bull. Chem. Soc. Jap.*, **43**, 3195, 3267 (1970).

114. Hearn, M.T.W., *Aust. J. Chem.*, **29**, 107 (1976).

115. Hill, H., *Topics in Carbon-13 NMR Spectroscopy*, Vol. 3, G.C. Levy, Ed., Wiley-Interscience, New York, 1979, Chap. 1, and references cited therein.

116. Höbold, W., R. Radeglia, and D. Klose, *J. Prakt. Chem.*, **318**, 519 (1976).

117. Höfle, G., *Tetrahedron*, **32**, 1431 (1976).

118. Holm, C.H., *J. Chem. Phys.*, **21**, 707 (1957).

119. Inamoto, N., S. Masuda, and K. Tokumaru, *Tetrahedron Lett.*, **1976**, 3707, 3711.

120. Jackman, L.M., and D.P. Kelley, *J. Chem. Soc. (B)*, **1970**, 102.

121. Jakobsen, H.J., and H. Beldsøe, *J. Magn. Resonance*, **26**, 183 (1977).

122. Jakobsen, H.J., S.A. Linde, and S. Sørensen, *J. Magn. Resonance*, **15**, 385 (1974).

123. Janowski, K., and W.T. Raynes, *J. Chem. Research (S)*, **66**, (1977) and references cited therein.

124. Johnson, L.F., and W.C. Jankoswki, *Carbon-13 Spectra*, Wiley-Interscience, New York, 1972.

125. Jokisaari, J., *Org. Magn. Resonance*, **11**, 157 (1978).

126. Jokisaari, J., J. Kuonanoja, and A.M. Häkkinen, *Z. Naturforsch.*, **33a**, 7 (1978).

127. Jones, A.J., T.D. Alger, D.M. Grant, and W.M. Litchman, *J. Am. Chem. Soc.*, **92**, 2386 (1970).

128. Jones, R.G., and J.M. Wilkins, *Org. Magn. Resonance*, **11**, 20 (1978).

129. Jozetowicz, M.L., I.K. O'Neill, and H.K. Prosser, *Anal. Chem.*, **49**, 1140 (1977).

130. Kaiser, R., *J. Magn. Resonance*, **3**, 28 (1970).

131. Kalinowski, H.O., and H. Kessler, *Org. Magn. Resonance*, **6**, 305 (1974).

132. Khami, K.S., and J.B. Stothers, *Can. J. Chem.*, **43**, 510 (1965).

133. Kingsbury, C.A., D. Draney, A. Sopchik, W. Rissler, and D. Durhan, *J. Org. Chem.*, **41**, 3863 (1976).

134. Koer, F.J., A.J. deHoog, and C. Altona, *Rec. Trav. Chim. Pays-Bas.*, **94**, 75 (1975).

135. Komoroski, R.A., and A. Allerhand, *Biochemistry*, **13**, 369 (1974).

136. Komoroski, R.A., and A. Allerhand, *Proc. Nat. Acad. Sci.* (USA), **69**, 1804 (1972).

137. Komoroski, R.A., I.R. Peat, and G.C. Levy, *Biochem. Biophys. Res. Commun.*, **65**, 272 (1975).

138. Komoroski, R.A., I.R. Peat, and G.C. Levy, *Topics in Carbon-13 NMR Spectroscopy*, Vol. 2, G.C. Levy, Ed., Wiley-Interscience, New York, 1976, Chap. 4.

139. Koole, N.J., and M.H.A. deBie, *J. Magn. Resonance*, **23**, 9 (1976).

140. Kowalewski, J., *Progress NMR Spectrosc.*, **11**, 1 (1977).

141. Kowalewski, J., A. Ericsson, and R. Vestin, *J. Magn. Resonance*, **31**, 165 (1978).

142. Kowalewski, J., G.C. Levy, L.F. Johnson, and L. Palmer, *J. Magn. Resonance*, **26**, 533 (1977).

143. Kramer, G.K., I.R. Peat, and W.F. Reynolds, *Can. J. Chem.*, **51**, 915, 2596 (1973).

144. Kunesch, G., and C. Poupat, in E. Buncel, and C.C. Lee, Eds., *Isotopes in Organic Chemistry*, Vol. 3, Elsevier, Amsterdam, 1977.

145. Lauterbur, P.C., *J. Chem. Phys.*, **21**, 217 (1957).

146. Leipert, T.K., and D.W. Marquardt, *J. Magn. Resonance*, **24**, 181 (1976).

147. Lepoivre, J.A., R.A. Dommisse, and F.C. Alderweireldt, *Org. Magn. Resonance*, **7**, 422 (1975).

148. Levy, G.C., Ed., *Topics in Carbon-13 NMR Spectroscopy*, Wiley-Interscience, New York, Volumes 1 (1974), 2 (1976), and 3 (1979).

149. Levy, G.C., J.D. Cargioli, and F.A.L. Anet, *J. Am. Chem. Soc.*, **93**, 1527 (1973).

150. Levy, G.C., and U. Edlund, *J. Am. Chem. Soc.*, **97**, 4482 (1978).

151. Levy, G.C., U. Edlund, and J.G. Hexem, *J. Magn. Resonance*, **19**, 259 (1975).

152. Levy, G.C., U. Edlund, and C.E. Holloway, *J. Magn. Resonance*, **24**, 375 (1976).

153. Levy, G.C., and J.M. Hewitt, *J. Assoc. Off. Anal. Chem.*, **60**, 241 (1977).

154. Levy, G.C., and R.A. Komoroski, *J. Am. Chem. Soc.*, **96**, 678 (1974).

155. Levy, G.C., R.L. Lichter, and G.L. Nelson, *Carbon-13 Nuclear Magnetic Resonance Spectroscopy*, 2nd Ed., Wiley-Interscience, New York, 1980.

156. Levy, G.C., and G.L. Nelson, *Carbon-13 Nuclear Magnetic Resonance for Organic Chemists*, Wiley-Interscience, New York, 1972.

157. Levy, G.C., and G.L. Nelson, *J. Am. Chem. Soc.*, **94**, 4897 (1972).

158. Levy, G.C., and I.R. Peat, *J. Magn. Resonance*, **18**, 500 (1975).

159. New computer techniques are being developed continuously. See, for example, *Computer Networks in The Chemical Laboratory*, G.C., Levy, and D. Terpstra, Eds., Wiley-Interscience, New York, 1980.

160. Lichter, R.L., and J.D. Roberts, *J. Phys. Chem.*, **74**, 912 (1970).

161. Lindeman, L.P., and J.Q. Adams, *Anal. Chem.*, **43**, 1245 (1971).

162. Lippma, E., T. Pehk, J. Paasivirta, N. Belikova, and A. Platé, *Org. Magn. Resonance*, **2**, 581 (1970).

163. Lyerla, J.R., C.S. Yannoni, D. Bruck, and C.A. Fyfe, *J. Am. Chem. Soc.*, **101**, 4770 (1979).

164. Lynch, B.M., *Can. J. Chem.*, **55**, 541 (1977).

165. Maciel, G.E., *J. Phys. Chem.*, **69**, 1947 (1965).

166. Maciel, G.E., *Topics in Carbon-13 NMR Spectroscopy*, Vol. 1, G.C. Levy, Ed., Wiley-Interscience, New York, 1974, Chap. 2.

167. Maciel, G.E., V.J. Bartuska, and F.P. Miknis, *Fuel*, **58**, 155 (1979).

168. Maciel, G.E., V.J. Bartuska, and F.P. Miknis, *Fuel*, **58**, 391 (1979).

169. Maciel, G.E., and D.A. Beatty, *J. Phys. Chem.*, **59**, 3920 (1965).

170. Maciel, G.E., and H.C. Dorn, *J. Magn. Resonance*, **24**, 251 (1976).

171. Maciel, G.E., P.D. Ellis, J.J. Natterstad, and G.B. Savitsky, *J. Magn. Resonance*, **1**, 589 (1969).

172. Maciel, G.E., J.W. McIver, Jr., N.S. Ostland, and J.A. Pople, *J. Am. Chem. Soc.*, **92**, 1, 11 (1970).

173. Mareci, T.H., and K.N. Scott, *Anal. Chem.*, **49**, 2130 (1977).

174. Markley, J.L., W.H. Horsley, and M.P. Klein, *J. Chem. Phys.*, **55**, 3604 (1971).

175. Marr, D.H., and J.B. Stothers, *Can. J. Chem.*, **43**, 596 (1965).

176. Marr, D.H., and J.B. Stothers, *Can. J. Chem.*, **45**, 225 (1967).

177. Marshall, J.L., and A.M. Ihrig, *Org. Magn. Resonance*, **5**, 235 (1973).

178. Marshall, J.L., D.E. Miller, S.A. Conn, R. Seinwell, and A.M. Ihrig, *Acc. Chem. Res.*, **7**, 333 (1974).

179. McDonald, I.A., T.J. Simpson, and A.F. Sierakowski, *Aust. J. Chem.*, **30**, 1727 (1977).

180. McFarlane, W., *Molec Phys.*, **10**, 603 (1966).

181. McInnes, A.G., J.A. Walter, J.L.C. Wright, and L.C. Vining, *Topics in Carbon-13 NMR Spectroscopy*, Vol. 2, G.C. Levy, Ed., Wiley-Interscience, New York, 1976, Chap. 3.

182. McInnes, A.G., and J.L.C. Wright, *Acc. Chem. Res.*, **8**, 313 (1975).

183. Miyajima, G., and K. Takahashi, *J. Phys. Chem.*, **75**, 331, 3766 (1971).

184. Miyajima, G., K. Takahashi, and K. Nishimoto, *Org. Magn. Resonance*, **6**, 413 (1974).

185. Morris, G.A., *J. Am. Chem. Soc.*, **102**, 428 (1980).

186. Morris, G.A., and R. Freeman, *J. Am. Chem. Soc.*, **101**, 760 (1979).

187. Müller, L., *J. Magn. Resonance*, **38**, 79 (1980).

188. Müller, L., A. Kumar, and R.R. Ernst, *J. Chem. Phys.*, **63**, 5490 (1975).

189. Naulet, N., M.L. Filleux, G.J. Martin, and J. Pornet, *Org. Magn. Resonance*, **7**, 326 (1975).

190. Nelson, G.L., unpublished results.

191. Nelson, G.L., and G.C. Levy, unpublished results.

192. Nelson, G.L., G.C. Levy, and J.D. Cargioli, *J. Am. Chem. Soc.*, **94**, 3090 (1972).
193. Nelson, G.L., and E.A. Williams, *Progr. Phys. Org. Chem.*, **12**, 229 (1976).
194. Nelson, S.F., and G.R. Wiseman, *J. Am. Chem. Soc.*, **98**, 3281 (1976).
195. Newton, M.D., and J.M. Schulman, *J. Am. Chem. Soc.*, **96**, 6295 (1974).
196. Newton, M.D., J.M. Schulman, and M.M. Manus, *J. Am. Chem. Soc.*, **96**, 17 (1974).
197. Nomura, Y., N. Masai, and Y. Takeuchi, *J. Chem. Soc. Chem. Commun.*, **1975**, 307.
198. Norton, R.S., A.O. Clouse, R. Addleman, and A. Allerhand, *J. Am. Chem. Soc.*, **99**, 79 (1977).
199. Olah, G.A., and P.W. Westerman, *J. Am. Chem. Soc.*, **95**, 3706 (1973).
200. O'Neill, I.K., and M.A. Pringuer, *Org. Magn. Resonance*, **6**, 398 (1974).
201. Opella, S.J., M.H. Frey, and T.A. Cross, *J. Am. Chem. Soc.*, **101**, 5856 (1979).
202. Pelletier, S.W., Z. Djarmati, and C. Pape, *Tetrahedron*, **32**, 995 (1976).
203. Perlin, A.S., and G.K. Hamer, in *Carbon-13 NMR in Polymer Science*, W.M. Pasika, Ed., ACS Symposium Series No. 103, 1979, p. 123.
204. Perlin, A.S., and H.J. Koch, *Can. J. Chem.*, **48**, 2639 (1970).
205. Piccinni-Leopardi, C., O. Fabre, D. Zimmerman, J. Reisse, F. Cornea, and C. Fulea, *Can. J. Chem.*, **55**, 2649 (1977).
206. Piccinni-Leopardi, C., O. Fabre, D. Zimmerman, J. Reisse, F. Cornea, and C. Fulea, *Org. Magn. Resonance*, **8**, 536 (1976).
207. Pihlaja, K., and T. Nurmi, *Finn. Chem. Lett.*, 141 (1977).
208. Pines, A., M.G. Gibby, and J.S. Waugh, *J. Chem. Phys.*, **59**, 569 (1973).
209. Plavac, N., I.W.J. Still, M.S. Chauhan, and D.M. McKinnon, *Can. J. Chem.*, **53**, 836 (1975).
210. Pugmire, P.J., J.C. Smith, D.M. Grant, B. Stanovnik, and M. Tisler, *J. Heterocyclic Chem.*, **13**, 1057 (1976).
211. Rabenstein, D.L., and T.L. Sayer, *J. Magn. Resonance*, **24**, 27 (1976).
212. Rae, I.D., *Aust. J. Chem.*, **32**, 567 (1979).
213. Rakita, P.E., J.P. Srebo, and L.S. Worsham, *J. Organomet. Chem.*, **104**, 27 (1976).
214. Randall, J.C., *Polymer Sequence Determination, Carbon-13 NMR Method*, Academic Press, New York, 1977.
215. Reynolds, W.F., I.R. Peat, M.H. Freedman, and J.R. Lyerla, *Can. J. Chem.*, **51**, 1857 (1973).
216. Rill, R.L., P.R. Hilliard, J.T. Bailey, and G.C. Levy, *J. Am. Chem. Soc.*, **102**, 418 (1980).
217. Roberts, J.D., F.J. Weigert, J.I. Kroschwitz, and H.J. Reich, *J. Am. Chem. Soc.*, **92**, 1338 (1970).
218. Rock, S.L., and R.M. Hammaker, *Spectrochim. Acta*, **27A**, 1899 (1971).
219. Rojas, A.C., and J.K. Crandall, *J. Org. Chem.*, **40**, 2225 (1975).
220. Rutar, V., M. Burgar, R. Blino, and L. Ehrenberg, *J. Magn. Resonance*, **27**, 83 (1977).
221. Sarneski, J.E., and C.N. Reilley, *Anal. Chem.*, **48**, 1303 (1976).
222. Sarneski, J.E., H.L. Surprenant, F.K. Molen, and C.N. Reilly, *Anal. Chem.*, **47**, 2116 (1975).
223. Sass, M., and D. Ziessow, *J. Magn. Resonance*, **25**, 263 (1977).
224. Sawhney, S.N., and D.W. Boykin, *J. Org. Chem.*, **44**, 1136 (1974).
225. Schaefer, J., *Topics in Carbon-13 NMR Spectroscopy*, Vol. 1, G.C. Levy, Ed., Wiley-Interscience, New York, 1974, Chap. 4.
226. Schaefer, J., and E.O. Stejskal, *J. Am. Oil Chem. Soc.*, **51**, 210 (1974).
227. Schaefer, J., and E.O. Stejskal, *J. Am. Oil Chem. Soc.*, **51**, 562 (1974).
228. Schaefer, J., and E.O. Stejskal, *J. Am. Oil Chem. Soc.*, **52**, 366 (1975).
229. Schaefer, J., and E.O. Stejskal, *Topics in Carbon-13 NMR Spectroscopy*, Vol. 3, G.C. Levy, Ed., Wiley-Interscience, New York, 1979, Chap. 4.

230. Schleich, T., B.P. Cross, B.J. Blackburn, and I.C.P. Smith, in *Structure and Conformation of Nucleic Acids and Proton-Nucleic Acid Interactions*, M. Sundaralingam, and S.T. Rao, Eds., University Park Press, Baltimore, 1975, pp. 223-251.
231. Schmidt, C.F., Y. Barenholz, C. Huang, and T.E. Thompson, *Biochemistry*, **6**, 3948 (1977).
232. Scott, A.I., and co-workers, unpublished results.
233. Servis, K.L., W.P. Weber, and A.K. Willard, *J. Chem. Phys.*, **74**, 3960 (1970).
234. Severini-Ricca, G., P. Manitto, D. Monti, and E.W. Randall, *Gazz. Chim. Ital.*, **105**, 1273 (1975).
235. Shapiro, Y.E., A.V. Viktorov, V.I. Volkova, L.I. Barsakov, V.F. Bystrov, and L.D. Bergelson, *Chem. Phys. Lipids*, **14**, 227 (1975).
236. Shaw, D., *Fourier Transform NMR Spectroscopy*, Elsevier, London, 1976.
237. Shoolery, J.N., *Progress. NMR Spectroc.*, **11**, 79 (1977).
238. Shoolery, J.N., and W.L. Buddle, *Anal. Chem.*, **48**, 1458 (1976).
239. Shoolery, J.N., and W.C. Jankowski, Varian Application Notes NMR-73-4 (1973).
240. Sibi, M.P., and R.L. Lichter, *J. Org. Chem.*, **44**, 3017 (1979) and references cited therein.
241. Sievers, R.E., Ed., *Nuclear Magnetic Resonance Shift Reagents*, Academic Press, New York, 1973.
242. Simonnin, M.P., M.J. Pouet, and F. Terrier, *J. Org. Chem.*, **43**, 855 (1978).
243. Simpson, T.J., *Chem. Soc. Rev.*, **4**, 497 (1975).
244. Smith, W.B., and D.L. Deavenport, *J. Magn. Resonance*, **6**, 256 (1972).
245. Sørensen, S., R.S. Hansen, and H.J. Jakobsen, *J. Magn. Resonance*, **14**, 243 (1974).
246. Spiesecke, H., and W.G. Schneider, *J. Chem. Phys.*, **35**, 731 (1961).
247. Srinivasan, P.R., and R.L. Lichter, *Org. Magn. Resonance*, **8**, 198 (1976).
248. Still, I.W.J., N. Plavac, D.M. McKinnon, and M.S. Chauhan, *Can. J. Chem.*, **54**, 280 (1976).
249. Stöcker, M., and M. Klessinger, *Org. Magn. Resonance*, **12**, 107 (1979).
250. Stothers, J.B., *Carbon-13 NMR Spectroscopy*, Academic Press, New York, 1972.
251. Stothers, J.B., and P.C. Lauterbur, *Can. J. Chem.*, **42**, 1563 (1964).
252. Stothers, J.B., and C.T. Tan, *Can. J. Chem.*, **53**, 581 (1975).
253. Stothers, J.B., C.T. Tan, and K.C. Teo, *Can. J. Chem.*, **54**, 1211 (1976).
254. Strong, A.B., D. Ikenberry, and D.M. Grant, *J. Magn. Resonance*, **9**, 145 (1973).
255. Sullivan, G.R., *J. Am. Chem. Soc.*, **98**, 7162 (1976).
256. Swain, C.G., and E.C. Lupton, *J. Am. Chem. Soc.*, **90**, 4328 (1968).
257. Tanabe, M., in *Specialist Periodical Reports, Biosynthesis*, Vol. 4, Chemical Society, London, 1976, p. 204.
258. Taskinin, E., *Tetrahedron*, **34**, 353 (1978).
259. Terpstra, D., *Topics in Carbon-13 NMR Spectroscopy*, Vol. 3, G.C. Levy, Ed., Wiley-Interscience, New York, 1979, Chap. 1, Section 6.
260. Thiault, B., and M. Mersseman, *Org. Magn. Resonance*, **8**, 28 (1976).
261. Torchia, D.A., J.R. Lyerla, Jr., and A.J. Quattrone, *Biochemistry*, **14**, 887 (1975).
262. Torchia, D.A., and D.L. Vanderhart, *Topics in Carbon-13 NMR Spectroscopy*, Vol. 3, G.C. Levy, Ed., Wiley-Interscience, New York, 1979, Chap. 5.
263. Ugurbil, K., T.R. Brown, J.A. denHollander, P. Glynn, and R.G. Shulman, *Proc. Nat. Acad. Sci.* (USA), **75**, 3742 (1978).
264. VanBinst, G., M. Biesemans, and A.O. Barel, *Bull. Soc. Chim. Belg.*, **84**, 1 (1975).
265. Vanderhart, D.L., and H.L. Retcofsky, *Fuel*, **55**, 202 (1976).
266. Vierhapper, F.W., and E.L. Eliel, *J. Org. Chem.*, **42**, 51 (1977).

267. Vierhapper, F.W., and E.L. Eliel, *J. Org. Chem.*, **44**, 1081 (1979).
268. Vitorović, D., D. Vučelić, M.J. Gasic, N. Juranić, and S. Macura, *Org. Geochem.*, **1**, 89 (1978).
269. Voelter, W., G. Jung, E. Breitmaier, and E. Bayer, *Z. Naturforsch.*, **26b**, 213 (1971).
270. Vögeli, U., and W. vonPhilipsborn, *Org. Magn. Resonance*, **7**, 617 (1975).
271. Vold, R.L., J.S. Waugh, M.P. Klein, and D.E. Phelps, *J. Chem. Phys.*, **48**, 3831 (1968).
272. Wasylishen, R.E., *Annu. Rep. NMR Spectrosc.*, **7**, 250*ff* (1977) and reference cited therein.
273. Wasylishen, R.E., K. Chum, and J. Bukata, *Org. Magn. Resonance*, **9**, 473 (1977).
274. Wasylishen, R.E., T.R. Clem, and E.D. Becker, *Can. J. Chem.*, **53**, 596 (1975).
275. Wasylishen, R.E., and H.M. Hutton, *Can. J. Chem.*, **55**, 619 (1977).
276. Wasylishen, R., and T. Schaefer, *Can. J. Chem.*, **50**, 2710 (1972).
277. Wehrli, F.W., *Topics in Carbon-13 NMR Spectroscopy*, Vol. 2, G.C. Levy, Ed., Wiley-Interscience, New York, 1979, Chap. 6.
278. Wehrli, F.W., and T. Wirthlin, *Interpretation of Carbon-13 NMR Spectra*, Heyden, London, 1976.
279. Weigert, F.J., and J.D. Roberts, *J. Am. Chem. Soc.*, **92**, 1338, 1347 (1970).
280. Wenkert, E., D.W. Cochran, E.W. Hagman, R.B. Lewis, and F.M. Schell, *J. Am. Chem. Soc.*, **93**, 6271 (1971).
281. White, D.M., and G.C. Levy, *Macromolecules*, **5**, 526 (1972).
282. Wiberg, K.B., G.M. Lampman, R.P. Ciula, D.S. Conner, P. Schertler, and J. Lavanish, *Tetrahedron*, **21**, 2749 (1965).
283. Wilkins, C.L., T.R. Brunner, and D.J. Thoennes, *J. Magn. Resonance*, **17**, 373 (1975).
284. Williams, E.A., J.D. Cargioli, and S.Y. Hobbs, *Macromolecules*, **10**, 682 (1977).
285. Wright, D.A., D.E. Axelson, and G.C. Levy, *Topics in Carbon-13 NMR Spectroscopy*, Vol. 3, G.C. Levy, Ed., Wiley-Interscience, New York, 1979, p. 150-156.
286. Wüthrich, K., S. Meiboom, and L.C. Snyder, *J. Chem. Phys.*, **52**, 230 (1970).
287. Yavari, I., and F.A.L. Anet, *Org. Magn. Resonance*, **8**, 158 (1976).
288. Ziessow, D., and B. Blümich, *Ber der Bunsen-Gessellschaft Phys. Chem.*, **78**, 1168 (1974).

Part I
Section I

Chapter 3

ELECTRON-SPIN-RESONANCE SPECTROSCOPY

BY IRA B. GOLDBERG, *Science Center, Rockwell International, Thousand Oaks, California*
AND ALLEN J. BARD, *Department of Chemistry, University of Texas at Austin, Austin, Texas*

Contents

I. INTRODUCTION

Electron-spin-resonance (ESR) spectroscopy (also sometimes called by the synonymous terms electron-paramagnetic-resonance, EPR, and electron-magnetic-resonance, EMR, spectroscopy) is based on the absorption of electromagnetic radiation (usually in the microwave region) which causes transitions between energy levels produced by the action of a magnetic field on an unpaired electron. The phenomenon is basically the same as that in nuclear magnetic resonance (NMR), except that the magnetic moment of an electron is about 658 times that of a proton, so that in a magnetic field the energy level separation for electrons is 658 times that of the proton. This accounts for the difference in the spectral regions of the techniques at equivalent magnetic fields (e.g., 60–100 MHz for NMR versus 9.5–35 GHz for ESR) and for the much greater sensitivity of the ESR technique. Thus, while many of the basic concepts are common to the two techniques, the instrumentation and range of applications are very different.

The discovery of the ESR phenomenon by E. Zavoisky in 1944 (121) actually predated the first NMR experiments by the physicists Purcell and Bloch in 1946. The ESR technique was mainly used by physicists in England and the United States in the late 1940s to study paramagnetic metal ions in crystal lattices. In the 1950s ESR techniques began to be applied to chemical problems, especially those involving free radicals and radical ions and relaxation (57,109). The availability of commercial ESR instrumentation, especially the Varian V4500 spectrometer in 1957 and the V4502 system a few years later, spurred vigorous research efforts in this field that have continued to the present time.

Since ESR requires the presence of unpaired electrons in the substance being analyzed, its range of application is narrower than many other analytical techniques. It is uniquely suited to the detection and study of free radicals and radical ions. It has also been used for the analysis of paramagnetic metal ions, especially those of transition metals, rare earths, and actinides. ESR signals are also exhibited by carbonaceous materials, defects in solids (e.g., F and V centers, impurity sites, and from radiation damage), for triplet states in solids, and for many atoms and several diatomic molecules in the gas phase. Solid, liquid, and gaseous samples, and even biological materials, can be used directly in the ESR spectrometer. At the time of the first edition of this series (1963), ESR was considered a relatively young technique. Numerous applications have appeared in the intervening years, and its high sensitivity, speed, and specificity have made it an important method for a special class of analytical problems.

II. PRINCIPLES OF ELECTRON-SPIN RESONANCE

Many books and reviews have been written about the use of ESR techniques for the study of chemical phenomena. Because a detailed description of ESR is not possible in this chapter, the reader is directed to references 3, 5, 7, 10, 23, 58, and 110 for additional information. These references were selected for their generality, the introductory nature of their treatment, and their detailed description of ESR techniques. This chapter will be devoted to a discussion of those features of ESR needed for analytical purposes.

Because ESR is utilized in many different research areas, magnetic parameters are cited in a great variety of units. To reduce the difficulty of converting parameters to

TABLE 3.I
Fundamental Constants Used in Magnetic Resonance Spectroscopy[a]

Constant	Symbol	Value
Free-electron g factor	g_e	2.002319313
Bohr magneton	β_e	9.274078×10^{-21} erg/Gauss
Planck's constant	h	6.626176×10^{-27} erg/sec
Velocity of light	c	2.9979246×10^{10} cm/sec
Boltzmann's constant	k	1.380662×10^{-16} erg/deg
Electron change	e	$4.8032424 \times 10^{-10}$ esu
Electron mass	m_e	9.109534×10^{-28} g
Electron charge/electron mass	e/m_e	5.272764×10^{17} esu/g
Nuclear magneton	β_N	5.050824×10^{-23} erg/Gauss
Proton g factor	g_H	5.585564

[a] Taken from E. R. Cohen and B. N. Taylor, *J. Phys. Chem. Ref. Data*, **2**, 663 (1973).

different units, the values of the relevant fundamental constants and some useful conversion factors are given in Tables 3.I and 3.II.

A. THE SPIN-RESONANCE PHENOMENON

The basic concepts of ESR are similar to those of NMR, although the applications and the instrumentation of the techniques are quite different. An electron may be pictured as a spinning, negatively charged, particle. By virtue of its charge and spin, the electron behaves as a magnet (just as a loop of wire carrying a current produces a magnetic field) and can interact with an external magnetic field. The magnetism of an electron can be expressed by saying that an electron has a magnetic moment, μ, which is proportional to e/m_e where e is the charge on the electron and m_e is its mass. The magnitude of the magnetic moment of a "classical" free electron would be β_e, the Bohr magneton, which has a value of 9.2732×10^{-21} erg/gauss. Because an electron is a subatomic particle, its momentum and energy are governed by quantum mechanical considerations. Each electron is assigned a spin quantum number M_s which is either $+1/2$ or $-1/2$ denoting that there are only two possible orientations of an electron in a magnetic field. The magnetic moment of the electron in the direction of the magnetic field, μ_z, is

$$\mu_z = -g\beta_e M_s \tag{1}$$

where g, called the g factor or spectroscopic splitting factor, depends upon the orbital and the electronic environment of the electron. For a free electron, g_e is about 2. When unpaired electrons are placed in a magnetic field, measured in gauss (or oersteds), the energy of the electrons will be changed by a certain number of ergs, given by

$$E = -\mu H = g\beta M_s H = \pm 1/2(g\beta H) \tag{2}$$

where H is the magnetic field strength. Those of spin $-1/2$ (pointing opposite to the

TABLE 3.II

Conversion Factors for Use in Electron Spin Resonance

To convert to	Multiply to				
	erg	cm^{-1}	MHz	eV	G
erg	1.0	5.0340×10^{15}	1.5092×10^{20}	6.2414×10^{11}	$5.38512 \times 10^{19} (g_e/g)$
cm^{-1}	1.9865×10^{-16}	1.0	2.9979×10^{4}	1.2398×10^{-4}	$1.06974 \times 10^{4} (g_e/g)$
MHz	6.6262×10^{-21}	3.3356×10^{-5}	1.0	4.1357×10^{-9}	$0.356828 (g_e/g)$
eV	1.6022×10^{-12}	8.0655×10^{3}	2.4180×10^{8}	1.0	$0.86280 \times 10^{14} (g_e/g)$
G	$1.85697 \times 10^{-20} (g/g_e)$	$0.93481 \times 10^{-4} (g/g_e$	$2.80247 (g/g_e)$	$1.15902 \times 10^{-14} (g/g_e)$	1.0

1 Gauss = 10^4 Tesla. The Gauss, a unit of magnetic induction is numerically equal to the Oersted, a unit of magnetic field intensity, in vacuum. The Tesla is the SI unit of magnetic field intensity.

direction of the magnetic field) will decrease in energy by about $1/2(g\beta H)$, while those with spin $+1/2$ (pointing in the direction of the field) will increase in energy by a like amount, so that the difference in energy in between the two levels is

$$\Delta E = g\beta H \tag{3}$$

Note that the magnetic moment of the electron points opposite to its spin. Just as in other spectroscopic experiments, there is a certain frequency of electromagnetic radiation that will induce transitions between these states. The frequency associated with this energy difference satisfies the equation

$$\Delta E = h\nu = g\beta H \tag{4}$$

In distinction to other types of spectroscopy, in which the energy levels are essentially fixed, in ESR spectroscopy the splitting of energy levels, and, therefore, the frequency capable of causing transitions between these levels, is a function of the magnetic field strength, H. Putting in values for the constants in equation 4 yields a value for ν/H of 2.8026 MHz/gauss for $g = g_e$.

It is helpful to compare ESR spectroscopy to NMR and absorption spectroscopy. For a magnetic-field strength of 3400 gauss (a field strength which is commonly employed) the energy level difference due to electron spins is 6.31×10^{-17} ergs (0.00091 Kcal/mole) and ν is 9.53 GHz. This frequency lies in the microwave region of the electromagnetic radiation spectrum, so that the most convenient instrumentation used will involve radar-type components such as waveguides, microwave cavities, and klystrons. Note that identical considerations hold for proton magnetic resonance, but because the hydrogen nucleus is about 1800 times heavier than the electron, and its g value is about 5.585, for NMR at 3400 gauss, ν would be about 14.5 MHz in the radiofrequency region. Because ΔE is relatively small for ESR in comparison to visible or infrared spectroscopy, differences in behavior other than a different frequency of the transition are also noted. The relative population of two energy levels separated by an energy difference ΔE is governed by the Boltzmann equation

$$\frac{n_{+1/2}}{n_{-1/2}} = e^{-\Delta E/kT} = e^{-h\nu/kT} \sim 1 - \frac{h\nu}{kT} \tag{5}$$

where k is the Boltzmann constant and T is the temperature. For a ΔE of 6.3×10^{-17} ergs, the relative population of the two energy levels is 0.9984 at room temperature.

It is sometimes possible to produce a sufficient rate of transitions to cause the population of the higher energy state to be equal to that of the lower one. This phenomenon, known as saturation, depends upon the intensity of the microwave radiation and upon the time required for a molecule in the upper level to fall back to the lower level. This time, related to the spin–lattice relaxation time, is a measure of the interaction of the unpaired electron with its environment (the lattice). When saturation begins, the signal level decreases and the signal broadens.

B. THE BASIC ESR EXPERIMENT

A very simple instrument for carrying out ESR spectroscopy at microwave frequencies and, for comparison, the familiar spectrometer operating in the visible region are shown in Fig. 3.1. The source of the microwave (or rf) radiation (often 9.5 GHz) is a klystron. The microwaves are conducted to the sample through a waveguide. The sample, contained in the sample tube, is held in a microwave cavity between the poles of a magnet operating in the region of 3400 gauss. The detector is a diode which produces a dc output related to the level of incident microwave power. The dc from the crystal is displayed on a recorder or oscilloscope. While it would be possible to operate in a mode analogous to visible spectrometers, by holding the magnetic field fixed and varying the frequency of the klystron, in practice it is much easier to hold the klystron frequency fixed and vary the magnetic field because then the cavity acts as a high-Q-tuned circuit. At a given klystron frequency there is a certain value of the field strength that satisfies equation 4. At this value of H, transitions are induced from the lower to the upper energy level, and microwave energy is absorbed by the sample. The microwave energy falling on the crystal is then

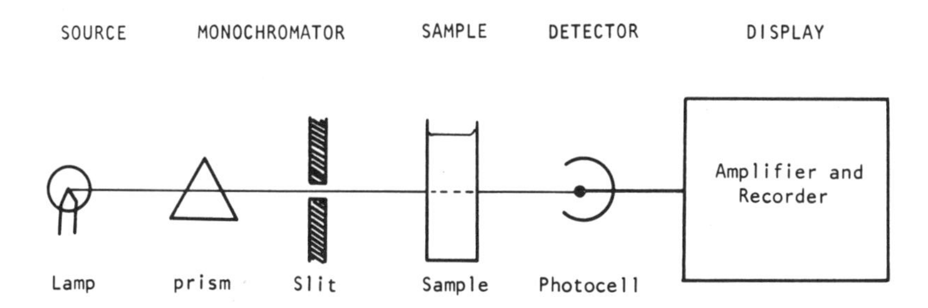

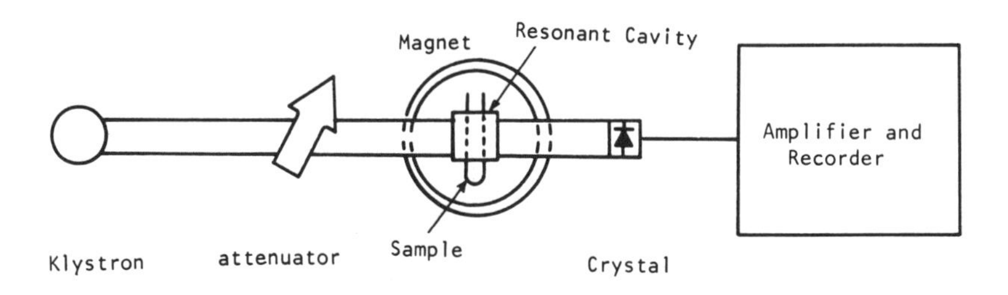

Fig. 3.1. Comparison between an optical spectrometer and a simple ESR spectrometer.

smaller, and the dc output from the crystal as a function of H shows the absorption band.

Although this method of operation of an ESR spectrometer is possible, it is not usually used because of its poor sensitivity. Various modifications involving modulation of the magnetic field and lock-in (phase-sensitive) or superheterodyne detection give much higher sensitivity and are used in most commercial instruments. These are described in Section III.

C. *g* FACTORS

1. The *g* Tensor

Equation 1 introduces the g factor as a single parameter. However, the value of g is highly dependent upon the environment of the electron. Unpaired electrons that are not in s orbitals will generate a magnetic field that either adds or subtracts from the applied field, and thus changes the value of the g factor from g_e. The actual value will depend upon the orientation of the molecule with respect to the magnetic field. In the gas or liquid phases, the observed g factor is the average of the three diagonal components of the g tensor so that the system is isotropic.

Analysis can be carried out for liquid systems as well as for single crystals and powders or randomly oriented solids. A crystal with low symmetry will have $g_x \neq g_y \neq g_z$ where the x, y, and z axes relate to molecular coordinates. A single crystal will then exhibit a g factor that is dependent upon its orientation in the magnetic field. A powder of the same material will exhibit a broad spectrum which covers the magnetic field range determined by the maximum and minimum g factors.

A molecule that is axially symmetric in a single crystal is characterized by a $g_\perp = g_x = g_y$ and $g_\parallel = g_z$. When the z axis is parallel to the field, the observed g factor is independent of the rotation of the molecule. In a powder, the spectra are characterized by an unusual shape which is basically the envelope of all possible orientations of the molecule with respect to the magnetic field. The magnetic field range of the spectrum would be determined by $g_\parallel$ and $g_\perp$. Various representative spectra resulting from powdered samples are shown in Fig. 3.2.

An electron will always have its g factor equal to g_e. The reason for deviations from g_e result from coupling of residual orbital angular momentum in the molecule (contributed for example from excited states) to the angular momentum or spin of the electron. When the orbital angular momentum is small, g approaches g_e. It is this property that makes the g factor diagnostic of particular materials.

2. Organic Radicals

Virtually all π-type organic radicals have g factors within a percent of g_e. For the most part, g factors are slightly greater than g_e, ranging from 2.0025 for hydrocarbons to about 2.0100 for some highly substituted or sulfur-centered radicals. The g factor depends upon the distribution of the unpaired electron among atoms in the radical or radical ion, and the spin–orbit coupling of those atoms. Heavier atoms have much larger spin–orbit coupling constants, and for a given aromatic nucleus,

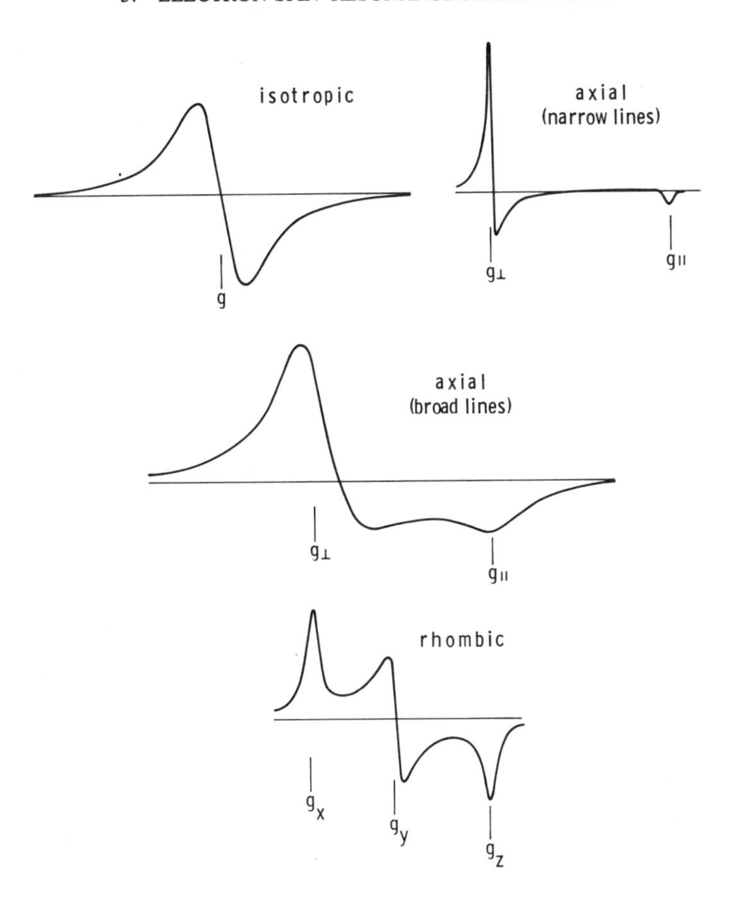

Fig. 3.2. Examples of derivative ESR spectra of solids. (*a*) Typical isotropic line; (*b*), (*c*) spectrum of species with axial symmetry $g_{\perp} > g_{\parallel}$ and narrow line width (*b*); broad line width (*c*); (*d*) spectrum of species with low symmetry $g_x \neq g_y \neq g_z$.

for example a semiquinone, halogen substitution will cause the g factors to increase in the order $g(\mathrm{H}) < g(\mathrm{F}) < g(\mathrm{Cl}) < g(\mathrm{Br}) < g(\mathrm{I})$. The same holds true for substitution of heterocyclic atoms.

One would think that g factors would be a good qualitative indication of the nature of a radical; however, there are other parameters (e.g., hyperfine splittings) which are often more easily measured can also serve this purpose. It is also important to mention that the ESR signal amplitude is quantitatively dependent upon the square of the g factor (vide infra). For most purposes, ESR spectrometers can be used to measure g with an accuracy better than ± 0.001 unit, making this error unimportant in most analyses.

At present, there is no set of comprehensive tables of ESR parameters. Some may be found in references 11, 16 and 70 for organic radicals. Solvent and medium effects may also affect the g factor, but in most cases these contributions are smaller than ± 0.001 unit.

3. Inorganic Radicals

Inorganic radicals, excluding transition metal ions, encompass a very broad range of structures. Systems which contain many atoms in asymmetric arrangements typically exhibit g factors near to g_e, but with larger deviations than exhibited by organic radicals. Radicals that contain symmetric arrangements of atoms; e.g., octahedral (AB_6), tetrahedral (AB_4), or linear (ABC or AB), can exhibit g factors very different from g_e. The g factors of inorganic radicals are discussed in reference 6.

Since most atoms and inorganic radicals are not stable at room temperature and are not usually detected in solution using ESR, they are of relatively little importance in analytical applications. Only certain atoms, namely those with S ground states can be detected in condensed phases. These include H, alkali metal, and N atoms. In these cases, the g factor is isotropic. Diatomic species cannot generally be detected at temperatures much higher than 100°K, and rarely in solution. This is because they have considerable angular momentum which allows rapid relaxation. At lower temperatures, spectra of species with axial symmetry are observed. Representative systems include $O_2^{\cdot}$ and NO adsorbed on surfaces, and halogen molecule ion defects in halide crystals. Generally, $g_\perp$ is near g_e, but $g_\parallel$ is very different than g_e. $O_2^{\cdot}$ has also been observed in solutions at room temperature, but is most likely strongly solvated or bound to a cation.

4. Transition Metal Ions

Transition metal ions also exhibit a wide range of g factors. Most transition metal ions have unpaired electrons in d levels. If it were not for the presence of a crystal field, all of the transition metal ions except d^5(Mn^{2+}, Fe^{3+}, etc.) states would exhibit large contributions of angular momentum from the orbital of the unpaired electron. The crystal field will quench or partially quench this angular momentum, and thus many transition metal ions yield strong signals at ambient temperatures and in solution: Small residual angular momenta usually increase the spin–lattice relaxation time and thus increase the intensity of the ESR signal.

g factors and linewidths of transition metal ions are very strongly dependent upon the environment. Strongly bonded complexing agents tend to quench the orbital angular momentum and shift the g factors toward g_e. The ligand field strength alone is insufficent to quench the orbital angular momentum. Ligands that are symmetrically bonded to a metal ion (e.g., six CN^- ligands) will cause splitting of the five d-electron levels into three lower and two upper states. Thus Cr^{3+} (d^3), which usually exhibits a broad signal and $g \neq g_e$ will now have one electron in each of three orbitals and is therefore nondegenerate so that $g \sim g_e$, and the linewidth also decreases. Ti^{3+} (d^1) under these conditions would still exhibit a broad signal; and the g factor will be different from g_e. If now one CN^- were replaced by an OH^- group, than a "tetragonal" crystal field would occur. This would further split the three lower-energy orbitals into a single energy level lower than the original energy level and two energy levels slightly higher than the original one. Thus, under these conditions Ti^{3+} would exhibit a strong signal.

A complete review of the effect of the crystal field environment on g factors and linewidths is beyond the scope of this chapter. For more information, the reader is directed to references 1, 5, 48, 68, 69, 85, and 110. The possibility of utilizing different ligands to cause shifts in ESR spectra and detect selectively specific components in the analysis of mixtures has hardly been exploited.

5. Gas-Phase Species

For gas-phase species, the electron spin angular momentum is coupled to the orbital angular momentum as well as rotational angular momentum (for molecules). Typically, atoms yield very intense spectra, but diatomic and linear triatomic molecules yield spectra of considerably lower intensity because there are many occupied rotational states each exhibiting different lines. Nonlinear triatomic molecules can often be detected, but their spectra are relatively weak and difficult to analyze. The details of ESR spectra of gas-phase species can be found in reference 20, 24, and 25.

The g factor of an atom in the gas phase is given by the equation

$$g_J = 1.0023 + \frac{J(J+1) + S(S+1) - L(L+1)}{2J(J+1)} \tag{6}$$

where $J = L+S, L+S-1, \ldots, L-S$, depending on the particular state of the atom. Note that when $L = 0$, $g_J = g_e$. This is true of H and N in their ground states. Oxygen exists in the thermally accessible states $^3P_{2,1,0}$, where the superscript is $2S+1$ (spin multiplicity) and the subscripts are the values of J. The letter represents the value of L (i.e., S for $L = 0$; P for $L = 1$; D for $L = 2$, etc.). For both the $^3P_2$ and $^3P_1$ states, $g_J = 1.5$, but for $J = 0$, there is no net angular momentum (g_J is undefined), and even though there are two unpaired electrons, this state is diamagnetic.

A discussion of the coupling of rotation to the spin of diatomic molecules is beyond the scope of this chapter, but sufficient information is available in the literature so that, if the analytical need arise, the spectrum can be found.

D. HYPERFINE INTERACTIONS

1. Line Positions

Well-resolved ESR spectra of many substances contain hyperfine structure. This hyperfine structure allows the identification of the paramagnetic substance in many cases, and also provides information about the environment of the molecule and the distribution of the electron density within the molecule.

The resonance frequency of an electron depends upon the magnetic field at the electron. Previously, the applied magnetic field was assumed to be the field at the electron. Actually, the electron is affected by both the applied magnetic field, H_0, and any local fields due to the magnetic fields of nuclei, or other effects, H_{local}, so that equation 7 may be written

$$h\nu = g\beta(H_0 + H_{\text{local}}) \tag{7}$$

The effect of the magnetic moments of nuclei on the ESR spectrum is called hyperfine interaction, and leads to a splitting of the ESR line (hyperfine structure). Consider a hydrogen atom, composed of an unpaired electron associated with a proton. Since the proton is a charged spinning particle with a nuclear spin, I, of 1/2, it has a magnetic moment, and the electron will be affected by the magnetic field of the proton as well as that of the applied magnetic field. The proton can be oriented parallel or antiparallel to the direction of the applied field, $M_I = +1/2$ or $M_I = -1/2$. The relative orientation of the nuclear magnetic moment in the magnetic field is unchanged during electron magnetic transitions, so that only two allowed transitions can occur. The result is a splitting of the original line into a doublet (Fig. 3.3).

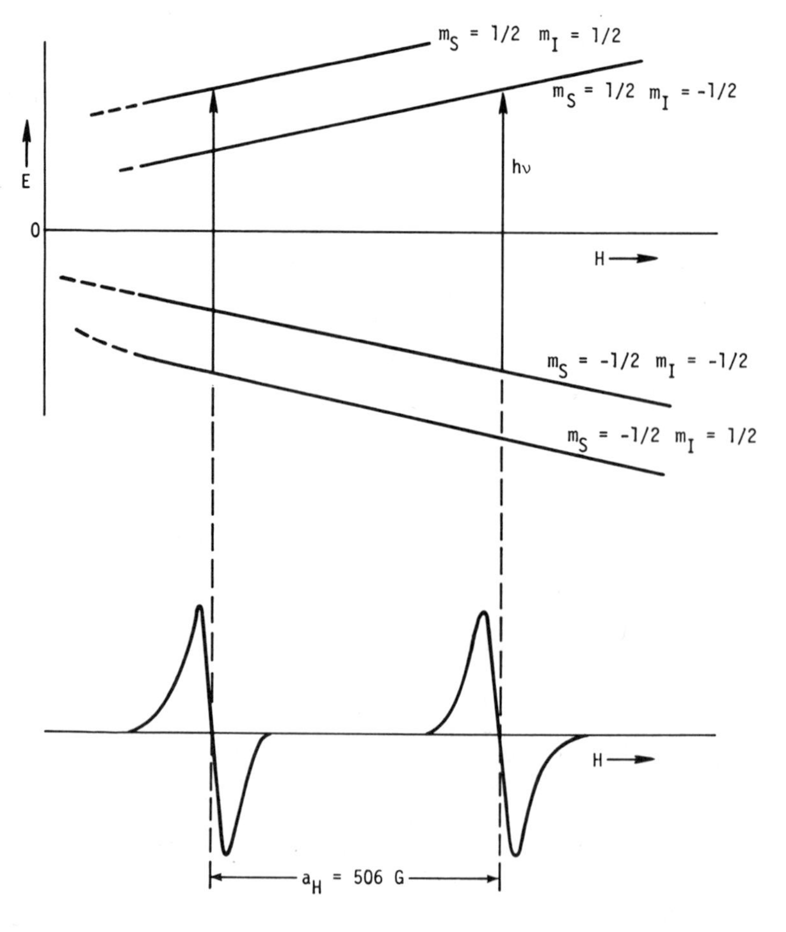

Fig. 3.3. Energy level diagram of the hydrogen atom and typical spectrum.

Another way of visualizing the resulting spectrum is as follows: When the proton points in the same direction as the applied field, the electron finds the appropriate resonance H at the lower value of H_0. When the proton magnetic moment opposes the field, a higher value of H_0 is needed for resonance. The magnitude of the splitting, usually given in gauss, is called the hyperfine coupling constant, a. For the hydrogen atom, a_H is about 506 gauss. In general, a single nucleus of spin I, will cause a splitting into $(2I + 1)$ lines, so that interaction with a single nitrogen nucleus (N^{14}, $I = 1$) will cause a splitting into three lines, and unpaired electrons in manganese(II) (Mn^{55}, $I = 5/2$) interacts with the nucleus to form a six-line spectrum. Many common nuclei, such as carbon-12 and oxygen-16, have zero spin and do not interact with the electron.

The $(2I + 1)$ line positions H_k for an equivalent set of nuclei are given by equation 8,

$$H_k = \frac{h\nu}{g\beta_e} - a_i M_i - \frac{a_i^2}{H}[I(I + 1) - M_I^2] \quad (8)$$

where $a_i = hA_i/g\beta$, and A_i is the isotropic value of the hyperfine splitting constant, and M_I is the quantum number of the component of nuclear spin in the direction of the magnetic field. The bracketed term on the right-hand side of equation 8 accounts for a second-order correction. This is usually small enough to be neglected when the hyperfine splittings are small at large applied fields. In that case, the ESR spectrum of $2I + 1$ evenly spaced lines will be centered around the field $h\nu/g\beta_e$. Deviations from first order result because the electron and nuclear spins are not decoupled by the magnetic field; i.e., the values of M_I and M_S are not exact quantum numbers and cannot be treated independently.

2. Relative Intensities

When the high-field approximation is valid, the relative intensities of the lines of an equivalent set of nuclei are determined by the multiciplicity or degeneracy of specific nuclear orientations with respect to magnetic field. For example, the degeneracies of lines due to splittings of nuclei with spins of 1/2 are given by coefficients of the binomial expansion $(a + b)^n$, where n is the number of equivalent nuclei. For a spectrum with more than one equivalent set of nuclei, the relative intensity of a line is proportional to the product of the degeneracies from each set of nuclei. Examples of the analyses of ESR spectra are given below.

For a system with the unpaired electron interacting with two equivalent $I = 1/2$ nuclei (protons), the analysis shows that formation of a triplet occurs with relative intensities 1:2:1 (Fig. 3.4). The two protons may either both oppose or both act in the direction of the magnetic field, causing the two extreme lines, or they may act in opposite directions from one another, essentially canceling their effect on the electron, resulting in the line located at the same position as the unperturbed line. Since this latter condition can occur in either of two ways, this center line is twice as intense as the two extreme lines. Continuing an analysis in this way results in the general conclusion that n equivalent protons cause a splitting into $n + 1$ lines; the relative intensities of these lines follow the binomial expansion. For example, the *p*-benzosemiquinone ion has four equivalent protons, so that its ESR spectrum shows five lines with relative intensities of 1:4:6:4:1, and a coupling constant of 2.4 gauss. For the general case of n equivalent nuclei of spin I, the resulting spectrum exhibits $(2nI + 1)$ lines. For example, two equivalent N^{14} nuclei produce a five-line spectrum of relative intensities 1:2:3:2:1.

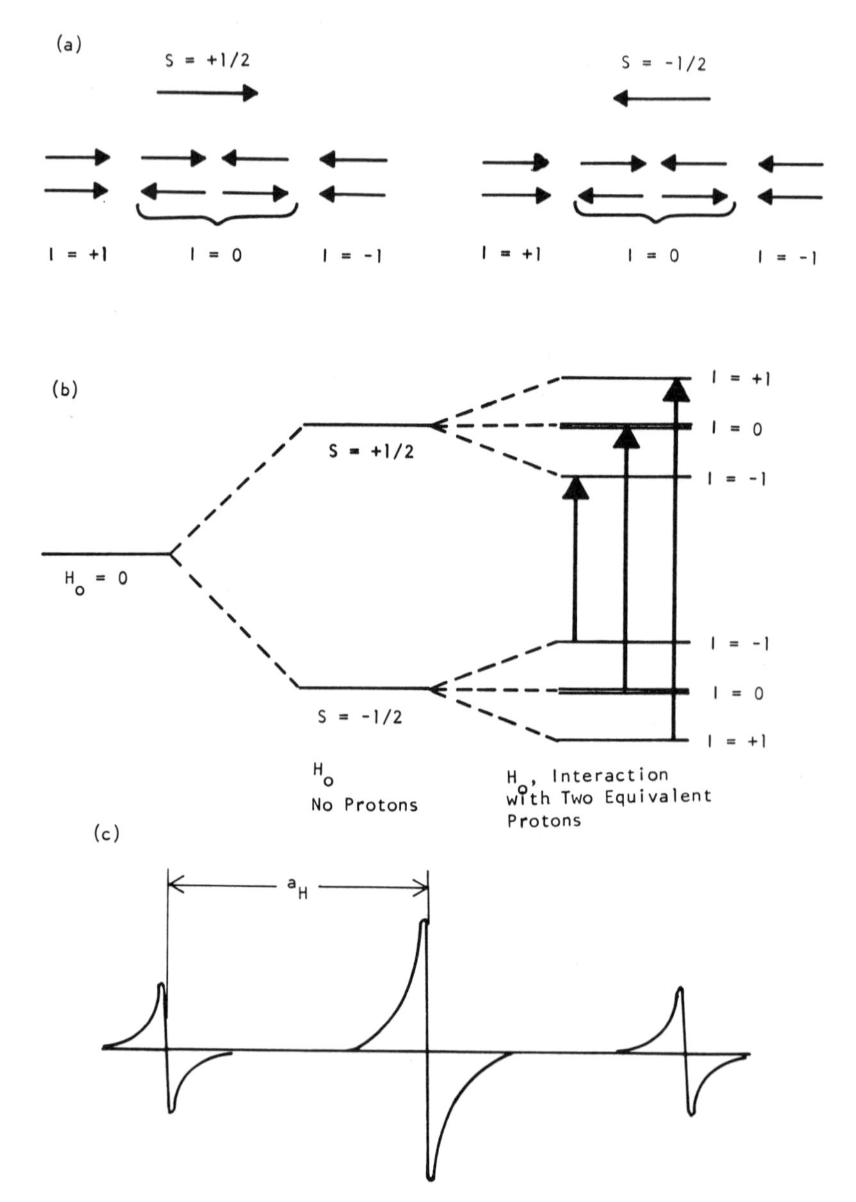

Fig. 3.4. Interaction of the unpaired electron with two equivalent protons.

For nonequivalent nuclei, each will interact with the electron with a different coupling constant to produce a hyperfine splitting. For example, if the electron interacts with one proton (H_1) with a coupling constant a_1, and another (H_2) with a

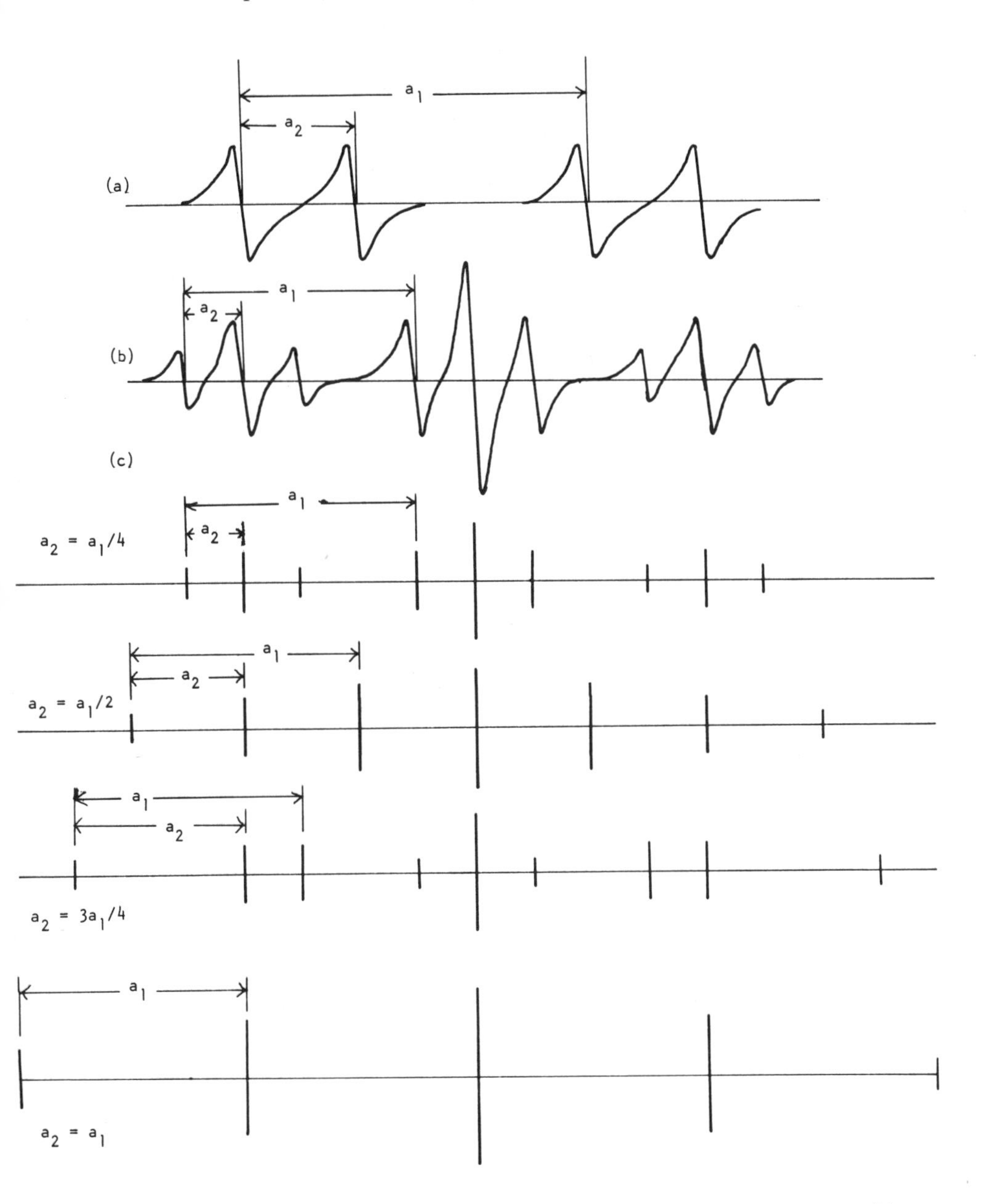

Fig. 3.5. A spectra due to interaction of unpaired electrons with: (*a*) two nonequivalent protons; (*b*) two nonequivalent sets of two equivalent protons; (*c*) Variation of spectra for interaction of two sets of equivalent pairs of protons for different values of the coupling constants a_1 and a_2.

coupling constant a_2, assuming a_1 is much larger than a_2, a doublet due to the interaction of H_1, split into a doublet due to H_2, will result (Fig. 3.5*a*). Similarly, the interaction of two equivalent hydrogens with a second pair of equivalent hydrogens results in a triplet of triplets for very different coupling constants (Fig. 3.5*b*). However, as the values of the coupling constants approach one another in magnitude, the spectra may appear to differ from the above simple behavior (Fig. 3.5*c*). Further details on the interpretation of ESR spectra can be found in references 3, 5, 7, 10, 23, 58, and 110.

3. Second-Order Effects

Although second-order effects are small and can usually be neglected, and they contribute no additional information to the ESR spectrum that would not be obtained in the first-order spectrum, they can cause distortions in the spectrum of which the user should be aware. These distortions take the form of shifts in the lines and splittings of lines which would be degenerate in the first-order case.

For the case of a single nucleus, which exhibits a large splitting, equation 8 shows that the lines will no longer be uniformly spaced. A representative example is shown in Fig. 3.6 for a V^{4+} complex in CS_2 which exhibits an isotropic spectrum. In this spectrum there is no line, for $M_I = 0$. If there were, this line would not be located

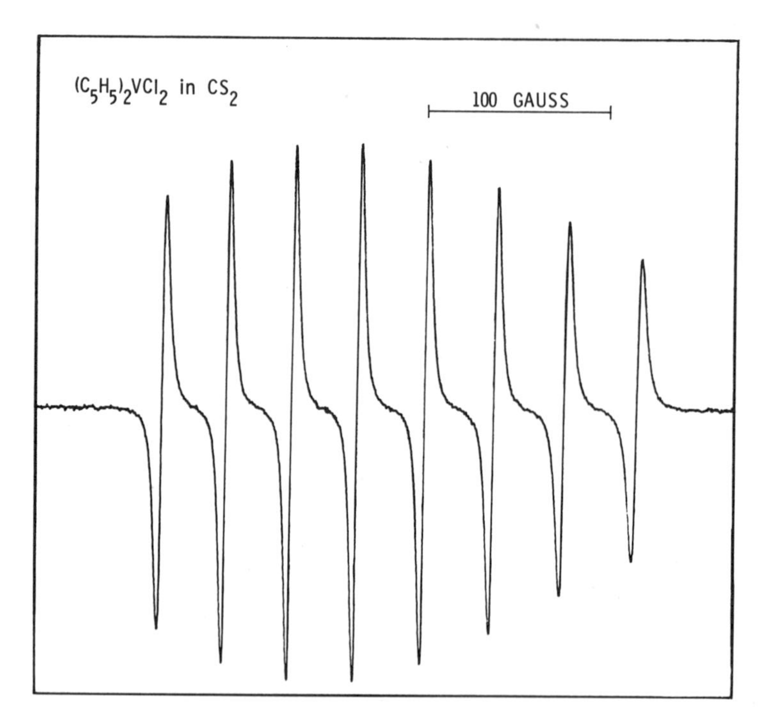

Fig. 3.6. Spectrum of V^{4+} showing unequal line spacings due to second-order shifts, and broadening effects.

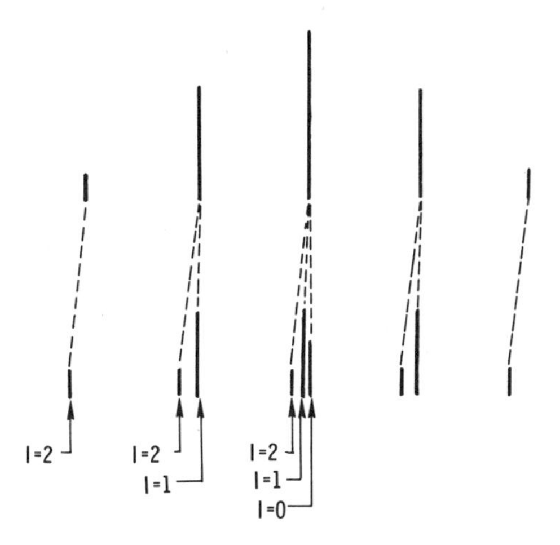

Fig. 3.7. Effect of second-order splittings on the spectrum from three equivalent $I = 1/2$ particles, with an unpaired electron.

at the field $h\nu/g\beta_e$ in contrast to first-order spectra. Also notice that the lines are not uniformly spaced. Incomplete averaging of the g- and/or hyperfine tensor causes nonuniform amplitudes. This phenomenon is typical for Mn^{2+}, V^{4+}, and VO^{2+} in solutions.

An additional second-order effect occurs when the spectra are produced by a molecule with a set of several equivalent nuclei. Usually the total nuclear spin $I = n \cdot I_x$ where I_x is the spin of the single nucleus. However, according to the rules of addition of quantized angular momentum, I can also be less than nI_x. For example, for three nuclei of $I_x = 1/2$, I can be 3/2, but it can also be 1/2. Thus, a second-order splitting can be observed according to equation 8 when the hyperfine splitting is large. Figure 3.7 illustrates this process for four "equivalent" protons, where $I = 2, 1$, and 0. Alkyl radicals and many F-substituted radicals exhibit second-order splittings.

4. Mechanism of Hyperfine Interactions

Hyperfine interactions can be classified into isotropic and anisotropic components as in the case of g factors. In solutions, where the molecule is free to rotate, only the isotropic portions of the hyperfine splittings are observed. The anisotropic part averages to zero. In spectra of single crystals or powders, both the isotropic and anisotropic parts can be separated. As in the case of g factors, the hyperfine splitting can be written in tensor form. Details of the analyses of these spectra can be found in references 1, 5, 6, 23, 48, 85, 100, and 110.

Contact Interactions: Contact interactions are dependent upon unpaired electron density at the nucleus, according to equation 9

$$A_i = \frac{8\pi}{3h} g\beta_e g_N\beta_N \mid \psi(0) \mid^2 \tag{9}$$

where g_N and β_N are the nuclear g factor and nuclear magneton, respectively, and $|\psi(0)|^2$ represents the probability of the unpaired electron being at the nucleus. Only s orbitals have a finite probability at the nucleus; for all others $|\psi(0)|^2 = 0$ at $r = 0$. However, many systems with unpaired electrons in p or d orbitals do exhibit isotropic splittings. In these cases, the p orbital will polarize the inner-shell electrons, as illustrated for the CH fragment of an aromatic system in Fig. 3.8. Note that the p orbital contains the unpaired electron, but the $C(sp^2)-H(s)$ bond is polarized such that the spin near the C atom is the same as that of the unpaired electron in the p orbital, while that near the H atom is opposite to that of the unpaired electron. Splittings that result when the spin at the nucleus is parallel to the unpaired electron are said to be positive, while those that result when the spin at the nucleus is opposite to the unpaired electron are said to be negative. Alternatively, spin can be transferred to another atom such as between the p orbital of a carbon atom and the proton of an adjacent CH_3 group. The spin on the proton in this case is of the same direction as the p orbital. This is often called hyperconjugation.

In most systems containing unpaired electrons there are local variations of the spin direction (references 1, 5, 6, 7, 10, 14, 23, 38, 48, 58, 95, and 110). Thus, the net spin at a given nucleus is the difference between positive and negative contributions. In many transition metal complexes, the dominant hyperfine splitting is due to the nucleus of the ion. In most organic radicals, and inorganic radicals, the unpaired electron is delocalized over the molecule, and significant hyperfine splittings will be observed for many atoms.

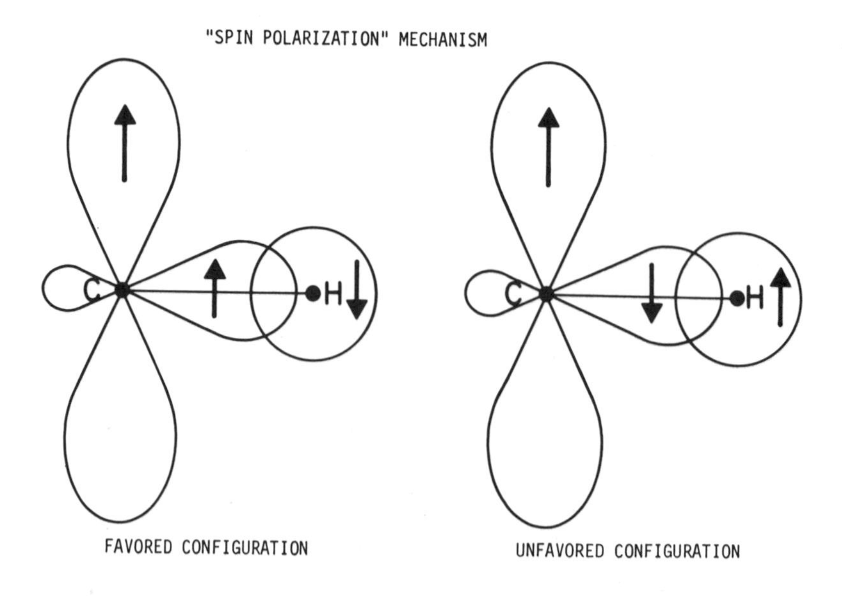

Fig. 3.8. Mechanism of spin polarization mechanism.

Dipolar Interactions: Hyperfine splittings that arise from dipolar interactions between the nucleus and the electron can only be resolved for rigid samples, and are therefore not as important in analytical applications as hyperfine splittings that arise from contact interactions. Equation 10 governs the behavior of the dipolar interaction as a function of orientation of the radical in the magnetic field,

$$E(\theta) = g\beta_e g_N\beta_N \frac{1 - 3\cos^2\theta}{\langle r^3 \rangle} \cdot S \cdot I \tag{10}$$

where r is the electron-nuclear separation and θ is the angle between the applied field and the line joining the dipoles. The average of equation (10) over all orientations θ is equal to zero.

E. ELECTRON-ELECTRON DIPOLAR INTERACTIONS

Triplet states of organic molecules and high spin states of transition metal ions have more than one unpaired electron. These electrons may interact with each other by virtue of their dipole moments. The energy of interaction is similar to equation 10, and is

$$E(\theta) = \left[g^2\beta^2 \frac{1 - 3\cos^2\theta}{\langle r^3 \rangle} \right] \cdot S_1 \cdot S_2 \tag{11}$$

When this coupling is strong, the system is very anisotropic, and solution spectra cannot be observed. If the term in brackets is smaller than the resonant frequency, then for powdered, frozen, or crystalline samples "forbidden transitions" can also be observed at about one-half of the normal applied field. In single crystals, the $g = 2$ resonance lines are strongly orientation-dependent, so that in powdered samples the resonance profile is spread over a large field. In addition to the $\Delta M_s = 1$ transitions, a forbidden "$\Delta M_s = 2$" resonance can be observed at about one-half of the normal resonance field. This line is only slightly orientation dependent. It is weak relative to the $\Delta M_s = 1$ lines exhibited by a single crystal, but strong relative to typical spectra exhibited by a powder. The transition probabilities of the $\Delta M_s = 2$ lines are difficult to calculate so that this line can only be used to determine the relative amounts of the paramagnetic sample, and has often been used to determine the population of states in which there may be a small separation of singlet and triplet energies.

F. LINESHAPES AND RELAXATION

The lineshapes in electron-spin resonance are typically Lorentzian or Gaussian. The equations of the normalized lines (integral of the absorption between $H = 0$ to $H = \infty$) are given in Table 3.III. Lorentzian lineshapes reflect the fundamental magnetic resonance and relaxation of a single transition, while Gaussian lines are inhomogeneously broadened envelopes of narrower unresolved lines. As will be seen, it is important to determine whether a derivative curve is either Gaussian or Lorentzian. Table 3.IV shows that the Gaussian curve decreases much more rapidly

TABLE 3.III

Comparison of Normalized Lorentzian and Gaussian Lines[a]

	Lorentzian lineshape	Gaussian lineshape
Equation for normalized absorption	$Y = \frac{1}{\pi\Gamma} \cdot \frac{\Gamma^2}{\Gamma^2 + (H-H_0)^2}$	$Y = \left(\frac{ln2}{\pi}\right)^{1/2} \cdot \frac{1}{\Gamma} \exp\left[-\frac{ln2(H-H_0)^2}{\Gamma^2}\right]$
Half width at half height	Γ	Γ
Equation for first derivative	$Y' = -\frac{1}{\pi\Gamma} \cdot \frac{2\Gamma^2(H-H_0)}{[\Gamma^2 + (H-H_0)^2}$	$Y' = -\frac{2(ln2)^{3/2}}{\pi^{1/2}\Gamma^3} \cdot (H-H_0) \exp \left[\frac{ln2(H-H_0)^2}{\Gamma^2}\right]$
Peak-to-peak amplitude	$2Y'_{max} = \frac{3\sqrt{3}}{4\pi} \cdot \frac{1}{\Gamma^2}$	$2Y'_{max} = 2\left(\frac{8}{\pi e}\right)^{1/2} \frac{ln2}{\Gamma^2}$
Peak-to-peak width	$\Delta H_{pp} = \frac{2}{\sqrt{3}}\Gamma$	$\Delta H_{pp} = \left(\frac{2}{ln2}\right)^{1/2} \Gamma$

[a] Taken from reference 21

in the wings of the line than does the Lorentzian curve. However, in the field between the magnetic resonance center and ± 0.7 times the peak-to-peak width, there is little distinction between both lineshapes. It is therefore important to analyze the lineshape away from the central portion.

It was mentioned earlier that many ESR transitions can be saturated by the available power in an ESR spectrometer. A solution of the Bloch equations which govern the relaxation behavior shows that the power absorbed, P_a, is proportional to the imaginary part of the complex susceptibility χ'',

$$P_a \propto \chi'' H_1^2 = \frac{\omega_0 \omega \gamma \chi_0 T_2 H_1^2}{1 + (\omega_0 - \omega)^2 T_2^2 + \gamma H_1^2 T_1 T_2} \tag{12}$$

where the equation is given in terms of frequency ($2\pi\omega$) units rather than magnetic field. The parameter χ_0 is the static magnetic susceptibility. H_1^2 is the rf magnetic field and is proportional to the incident microwave power, γ is the magnetogyric ratio $g\beta/\hbar$, and T_1 and T_2 are the spin–lattice and spin–spin relaxation times. The means by which energy is dissipated is strictly through spin–lattice relaxation at the rate governed by T_1. Thus, as long as $\gamma H_1^2 T_1 T_2 \ll 1$ the population of the spins will be undisturbed from that given in equation 5, and the power pumped into the system is effectively dissipated. However, when $\gamma H_1^2 T_1 T_2 \sim 1$, then the spins are being pumped to a higher level at a rate faster than they can relax to the level established by thermal equilibrium. The ratio of the numerator and denominator determine the net power absorbed. An increase in the denominator reflects a decrease of the population difference between spins in the upper and lower states, while the absorption of power by the remaining spins is still linear with H_1^2. At this point, P_a will increase more slowly than linearly with incident power. At the limit where $\gamma H_1^2 T_1 T_2 \gg 1$, the power absorbed will approach a constant value because sufficient energy will be put

TABLE 3.IV

Diagnostic Criteria for Derivatives of Gaussian and Lorentzian Lineshapes

$\frac{(H - H_0)}{1/2\Delta H_{pp}}$	Height Above baseline/maximum height	
	Lorentzian	Gaussian
1.00	1.000	1.000
1.40	0.911	0.849
1.50	0.870	0.808
1.60	0.828	0.733
1.70	0.784	0.661
1.80	0.739	0.587
1.90	0.696	0.515
2.00	0.653	0.446
2.20	0.572	0.331
2.40	0.500	0.222
2.60	0.436	0.146
2.80	0.379	0.092
3.00	0.333	0.055
3.50	0.240	0.012
4.00	0.177	0.002
4.50	0.133	
5.00	0.102	
6.00	0.063	
7.00	0.041	
8.00	0.029	
10.00	0.015	

into the system to maintain an equal spin population. This is known as saturation. In the absence of saturation, the linewidth is governed only by T_2.

A "rule of thumb" can be given to predict the effect of incident power which is typically valid above 77°K. If W is the microwave energy in the cavity, and P is the incident power

$$W = P\,\frac{2Q_L}{\omega} \tag{13}$$

where Q_L is Q of the cavity containing the sample. Usually the sample is placed in the region where H_1 is a maximum. In this case, one can neglect the electric field component of the rf field, and it can be shown that

$$2P\,\frac{Q_L}{\omega} \sim \frac{1}{4\pi}\,H^2_{1(\max)}\,(V_c/4) \tag{14}$$

where V_c is the cavity volume. The component of the field that causes resonance is actually 1/2 of H_1^2 calculated in equation 14. Combining all of the appropriate constants,

$$H_1(G) = 6.3\left[\frac{P(\text{watts})Q_L}{\nu(\text{MHz})\,V_c(\text{cm}^3)}\right]^2 \tag{15}$$

Typical values of Q_L for a rectangular cavity are often between 1000 and 5000, depending upon the sample. If the value calculated for H_1^2 is in the order of, or greater than the peak-to-peak linewidth, then the spectrometer is invariably set in a region that can saturate the sample. In that case, the investigator should determine carefully whether the sample is being saturated. Typical values can be used to illustrate this process. Assume $Q_L \sim 4000$, $\nu = 10^4$ MHz, $P = 1$ mW, and $V_c = 10$ cm^3. Equation 15 gives $H_1 \sim 0.04$ G. This is probably large enough to begin to saturate many organic radicals which exhibit linewidths in the order of 0.06 G or less, but will not saturate transition metal ions in which the linewidths are usually greater than about 2 G.

It is important therefore to determine that in an analytical experiment, the incident power does not cause saturation of the spin system. Before this can be ascertained, the behavior of detectors used in ESR must be understood. There are two regions in detector response. These are the square-law region where the output voltage is linear with power, and the linear region where the output voltage is proportional to the square root of the incident power. This is confusing unless it is realized that $E_{in}^2 \propto P_{in}$, so that in the linear region $E_{out} \propto E_{in}$ while in the square-law region $E_{out} = E_{in}^{1/2}$. For reasons of sensitivity, the ESR spectrometer is usually set so that the detector is in the linear region. Thus, in the absence of saturation, the ESR signal amplitude will be proportional to the square root of the incident power.

III. EXPERIMENTAL TECHNIQUES

A. INSTRUMENTATION

1. Basic Spectrometer

Since the radiation used in ESR spectroscopy is in the microwave region, an understanding of ESR instrumentation requires some knowledge of the operation of microwave components. A brief description of these and other spectrometer components and the design of a typical commercial ESR spectrometer will be given. Detailed discussions of the design and operating techniques of ESR spectrometers can be found in the general references, particularly in the books by Alger (3), Poole (87), Wilmshurst (115), and Ingram (58). A typical spectrometer (Fig. 3.9) consists of the radiation source, the klystron, and the wave guide, attenuators, isolators, and couplers for conveying the radiation to the sample contained in a resonant cavity. The microwave radiation is maintained monochromatic at the resonant frequency by an automatic frequency control (AFC) feedback circuit. The sample cavity is usually contained on one arm of a balanced hybrid T or circulator bridge. Absorption of energy by the sample causes a change in the microwave energy reflected to the diode detector. The magnetic field at the sample is produced by a large electromagnet which is slowly (several minutes to several hours) scanned and by modulation coils which cause a sinusoidal variation of the field at the modulation frequency (typically 35 Hz–100 KHz). This field modulation causes the signal reaching the crystal

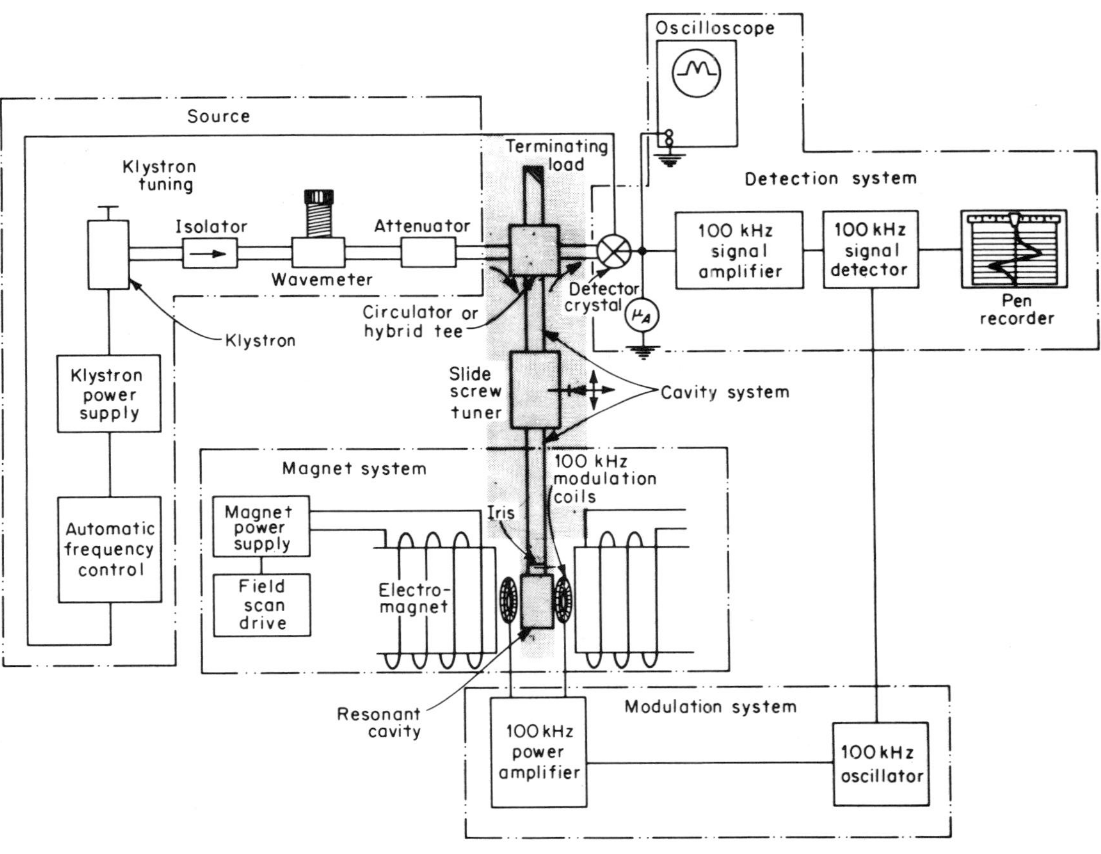

Fig. 3.9. Block diagram of typical ESR spectrometer employing 100-KHz field modulation. (Courtesy of J. E. Wertz and J. R. Bolton.)

detector to be modulated. The detected signal is amplified, demodulated with a phase-sensitive detector, and the resulting dc voltage displayed on a recorder. As a result of the use of field modulation, the demodulated signal is approximately equal to the derivative of the absorbed power as a function of magnetic field.

2. Spectrometer Components

a. Klystrons

The source of radiation is a klystron. These are available for frequencies of 2.5–220 GHz, although most analytical work is performed in the X-band (3 cm) region at about 9 GHz. A klystron can be tuned over about $\pm 3\%$ of its central frequency by varying the dimensions of the resonant cavity inside the tube. The output frequency is also a function of the resonator and reflector voltages applied to the klystron by the power supply. It is usually stabilized against temperature variations by immersion in an oil bath or by forced air or water cooling. A feedback AFC circuit, utilizing a portion of the signal fed back from the sample cavity, stabilizes the klystron frequency to about 1 – 10 ppm. The output power of typical klystrons used in ESR spectrometers is about 150–700 mW. Recently, solid-state Gunn diodes have also been used for ESR.

b. Waveguides, Attenuators, Isolators

The microwave radiation is conducted to the bridge, sample, and detector by a waveguide, which is a hollow, rectangular tube of copper or brass often plated with silver or iridium. The dimensions of the waveguide depend upon the wavelength of the microwave radiation employed. For X band, the cross-sectional dimensions are 0.9×0.4 in. The microwave power propagated down the waveguide can be decreased by inserting a variable attenuator, consisting of a piece or resistive material, into the waveguide. The power at the sample may thus be attenuated by a factor of 10^2–10^6 (20–60 dB). Reflection of microwave power back to the klystron is prevented by an isolator, which is a strip of ferrite material that passes microwaves in only one direction. This helps to stabilize the klystron frequency.

c. Cavities

The sample is contained in a resonant cavity (Fig. 3.10). This is essentially a piece of wave guide one or more half-wave lengths long in which a standing wave is set up. The cavity is analogous to a tuned circuit (e.g., a parallel R-L-C combination) used at lower frequencies. A measure of the quality of the cavity, which directly affects spectrometer sensitivity, is its Q value or Q factor, defined as

$$Q = \frac{\text{energy stored in cavity}}{\text{energy lost per cycle}} \tag{16}$$

The standing wave in the cavity is composed of both magnetic and electric fields at right angles to each other. Two cavities which are frequently employed are the rectangular TE_{102} cavity and the cylindrical TE_{011} cavity. The rectangular cavity used on Varian spectrometers is shown in Fig. 3.11. Typically the Q of a cavity without a

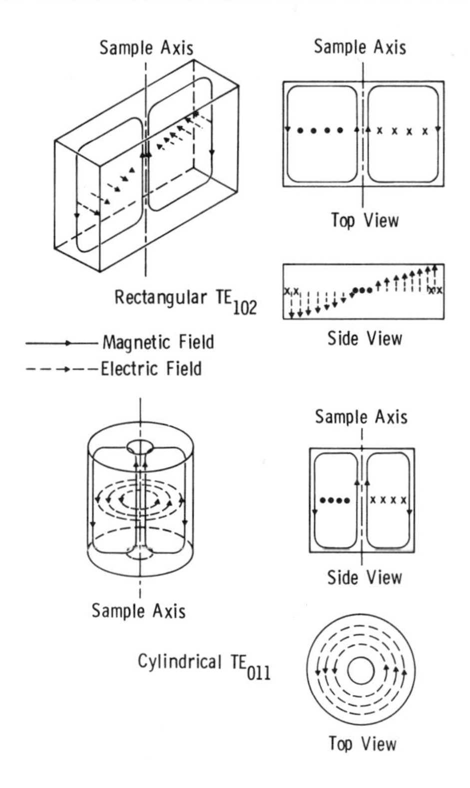

Fig. 3.10. Resonant cavities and the electric and magnetic fields of the standing waves.

sample (the "unloaded Q") is about 7000 for such a cavity. Since the component of the microwave radiation that interacts with the sample is the rf magnetic field, the sample is located in the cavity where this field is at its highest. The rf electric field also interacts with the sample, however, and if the sample has a high dielectric constant (i.e., is "lossy") the Q of the cavity may be decreased significantly. Therefore, the sample is usually located in the cavity in a position of maximum rf magnetic field and the minimum rf electric field. Rectangular flat cells with a thickness of about 0.25 mm and sample volume of 0.05 ml are often used for aqueous samples (which are particularly lossy) in rectangular cavities. With such a cell, the loaded Q of the cavity is about 2250. Tubing of 3–5-mm i.d. with sample volumes of about 0.15–0.5 ml can be used with samples that are not lossy.

Other variations of the basic cavity design include rotatable cavities for studying anisotropic effects in single-crystal and solid-sample studies. Dual cavities, consisting of two TE_{102} cavities joined to form a TE_{104} cavity, can be employed for simultaneous spectroscopic observation of a sample and a standard. These are particularly useful for precise g-value determinations and quantitative measurements. Slots can be machined into the walls of the cavity (parallel to the current direction in the walls) without degrading appreciably the cavity Q factor. Observa-

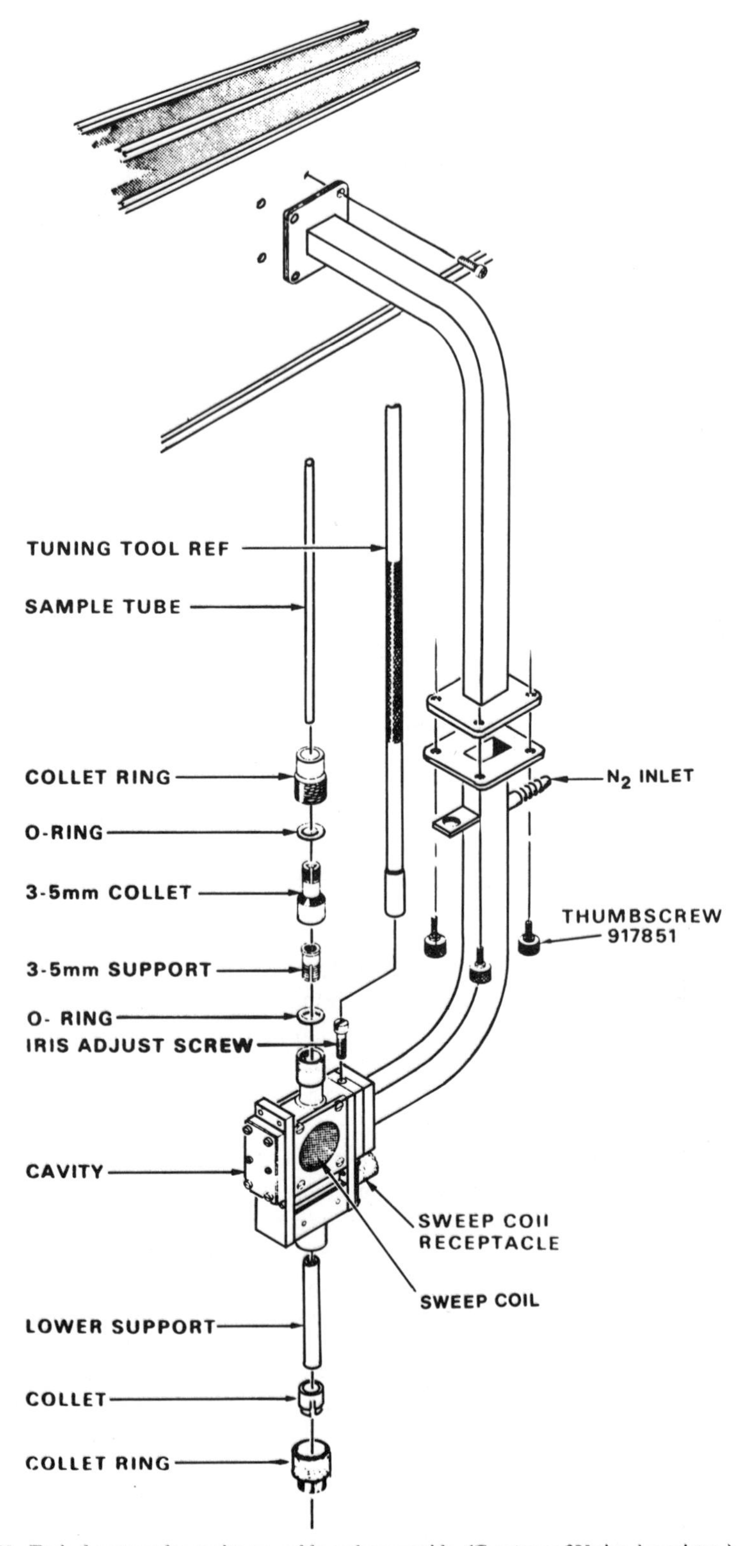

Fig. 3.11. Typical rectangular cavity assembly and waveguide. (Courtesy of Varian Associates.)

tion of species generated by irradiation of the sample with UV or visible light can thus be carried out. Cells are also available for electrogeneration of paramagnetic species directly in the cavity and for chemically producing them by flow-mixing reagent streams near the sensitive portion of the cavity. Recently TE_{101} mode microwave cavities have been made available for very lossy samples. This cavity geometry provides a two-fold increase in the signal to noise ratio over TE_{102} mode cavities when used with aqueous samples (70a).

The components of the microwave assembly can be coupled together by a variety of methods. Frequently, irises or slots of various sizes are used. For example, the resonant cavity can be attached to the wave guide with a standard-sized flange and coupled by an iris. Matching of wave guide elements (which is analogous to impedance matching in conventional circuits) is accomplished by screws or stubs, which can be positioned to the wave guide or across the coupling iris (see Fig. 3.11).

d. DETECTORS

Detectors used for microwave radiation for frequencies of 3 GHz and above are diodes that convert the rf radiation to dc. These diodes are designed so that they do not rectify the modulation frequencies (100 kHz or smaller). Thus a detected 100 kHz modulated rf wave consists of a 100 kHz signal. Initially, silicon-crystal detectors, usually selected for low noise characteristics, were used. To minimize $1/f$ noise, which is fairly high for this type of detector, 100 kHz magnetic-field modulation is used. A problem with 100 kHz modulation is that sidebands at about 36 milligauss on either side of the absorption line may arise; these are troublesome if the signal linewidth is less than about 60 milligauss, which is typical of organic radicals in solution. Superheterodyne methods can be employed in such cases, but such spectrometers, which require two klystrons, usually do not permit high-incident powers and are sometimes difficult to tune and lock so that they have not found wide general application. More recently Schottky-barrier diodes and backward diodes have been used as detectors. They do not require as much incident rf power to bias them in their most sensitive regions and they have lower $1/f$ noise characteristics. Microwave transistors recently developed also show promise as detectors for some ESR applications.

e. MAGIC-T AND CIRCULATOR BRIDGES

Rather than employ a detection technique that requires the observation of a small decrease in a large zero signal, which prevents very high amplification, a bridge arrangement is generally used. This permits the observation of a small microwave signal and is analogous to the resistance-bridge arrangement used in gas chromatography. Microwave bridges (which are analogous to impedance bridges in conventional circuits) can be of the "magic-T" or "hybrid-T," or circulator variety. A magic-T bridge is shown in Fig. 3.12. Power from the klystron and attenuator entering arm A will divide between arms B and C if the impedance of B and C is the same; no power will enter arm D, to the detector. Under these conditions the bridge is said to be balanced. If the impedance of arm B changes, say because the Q of a resonant cavity coupled to the end of arm B changes when ESR absorption by the sample occurs, the bridge becomes unbalanced and the difference between the power reflected from arms B and C enters into arm D, and is detected.

Circulators behave in a similar way to hybrid T bridges. As before, microwave power enters, but is divided between the clockwise and counterclockwise paths. Since the reflected power from each arm must complete a loop, twice as much power is incident on the cavity. As before, the difference between the power reflected from arms analogous to C and B of the hybrid T enters arm D, which houses the detector. This bridge geometry (whether wave guide, ferrite or strip line) has an advantage over the hybrid-T arrangement because it allows twice as much power at the detector and twice as much power into the cavity; thus a better signal-to-noise ratio (S/N) is obtained with a circulator element as the microwave bridge. A magic-T bridge is used in the Varian V-4502 spectrometer while a circulator is employed in the E-line series. In practice, the bridge is usually slightly unbalanced initially (i.e., with no ESR signal) to bias the detector into its most sensitive region. Alternatively, power is passed around the bridge element to the detector. This permits the cavity to be coupled to the wave guide for optimum sensitivity.

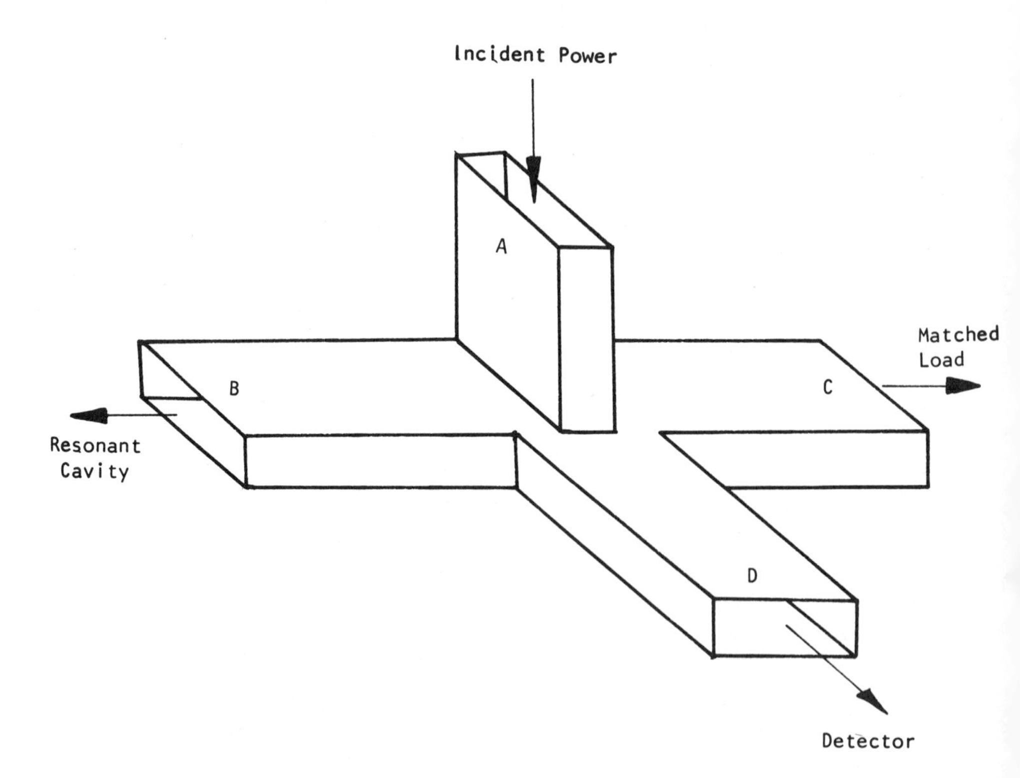

Fig. 3.12. Hybrid-T (or Magic-T) bridge element.

f. Magnets

An electromagnet capable of producing fields of at least 5000 gauss is required for X-band ESR. The homogeneity of the field for solution studies should be about 50 milligauss for studies of organic radicals and 1–2 gauss for most transition metals over the ESR sample region. This homogeneity and stability is much less than that required for high-resolution NMR studies, so that the stabilizing systems, "shimming" coils, and sample spinning techniques used in NMR are not necessary. Thus, magnets with 4-in. or 6-in. pole-piece diameters are often employed in ESR spectrometers, although 9-in. or larger magnets offer wider gaps, which are convenient in many types of ESR work, and allow easier attainment of the required homogeneity over a larger volume. The ESR spectrum is recorded by slowly varying the magnetic field through the resonance condition by sweeping the current supplied to the magnet by the power supply; this sweep is generally accomplished with a variable-speed motor drive. Both the magnet and the power supply may require water cooling. In most cases, the magnetic field is regulated by using a feedback circuit to sense changes in the magnetic field and correct for these changes. This approach involves the use of a field-strength sensor (e.g., a Hall probe or rotating coil), which generates a signal that is proportional to the field.

g. Modulation Coils

The modulation of the signal at a frequency consistent with good S/N in the crystal detector is accomplished by a small alternating variation of the magnetic field. Modulation amplitudes from 0.05 to 40 gauss peak-to-peak are frequently used. This variation is produced by supplying an ac signal to modulation coils oriented with respect to the sample in the same direction as the magnetic field. For low-frequency modulation (400 Hz or lower) the coils can be mounted outside the cavity and even on the magnet pole pieces. Higher modulation frequencies (1 KHz or higher) cannot penetrate metal effectively, and either the modulation coils must be mounted inside the resonant cavity or cavities constructed of a nonmetallic material (e.g., quartz or ceramic) with a thin silver plating must be employed.

At high modulation amplitudes, (e.g., larger than 10 gauss) heating of the cavity can occur. This causes dimensional changes in the cavity and may result in baseline drift.

3. Commercial Spectrometers

A typical commercial ESR spectrometer is shown in Fig. 3.13 and several types which are currently in use and their characteristics are shown in Table 3.V. Although the systems involving low frequencies (200–800 MHz) can employ a small permanent magnet or Helmholtz coils and are relatively inexpensive, their sensitivity is low, and they are mainly employed for educational and demonstration purposes. The bulk of the analytical methods and research studies have employed X-band spectrometers. In recent years, dedicated minicomputers have been incorporated into the spectrometer and signal-averaging devices (e.g., computers-of-average transients, CATs) have been used to improve the signal-to-noise in very weak spectra. Currently, high quality

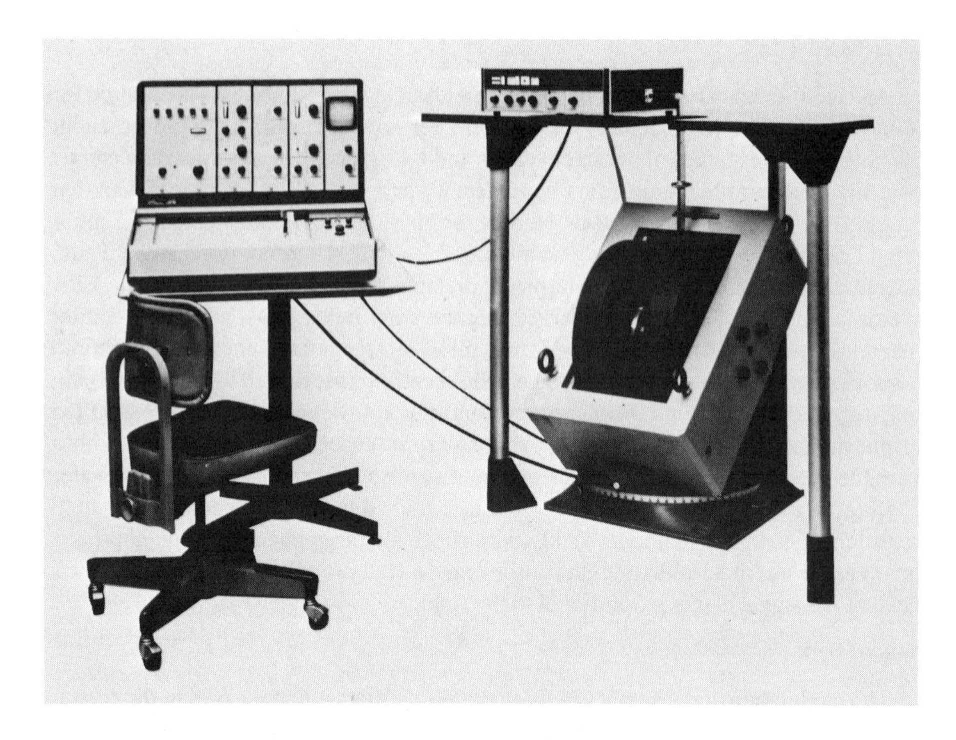

Fig. 3.13. A modern ESR spectrometer. The electromagnet and microwave bridge are at the right, the power supplies, amplifiers, and recorder are contained on the console on the left. (Courtesy of Varian Associates.)

spectrometers which operate at selected frequencies between 1.1 and 35 GHz are commercially available.

B. SENSITIVITY

The sensitivity of a spectrometer, in terms of the smallest detectable concentration of a given species, depends upon spectrometer parameters (e.g., microwave power, cavity Q, bandwidth and detector, klystron and amplifier noise), signal parameters (e.g., linewidth, relaxation times), and sample characteristics (e.g., solvent). The absorption signal depends on the magnetic susceptibility of the sample, which, in turn, is a function of the number of unpaired electrons or spins contained in the cavity. In general, the minimum number of detectable spins $N_{\min}$, is proportional to several variables (39,87)

$$N_{\min} \propto \frac{\Delta\nu}{Q_0 \eta \nu_0} \left(\frac{\Delta f k T}{2P_0} \right)^{1/2} \tag{17}$$

where $\Delta\nu$ is the linewidth of the absorption, Q_0 is the Q of the unloaded cavity, η is the filling factor described in equation 18, ν_0 and $\Delta\nu$ are the frequency and linewidth of

TABLE 3.V
Typical Characteristics of ESR Spectrometers

Frequency	Sensitivity[a] (unpaired spins)	Cavity	Magnetic field at frequency for $g = 2$ (gauss)	Manufacturer[b]
335 MHz	$10^{14}\,\Delta H$	Helix	120	Alpha, Ealing
200–800 MHz		Helix	70–290	Bruker
1.1 GHz (*L* band)		Helix		Bruker
3 GHz (*S* band)	$2 \times 10^{13}\,\Delta H$		1,100	Bruker, Micro-now
9.5 Ghz (*X* band)	2×10^{11} to $5 \times 10^{10}\,\Delta H$	TE_{102}, TE_{104}, TE_{011}, TE_{101}	3,400	Bruker JEOL Micro-now Varian Decca Thompson-CSF Ventron Hilger and Watts
24 GHz (*K* band)	$10^{10}\,\Delta H$	TE_{012}, TE_{011}	8,600	Bruker JEOL Ventron
35 GHz (*Q* band)	5-6 $\times 10^{9}\,\Delta H$	TE_{011}, TE_{012},	12,500	JEOL Varian Ventron
70 GHz (*V* band)	$10^{9}\,\Delta H$	TE_{012}, TE_{011}	25,000	Ventron

[a] Approximate; represents the minimum number of detectable spins, where ΔH is the signal linewidth in gauss at half-maximum absorption with a 1-sec time constant.

[b] Alpha Scientific Laboratories, Inc., no longer manufactured; Bruker Instruments, Inc., sold in the U.S. by IBM Instruments, Danbury, Connecticut; Decca, no longer manufactured; The Ealing Corp., South Natic, Massachusetts; Hilger and Watts, London, England; JEOL Analytical Instruments, Inc., Cranford, New Jersey; Micro-now Instrument Co., Inc., Chicago, Illinois; Thompson-CSF, no longer manufactured; Varian Instrument Division, Palo Alto, California; Ventron Instruments, Corp., no longer manufactured.

the absorption, Δf is the bandwidth of the response of the spectrometer, P_0 is the incident power, k is Boltzmann's constant, and T is the detector temperature in degrees Kelvin. The filling factor measures the effectiveness of coupling of the sample to the microwave magnetic field and is essentially the integral of the power density over the sample divided by the integral of the power density over the volume of the cavity,

$$\eta = \frac{\int_{V_s} H_1^2 \, dV_s}{\int_{V_c} H_1^2 \, dV_c} \tag{18}$$

where H_1 is the rf magnetic field, V_s is the sample volume, and V_c is the cavity volume.

To maximize the sensitivity of the spectrometer, several parameters in equation 17 can be optimized. For example, the incident power can be increased, assuming that the absorption is not saturated, and the extent of filtering can be increased (i.e., the Δf decreased). Often it is possible to use a higher microwave frequency (and thus a higher H) to increase the population difference between the two energy levels. For a given quantity of sample, where the geometry is kept the same (except that the dimensions of the cavity are made proportional to the frequency), it can be shown that the sensitivity is proportional to $\nu_0^{7/2}$. In practice, however, the cavity Q decreases at higher frequencies, and klystrons and detectors are more noisy. Theoretically, a hundredfold increase in sensitivity would be expected on increasing the frequency from 9 to 35 GHz, but in fact only a tenfold gain is usually realized. On the other hand, if the sample size is unlimited, the sensitivity per unit volume of sample actually decreases at higher frequencies. It can be shown that if the same geometry is maintained, the minimum detectable concentration is theoretically proportional to $\nu_o^{-1/2}$.

Finally, one must consider the filling factor and the cavity Q. For a given sample the best S/N is obtained when the cavity Q is reduced to two-thirds of that of the unloaded cavity by introducing the sample. This puts some constraint on the filling factor for a given sample with a minimum reduction of Q. As a general rule, at X-band frequencies, rectangular cavities provide better performance for lossy samples, while cylindrical cavities show better sensitivity for small-volume or solid samples. For X-band spectrometers concentrations of about 10^{-9} M can probably be determined in samples with very low dielectric losses. For aqueous solutions, 10^{-7} M probably represents a realistic estimate of the lower limit of detection. Of course, signal-averaging techniques can improve the signal-to-noise and increase the effective sensitivity, assuming that the magnetic field and spectrometer parameters can be maintained constant over the times necessary to accumulate the signals. Signal averaging will probably be more useful in the detection and identification of species than in their quantitative determination.

C. TREATMENT OF DATA FOR ANALYTICAL APPLICATIONS

1. Spectroscopic Analysis

The ESR spectrum is usually recorded directly from a phase-sensitive detector as the first derivative of the absorbed power as a linear function of the applied magnetic field. Spectroscopic analysis; i.e., determination of g factors, hyperfine splittings, and zero-field splittings, useful for qualitative analysis can be derived directly from this data. This is discussed in all of the general texts.

2. Quantitative Analysis

There are two general types of quantitative analysis that can be carried out using ESR. The first of these is general to analytical spectroscopy and involves the preparation of standards of a similar nature to the unknown, and a comparison of the amplitude of response of the unknown with those of the standard. The second method is essentially an absolute calibration which relies on the fact that the transition probability of the unpaired electron can be calculated accurately from the spectroscopic data. This allows any well-characterized material to be used as a standard.

3. Method of Direct Comparison

When a direct calibration of the signal response is carried out utilizing known samples, generally the peak-to-peak amplitudes of the ESR signal are used as a measure of concentration. For most species, the amplitude of response for a given molecule will be linear with concentration over several orders of magnitude. As the concentration increases, however, the lines will begin to broaden so that the amplitude will become less than linear with concentration. This broadening is due to exchange of unpaired electrons between molecules that tends to average the hyperfine fields to yield one line at very fast exchange rates. Nevertheless, this broadening will not affect the analysis provided the concentration of the unknown is within the limits of the standards.

There are also several instrumental conditions that must be maintained during the analysis. The ESR spectrometer output is the derivative of the power absorbed, as opposed to optical spectrometers in which the output signal is a function of the relative power absorbed by the sample. Since the ESR signal is usually proportional to the square root of the incident power (Section II.F), it is important the incident power be accurately set. It is also important that the iris on the ESR cavity be reset to the same point. This is easily done by keeping the current through the detector (crystal current or leakage current) the same. Finally, even under conditions of partial power saturation of the ESR signal, analyses can be carried out because the response of the unknown will be identical to the standard. The linearity of the ESR signal is dependent upon a small change in the Q factor of the cavity during resonance. For very intense signals, the Q factor will change significantly and again the ESR signal will increase less than linearly with concentration.

4. Sample Handling

For accurate analyses, several precautions must be taken to insure that reproducible results are obtained. These are listed below.

1. For liquid and gaseous samples the same cells must be used, and these must be positioned reproducibly in the cavity. Unlike UV spectroscopy, ESR samples are much smaller than the wavelength so that they must be placed at the same region of oscillating magnetic field.

2. For a liquid sample, the same solvent should be used for the standards and samples. Different solvents have different losses, and thus changes in the cavity Q may result if different solvents are used.

3. Solution concentrations of electrolytes greater than ca 0.5 M will alter the cavity Q factor, and thus change the calibration factor.

4. If powdered samples are used, they must also be of the same geometry as the calibration sample, and preferably of the same matrix. Typically, powdered samples do not affect the Q of the cavity as much as liquids unless they are good conductors.

Several useful techniques to assist analyses have been reported in the literature. Often "flat cells" are used for liquid samples, however, it was found that a cylindrical 3-mm tube held in place by Teflon plugs which fit firmly into the collet rings of the ESR cavity (Fig. 3.11) insure reproducible repositioning of liquid samples (28).

Often a dual cavity is used to assure reproducible ESR settings. In this application, an arbitrary reference sample is used, so that if the ESR settings are changed between experiments, the amplitude of the reference will also be altered, and a correction can therefore be made. It has also been pointed out that a double internal standard, i.e., two equal secondary standards placed on opposite sides of the unknown sample in the ESR cavity permit an internal calibration for a single-sample cavity. Dilute Mn^{2+} in a host lattice gives no signal within ± 40 G of $g = 2$ since it exhibits no center line (73). This technique was also found to reduce the error caused by varying amounts of water in biological samples which change the cavity Q.

5. Theory of Absolute Measurements

a. General Case

Part of the versatility of the ESR technique is due to the ability to calibrate the instrument with any standard for the determination of an unknown within the constraints of geometry. Most of the theory that relates the ESR signal amplitude to concentration has been developed for use in gas-phase work (17,112–114). This approach provides the most general treatment, so that the same approach will be followed here. The general case can be simplified for many specific applications, some of which will be considered. Finally, techniques of determining the double integral of the ESR derivative spectrum and errors involved with it will be discussed.

The imaginary part of the susceptibility which gives rise to power absorption for the transition between states i and j is given by

$$\chi''_{ij} = \frac{n_{ij}}{h}\mu_{ij}^2 f'(\nu - \nu_0) \tag{19}$$

where n_{ij} is the difference in the number of molecules in the upper and lower spin states (e.g., $N_{-1/2} - N_{1/2}$), $f'(\nu - \nu_0)$ is a normalized lineshape such that its integral between $\nu = 0$ and $\nu = +\infty$ is unity, and μ_{ij}^2 is the square of the transition probability matrix element for the transition between states i and j. First $f'(\nu - \nu_0)$ must be converted to a field-dependent function to be consistent with ESR instrumentation. An effective g factor, g_{eff}, is defined as

$$g_{\text{eff}} = \left(\frac{h}{\beta}\right)\frac{d\nu}{dH} \tag{20}$$

which is analogous to equation 4. Usually $g_{\text{eff}} = g$, except in systems in which the hyperfine splitting or dipolar interaction is of the order of the applied field, so that the high-field approximation is no longer valid.

The population n_{ij} is given by

$$n_{ij} = \frac{h\nu_0}{kT}\cdot\frac{\exp(-E_i/kT)}{Z}\cdot N \tag{21}$$

where E_i is the energy of the lower state of the transition excluding the contribution from the electron-magnetic field interaction, N is the total number of molecules in the sample, and Z is the partition function. The term $h\nu_0/kT$ arises from the electron-magnetic field interaction, equation 5, for an unsaturated sample. The second term arises from thermally accessible energy levels. For most species, $\exp(-E_i/kT)$ is equal to unity; however, it is important for gas-phase systems and for systems such as those which have low energy excited states such as a singlet-triplet state equilibrium or systems which exhibit very large electron-electron dipolar interactions.

Combining equation 19, 20, and 21, the total number of molecules is given by

$$N = \left(\frac{kT}{h\nu_0}\right)\left[\frac{Zg_{\text{eff}}}{\mu_{ij}^2\exp(-E_i/kT)}\right]\int_0^\infty \chi''_{ij}\,dH \tag{22}$$

for a system that exhibits a single transition. In most cases, a system will exhibit many transitions. Often some of these transitions are degenerate or unresolved so that it is easier to determine the integral of χ_{ij}^{11} over several transitions. Equation 22 then becomes

$$N = \left(\frac{kT\beta}{h\nu_0}\right)\sum_i\left[\frac{Z_i(g_{\text{eff}})}{\mu_{ij}\exp(-E_i/nT)}\right]\int_0^\infty \chi''_{ij}\,dH \tag{23}$$

At this point the partition function and the transition probabilities should be determined. The partition function Z is given by equation 24

$$Z = \sum_i (2J + 1)\Pi_k(2I_k + 1)^{n_k}\exp(-E_i/kT) \tag{24}$$

where n_k is the number of nuclei of spin I_k. J can have the values $|L + S|$, $|L + S - 1|$, ..., $|L - S|$ as described in Section II.C.5. (For condensed-phase species $L = 0$ so that $J = S$.) The product of $(2I_k + 1)^{n_k}$is the degeneracy due to hyperfine splitting while $(2J + 1)$ is the degeneracy of the spin state.

The transition probability, μ_{ij}^2 is only dependent upon the g factor and spin, and is given by equation 25 (5,23,58,110).

$$\mu_{ij}^2 = \frac{1}{2}g^2\beta^2(J - M_J)(J + M_S + 1) \tag{25}$$

so that equation 24 becomes

$$N = \left(\frac{2kT}{h\nu_0\beta}\right)\left[\sum\frac{g_{\text{eff}}Z}{g_J^2(J - M_J)(J + M_J + 1)\exp(-E_{J,MJ}/kT)}\right]\int_0^\infty \chi''_{JM_J}\,dH \tag{26}$$

Equation 26 can be used for all analyses; however, simpler forms can be determined for specific applications.

b. Condensed Phases

Consider, for example, a radical in a solution or a solid with $S = 1/2$, which exhibits only one line with a transition from $M_S = -1/2$ to $M_S = +1/2$. Most radicals have no near-lying electronic energy levels so that $\exp(-E_{S,M_S}/kT) = 1$. The value of Z is therefore $2S + 1 = 2$, and $(S - M_s)(S + M_s + 1) = 1$. In this case, equation 26 becomes

$$N = \left(\frac{2kT}{h\nu_0\beta}\right) \cdot \frac{2}{g} \int_0^{\infty} \chi'' dH \tag{27}$$

If the spectrum of the radical anion was split by four equivalent protons, the degeneracies of each of the five lines are 1:4:6:4:1. The term $(2I + 1)^{nk} = 16$. If the center line of degeneracy 6 is used, equation 26 then becomes

$$N = \left(\frac{2kT}{h\nu_0\beta}\right) \cdot \frac{32}{6g} \int_0^{\infty} \chi'' dH \tag{28}$$

It can be shown that by substitution of $g\beta H_0$ for $h\nu_0$, and by summing all of the transitions used for the analysis

$$N = \frac{3kT}{g^2\beta^2 H_0[S(S + 1)]} \cdot \frac{\Pi[(2I_k + 1)^{n_k}]}{\Sigma_l D_l} \int_0^{\infty} \chi'' dH \tag{29}$$

where $\Sigma_l D_l$ is the sum of the degeneracies of all lines l which are used in the analysis. This form of the equation has been used for most analytical work.

In solids when $S > 1$, the crystal fields may cause splittings of the lines with different values of M_s. A typical example would be Fe^{3+} in a cubic host (MgO, for example), where $S = 5/2$, so that five transitions are observed: $M_s = -5/2 \rightarrow M_s = -3/2$, $M_s = -3/2 \rightarrow M_s = -1/2$, etc. The hyperfine splitting due to Fe^{3+} is not resolved so that $Z = 6$. In this case if all of the lines are utilized in the analysis, equation 29 may be used to determine N, but if only one of the transitions is utilized, equation 26 must be used because $(S - M_s)(S + M_s + 1)$ will depend upon which transition is selected. In solution, only one line will be observed, so that equation 29 will again be valid.

In solution and under most conditions in solids, and for S-state atoms in the gas phase, equation 29 can be used to determine analytical concentrations. Upon the condition that there are large fine splittings due to dipolar interactions, in a solid, the more elaborate form, equation 26, must be used.

c. Gas Phase

A typical case in which the exponential in equation 21 is not unity is in the analyses of gas phase species. This subject is beyond the scope of this chapter, but has been treated in detail (112). For atoms, there are many cases in which there are thermally accessible states. For example, the oxygen atom (Section II.C.5) has three states, $^3P_2$, $^3P_1$, and $^3P_0$, with energies of 0, 453, and 647 cal/mole, respectively. The $^3P_2$ state exhibits four closely spaced lines, and the $^3P_0$ state exhibits two lines, one on either side of the $^3P_2$ spectrum. The $^3P_0$ state is diamagnetic. Depending on which lines are used for the determination of 0, equation 26 will have different parameters.

Diatomic species also have rotational components, thus Z must include the rotational partition function as well. In addition, some of the transitions of diatomics such as NO are not only magnetic, but also electric by virtue of the dipole moment (17). It is therefore convenient in the case of gas phase atoms and molecules to define a term Q given in equation 30

$$Q = \frac{\beta^2 Z}{2} \sum_i \frac{g_{(\mathrm{eff})}}{\mu_{ij}^2 \exp(-E_i/kT)} \tag{30}$$

such that

$$N = \left(\frac{kT}{h\nu_0}\right) Q \int_0^\infty \chi_{ij}'' \, dH \tag{31}$$

Westenberg (112) has calculated an extensive list of values of Q for atoms and diatomics at 300°K which can be used for analysis. Most applications of gas phase analysis have been applied to kinetic studies.

d. Double Integration

The value of $\int \chi'' dH$ must be determined from the derivative spectrum. This requires double integration of the ESR signal or similar handling of the data. For any absorption lineshape $f(H - H_0)$,

$$\int_0^\infty \chi'' dH = \frac{K}{A \cdot H_m \cdot p^{1/2}} \int_{-\infty}^{\infty} \int_{-\infty}^{H} \left[\frac{df(H - H_0)}{dH} dH\right] dH \tag{32}$$

where the ESR data are presented as $df(H - H_0)/dH$ and K is an instrumental constant that must be determined by calibration for a constant sample geometry. A is the amplification, and p is the incident power. Equation 32 requires that the signal must be doubly integrated. However, it can also be shown that

$$\int_{-\infty}^{\infty} \int_{-\infty}^{H} \left[\frac{df(H - H_0)}{dH} dH\right] dH = \int_{-\infty}^{\infty} (H - H_0) \frac{df(H - H_0)}{dH} dH \tag{33}$$

That is, the double integral in the limit of integration from $+\infty$ to $-\infty$ can be written as the integral of the product $(H - H_0)[df(H - H_0)/dH]$, equation 33. The advantage of this form is that it converges more rapidly than does the double integral (29,90).

Although it is important to double integrate a line over the field range from $+\infty$ to $-\infty$ it is never possible to approach this condition because of interferences due to other spectral lines or instrumental drift. Nevertheless, it is a relatively simple matter to correct the integrated portion of the line to approach the limit.

Figure 3.14 shows the discrepancy between the line integrated between the limits of $+\infty$ and $-\infty$, and the truncated scan. Notice that the first moment (M) converges more rapidly than does the double integral (I) of the derivative curve. In either case, if the lineshape is known (e.g., Gaussian, Lorentzian, or an analytical equation) it is a simple matter to correct the integral. For example, for a truly Lorentzian line, the double integral of the derivative of a Lorentzian absorption is given by equation 34 for a spectrum recorded between $H_0 - H_a$ and $H_0 + H_a$.

$$I_L = -\frac{2}{\pi\Gamma}\int_{H_o - H_a}^{H_o + H_a}\int_{H_o - H_a}^{H}\frac{(H - H_0)dH'}{[\Gamma^2 + (H - H_0)^2]^2}dH =$$

$$\frac{2}{\pi}\left\{\tan^{-1}\left[\frac{(H - H_a)}{\Gamma}\right] - \frac{(H - H_a)/\Gamma}{1 + \left(\frac{H - H_a}{\Gamma}\right)^2}\right\} \tag{34}$$

On the other hand, the first moment is given by equation 35

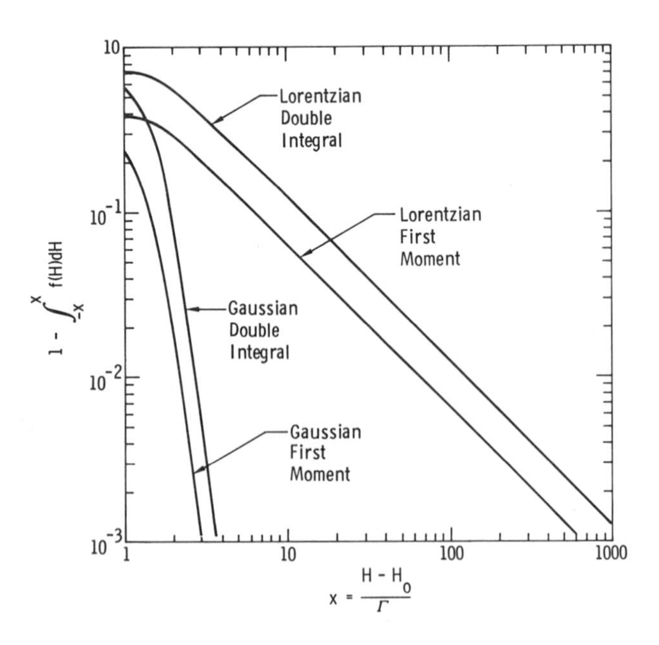

Fig. 3.14. Errors due to truncated integration of ESR lines.

$$M_L = -\frac{2}{\pi\Gamma}\int_{H_0 - H_a}^{H_0 + H_a} \frac{(H - H_0)^2 dH}{[\Gamma^2 + (H - H_0)^2]^2} = \frac{2}{\pi}\tan^{-1}\left[\frac{(H - H_a)}{\Gamma}\right] \tag{35}$$

Notice that the I_L is smaller than M_L by an amount approximately proportional to $1/(H - H_0)$ for large scans. Since the scan divided by Γ can be determined from ΔH_{pp} (Table 3.III), a correction can be made. This procedure is illustrated for the double integral of a line due to ground state 0_2 at 298°K in Table 3.VI (44a). The oxygen line selected, line C of reference 114, was shown to be Lorentzian. All spectra were recorded over the same magnetic field range but due to pressure broadening, the ratio of scan to linewidth decreases with pressure. Even though the ratio of the scan to the linewidth changes by nearly a factor of 2, the correction term does not change by more than a few percent because the spectrum was recorded well into the wings. Yet

TABLE 3.VI

Double Integral and Corrected Double Integral of a Line of Ground State 0_2

Pressure	scan / ΔH_{pp}	I_L^a / Gain · pressure	Correction factor	I_L (corrected) / Gain · pressure
0.096	21.3	4.518	1.115	5.038
0.096	21.3	4.453	1.115	4.965
0.096	21.3	4.414	1.115	4.922
0.175	18.3	4.317	1.136	4.904
0.175	18.3	4.400	1.136	4.998
0.175	18.3	4.309	1.136	4.895
0.312	17.1	4.403	1.147	5.050
0.312	17.1	4.391	1.147	5.036
0.312	17.1	4.372	1.147	5.015
0.414	16.7	4.323	1.151	4.976
0.414	16.7	4.312	1.151	4.963
0.414	16.7	4.281	1.151	4.927
0.518	15.7	4.409	1.162	5.123
0.518	15.7	4.290	1.162	4.985
0.518	15.7	4.420	1.162	5.136
0.697	15.0	4.395	1.170	5.142
0.697	15.0	4.322	1.170	5.056
0.697	15.0	4.402	1.170	5.150
0.857	14.5	4.348	1.177	5.118
0.857	14.5	4.233	1.177	4.982
0.857	14.5	4.217	1.177	4.963
0.999	13.6	4.230	1.191	5.038
0.999	13.6	4.130	1.191	4.919
0.999	13.6	4.202	1.191	5.004
Mean		4.337		5.013
σ		0.091		0.078
Relative error		2.1%		1.55%

[a] Experiments were carried out at the same incident power, modulation amplitude, and scan rate.

this correction would allow for O_2 to be used as a calibration for any other gases in the same container. Correction factors of up to about 35% have been used in a number of samples with good results (44). Similar methods have been used for powdered isotropic as well as liquid samples.

e. SOURCES OF ERROR IN DATA HANDLING AND RECORDING

Various sources of error in data handling have been discussed in the literature, and quantitative estimates of some of these errors have been provided (63,67,90). Some authors include in their errors the assumption that the first derivative data starts at an amplitude of zero. We have made the assumption in the correction factors that the true zero point can be determined.

Baseline: Exact determination of the zero point of the derivative signal amplitude is important. Either an offset, α, or a drift $\beta \cdot H$ (assumed linear with field) will cause significant errors. The error due to a fixed offset will cause an error proportional to the square of the scan range for the first moment calculation, while in the double integral calculation a linear term must also be added. For a baseline drift, the error in the first moment will be proportional to the cube of the scan range while for the double integral, a quadratic term must also be added. Thus the first moment offers another advantage in that baseline errors are minimized. Equations have been derived for double integration in the presence of baseline drift (72).

We have found that if a spectrum is symmetrical, or if it is recorded into the wings and there is no drift problem, the average value of all of the data points, equally spaced in field, gives an accurate value of the baseline. When there is a drift problem, the spectrum must be recorded far out into the wings, and a line drawn. A good check to find whether the spectrum does approach the baseline for a Lorentzian curve is to plot the derivative signal amplitude against $1/(H - H_0)^3$. This line should extrapolate to the baseline as $H \rightarrow \infty$, and drifts will become evident. This technique has been utilized in ferromagnetic resonance but it has not been applied to paramagnetic resonance.

Power saturation: The effect of power saturation has been discussed in Section II.E. Since values of T_1 and T_2 of equation 12 are not well known, it is difficult to correct for errors caused by saturation. Even in this case, if standards of the same material can be obtained and the double integral recorded as a function of power, then one can correct for saturation. However, if these standards are available, the peak-to-peak amplitudes can then be used for analysis.

Modulation broadening: Up to this point we have assumed that the lineshape is the derivative of the absorption. However, a finite modulation amplitude will cause line broadening. Modulation-broadened lineshapes have been calculated for Gaussian and Lorentzian lines (22,99,108), and the relative peak-to-peak linewidth $[(\Delta H_{pp}/\Delta H_{pp}(o)]$ of a Lorentzian line is shown as a function of modulation amplitude in Fig. 3.15. H_m is defined by the equation $H_m(t) = H_m\cos(2\pi f_m t)$ where f_m is the modulation frequency. (In some papers

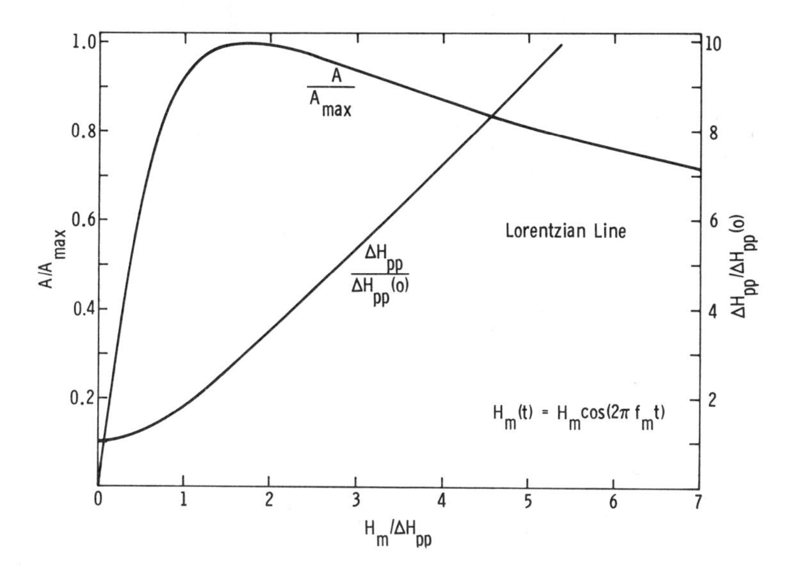

Fig. 3.15. Peak-to-peak width and relative signal amplitude of a Lorentzian line as a function of modulation amplitude.

H_m is defined as twice this value in order to be consistent with spectrometers in which the modulation amplitude is given as the peak-to-peak value.) Notice that the maximum signal amplitude occurs when the peak-to-peak linewidth is three times that of the width in the absence of modulation. Even at about half of the maximum sensitivity, ΔH_{pp} is 12% larger than the natural value. Randolph (91) has shown experimentally, and Buckmaster has shown mathematically (22) that the total double integral of an ESR curve is linear with modulation amplitude even though the lines are broadened. Thus equation 32 is valid. Since the wings of the experimental derivative line are not greatly affected by modulation amplitude even for a broadened line, if ΔH_{pp} at a very low modulation amplitude is known, a correction can still be applied to the lines if the scan is outside of $\pm 5 \cdot \Delta H_{pp}$ (experimental).

Cavity Q, Coupling, and Spin Quantity: The effect of coupling of the cavity is critical in obtaining reproducibly quantitative ESR determinations. Cavities can be undercoupled, critically coupled, or overcoupled to the waveguide. This coupling is adjusted by the iris (Fig. 3.11). In a critically coupled cavity, the cavity is matched to the waveguide so that there is a minimum signal at the detector for the spectrometer shown in Fig. 3.9. To bias the detector, rf power can be brought in from a path around the circulator or magic T. Alternatively, the cavity coupling can be changed to provide reflected power. Critical coupling provides maximum sensitivity, but because of bias requirements this is not always possible.

The sensitivity also depends on the Q factor of the cavity (equation 17). For small samples, the Q at resonance, Q_R, is approximately equal to Q_0. For large

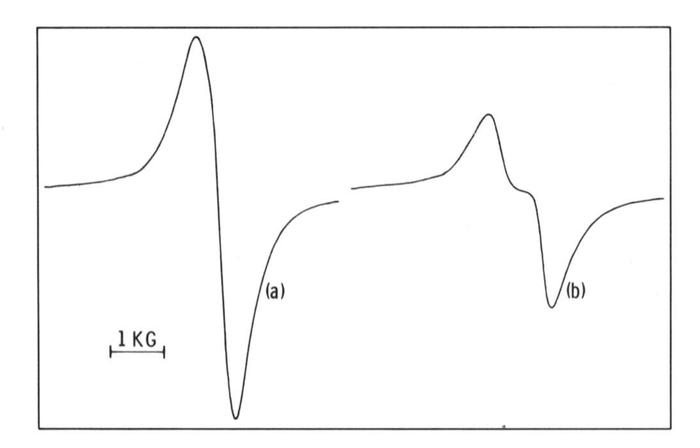

Fig. 3.16. ESR spectra of a large magnetic sample for a critically coupled (*a*) and overcoupled (*b*) cavity. Undercoupling results in a similar curve as (*a*) but with slightly lower intensity[(44a)].

spin concentrations, $Q_R < Q_O$, and the signal amplitude and double integral become nonlinear with concentration (46a,46b,107).

For large samples overcoupling and undercoupling yield different results, as shown for a large sample in Fig. 3.16. Overcoupling (Fig. 3.16*b*), because of changes when $Q_R < Q_O$, causes the reflected power to decrease while undercoupling causes the reflected power to increase. A very strong sample can cause a large change in the bias power on the detector causing a weaker signal. Clearly the undercoupled or critically coupled case, which are similar (Fig. 3.16*a*), are advantageous.

Bias Power: The bias power at a detector must be set reproducibly using the ammeter or voltmeter provided in the spectrometer. Since the signal does not depend strongly on this power level, it does not contribute a significant error, except at very low levels (< 100 μA using conventional detectors).

D. TRICKS TO IMPROVE ANALYSIS

1. It is a fairly straightforward matter to utilize a desk calculator for double integral or first-moment computations (29). However, it is advantageous to utilize a laboratory computer or a device that permits paper tape output. In this way, more data points can be used and thus errors are reduced by the averaging process and a better value of the baseline can be determined.

2. The effective S/N of an ESR signal is improved by integration (88) (Fig. 3.17), so that even if a signal is scarcely detected on the first derivative presentation it can yield a significant double integral if at least 20 data points per linewidth are used.

3. Often an inhomogeneous magnetic field will produce an asymmetric line. If the ESR spectra can be recorded sufficiently far out into the wings, either the average of

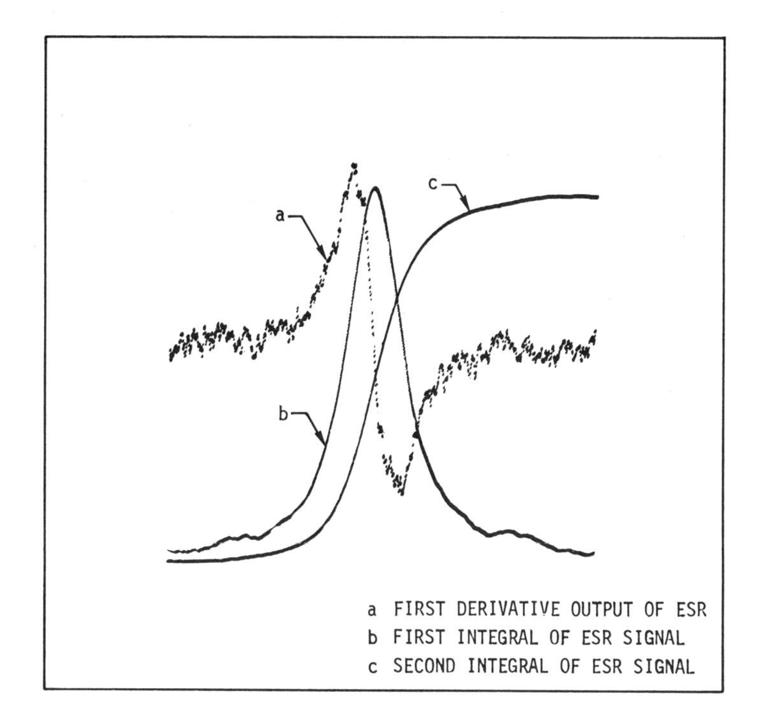

Fig. 3.17. First derivative curve, integral and double integral of a line of the F-atom signal or a concentration of 3×10^{-13} moles/cm^{2}(47).

all data points or an extrapolation is still valid to determine the baseline. Westenberg (112) has pointed out that each half of the double integral will yield different values. This is true only if a truncated scan is used. Since the correction factors do not change rapidly with linewidth, the experimental linewidth in the absence of additional modulation broadening can be used for ΔH_{pp}. Auxiliary coils have also been designed to improve field homogeneity (46).

4. In the case of baseline drift and interference due to adjacent lines, truncating the double integral and applying a large correction factor minimizes errors resulting from the baseline determination. Up to a 35% correction has been used in our laboratories with reproducibility of $\pm 2\%$ or better.

5. The treatment above has been applied to lines in which $H_0 \gg \Delta H_{pp}$. When the line is very broad, equation 22 must also be multiplied by a factor H/H_0 (34). If the entire analysis is carried out, it is found that for $\Delta H_{pp} < 0.2\ H_0$, the lineshape is still sufficiently symmetric for most analyses except that the absorption maximum (derivative zero) shifts slightly upfield from the predicted value.

6. The 100 KHz modulation-detection units of Varian V4502 and E3 ESR spectrometers are constructed using 5% resistors for gain and modulation amplitude settings. For accurate analyses these may be calibrated or rebuilt using precision resistors (44a).

7. Dual cavities have been used for a sample and quantitative reference (55,66). However, changing the sample geometry will alter the microwave field at the reference material. Thus, using a dual cavity, comparisons are only valid for nearly constant samples and geometries (27,90).

8. In addition to utilizing analog computers, digital computers, and desk calculators for double integral and first-moment determinations, first-moment balances have been constructed in which the spectrum is cut out and placed along an arm of the balance (67).

IV. ANALYTICAL APPLICATIONS

A. GENERAL CONSIDERATIONS

Direct analysis by ESR is limited to species containing unpaired electrons. Transition metal ions and their complexes, e.g., Mn(II), Cu(II), and V(IV), have been investigated and a number of analytical methods have been proposed. Paramagnetic organic species, i.e., radical ions ($R^{\overline{\cdot}}$ and R^{+}) and free radicals ($R\cdot$) have been investigated extensively. While most of this work has been concerned with theoretical treatments or the elucidation of structure or electron distributions, ESR is also very valuable in the identification and determination of such species. When the species of interest is itself not paramagnetic, it frequently can be converted to a paramagnetic species by an appropriate treatment; such techniques are discussed in Section IV.B.2. Transient radicals can sometimes be stabilized using a "spin trapping" technique. In addition to these direct methods, indirect determinations, where the effect of the component of interest on the ESR signal of another species is observed, are possible. These may involve changes in the amount of the paramagnetic species, e.g., when the compound of interest "titrates" the detected species to form one that does not produce a signal, or changes in line shape or intensity when the added species affects the relaxation processes of the paramagnetic one (Section IV.B.3). Qualitative analysis, such as the identification of radical species, is usually based on an analysis of the hyperfine structure of the spectrum (Section II.C). The fine detail and wealth of information in the patterns of organic radical species in solution makes this technique particularly useful, especially where only a single paramagnetic species is present. Mixtures of radicals present greater problems because their g values are often quite close, and complicated overlapping spectra result. However, the resolution is often sufficient to allow separate lines to be detected for each component. Computer simulation methods can also be employed for resolution of such spectra. Although the g values for most organic species are within about 0.5% of the free electron value (Section II.B), precise measurement of the g value can sometimes be used to distinguish between different classes of species (e.g., between hydrocarbon and peroxy radicals) or between different forms of a given metal ion (e.g., V-containing species in petroleum). We will subdivide our discussion of applications according to the nature of the medium, since experimental considerations and the results are somewhat different in liquid, solid, and gaseous media. Typical examples of analyses by ESR will be given, but an exhaustive coverage of the literature will not be attempted. The analytical uses of ESR before 1965 have been reviewed (8,43) and reviews of more recent work are also available (35,60,61 108a, 108b, 108c).

B. LIQUIDS

1. Metal Ions in Solution

a. Analysis of Vanadium in Petroleum

Vanadium exists in the paramagnetic +4 state as the VO^{2+} ion in the ppm range in petroleum oils, predominantly as porphyrin complexes. The interaction of the unpaired $3d$ electron with the nucleus ($I = 7/2$) produces an eight-line ESR spectrum which shows a more complex pattern in solution (Fig. 3.6) because of the anisotropy of the g and hyperfine tensors. Thus, ESR can be used for the direct and rapid determination of V without preliminary separations (32,33,92,96,106). The peak height of the first derivative of a given hyperfine line was used for the analysis with vanadyl etioporphyrin I dissolved in a heavy oil distillate as a standard (96). Special quartz tubes of 3-mm i.d. were used for the samples. The signal intensity was shown to vary with the square root of rf power. The modulation amplitude (ca. 5 gauss) was maintained constant for all samples. Amounts of V in the range of 0.1 to several hundred ppm were determined. ESR has also been employed to demonstrate the existence of V(IV) in several different forms (15) in petroleum (32,33). This was accomplished by showing that different fractions of the petroleum showed different isotropic g values with shifts of 22.3–25.8 parts per thousand from the free electron value.

b. Determination of Other Metal Ions in Solutions

A number of studies have been concerned with the effect of instrumental (e.g., receiver gain, G, modulation amplitude, M, and rf power, P) and procedural variables on quantitative analysis of transition metal ions by ESR (21,50–53,60,76,79,80,108d). A typical analysis is that of Mn(II), which shows a strong, six-line spectrum from interaction of the unpaired electron with the nucleus with $I = 5/2$ and $g = 2$ (Fig. 3.18). By reproducible tuning of the spectrometer and use of an aqueous quartz flat cell good precision (standard deviation of 0.4%) could be obtained (50). At constant rf power a plot of $S/G \cdot M$ (where S is the peak-to-peak signal amplitude of the fourth downfield line) was linear with concentration (on a log-log plot) between 10^{-3} and 10^{-6} M Mn^{2+}. An accuracy of 2% was obtained for a useful analysis range of 0.1 to 10^{-6} M. The variation of S with power (P) was also investigated (21,80). In general the signal varies nonlinearly with P (Fig. 3.19). It was suggested that the power dependence could be removed by using as the parameter on the analytical curves $S/G \cdot M \cdot \log P$ (80) or $S/G \cdot M \cdot P^k$ (where k is a constant near 1/2 which differs slightly for each metal ion; for Mn(II), $k = 0.55$) (21). Similar techniques have been reported for other transition metal ions; typical results are given in Table 3.VII.

The effect of addition of possible interferences on the ESR results was also investigated (51,76). Additives can affect the ESR signal by causing a broadening, by

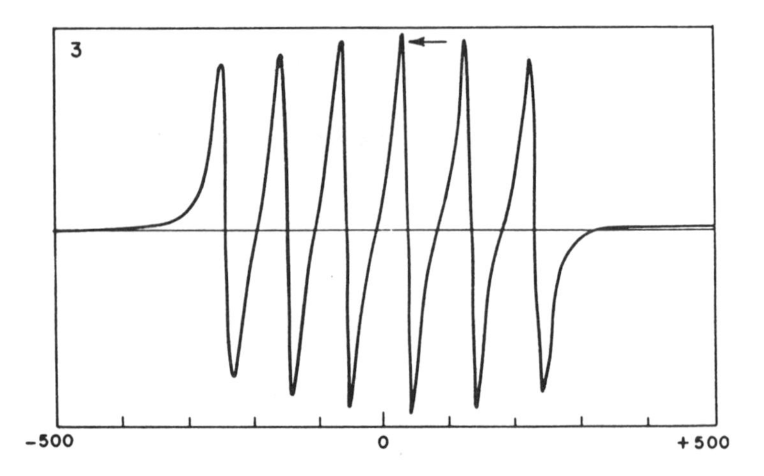

Fig. 3.18. ESR spectrum of Mn(II) sulfate solution, 10^{-4} *M* (Varian Model E-3 spectrometer; time constant, 0.3 sec; rf power, 10 mW; modulation amplitude, 10 G)[50]. (Courtesy of *Analytical Letters*.)

TABLE 3.VII

ESR Determination of Various Ions

Ions[a]		Upper limit (*M*)	Lower limit (*M*)	Mean k^5	Precision (%)	Error (%)
Mn(II)	(1)	0.1	1×10^{-6}		±0.4	2
	(2)	0.005	5×10^{-8}			
	(3)	0.1	1×10^{-6}			
	(5)	0.1	9×10^{-8}	0.55	±0.4	
Cu(II)	(1)	0.1	1×10^{-5}		0.4	1.5
	(2)	0.1	3×10^{-6}			
	(3)	0.02	5×10^{-6}			
	(5)	1	2×10^{-7}	0.48	±0.2	
Cr(III)	(1)	0.1	8×10^{-5}		±0.4	2.2
	(2)	0.01	1×10^{-5}			
	(3)	0.2	2×10^{-4}			
VO(II)	(1)	0.05	1×10^{-5}	0.52	±0.6	1.8
	(5)	0.02	2×10^{-7}		±0.7	
Fe(III)	(4)	0.1	8×10^{-6}		±0.4	2.5
	(3)	0.02	7×10^{-5}			
Co(II)	(1)	0.5	1×10^{-6}			
Gd(III)	(2)	0.04	4×10^{-5}			
	(3)	0.1	1×10^{-4}			

[a] (1) In aqueous solution taken from references 50, 51, and 76; (2) in aqueous solution taken from reference 80; (3) in ethanol solution taken from reference 80; (4) in acetone solution taken from reference 76; (5) in aqueous solution taken from reference 21.

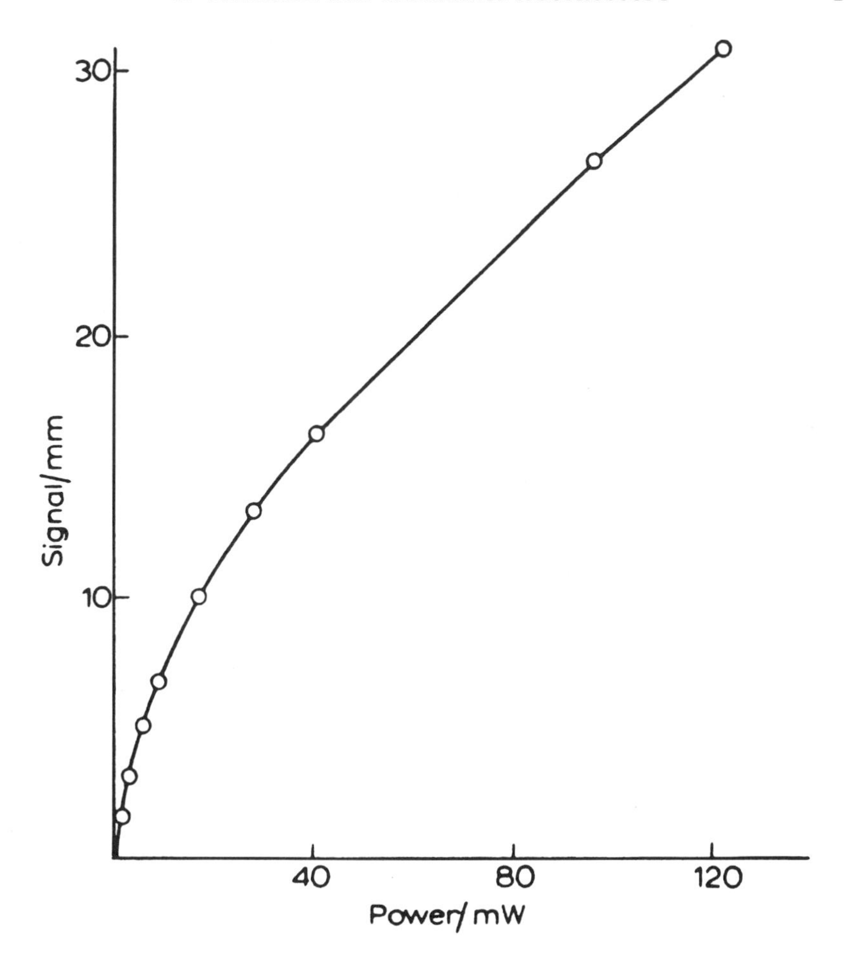

Fig. 3.19. Variation of signal (S) with rf power for a 3.5×10^{-3} *M* $MnCl_2$ solution[21]. (Courtesy of *Analytica Chimica Acta*.)

shifting the effective *g* factor, or by causing the lines in the signal to split or coalesce, e.g., through changes in complexation of the metal ion. Typical results compiled by Janzen (60) are shown in Table 3.VIII. Consider the case of Cu(II). In noncomplexing media, Cu(II) shows a broad structureless signal (linewidth 800–900 gauss). Upon the addition of ethylenediamine, the broad signal decreases and a four-line spectrum resulting from hyperfine coupling with the ^{63}Cu and ^{65}Cu nuclei ($I = 3/2$) appears (51). This quartet spectrum, with interline spacings of 170 G, is more convenient for determining Cu(II) in the presence of species which would show overlapping absorptions. Similarly the $Cu(DDC)_2$ complex (DDC = diethyldithiocarbamate) extracted into benzene showed a quartet useful for Cu(II) analysis (117). The signal intensity depended upon the solvent used for extraction. The intensity in toluene, for example, was only 0.6 that in benzene. While a number of metal ions (Mg^{2+}, Fe^{2+}, Ba^{2+}, Mn^{2+}, Zn^{2+}, Co^{2+}, Po^{2+}, Ni^{2+}) and anions (I^-, Br^-, NO_3^-) did not interfere with the determination, Mg(II) and CN^- did, probably by destruction of the $Cu(DDC)_2$ complex, e.g.

$$Cu(DDC)_2 + Hg^{2+} \rightarrow Hg(DDC)_2 + Cu^{2+}$$

This type of reaction was used as the basis of an indirect method for the determination of Hg^{2+} by noting the decrease in the height of the ESR signal for the $Cu(TET)_2$ complex (TET = tetraethylthiuram disulfide) upon addition of Hg^{2+} (116).

Chelates are able to alter the stability of various ions. For example, the paramagnetic Ag^{2+} ion is stabilized in the complex with tetraethylthiuram disulfide which permits analysis to the 0.1 ppb level (118).

TABLE 3.VIII

Maximum Tolerable Concentrations (*M*)[a] of Diverse Substances on ESR Spectra of Ions Indicated in Aqueous Solution (From reference 60) (Courtesy of *Analytical Chemistry*)

Substance	Mn(II)[b]	Cu(II)[c]	Cu-En[d]	Cr(III)[e]	VO(II)[f]
HCl	0.10	0.02	0.20	0.05	0.05 (or 0.04)
HNO_3	0.02	0.1	0.1	0.1	0.06
$HClO_4$	0.05	0.2	0.05	0.1	0.1
H_2SO_4	0.005	0.01	0.02	0.002	0.02
H_3PO_4	0.005	1.0	0.04	0.02	0.02
KH_2PO_4	0.005	0.02	0.04	0.001	
Formic Acid	0.2	1.8			
Acetic Acid	0.2	0.10	0.2	0.1	0.1
K_2HPO_4				0.0008	0.1
KCl	0.10				
NaCl	0.10				
$CaCl_2$	0.10	0.20	0.02	0.05	0.2
$MgCl_2$	0.10			0.05	0.1
$Zn(NO_3)_2$	0.02	0.5	0.02	0.2	0.5
KNO_3	0.02	0.1	0.1	0.1	0.3
$Ca(NO_3)_2$	0.02		0.02	0.2	0.5
$NaClO_4$	0.05	0.2	0.05	0.1	0.4
K_2SO_4	0.005				
Na Acetate	0.002	0.10	0.2	0.0005	0.0005
NaF	0.01	0.02			
KBr	0.80	0.20	0.05	0.08	0.3
KI	0.50			0.1	0.5
$Na_2S_2O_3$	0.005				
KCN	5×10^{-5}	0.0001[i]		0.0002	0.0002[k]
Na Citrate	5×10^{-5}	0.0001	0.02	0.0001	0.0002[j]
K Oxalate	0.0001	0.01[h]	0.005[g]	0.0005	0.0004[j]
K, Na Tartrate	0.0002	0.0005	0.005	0.0005	0.0001[j]
EDTA	2×10^{-5}	0.0001[h]	0.0001	0.0002	0.0002[k]
KCSN	0.002	0.02[g]	0.02[g]	0.2	0.05
Na_2SO_3	0.0005				
$ZnSO_4$	0.0002	0.002		0.0008	0.02

TABLE 3.VIII *(Cont.)*

Substance	Mn(II)[b]	Cu(II)[c]	Cu-En[d]	Cr(III)[e]	VO(II)[f]
Na_2SO_4	0.0002	0.02	0.04	0.0005	0.02
$MgSO_4$	2×10^{-5}	0.002		0.0008	0.02
$C_2H_4(NH_2)_2$	2×10^{-5}	0.001[h]		0.0001	0.0005
CH_3CN	2% (by volume)	2–3%	2–3%	2%	4–5%
CH_3OH	1% (by volume)			3%	2%
DMF	1% (by volume)	10%	10%	1%	1%
$MnCl_2$		0.20	0.02		
Acetone	1% (by volume)	2–3%	2–3%	2%	2%
Glycerine	1% (by volume)	2–3%	2–3%		
NH_4OH		0.0001[h]	0.20	0.005	0.0005[j]
Ethanol		2–3%	2–3%		
Dioxane		2–3%	2–3%	1%	2%

[a] Maximum concentration that can be added with no effect on the ESR spectra.
[b] 1×10^{-3} *M* $Mn(ClO_4)_2$ (51).
[c] 1×10^{-2} *M* $CuSO_4$ (51).
[d] 1×10^{-2} *M* $CuSO_4$; 1.28 *M* $C_2H_4(NH_2)_2$ (51).
[e] 1×10^{-2} *M* $Cr(NO_3)_3$ (76).
[f] 1×10^{-2} *M* $VOSO_4$ (76).
[g] Concentrations greater than this cause precipitation.
[h] Quartet observed.
[i] Reduction of Cu(II) to Cu(I) occurs and no signal is observed.
[j] Some new features appeared in spectrum in presence of excess of ligand.
[k] Unusually narrow spectrum with relatively intense peaks.

When the relaxation time of an ion is very short, the ESR signal may be so broadened that it extends over a very wide magnetic field range and hence is difficult or impossible to observe. For example, Fe^{3+} in aqueous media shows an extremely broad (ca. 1000 G) signal. However, since the relaxation time depends upon the environment of the ion (solvent, ligand), changes in this can sometimes lead to improved signals. Thus Fe(III) in acetone or alcohol in the presence of excess chloride ion gives a sharper signal (76,80). Similarly Ni(II) often does not give a useful ESR signal, but the chloroform extract of an aqueous solution containing the toluene-3,4-dithiol complex of Ni(II) can be used for analysis (19). Numerous possibilities exist for ESR analysis of metal ions in solution (120) and new applications wil undoubtedly appear during the coming years.

Titrations using ESR detection offer the advantages that diamagnetic materials can be detected by indirect methods and that precise repositioning of the sample is not required.

The titration of As_2O_3 by $KMnO_4$ was carried out in a recirculating cell, utilizing the detection of the Mn^{2+} ion (2), with a precision of better than $\pm 1\%$. The amplitude of the ESR signal of Mn^{2+} was sufficiently linear with concentration to determine the end point of the titration. $K_2Cr_2O_7$, which produces Cr^{3+}, may also be a useful titrant.

Determinations of phenanthroline and 2,4,6-tri-pyridyl-*s*-triazine were carried out using a photochemical titration (40). Fe^{2+} was generated photochemically from Fe^{3+}-citrate. The Fe^{2+} then complexes the organic ligand, and the chelate is detected by EPR. The ligand *o*-bipyridyl was found to complex too slowly with Fe^{2+} to be useful with EPR detection. Similarly, Cu(II) was determined by titration with the photochemically generated anthraquinone radical anion (41).

2. Organic Species in Solution

Since most organic compounds do not exist as radical species, the analysis of these by ESR requires that they be converted quantitatively or at a constant yield to radicals or radical ions that are stable for a long enough time for an ESR spectrum to be obtained. While a variety of methods exist for producing paramagnetic species and hundreds of radicals and radical ions have been investigated by ESR, there have been very few studies of the quantitative aspects of the radical production or actual quantitative determinations. A number of chemical methods exist for the production of radical ions (9,64,102). Radical anions of aromatic compounds are frequently produced by treatment of solutions in aprotic solvents (e.g., dimethoxyethane or tetrahydrofuran) with an alkali metal. Radical cations can be produced by dissolution in concentrated sulfuric acid or by various chemical oxidants. Electrochemical reduction and oxidation is also frequently employed, using external generation in a conventional electrolysis cell and transfer to a sample tube for ESR examination or direct generation in a small electrolytic cell contained in the resonance cavity (the *intra muros* technique) (45,65,75). Radicals can also be produced by photolysis, pyrolysis, and adsorption on various catalyic surfaces.

a. Determination of Polynuclear Aromatic Hydrocarbons

Treatment of low concentrations of certain hydrocarbons with dehydrated silica-alumina catalyst leads to quantitative conversion to the radical cation. This reaction was used as the basis of a method for the detection and estimation of anthracene, perylene, 9,10-dimethylanthracene and naphthacene (42). Solutions of the hydrocarbon in amounts of ca. 5 to 200 μg in CS_2 or benzene were treated with activated catalyst (10% Al_2O_3, 89.8% SiO_2, 0.05–0.10% Fe_2O_3) and the ESR spectra observed in selected uniform bore 6-mm glass tubes (Fig. 3.20). Note that the spectra of these adsorbed radical cations are more poorly resolved than those obtained by chemical or electrochemical oxidation. Since the *g* values of all of these species are similar, determination of mixtures of these would probably not be possible. The radicals produced by this procedure were stable for at least 20 min, even when the sample tubes were not protected from oxygen and moisture. Linear plots of signal amplitude (peak-to-peak) versus amount of hydrocarbon were obtained. The precision and accuracy of the determination was about $\pm 6\%$. Benzene and naphthalene, which do not form radical cations under these conditions, do not interfere.

With care, alkali metals can be used to reduce successively aromatic hydrocarbons (44a). A solution of 50 μg each of perylene and pyrene in dimethoxyethane was reacted in steps using a Pyrex tube with a 3-mm o.d. side arm (Fig. 3.21). Initially, the spectrum of the perylene radical anion appears and reaches a steady state. Gradually as the perylene is reduced to the dianion, the initial spectrum is replaced by the spectrum of the pyrene anion. Thus, independent spectra of both species can be

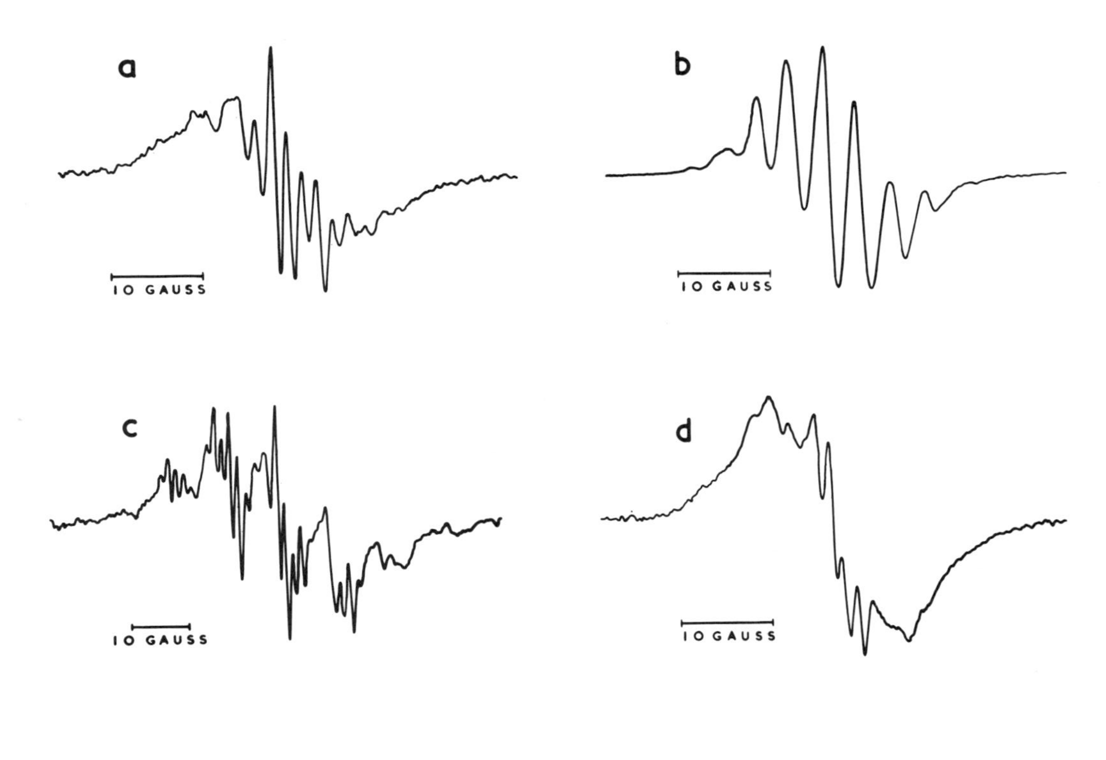

Fig. 3.20. ESR spectra of (*a*) anthracene, (*b*) perylene, (*c*) 9,10-dimethylanthracene (adsorbed from CS_2), and (*d*) naphthacene (adsorbed from benzene) obtained by treatment with activated silica-alumina(42). (Courtesy of *Talanta*.)

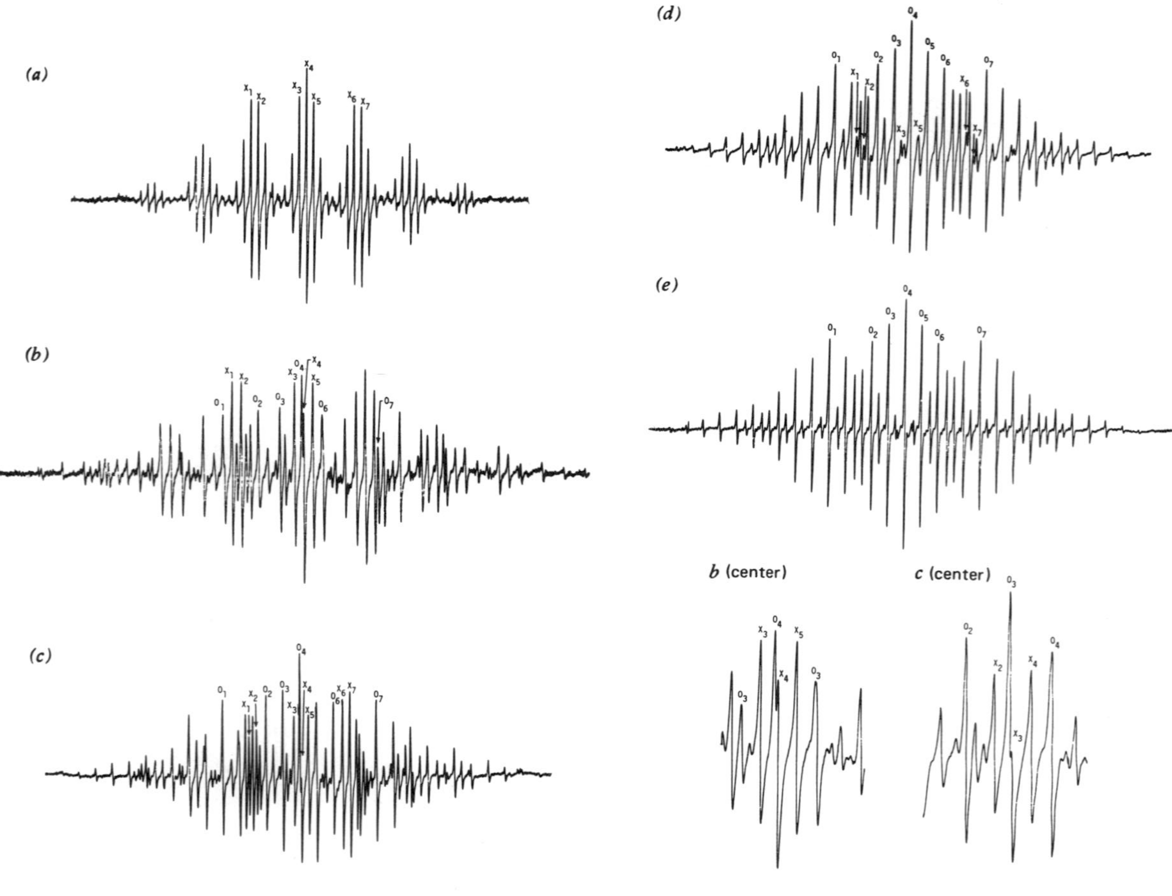

Fig. 3.21. Spectra of a mixture of perylene and pyrene successively reduced with alkali metals. (*a*) Perylene anion; (*b*–*d*) further reduction of the perylene anion to the dianion and appearance of the pyrene anion; (*e*) pyrene anion showing no perylene anion[44a].

observed. Several experiments gave accuracies of ±5% for the relative amounts of perylene to pyrene.

ESR has also been used to study deuterium exchange in homogeneous and heterogeneous systems (30,31). Radical anions were generated from the reaction products. The positions and the relative amounts of deuteration can be determined unambiguously from the hyperfine patterns in ESR spectrum.

b. DETERMINATION OF QUINONES

ESR signals obtained from the radical anions of quinones (semiquinones) are easily obtained, for example, by atmospheric oxidation of alkaline aqueous solutions, and are reasonably stable. Although quantitative studies have not been carried out, the effect of pH on the *p*-benzoquinone radical signal during formation from quinhydrone in deaerated aqueous was investigated (83). The absolute concentration of radical ion was determined by double integration of the ESR signal and comparison to that obtained from peroxylamine disulfonate. At an estimated concentration of *p*-benzosemiquinone of $6 \times 10^{-6}\,M$ the standard deviation was 4%.

Deuteration of hydroquinones was also studied by ESR (119). This paper also provides a good description of the treatment of the data.

c. SPIN-TRAPPING TECHNIQUES

One difficulty with the ESR analysis of organic species is that the radical species are often unstable, especially in the presence of oxygen and moisture. This may necessitate rather elaborate measures for their formation and transfer to the spectrometer and may discourage quantitative analytical applications of many types of compounds. Nitroxide radicals, e.g., $R_2NO\cdot$, are very stable, however, and conversion of an organic species to the nitroxide allows the detection of this species. This technique is called "spin-trapping" and involves reaction of the radical with a nitrone or nitroso compound. For example, phenyl radical ($Ph\cdot$) can be trapped by α-phenyl-*N*-*tert*-butylnitrone (59).

$$Ph\cdot + C_6H_5CH = \overset{\overset{\displaystyle O}{\uparrow}}{N} - C(CH_3)_3 \rightarrow (C_6H_5)_2CH\overset{\overset{\displaystyle O}{|}}{N}C(CH_3)_3$$

While few quantitative studies using spin-trapping have been reported, it is a possible analytical method for the determination of unstable radicals.

d. ESR DETECTOR FOR LIQUID CHROMATOGRAPHY

Separation of nitroxide radicals by liquid chromatography (LC) on a silica column was monitored by ESR (93). Successive 30-sec sweeps were used to record the spectra. Similarly the separation of DPPH from its reaction products by LC was monitored by ESR.

3. Methods Based on Relaxation Times and Reaction Rates

Relaxation methods are based on changes in the shape of the ESR spectrum (e.g., line broadening, signal intensity) because of variations in the environment of the detected radical. A number of methods employ nitroxide radicals as probes. Most nitroxides (Fig. 3.22) in nonviscous media show a well-defined three-line spectrum with a splitting of ca. 14–17 G attributable to interaction of the unpaired electron with the ^{14}N nucleus ($I = 1$) Fig. 3.23*a*). The nearest hydrogen atoms are in the γ-position, so hyperfine splitting by these is usually not observed, although they may contribute to the line width (ca. 1 G). When the nitroxide is in a crystal, in a very viscous medium or in any environment where its motion is restricted, a broadened spectrum (known as the rigid glass, powder, polycrystalline, or strongly immobilized spectrum) results (Fig. 3.23*b*). This broadening is due to the anisotropic g value (i.e., a g value that depends upon the orientation of the nitroxide with respect to the magnetic field) and N hyperfine splitting. These changes in shape from the "free" spectrum to one with various degrees of broadening is the basis of the "spin-labeling" technique for studying the polarity of the environment, the molecular motion, and the orientation of a free radical probe introduced into a molecule or membrane (49,74). Analytical methods are also based on changes in the nitroxide spectrum.

a. ESR IMMUNOASSAY OF DRUGS–FRAT

The analytical application of spin-labeling is based upon generation of the "free" nitroxide spectrum from the "immobilized" one by reaction of a labeled reagent with the compound of interest. Consider, for example, the method for the determination of morphine in urine, saliva, and other biological fluids (71). A spin-labeled morphine analog (Fig. 3.24*c*) (XNO ·) is prepared. The ESR of this substance shows the typical "free" nitroxide pattern. An antigen form of morphine is also prepared by coupling morphine to bovine serum albumin (BSA) (Fig. 3.24*b*). When this antigen is injected into rabbits, antibodies against it are formed; these antibodies (AB) will bind strongly to morphine itself or its spin-labeled form

$$\mathrm{AB} + \mathrm{XNO}\cdot \rightarrow [\mathrm{AB{-}XNO}\cdot]_{\mathrm{immob}}$$

Thus, a reagent prepared by mixing the antibody and labeled drug in such amounts that no free XNO · remains will show an ESR spectrum with the "immobilized" or broadened signal. When the sample presumed to contain morphine (M) is added to the reagent, some of the antibody-bound XNO · will be exchanged by M, thus liberating free XNO · and producing the sharp-lined nitroxide spectrum:

$$[\mathrm{AB}\cdots\mathrm{XNO}\cdot]_{\mathrm{immob}} + \mathrm{M} \rightleftharpoons [\mathrm{AB}\cdots\mathrm{M}]_{\mathrm{immob}} + \mathrm{XNO}\cdot$$

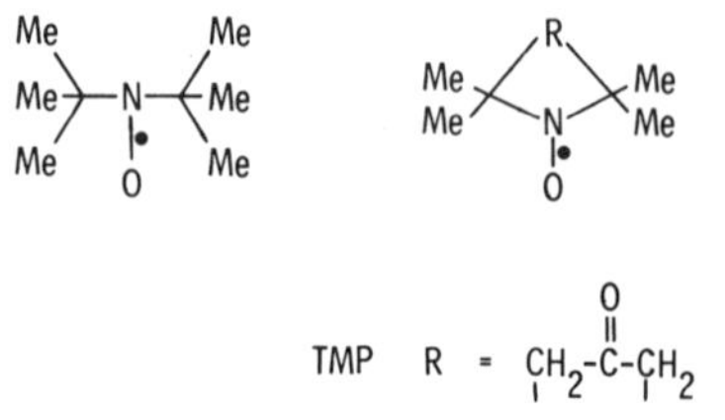

TMP R = CH_2-C(=O)-CH_2

Fig. 3.22. Stable nitroxides.

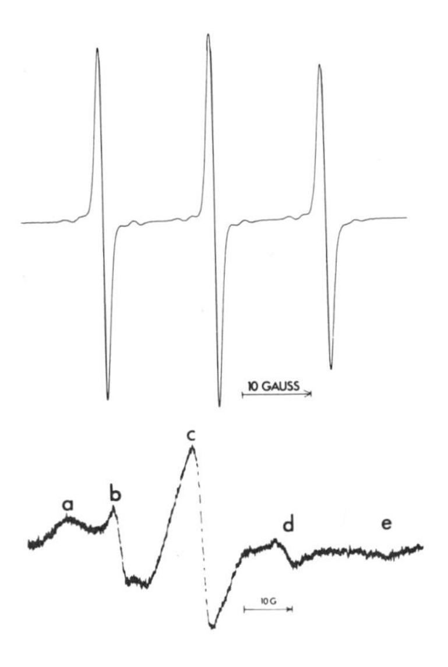

Fig. 3.23. (*a*) Typical ESR spectrum of "free" nitroxide radical in aqueous solution; (*b*) typical ESR spectrum of immobilized nitroxide; the extent of broadening and spectrum in any given case depends upon environment, orientation and motion of the spin label. (Courtesy of Aldrich Chemical Co.)

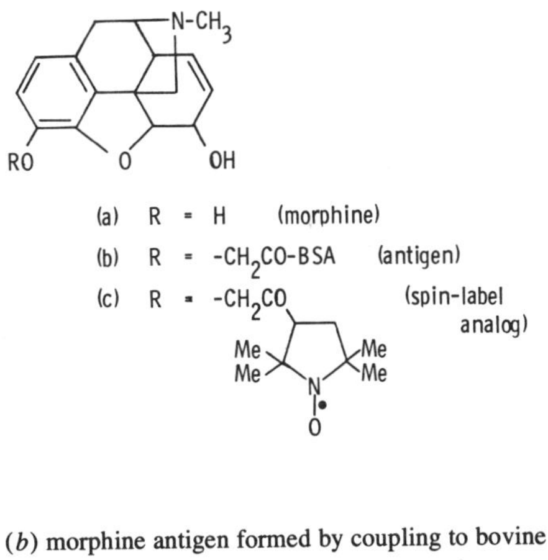

$N-CH_3$

RO O OH

(a) R = H (morphine)

(b) R = $-CH_2CO-BSA$ (antigen)

(c) R = $-CH_2CO$ (spin-label analog)

Me Me Me Me N O

Fig. 3.24. (*a*) Morphine; (*b*) morphine antigen formed by coupling to bovine serum albumin; (*c*) spin-labeled morphine antigen.

Because of the specificity of the antibody, appearance of the "free" ESR spectrum is evidence of morphine or its close analogs (e.g., codeine, ethyl morphine); other drugs of abuse, such as barbiturates, amphetamines, methadone, etc., do not interfere. Calibration curves of free signal height against amount of drug can be used to determine the quantities of drug in the sample. Prepared reagent for morphine and other drugs is commercially available. This technique, call FRAT (an acronym for "free radical assay technique") uses very small samples (e.g., 20 μl of urine), is very rapid (ca. 1 min), and is specific. The reagent is expensive, however. The use of similar methods applied to enzymes and interferences are discussed in a recent review (61).

b. Determination of Dissolved Oxygen

Oxygen is a paramagnetic triplet. Dissolved oxygen leads to line broadening in the ESR spectrum of a radical, because the oxygen–radical interaction contributes to the relaxation of the free radical. The extent of this broadening depends upon the oxygen concentration, so that the method can be used to measure dissolved oxygen (89). The linewidth of the nitroxide radical 2,2,6,6-tetramethylpiperidinoxy (TMP) (Fig. 3.22) at a concentration of 0.34 mM varied from its oxygen-free value of 0.55 G to up to 4 G; the results showed a fair amount of scatter, so the method is only semiquantitative.

A method based on oxygen's effect on the spin–lattice relaxation time has also been proposed (56). The ESR spectrum of a free radical (of an unspecified nature) which originates in a terphenyl coolant of a reactor was studied as a function of incident microwave power. Because the radical has a long spin–lattice relaxation time, the ESR signal saturates fairly easily, and the signal height decreased with increasing rf power in the absence of oxygen. Addition of oxygen decreases the relaxation time and changed the nature of the peak height versus rf power relation. This change was used to estimate the oxygen content and was proposed for oxygen concentrations as low as 10^{-5} M.

c. Determination of Hydroperoxides by Reaction with DPPH

The effect of substances that react and decrease the signal of a stable free radical can be employed as an indirect method of analysis. Thus the measurement of the rate of reaction of the stable radical diphenylpicrylhydrazyl (DPPH) with several hydroperoxides was proposed as a method of determination of the hydroperoxide (104). The ESR signal of DPPH or that of a new species formed on reaction was monitored with time and the hydroperoxide concentration was calculated based on the measured rate of reaction. Although different hydroperoxides could be distinguished based on their different reactions rates, the specificity of the method is probably not very good and interferences and temperature control may be a problem.

A similar technique utilizing diphenylamine which reacts with the hydroperoxide to form diphenylnitroxide has also been reported (65a). This technique involves the increase in signal due to the free radical.

C. GASES

Most gaseous systems that contain paramagnetic atoms or molecules are of interest in studies of either kinetics or flames. This subject has been reviewed in detail (112). Rarely are atoms obtained in an environment which needs to be analyzed, and most paramagnetic molecules are not stable. Exceptions to this rule are O_2, NO, and NO_2, although O_2 itself is typically not of interest. O_2 in the metastable $^1\Delta_g$ state is a unique case for which few techniques are available for analysis.

1. Analysis of NO and NO_2

O_2 usually interferes with EPR determinations in the gas phase. At low pressures, the X-band spectrum of O_2 exhibits several thousand lines which extend from about 2 KG to beyond 20 KG. At high pressures, O_2 exhibits several broad lines. However, electric field modulation (Stark modulation), in addition to magnetic field modulation can eliminate the spectrum of nonpolar molecules (26). Using this technique, Uehara and Arimitsu (105) developed an ESR cavity and sampling system which permits air at ambient pressure to be monitored for NO_2 with a sensitivity of 10 ppm. They also utilized a low-temperature trapping technique in which NO and NO_2 are deposited in a trap while the air is pumped out. This technique provided a sensitivity of about 30 ppb for both NO and NO_2. They suggest the utilization of this technique for exhaust gases.

2. Singlet Molecular Oxygen

O_2 can be formed in the metastable singlet state ($^1\Delta_g$) which is paramagnetic by virtue of its orbital angular momentum (37,77). It is of current interest in air pollution studies as well as in synthetic procedures. Optical and infrared absorption cannot be used with this molecule. However, a red emission that is proportional to the square of the concentration can be easily detected but needs to be calibrated for each system. EPR provides a useful quantitative technique for this material (36), and can be used with accuracies of several percent (44b). A calibration of the optical emission versus the square of the pressure determined by ESR is shown in Fig. 3.25.

D. SOLIDS

The analysis of solids has not received much attention. This may be due in part to difficulties in sample placement. However, ESR offers a nondestructive tool that, once calibrated, can provide rapid analyses.

1. Mn^{2+} in Calcium Carbonate

Mn^{2+} in calcium carbonate exhibits a spectrum that consists of six intense lines as well as some weaker transitions between those lines. Samples of barnacle shells were analyzed for Mn^{2+} (12) by grinding the material, and positioning a sample of about 10 mg in the center of the cavity. The amplitude of the signal was used as a measure of concentration, and was calibrated against atomic absorption measurements. The

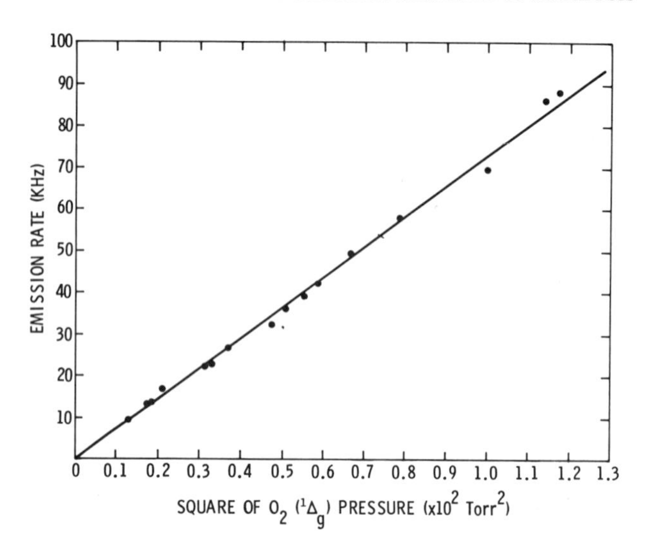

Fig. 3.25. Optical emission rate for 0_2 in the $^1\Delta_g$ state versus square of the $0_2(^1\Delta_g)$ pressure determined by EPR spectroscopy.[44b].

limits of detection were about 20 ppb, and agreed to within $\pm 3\%$ of the atomic absorption determination.

2. Active Surface Area

It was mentioned in Section IV.3 that metal oxide catalysts can oxidize many aromatic molecules. To determine the surface oxygen content of MnO_2 as an oxidizing agent (84) two methods were used. In the first, diphenylpicrylhydrazine ($DPPH_2$) was reacted with the MnO_2/O_2 system to yield the diphenylpicrylhydrazyl radical (DPPH). The concentration of the DPPH was then determined by ESR. The second technique involved the direct absorption of DPPH from a solution in benzene, in which a reduction of the DPPH concentration is determined.

3. Ferromagnetic Systems

A discussion of ferromagnetic resonance is beyond the scope of this chapter. However, for many systems, the spins of the sample may be considered to be completely aligned in the magnetic field. Thus the magnetization (M_F) of a ferromagnetic material is given by equation 36

$$M_F = n_{\text{eff}} g \beta N_F \tag{36}$$

where n_{eff} is the effective number of spins per atom and N_F is the number of ferromagnetic molecules. In contrast, the magnetization of a nondilute paramagnetic material (M_P) is given by equation 37

$$M_P = \frac{g^2\beta^2 S(S + 1)H_0}{3k(T - \theta)} N_P \qquad (37)$$

where θ is the Curie temperature, and N_P is the number of paramagnetic molecules. Thus, to utilize ESR to determine the quantity of ferromagnetic material, equation 29 can be used if the first term on the right-hand side is replaced by $1/(n_{\mathrm{eff}}\beta)$.

This technique was used to determine the amount of fine-grained metallic iron in lunar return samples (54). The ESR spectrometer was calibrated with $MnSO_4 \cdot H_2O$ as a paramagnetic stand, where $\theta = -26°$,[4] $g = 2.000$, and $S = 5/2$. Often θ is sufficiently close to 0 for even pure materials, so that it can be neglected, however, $MnSO_4 \cdot H_2O$ is an exception.

D. STANDARDS

The usual criteria for the selection of a primary or secondary standard such as ease of preparation, availability, stability, and reproducibility also apply to ESR. Two types of samples are typically used, those which provide a quantitative standard and those which provide a magnetic field standard.

1. Magnetic Field Standards

Standards to calibrate the magnetic field must exhibit sharp lines and have spacings between lines that are known exactly. In addition, the spectrum of the standard should extend over the range of magnetic field of the sample. There are many of these standards in use, but several examples will be given below.

Wursters Blue Perchlorate in Ethanol (111): The perchlorate salt of *N,N,N',N'*- tetramethyl-*p*-phenylenediamine is easily prepared. The spectrum of the material dissolved in deoxygenated absolute ethanol exhibits an apparent g factor of 2.00305±0.00001 and hyperfine splittings due to two equivalent nitrogen atoms of 7.051±0.009 G, 12 equivalent protons of 6.773±0.005 G, and 1.989±0.009 G. An intense spectrum is observed over a range of about 80 G.

Mn^{2+} Standard (94): Usually SrO powder contains sufficient Mn^{2+} to give a strong ESR signal which exhibits six lines of a width of 1.5 G. The g factor is 2.0012±0.0002 and the hyperfine splitting is 83.71 G. The line positions should be calculated to at least second order (equation 8) to be used as a standard.

Perylene Radical Cation: The perylene radical cation is prepared by adding a small amount of perylene to 98% sulfuric acid, and degasing the solution. The spectrum is characterized by a g factor of 2.002569±0.000006 (78,97) and three sets of hyperfine splittings 4.053, 3.054, and 0.446 G (13) each for four equivalent protons.

2. Quantitative Standards

a. LIQUIDS

A number of dissolved paramagnetic materials provide good quantitative standards. The following have been utilized:

α,α'-diphenylpicrylhydrazyl can be weighed and dissolved readily in many polar and nonpolar solvents. However, this material is not stable for long periods of time (18).

Potassium peroxylamine disulfonate $[K_2NO(SO_3)_2]$ is a useful standard for aqueous solutions, although it is not stable in solution for more than about a day. However, the concentration can be determined optically (62). The molar extinction coefficient is 1690 at 248 mμ and 2.08 at 545 mμ (81); 0.05 *M* carbonate retards decomposition. The spectrum consists of three lines of equal intensity with a hyperfine splitting of 13.05$\pm$0.03 G.

Nitroxide Radicals: Several tetramethyl piperidyl nitroxides have been found to form stable radicals which can be used in aqueous on nonaqueous solvents (61,111).

$CuSO_4 \cdot 5H_2O$ can be dissolved in aqueous media; however, the signal is quite broad. Similarly, $MnSO_4 \cdot H_2O$ ($S = 5/2$) exhibits a six-line spectrum. These have been used extensively for quantitative analysis ($CuCl_2 \cdot 2H_2O$, and $MnCl_2 \cdot 4H_2O$ have also been used (34).

A brief note describes the use of quinhydrone in buffered media to obtain a constant concentration of the semiquinone radical anion (83).

b. SOLIDS

Numerous solid single crystals and powders have been used for analysis. Standards include DPPH, $CuSO_4 \cdot H_2O$, and *F* centers in alkali halides (55) which can be determined by optical absorption. $MnSO_4 \cdot H_2O$ can be weighed directly (54,98). Single crystals of $CuSO_4 \cdot 5H_2O$, which can also be weighed are anisotropic, so that the calibration factor must include the correct value of g. The powder exhibits an irregularly shaped spectrum. Many pure materials can be used; however, it is important to determine the Curie temperature of the material. Dilute materials, such as *F* centers of transition metals in host matrices can be assumed to have a Curie temperature of 0°K. If transition metal ions in foreign matrices are to be used, the concentration must be determined by optical or chemical means.

c. GASES

In studies of kinetics using flow techniques, the ESR spectrometer is often calibrated by titration of one of the atoms of interest, with a titrant that reacts rapidly with the atoms on the time scale of the flow reactor or with O_2 or NO gas. For sampling measurements, either O_2 (47a,103,114) or NO (17) must be used. Typically, total pressures between 0.5 and 4 torr are used to keep exchange and pressure broadening to sufficiently low levels.

ACKNOWLEDGMENTS

This work was supported in part by the Office of Naval Research, and by the National Science Foundation and the Robert A. Welch Foundation. We thank Dr. T. McKinney for helpful criticisms on this manuscript.

ADDENDUM

Since this article was submitted for publication, several relevant articles have been written. Analytical Chemistry has extended its review through 1982 (108a,108b,108c). Two critical reviews were also written (35a,44c). A summary of different quantitative and magnetic field standards are given in ref. 44c. A method of correcting for certain saturation effects was applied to low temperature measurements (75a). An analysis of the contributing effects of various experimental parameters, including microwave power, modulation amplitude, and signal gain for the E-3 spectrometer, was reported (108d).

Several novel analytical methods have also been reported recently. Low temperatures (<4.2 K) used in order to quantitatively determine both Fe^{2+} and Fe^{3+} in MgO (78a). Fe^{2+} cannot be observed at higher temperatures. High pressure liquid chromatography was coupled with ESR to separate paramagnetic products formed after irradiation of aminoacids and subsequent reactions with spin traps (Sec. 3.IV.B.2.c). In this application, ESR was used as a detector for chromatography; however, it also provided positive structural identification of the paramagnetic materials. The quantitative production of $O_2(^1\Delta)$ in liquids was determined by indirect means (22a,71a). Excited oxygen reacts competitively with 2,2,6,6-tetramethylpiperidine to form the corresponding nitroxide which is relatively stable against subsequent reactions, and provides a time dependent measure of the integrated $O_2(^1\Delta)$ production. Effort (100a,122) has also been placed on developing paramagnetic complexing agents which interact with either paramagnetic or diamagnetic heavy metals to yield a new paramagnetic complex which may either be extracted from the original complexing agent or detected as part of a mixture. Both semiquinone-amine and nitroxide-amine complexing agents have been used, and Pb^{2+}, Zn^{2+}, Cd^{2+}, Et_2Tl^+, Cu^{2+}, Hg^{2+}, Co^{2+}, Co^{3+}, Fe^{3+}, Ni^{2+}, K^+, Na^+, Pr^{3+}, Pd^{2+}, Pt^{2+} and Ag^+, have been studied. The technique is new, and as yet limits of detection and accuracy have not been defined.

REFERENCES

1. Abragam, A., and B. Bleaney, *Electron Paramagnetic Resonance of Transition Ions*, Clarendon Press, Oxford, 1970.
2. Agerton, M., and E. G. Janzen, *Anal. Lett.*, **2**, 457 (1969).
3. Alger, R. S., *Electron Paramagnetic Resonance. Techniques and Applications*, Wiley-Interscience, New York, 1968.
4. Allain, Y., J. P. Krebs, and J. de Gunzbourg, *J. Appl. Phys.*, **39**, 1124 (1968).
5. Atherton, N. M., *Electron Spin Resonance*, Halsted, London, 1967.

6. Atkins, P. W., and M. C. R. Symons, *The Structure of Inorganic Radicals*, Elsevier, Amsterdam, 1967.
7. Ayscough, P. B., *Electron Spin Resonance in Chemistry*, Methuen, London, 1967.
8. Bard, A. J., "Electron Spin Resonance" in F. J. Welcher, Ed., *Standard Methods of Chemical Analysis*, Vol. 3, D. Van Nostrand, Princeton, New Jersey, 1966, pp. 616–635.
9. Bard, A. J., A. Ledwith, and J. J. Shine, *Adv. Phys. Org. Chem.*, **13**, 155 (1976).
10. Bersohn, M., and J. C. Baird, *Electron Paramagnetic Resonance*, Benjamin, New York, 1966.
11. Bielski, B. H. J., and J. M. Gebecki, *Atlas of Electron Spin Resonance Spectra*, Academic, New York, 1967.
12. Blanchard, S. C., and N. D. Chasteen, *Anal. Chim. Acta*, **82**, 113 (1976).
13. Bolton, J. R., J. *Phys. Chem.*, **71**, 3702 (1967).
14. Bolton, J. R., "Electron Spin Densities" in E. T. Kaiser and L. Keven, Eds., *Radical Ions*, Interscience, New York, 1968, Chapter 1.
15. Boucher, L. J., E. C. Tynan, and T. F. Yen, in T. F. Yen, Ed., *Electron Spin Resonance of Metal Complexes*, Plenum, New York, 1969.
16. Bowers, K. W., *Adv. Magnet. Resonance*, **1**, 317 (1965).
17. Breckenridge, W. H., and T. A. Miller, *J. Chem. Phys.*, **56**, 475 (1972).
18. Bridge, N. K., *Nature*, **185**, 31 (1960).
19. Brinkman, W. J., and H. Freiser, *Anal. Lett.* **4**, 513 (1971).
20. Brown, J. M., "Electron Resonance of Gaseous Free Radicals" in *Magnetic Resonance*, Vol. 4, M. T. P. International Reviews of Science; Physical Chemistry, C. A. McDowell, Ed., Butterworth, London, 1973, Chapter 7.
21. Bryson, W. G., D. P. Hubbard, B. M. Peake, and J. Simpson, *Anal. Chim. Acta*, **77**, 107 (1975).
22. Buckmaster, H. A., and J. C. Dering, *J. Appl. Phys.*, **39**, 4486 (1968).
22a. Cannistraro, S, A. van der Voorst, and G. Jori, *Photochem. Photobiol.*, **28**, 257 (1978).
23. Carrington, A., and A. D. McLachlan, *Introduction to Magnetic Resonance*, Harper and Row, New York, 1967.
24. Carrington, A., *Microwave Spectroscopy of Free Radicals*, Academic, London, 1974.
25. Carrington, A., D. H. Levy, and T. A. Miller, *Adv. Chem. Phys.*, **18**, 149 (1970).
26. Carrington, A., D. H. Levy, and T. A. Miller, *Rev. Sci. Instrum.*, **38**, 1183 (1967).
27. Casteleijn, G., J. J. ten Bosch, and J. Smidt, *J. Appl. Phys.*, **39**, 4375 (1968).
28. Chang, R., *Anal. Chem.*, **46**, 1360 (1974).
29. Cope, F. W., "A Method for Digital Calculation of Absolute Free Radical Concentrations," Report NADC-MR-6804 (AD 671 809) May, 1968.
30. Davis, K. P., and J. L. Garnett, *J. Phys. Chem.*, **75**, 1175 (1971).
31. Davis, K. P., J. L. Garnett, and J. H. O'Keefe, *Chem. Comm.*, 1672 (1970).
32. Dickson, F. E., C. J. Kunesh, E. L. McGinnis, and L. Petrakis, *Anal. Chem.*, **44**, 978 (1972).
33. Dickson, F. E., and L. Petrakis, *Anal. Chem.*, **46**, 1129 (1974).
34. Dohrmann, J. K., Ber. Bunsenges, *Phys. Chem.*, **74**, 575 (1970).
35. Eargle, D., *Anal. Chem.*, **40**, 303R (1968).
35a. Eaton, S. S. and G. R. Eaton, *Bull. Magn. Reson.*, **1**, 130 (1980).
36. Falick, A. M., "An EPR Study of the O_2 ($^1\Delta_g$) Molecules," University of California Radiation Laboratory, Berkeley, California, Report UCRL-17453 (1967).
37. Falick, A. M., B. H. Mahan, and R. J. Myers, *J. Chem. Phys.*, **42**, 1837 (1975).
38. Farach, H. A., and C. P. Poole, Jr., *Adv. Magnet. Resonance*, **5**, 229 (1973).
39. Feher, G., *Bell System Tech. J.*, **36**, 449 (1957).

40. Fitzgerald, J. M., and J. L. Beck, *Anal. Lett.*, **3**, 531 (1970).
41. Fitzgerald, J. M., and D. C. Warren, *Anal. Lett.*, **3**, 623 (1970).
42. Flockhart, B. D., and R. C. Pink, *Talanta*, **9**, 931 (1962).
43. Flockhart, B. D., and R. C. Pink, *Talanta*, **12**, 529 (1965).
44. (a) Goldberg, I. B., unpublished results; (b) Goldberg, I. B., and A. T. Pritt, unpublished results.
44c. Goldberg, I. B., *Spec. Period. Rep.: Electron Spin Reson.*, **6**, 1 (1981).
45. Goldberg, I. B., and A. J. Bard, "The Application of ESR Spectroscopy to Electrochemistry" in J. N. Herak and K. J. Adamic, Eds., *Magnetic Resonance in Chemistry and Biology*, Dekker, New York, 1975, Chapter 10.
46. Goldberg, I. B., and H. R. Crowe, *J. Magnet. Resonance*, **18**, 497 (1975).
46a. Goldberg, I. B. and H. R. Crowe, *Anal. Chem.*, **49**, 1353 (1977).
46b. Goldberg, I. B., H. R. Crowe, R. M. Housley and E. H. Cirlin, *J. Geophys. Res.*, **84B**, 1025 (1979).
47. Goldberg, I. B., H. R. Crowe, and D. Pilipovich, *Chem. Phys. Lett.*, **33**, 347 (1975).
47a. Goldberg, I. B. and H. O. Laeger, *J. Phys. Chem.*, **84**, 3040 (1980).
48. Goodman, B. A., and J. B. Raynor, *Adv. Inorg. Chem. Radiochem.*, **13**, 135 (1970).
49. Griffith, O., and A. Waggoner, *Account Chem. Res.*, **2**, 17 (1969).
50. Guilbault, G. G., and G. J. Lubrano, *Anal. Lett.*, **1**, 725 (1968).
51. Guilbault, G. C., and T. Meisel, *Anal. Chem.*, **41**, 1100 (1969).
52. Guilbault, G. G., and T. Meisel, *Anal. Chem. Acta*, **50**, 157 (1970).
53. Guilbault, G. G., and E. S. Moyer, *Anal. Chem.*, **42**, 471 (1970).
54. Housley, R. M., E. H. Cirlin, I. B. Goldberg, and H. R. Crowe, 7th Lunar Science Conference Proceedings, Houston, Texas, 1976
55. Hyde, J. S., "Experimental Techniques in EPR" in *Proc. 6th Annual NMR-EPR Workshop*, Varian Associates Instrument Division, Palo Alto, California, November 5–9, 1962.
56. Ingalls, R. B., and G. A. Pearson, *Anal. Chim. Acta*, **25**, 566 (1961).
57. Ingram, D. J. E., *Free Radicals, as Studied by Electron Spin Resonance*, Butterworths, London, 1958.
58. Ingram, D. J. E., *Spectroscopy at Radio and Microwave Frequencies*, 2nd Ed., Plenum, New York, 1967.
59. Janzen, E. G., *Accounts Chem. Res.*, **4**, 31 (1971).
60. Janzen, E. G., *Anal. Chem.*, **44**, 113R (1972).
61. Janzen, E. G., *Anal. Chem.*, **46**, 478R (1974).
62. Jones, M. T., *J. Chem. Phys.*, **38**, 2892 (1963).
63. Judeikis, H., *J. Appl. Phys.*, **35**, 2615 (1964).
64. Kaiser, E. T., and L. Kevan, eds., *Radical Ions*, Interscience, New York, 1968.
65. Kastening, B., "Joint Application of Electrochemical and ESR Techniques" in H. W. Nürnberg, Ed., *Electronanalytical Chemistry*, Vol. 10, Wiley, New York, 1974.
65a. Klyvera, N. D., G. E. Muratova, A. I. Kashlinskii, and A. V. Sokolov, *J. Anal. Chem. USSR*, **22**, 235 (1967).
66. Kohnlein, W., and A. Müller, "A Double Cavity for Precision Measurements of Radical Concentrations" in M. Blois, Ed., *Free Radicals in Biological Systems*, Academic, New York, 1961, Chapter 7.
67. Kohnlein, W., and A. Müller, *Phys. Med. Biol.*, **6**, 599 (1961).
68. Kokoszka, G. F., and G. Gordon, *Techn. Inorg. Chem.*, **7**, 151 (1968).
69. Kuska, H. A., and M. T. Rogers, "Electron Spin Resonance of First Row Transition Metal Complex Ions" in E. T. Kaiser and L. Keven, Eds., *Radical Ions*, John Wiley, New York, 1968, Chapter 13.

70. *Landolt-Bornstein Numerical Data*, Group II. Atomic and Molecular Physics: Vol. I. Magnetic Properties of Free Radicals; Vol. II. Magnetic Properties of Transition Metal Ions.

70a. Leniart, D., *Varian Instrum. Applic.*, **10**(2), 8 (1976).

71. Leute, R. K., E. F. Ullman, A. Goldstein, and A. Herzenberg, *Nature New Biol.*, **236**, 93 (1972).

71a. Lion, Y., M. Delmelle and A. van der Vorst, *Nature*, **263**, 442 (1976).

72. Loveland, D. B., and T. N. Tozer, *J. Phys.*, **E5**, 535 (1972).

73. Lukiewicz, S., and T. Sarna, *Folia Histochem. Cytochem.*, **9**, 127, 203 (1971).

74. McConnell, H. M., *Magnetic Resonance in Biological Systems*, Pergamon, New York, 1967.

75. McKinney, T., "Electron Spin Resonance and Electrochemistry," Vol. 10, A. J. Bard, Ed., Dekker, New York, 1976.

75a. Mailer, C., T. Sarna, H. M. Swartz and J. S. Hyde, *J. Magnet. Reson.*, **25**, 210 (1977).

76. Meisel, T., and G. G. Guilbault, *Anal. Chim. Acta*, **50**, 143 (1970).

77. Miller, T. A., *J. Chem. Phys.*, **54**, 330 (1971).

78. Möbius, K., *Z. Naturforschg.*, **20A**, 1102 (1965).

78a. Modine, F. A., E. Sonder and R. A. Weeks, *J. Appl. Phys.*, **48**, 3514 (1977).

78b. Moriya, F., K. Makino, N. Suzuki, S. Rokushika and H. Hatano, *J. Phys. Chem.*, **84**, 3085 (1980).

79. Moyer, E. S., and G. G. Guilbault, *Anal. Chim. Acta*, **52**, 281 (1970).

80. Moyer, E. S., and W. J. McCarthy, *Anal. Chim. Acta*, **48**, 79 (1969).

81. Murib, J. H., and D. M. Ritter, *J. Am. Chem. Soc.*, **74**, 3394 (1952).

82. Myers, R. J., *Molecular Magnetism and Magnetic Resonance Spectroscopy*, Prentice-Hall, New Jersey, 1973.

83. Narni, G., H. S. Mason, and I. Yamazaki, *Anal. Chem.*, **38**, 367 (1966).

84. Oei, A. T. T., and J. L. Garnett, *J. Catal.*, **19**, 176 (1970).

85. Orton, J. W., *Electron Paramagnetic Resonance*, Iliffe, Ltd., London, 1968.

86. Pake, G. E., *Paramagnetic Resonance*, W. A. Benjamin, New York, 1962, Chapter 2.

87. Poole, C. P., *Electron Spin Resonance*, Interscience, New York, 1967; 2nd Ed, 1982.

88. Posener, D. W., *J. Magnet. Resonance*, **14**, 129 (1974).

89. Povich, M. J., *Anal. Chem.*, **47**, 346 (1975).

90. Randolph, M. L., "Quantitative Considerations in Electron Spin Resonance Studies of Biological Materials" in H. M. Swartz, J. R. Bolton, and D. C. Borg, Eds., *Biological Applications of Electron Spin Resonance*, John Wiley, New York, 1972.

91. Randolph, M. L., *Rev. Sci. Instrum.*, **31**, 949 (1960).

92. Roberts, E. M., R. L. Rutledge, and A. P. Wehner, *Anal. Chem.*, **33**, 1879 (1961).

93. Rokushika, S., H. Taniguchi, and H. Hatano, *Anal. Letters*, **8**, 205 (1975).

94. Rosenthal, J., and L. Yarmus, *Rev. Sci. Instrum.*, **37**, 381 (1966).

95. Sales, K. D., *Adv. Free Radical Chem.*, **3**, 139 (1969).

96. Saraceno, A. J., D. T. Finale, and N. D. Coggeshall, *Anal. Chem.*, **33**, 500 (1961).

97. Segal, B. G., M. Kaplan, and G. K. Fraenkel, *J. Chem. Phys.*, **43**, 4191 (1965) with correction in R. Allendoerfer, *J. Chem. Phys.*, **55**, 3615 (1971).

98. Singer, L. S., *J. Appl. Phys.*, **30**, 1463 (1959).

99. Smith, G. W., *J. Appl. Phys.*, **35**, 1217 (1964).

100. Sorin, L. A., and M. V. Vlasova, *Electron Spin Resonance of Paramagnetic Crystals*, Plenum, New York, 1973.

100a. Stegmann, H. B., M. Schnabel and K. Scheffler, *Angew. Chem. Int. Ed.*, **18**, 943 (1979).

101. Swartz, H. M., J. R. Bolton, and D. C. Borg, eds., *Biological Applications of Electron Spin Resonance*, John Wiley, New York, 1972.

102. Szwarc, M., *Carbanions, Living Polymers and Electron Transfer Processes*, Wiley-Interscience, New York, 1968.
103. Tinkham, M., and M. W. P. Strandberg, *Phys. Rev.*, **97**, 937, 951 (1955).
104. Ueda, H., *Anal. Chem.*, **35**, 2213 (1963).
105. Uehara, H., and S. Arimitsu, *Anal. Chem.*, **45**, 1897 (1973).
106. Ulbert, K., *Coll. Czech. Chem. Commun.*, **27**, 1438 (1962).
107. Vigouroux, B., J. C. Gourdon, P. Lopez, and J. Pescia, *J. Phys.*, **E6**, 557 (1973).
108. Wahlquist, J., *J. Chem. Phys.*, **35**, 1708 (1961).
108a. Wasson, J. R. and P. J. Corvan, *Anal. Chem.*, **50**, 478R (1978).
108b. Wasson, J. R. and J. E. Salinas, *Anal. Chem.*, **52**, 50R (1980).
108c. Wasson, J. R., *Anal. Chem.*, **54**, 121R (1982).
108d. Warren, D. C. and J. M. Fitzgerald, *Anal. Chem.*, **49**, 250 (1977).
109. Wertz, J. E., *Chem. Rev.*, **55**, 829 (1955).
110. Wertz, J. E., and J. R. Bolton, *Electron Spin Resonance: Elementary Theory and Practical Applications*, McGraw-Hill, New York, 1972.
111. Wertz, J. E., and J. R. Bolton, pp. 463–467.
112. Westenberg, A. A., *Prog. Reaction Kinetics*, **7**, part 1, 23 (1973).
113. Westenberg, A. A., *J. Chem. Phys.*, **43**, 1544 (1965).
114. Westenberg, A. A., and N. deHaas, *J. Chem. Phys.*, **40**, 3087 (1964).
115. Wilmshurst, T. H., *Electron Spin Resonance Spectrometers*, Adam Hilger, Ltd., London, 1967.
116. Yamamoto, D., and N. Ikawa, *Bull. Chem. Soc. Jpn.*, **45**, 1405 (1972).
117. Yamamoto, D., T. Fukumoto, and N. Ikawa, *Bull. Chem. Soc. Jpn.*, **45**, 1403 (1972).
118. Yamamoto, D., and F. Ozeki, *Bull. Chem. Soc. Jpn.*, **45**, 1408 (1972).
119. Yao, H. C., and H. C. Heller, *Anal. Chem.*, **41**, 1540 (1969).
120. Yen, T. F., ed., *Electron Spin Resonance of Metal Complexes*, Plenum, New York, 1969.
121. Zavoisky, E., *J. Phys., U.S.S.R.*, **9**, 211 (1945).
122. Zolotov, Yu. A., O. M. Petrukhin, V. Yu. Nagy, and L. B. Volodarskii, *Anal. Chim. Acta*, **115**, 1 (1980).

Part I
Section I

Chapter 4

NUCLEAR QUADRUPOLE RESONANCE SPECTROMETRY

BY H. G. FITZKY, *Bereich Angewandte Physik, Fa. Bayer AG, 509 Leverkusen, West Germany*

Contents

I. INTRODUCTION

Since nuclear quadrupole resonance was introduced by Dehmelt and Krueger (20–23) in 1950, it has been used to an increasing extent in the investigation of chemical bonds in crystallized solids. In particular through the work of Townes and Dailey (80–83) it has become possible to use data obtained by spectroscopic methods for elucidating the electronic structure of the chemical bond. In the last few years, due to the development of commercial spectrometers, the analytical application of nuclear quadrupole resonance has been extended to include the identification of functional groups and stereochemical characteristics (8,9,31,32,53,55,73,75). This method is suitable for elucidating compounds with roughly 130 isotopes, the nuclear spin of which is greater than 1/2. Nuclear quadrupole resonance thus supplements nuclear magnetic resonance (NMR), as it can be used for detecting precisely those nuclei which cannot be measured by means of NMR due to excessive line widening caused by nuclear quadrupole coupling. Chlorine quadrupole resonance in particular has gained greater analytical significance, but nitrogen, bromine, iodine, and antimony resonance have also been used to a lesser extent. The resonance frequencies of the various quadrupole nuclei cover a relatively wide range from about 1 to 1000 MHz (Fig. 4.1). For recording spectra in the middle range of about 3–300 MHz, spectrometers that are relatively simple to operate are commercially available. Nuclear quadrupole resonance spectroscopy differs from other methods of spectroscopic analysis in that it requires fairly large amounts of well-crystallized sample material, i.e., compounds of maximum possible purity. Liquids are converted into the crystallized form by freezing. Consequently, only monomers, and not polymers, could be studied with this method. Another important difference lies in the relatively low intensity of the spectra, which take a fairly long time to be recorded. The routine method of elucidating the molecular structure consists as a rule in combining different methods, such as infrared and mass spectroscopy and, if required, ^{13}C nuclear resonance, above all to reduce the number of structural possibilities to be considered.

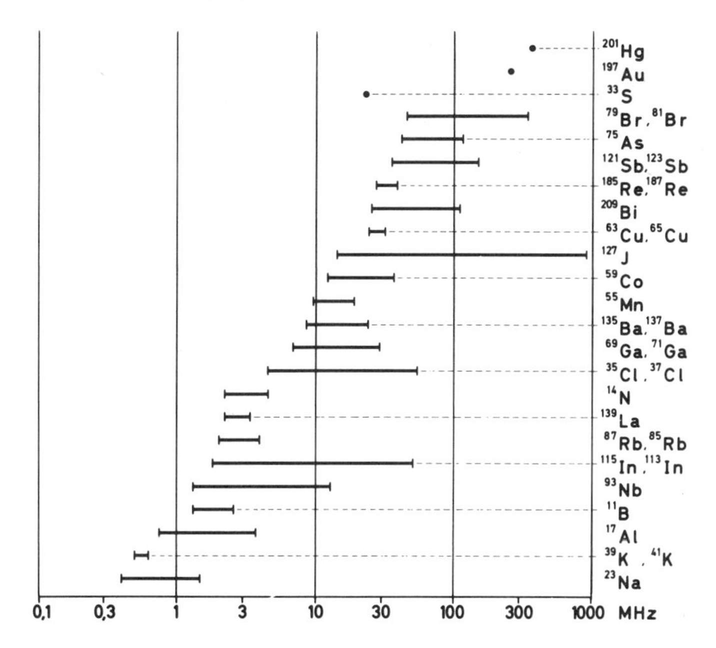

Fig. 4.1. Nuclear quadrupole resonance frequencies of the most important nuclei.

II. PRINCIPLES

A. THE RESONANCE PHENOMENON

1. The Nuclear Quadrupole and Its Environment

Nuclei with a nuclear spin of $I > 1/2$ have a nuclear charge distribution in the form of a rotational ellipsoid and consequently a nuclear quadrupole moment, the size and sign of which depend on the nature and degree of flattening of the ellipsoid (Fig. 4.2). Thus the nucleus ^{14}N (natural abundance 99.635%) has the nuclear spin $I = 1$ and the relatively small nuclear quadrupole moment, $eQ = +0.016$

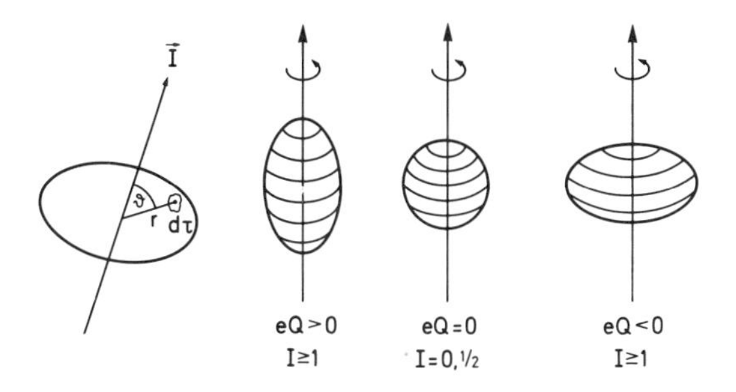

Fig. 4.2. Electric nuclear quadrupole moment. Definition of the scalar electric quadrupole moment: $eQ = \int \rho r^2(3\cos^2\vartheta - 1)d\tau$.

TABLE 4.I

Nuclear Quadrupole Moments eQ, natural Abundances, and Coupling Constants for a single p Electron e^2q_pQ/h

Nucleus	Isotope	Spin	Abundance (%)	eQ ($10^{-24}cm^2$)	e^2q_pQ/h (MHz)
H	2	1	0.015	0.0028	0.2
N	14	1	99.6	+0.016	8.4
Cl	35	3/2	75.4	−0.080	109.7
Cu	63	3/2	69.09	0.16	
Ga	69	3/2	60.4	0.19	125.0
As	75	3/2	100	0.29	380
Br	79	3/2	50.6	0.33	769.7
In	115	9/2	95.72	1.161	899.1
Sb	121	5/2	57.25	−0.53	1300
I	127	5/2	100	−0.79	2292.7

$\times 10^{-24}$ cm^2. Table 4.I contains data on some more frequently studied nuclei. Depending on the direction of the nuclear spin moment I in the inhomogeneous electrical field within the molecule, the nuclear quadrupole may take up energetically different positions. The energy levels are determined by the size of the nuclear quadrupole moment eQ and the inhomogeneity of the field at the nuclear site q, with the spin quantum number I determining the number of orientations of the nucleus, i.e., the number of energy terms. Its field gradient q supplies the chemical information. Its size is determined mainly by the bonding electrons in unfilled shells, principally p electrons. Filled shells and s electrons have no effect because of their spherical symmetry. The effect of d electrons is approximately 10 times less than that of p electrons. In the condensed phase, small additional components, e.g., from the crystal field, which do not exceed about 1%, also have an effect (45,52).

2. The Nuclear Quadrupole in the Field Gradient with Axial Symmetry

If the electric field gradient at the nuclear site is characterized by the second derivatives of the electrostatic potential, Laplace's relation is obtained

$$\frac{\delta^2 V}{\delta x^2} + \frac{\delta^2 V}{\delta y^2} + \frac{\delta^2 V}{\delta z^2} = 0 \qquad (1)$$

with the gradients

$$q_x = \frac{\delta^2 V}{\delta x^2},\ q_y = \frac{\delta^2 V}{\delta y^2},\ q_z = \frac{\delta^2 V}{\delta z^2} \qquad (2)$$

In the case of axial symmetry of the field gradient about the internuclear axis z, is $q_x = q_y$ and $q_z = -2q_x$.

From the classical relation for the energy of the quadrupole nucleus in the field of axial symmetry

$$E = 1/8\, e^2 q_z Q\, (3 \cos^2\phi - 1) \qquad (3)$$

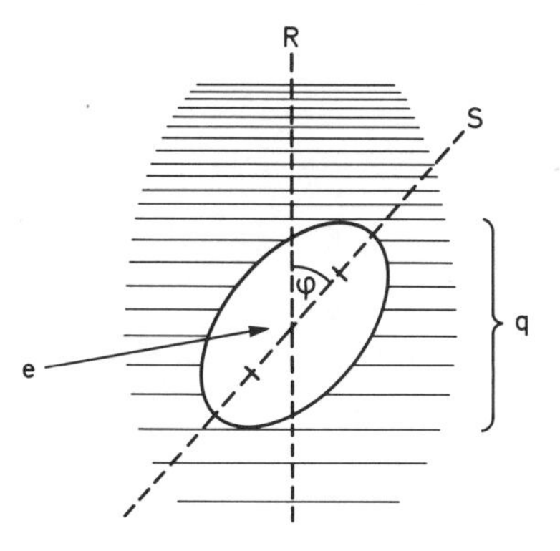

Fig. 4.3. Nuclear quadrupole in an inhomogeneous electric field. Q quadrupole moment, e electronic charge, q average value of the field gradient at the nuclear site, R direction of the field gradient (also quantization axis), S spin axis $(+Q)$.

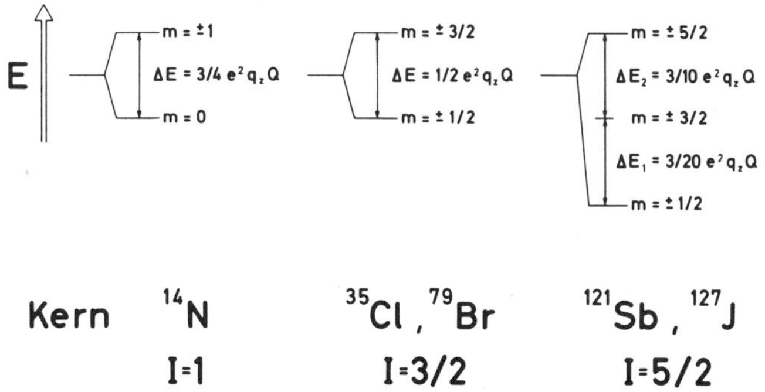

E

m = ±1
ΔE = 3/4 $e^2 q_z Q$
m = 0

m = ± 3/2
ΔE = 1/2 $e^2 q_z Q$
m = ± 1/2

m = ± 5/2
ΔE_2 = 3/10 $e^2 q_z Q$
m = ± 3/2
ΔE_1 = 3/20 $e^2 q_z Q$
m = ± 1/2

Kern ^{14}N | ^{35}Cl, ^{79}Br | ^{121}Sb, ^{127}J

I=1 | I=3/2 | I=5/2

Fig. 4.4. Energy levels of some quadrupole nuclei in the field gradient with axial symmetry.

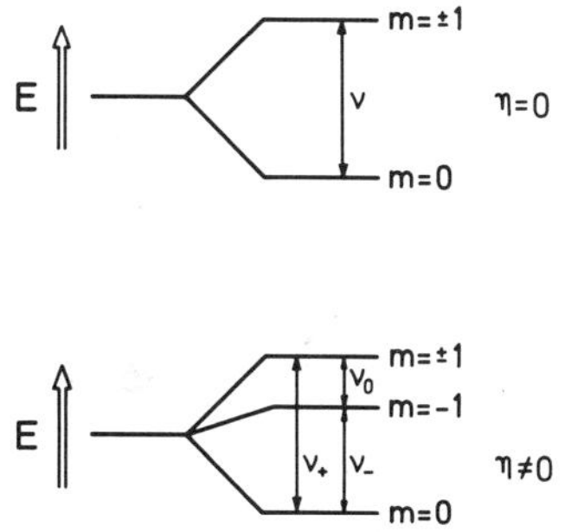

E
m=±1
ν
m=0
$\eta=0$

E
m=±1
ν_0
m=-1
ν_+ ν_-
m=0
$\eta\neq0$

Fig. 4.5. Energy-level diagram of the ^{14}N quadrupole nucleus in axially symmetrical (*top*) and nonaxially symmetrical field gradients (*bottom*).

TABLE 4.II

Transition Frequencies between Quadrupole Energy Levels for Nuclei with Spin $I = 1, 3/2$, and $5/2$

Spin	Transition		Frequencies
1	0	+1	$\nu_+ = 3/4(e^2qQ/h)(1 + \eta/3)$
	0	−1	$\nu_- = 3/4(e^2qQ/h)(1 - \eta/3)$
			$\nu_0 = \nu_+ - \nu_-$
3/2	3/2	1/2	$\nu = 1/2(e^2qQ/h)(1 + \eta\ 2/3)^{1/2}$
5/2	1/2	3/2	$\nu_1 = 3/20(e^2qQ/h)(1 + 1.0926\,\eta^2 - 0.6340\,\eta^4)$
	3/2	5/2	$\nu_2 = 6/20(e^2qQ/h)(1 - 0.2037\,\eta^2 + 0.1622\,\eta^4)$

the quantum mechanical energy that results by quantization of the angle ϕ between the spin axis S and the direction of the field gradient R (Fig. 4.3) is

$$E_m = 1/4\, e^2 q_z Q \frac{3m^2 - I(I + 1)}{I(2I - 1)} \tag{4}$$

with $m = I, I - 1, \ldots -I$.

For ^{14}N with $I = 1$, where there is axial symmetry, a transition (Fig. 4.4) with the frequency $\nu = 3/4\ e^2q_zQ/h$ corresponding to $\Delta m = \pm 1$ results. Similarly, in the case of axial symmetry, ^{35}Cl with $I = 3/2$ shows only one transition of the frequency $\nu = 1/2\ e^2q_zQ/h$, while on the other hand nuclei with a higher nuclear spin, such as ^{121}Sb and ^{127}I, give several transitions (11,45).

3. Asymmetric Field Gradient

In the case of an asymmetric field, i.e. when $q_x \neq q_y$, the m degeneracy of the terms is removed, so that further transitions occur. If the degree of asymmetry is characterized by the parameter η

$$\eta = \frac{q_x - q_y}{q_z} \text{ with } 1 > \eta > 0 \tag{5}$$

three possible transitions result when $I = 1$ (e.g., ^{14}N) (Fig. 4.5). Nuclei with $I = 3/2$ give only one transition, while two transitions take place when $I = 5/2$ and three when $I = 7/2$ (Table 4.II).

B. FIELD GRADIENT AND CHEMICAL SHIFT

1. General

The frequencies of the pure nuclear quadrupole transitions are proportional to the field gradient q at the nuclear site, which is determined almost exclusively by the number of p electrons in the valence shell. An approximate estimation of q was first undertaken by Townes and Dailey (11,80–83). The influence of the p electrons is represented by the relation

$$q = N_x q_x + N_y q_y + N_z q_z \tag{6}$$

taking into account the number N of the p_x, p_y, and p_z electrons. For a completely filled p shell with $N_x = N_y = N_z = 2$, $q = 0$; therefore $q_z = -2q_x = -2q_y$. If this is inserted into equation 6, the result is the basic equation of Townes and Dailey

$$q = q_z [N_z - 1/2 (N_x + N_y)] \tag{7}$$

In the case of one unpaired p electron, a maximum in the field gradient results. Thus, for instance, the chlorine atom with $N_x = N_y = 2$, $N_z = 1$ yields a maximum resonance frequency of $1/2\ e^2q_{at}Q/h = 54.87$ MHz. The other extreme is reached in the case of the chloride ion with its completely filled p shell and extremely low resonance frequencies. The ionic character of a bond influences the position of the resonance frequency to a very large extent. In simply coordinated compounds, there are three variables which have a substantial influence on the chemical shift of the NQR transition. These can be explained fairly simply: The first is the ionic bond character i, characterized, for example, by the electronegativity difference of the bond partners (39); the second is the degree of s hybridization s, which leads to a reduction of the field gradient because of the spherical symmetry of the s electrons; and the third is the multiple bond character π, which influences the magnitude and symmetry of the field gradient. These three factors can be represented as follows:

$$e^2q_zQ \,/\, (e^2q_{at}Q) = (1 - i)(1 - s) - \pi \tag{8}$$

The multiple-bond fraction of multiple bond character (conjugation effect) due to p_π bonding is

$$\pi = 2/3 \frac{e^2 q_z Q}{e^2 q_{at} Q} \eta \tag{9}$$

It usually causes a reduction in the transition frequency.

2. Nuclear Quadrupole Coupling Constants of Gases and Solids

The ionic-bond fraction of ionic character, s hybridization, and conjugation effect have the greatest effect on the chemical shift of the NQR frequencies. There is a considerable amount of experimental data, in particular on inorganic and organic halogen and nitrogen compounds. Most of the information derives from measurements of pure nuclear quadrupole transitions of crystallized solids, but a minor part also from investigations of the nuclear quadrupole hyperfine structures of the microwave rotation spectra of molecules in the gas phase (39,40,52,76). When comparing the values of the nuclear quadrupole coupling constants e^2q_zQ/h obtained by these methods, it must be borne in mind that there is a slight increase in the ionic share when measurements are carried out in the solid phase, due to the crystal field (76) (Table 4.III).

3. Structure and Chemical Shift

Compared with the shift ranges of NMR (^1H, ^{19}F), the shift range of nuclear quadrupole resonance (NQR) covers a considerably wider frequency range, owing

TABLE 4.III
Nuclear Quadrupole Coupling Constants e^2qQ/h in MHz in the Gas Phase and Solid State

Nucleus	Compound	e^2qQ/h Gas phase	e^2qQ/h Solid state	Ref.
^{35}Cl	CH_3Cl	74.77	60.4	(76)
	CH_2Cl_2	78.4	72.47	(76)
	$CHCl_3$	80.9	76.98	(76)
	CF_3Cl	78.05	77.58	(76)
	CF_2Cl_2	—	78.16	(39)
	$CFCl_3$	110.8	79.63	(76)
^{14}N	HC≡N	4.58	4.01	(52)
	ClC≡N	3.63	3.219	(52)
	BrC≡N	3.83	3.35	(39,52)
	JC≡N	3.80	3.40	(39,45)

mainly to the extremely strong coupling of the quadrupole nucleus to the molecular field. Thus the values of the quadrupole coupling constants for ^{35}Cl lie between 0.04 MHz in the case of KCl and 109.74 MHz in the case of Cl_2. The corresponding values for ^{14}N lie between a few kHz and 8.4 MHz.*

For reasons of intensity, transitions at higher frequencies, over about 500 kHz, are in particular accessible to pure NQR.

The electronic effects on the chemical shift that are most important for elucidating the structure are discussed in Sections II.B.3.a–f, using mainly chlorine and nitrogen compounds as examples. The observations can easily be applied to other nuclei.

a. IONIC BOND CHARACTER

A highly ionic bond ($i \longrightarrow 1$), as on the right side of equation 10

$$R^- Cl^+ \longleftrightarrow R{-}Cl \longleftrightarrow R^+ Cl^- \qquad (10)$$

yields frequencies tending to zero for the nuclear quadrupole resonance transitions, while predominantly covalent bonds such as N–Cl bonds (middle of equation 10) lead to approximately one unpaired electron on the chlorine nucleus and to very high frequencies, in extreme cases $1/2\,(e^2q_{at}Q/h) = 54.8$ MHz. If the electronegativity of the bond partner is higher than that of the Cl, the frequencies may be even higher ($R = F$). Two typical examples of increasing ionic bond character are given in Table 4.IV. It is known that, parallel to this, the rate of hydrolysis of these compounds also increases. Because of the greater difference in the electronegativities of the bond partners, the Si–Cl frequencies are considerably lower than those of the corresponding C–Cl bond.

*The value of 8.4 MHz is only an estimate, as measurement of the nuclear quadrupole coupling constant for an unpaired p electron is not possible because of the ^4S ground energy level of the N atom with $e^2q_0Q/h = 0$ (52,72).

TABLE 4.IV

Decreasing ^{35}Cl-NQR Frequencies with increasing Ionic Bond Character (32)

Compound	Frequency (MHz)	Compound	Frequency (MHz)
CH_3Cl	34.029	CH_3SiH_2Cl	—
$(CH_3)_2CHCl$	32.069	$(CH_3)_2SiHCl$	17.166[a]
$(CH_3)_3CCl$	31.065	$(CH_3)_3SiCl$	14.465

[a] Mean value.

TABLE 4.V

Reduction of ^{35}Cl NQR Frequency by Conjugative Effect (32)

Cl^+ (4-position), N	Cl (3-position), N	N, Cl^+ (2-position)
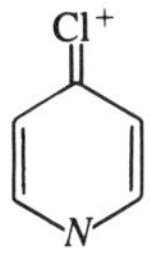 34.194 MHz	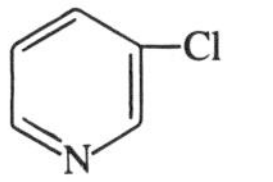 35.234	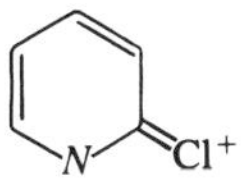 34.759
conjugative effect	no conjugative effect	conjugative effect

b. CONJUGATION EFFECTS

A conjugative effect is the participation of π electrons (double-bond character) in the C–Cl bond. A typical feature is the existence of limiting structures

$$C = C{-}Cl \longleftrightarrow C^-{-}C = Cl^+ \tag{11}$$

Because of the elimination of the rotational symmetry of the field gradient around the internuclear axis, the participation of π electrons in the C–Cl bond leads to the occurrence of an asymmetry parameter $\eta \neq 0$. According to Bersohn (5), the π-bond character is directly proportional to η (cf. equation 9). In the case of chlorine compounds, NQR Zeeman effect measurements on monocrystals or microwave gas spectroscopic measurements are essential to determine η. In the case of nitrogen compounds and other nuclei with spins greater than 3/2, it can be determined directly from the different NQR transitions. In general the shift value is reduced when there is a conjugative effect. Conjugative effects frequently occur when chlorine is linked to unsaturated C atoms. As an example, Table 4.V illustrates corresponding resonance structures in isomeric chloropyridines.

c. ELECTRONEGATIVE BOND PARTNERS

The factors that have the largest influence on the extent of the chemical shift are the electronegativies of the bond partners or groupings near the chlorine nucleus, with influences becoming effective also over several single bonds, as will be shown later, using N–Cl compounds as an example. These effects, which are familiar to the chemist as inductive effects, have been closely studied for the C–Cl σ bond and

TABLE 4.VI
^{35}Cl NQR Frequency, inductive Effect and Taft σ^* Parameter of the Compound $R-CH_2Cl$ (52)

R	NQR frequency (MHz)	σ^*
CH_3CH_2-	32.968	-0.100
CH_3-	32.985	0
$CH_3CH{=}CH-$	33.455	$+0.36$
H–	34.023	0.49
CH_3OCH_2-	33.453	0.52
C_6H_5-	33.627	0.60
$CH_3(CO)-$	35.278	1.65
$CH_3O(CO)-$	35.962	2.00
CF_3-	37.98	3.1

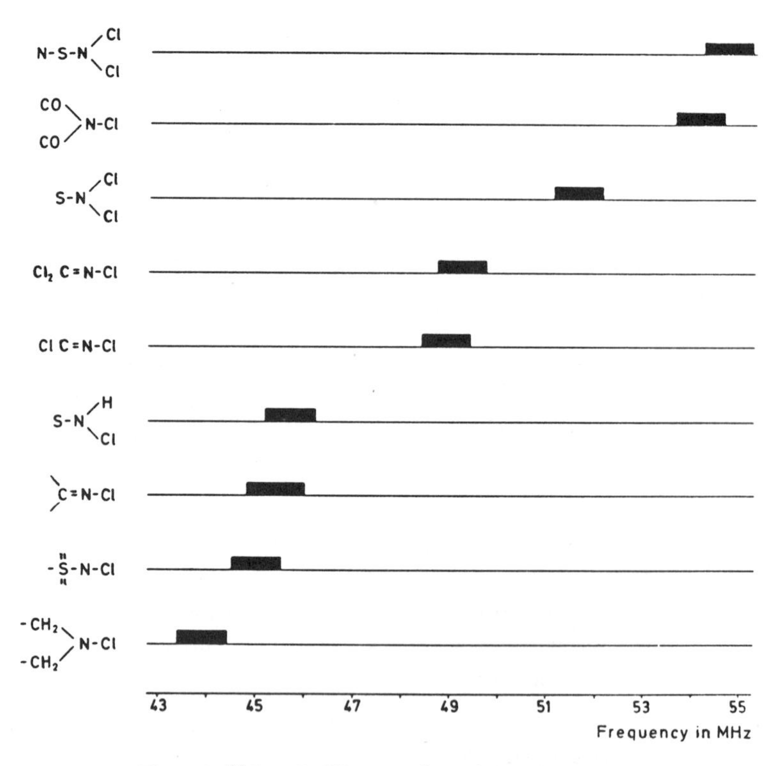

Fig. 4.6. ^{35}Cl NQR shift range of organic N–Cl compounds.

paralleled to Taft's σ^*-parameter (52,90) (Table 4.VI). An essential feature of the inductive effect is the electron shift along the C–Cl σ bond, which reduces the filling up of the p shell of the chlorine by increasing the electronegativity of the bond partner, resulting in correspondingly higher NQR frequencies.

Similar investigations have been carried out on 1-chloro-*n*-alkanes (52) and α-ω dichloroalkanes (44). Further studies of the correlations between the shift value and the electronegativity of trichloromethyl compounds (8,9,31,32), Si-chlorine compounds (51,44), carbonyl chlorides (10,90), and phosphoryl chlorides (54) give typical values for analytical purposes. For special groupings, shift tables for comparison purposes can easily be prepared from tables of the NQR frequencies (6). As an example, Fig. 4.6 gives a survey of the shift values of the N–Cl bond. The increase in shift here is caused not only by the proximity of electronegative groupings and the carbonyl group, but also by the accumulation of chlorine nuclei on the nitrogen. Thus, two neighboring carbonyl groups shift the frequency of the nitrogen-bound chlorine in *N*-chlorosuccinimide into the range of $-NCl_2$.

Fig. 4.7 shows the inductive influence of electronegative groupings which affects the chlorine over several bonds. It represents spectra of $\rangle$C = N–Cl compounds with the electronegativity of the adjacent grouping increasing from top to bottom (33). Other correlations between the shift value and the electronegativity of the bond partners can also be found in the spectra of other quadrupole nuclei. Thus Table 4.VII, which is a comparison of ^{14}N resonances of the CN-group with the ^{35}Cl resonance of the $-CCl_3$ group, shows that there is a considerable correlation between the shift and the electronegativity of the residual molecule. Other correlations of this type for bromium and iodine compounds are given in Table 4.III.

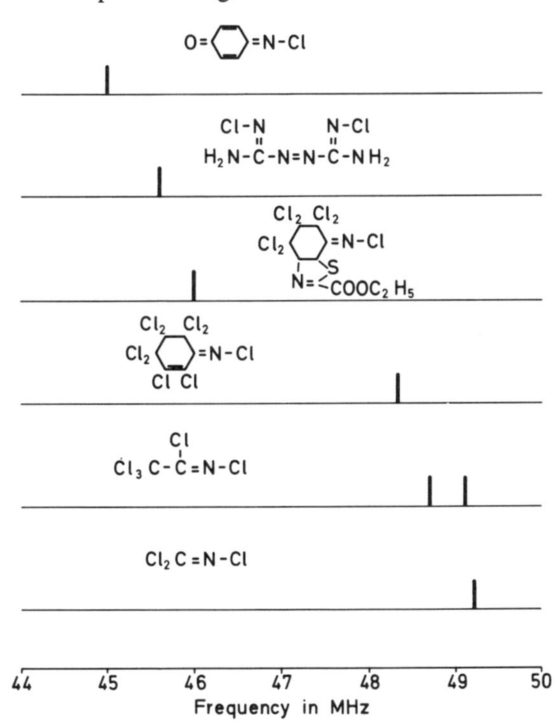

Fig. 4.7. ^{35}Cl NQR shift in $\rangle$C=N–Cl compounds with different electronegativities of the substituents.

TABLE 4.VII

^{14}N and ^{35}Cl NQR Shift and Electronegativity (34)

Compound	^{14}N NQR frequency (MHz)	Compound	^{35}Cl NQR frequency (MHz)
NC–CN	4.269	$NC-CCl_3$	41.46
Cl_3C-CN	4.052	Cl_3C-CCl_3	40.67
Cl–CN	3.219[a]	$Cl-CCl_3$	40.60
H–CN	4.018	$H-CCl_3$	38.28
C_6H_5-CN	3.885	$C_6H_5-CCl_3$	38.70
$CH_2=CH-CN$	3.800	$CH_2=CH-CCl_3$	38.30
C_2H_5-CN	3.775	$C_2H_5-CCl_3$	–
CH_3-CN	3.737	CH_3-CCl_3	37.94

[a] Because of a high conjugation component, which is expressed in the structure $Cl^+=C=N^-$ and leads to an excessively high π-electron density on the N, Cl–CN lies at a very low frequency.

d. Accumulation of Chlorine Nuclei

A frequent special case of the shift effect of electronegative bond partners is the accumulation of Cl nuclei at one and the same C (or Si, P, S, N) nucleus. The examples in Table 4.VIII show that, when further chlorine nuclei enter a certain functional group, approximately equal shift changes occur in each case which may permit an additive estimation of shift values. Experience has shown that with a purely additive precalculation of the spectra a sufficient degree of accuracy can be attained, as, for example, in the method suggested by Scrocco for chlorinated benzenes (74). According to this method, the ^{35}Cl resonances of the chlorobenzenes are

$$f\,(^{35}\mathrm{Cl}) = f\,(\text{chlorobenzene}) + a\,f_{ortho} + b\,f_{meta} + c\,f_{para} \tag{12}$$

with $f_{ortho} = 1.30, f_{meta} = 0.45$, and $c_{para} = 0.25$ MHz. a,b,c are the numbers of the nuclei concerned. With this formula, the shift contributions of further chlorine

TABLE 4.VIII

^{35}Cl NQR Shift and Shift Change (Δ) for Accumulation of Cl Nuclei (mean values of the frequency in MHz) (32)

Compound	Frequency	Δ	Compound	Frequency	Δ
CH_3Cl	34.023		C_6H_5Cl[a]	34.622	
CH_2Cl_2	35.991	1.968	$C_6H_4Cl_2$	35.273	0.651
$CHCl_3$	38.281	2.290	$C_6H_3Cl_3$	36.212	0.939
CCl_4	40.633	2.352	$C_6H_2Cl_4$	36.996	0.784
			C_6HCl_5	37.963	0.964
			C_6Cl_6	38.451	0.488

[a] Chlorobenzenes.

TABLE 4.IX

^{35}Cl NQR Frequency and Ring Strain (32)

$CH_3-CCl_2-CH_3$	1,1-dichlorocyclohexane	1,1-dichlorocyclopentane	1,1-dichlorocyclobutane	1,1-dichlorocyclopropane	$H_2C{=}CCl_2$
34.88 MHz	34.99 34.90	35.06 34.98	35.35[a]	36.61	36.54

[a] Estimated.

substituents, starting from chlorobenzene, can be determined with sufficient accuracy. It is also apparent that the effect of the newly entering chlorine nuclei decreases with the distance. This method of additive shift calculation can be easily modified for special cases, if, for example, no suitable comparison substances are available (31,32). Values of Taft's polarity parameter σ^* can also be used for precalculation, as was shown using $RSiCl_3$ compounds as an example (52). Correlations between Hammett σ values and ^{14}N shift values of substituted pyridines have also been investigated for this purpose (72).

e. RING STRAIN

Cyclic compounds show an increase in NQR frequencies which rises with increasing ring strain. Increased electronegativity of the carbon σ orbital can be assumed to be at least partially responsible for this, with the increase rising with the deviation from the sp^3 configuration (77). The relationships at the doubly chlorinated carbon of saturated rings with inclusion of the corresponding straight-chain compounds are demonstrated in Table 4.IX.

f. STEREOCHEMICAL EFFECTS

Certain stereochemical questions can presumably also be investigated with the aid of NQR spectroscopy (45,53). The results obtained so far show that the difference between the axial and equatorial Cl nuclei is often relatively small, while in the case of *cis*/*trans* isomers larger differences are to be expected. *Cis*- and *trans*-1,2-dichloroethylene give frequencies of 34.864 (mean value) and 34.497 MHz, distinctly different values. The shifts for the *cis*-arrangement are always somewhat higher (Table 4.X). The frequencies for equatorial and axial Cl in the four isomeric hexachlorocyclohexanes, however, are respectively 36.991 MHz (mean value) with a standard deviation of 0.259 and 36.422 with a standard deviation of 0.466, so that it is not always possible to distinguish safely between the two (53). On the other hand, results obtained with 2,3-dichlorodioxane (1,50) and various chlorinated sugars (16) show that equatorial Cl in the α-position to the ether oxygen yields values which are 2–3 MHz higher than those obtained with axial Cl. Table 4.X contains a survey of *cis*/*trans* effects that have so far become known and of data on syn- and anti-7-chloronorbornene (14). Other NQR measurements that are important for determining

TABLE 4.X

Configuration and ^{35}Cl NQR Frequency Change in MHz at 77°K

Compound		Frequency (MHz)	Change *cis*/*trans*	Reference
Cl CH=CH Cl	*cis*	34.865[a]	0.368	6
	trans	34.497		
Cl, O, =O, O, Cl	*trans*	35.970		[b]
Cl, Cl, O, =O, O	*cis*	36.291	0.321	[b]
Cl, Cl, COOMe, COOMe	*cis*	35.284		53
Cl, Cl, COOMe, COOMe	*trans*	34.546	0.738	53
Cl, Cl	*cis*	35.727		53
	trans	33.323	2.404	53
Cl, Cl	*cis*	34.705		53
	trans	33,936	0.769	53
Cl	*syn*	33.021	0.836	14
	anti	32.185		14

[a] Mean value.

[b] H.G. Fitzky, H. Kalkoff, unpublished results.

the structure of chloronorbornenes and chloronorbornanes have been published by M.J.S. Dewar et al. (25). Studies have also been made of similarly constructed insecticides and pesticides (37,69) and there have also been some conformational investigations with, for example, benzotrichloride (2,46), benzyl chloride (26), and 1,2-dichloroethane (7). Numerous chlorocyclopropanes have been investigated by

Lucken et al. with regard to the stereochemical effects of neighboring substituents (3).

III. PRACTICE

A. NQR SPECTROMETER

1. General

Transitions between the energy levels of the nuclear quadrupole are induced by the magnetic coupling of the nuclear spin with the high-frequency magnetic field (29,41,47). A spectrometer therefore consists essentially of a tuneable high-frequency generator producing a high-frequency magnetic field which is suitable with regard to frequency and intensity and in which the substance to be investigated is placed. Nuclei with short spin-lattice relaxation times T_1, such as ^{35}Cl, for example, require higher rf power, which is preferably produced by superregenerative oscillators. On the other hand, for nuclei with long spin-lattice relaxation times, such as ^{14}N, measuring oscillators are required that have sufficient detection sensitivity even with low intensity, such as the so-called marginal oscillators. A third type that has established itself in the past few years is the pulse Fourier transformation (FT) spectrometer which has greater detection sensitivity than with continuous wave (cw) spectrometers available, but is also considerably more expensive. Continuous wave spectrometers preferably measure the NQR absorption on the basis of a change in the damping of the oscillator circuit, while in the pulse FT spectrometer measurement is based on the spin echo (free induction decay).

2. Types of Spectrometer

a. SUPERREGENERATIVE OSCILLATOR SPECTROMETER (SRO)

SRO spectrometers have been used successfully since the beginnings of NQR spectroscopy because they are simply constructed (19,57,88), easy to operate, and are highly suitable for spectroscopy over large frequency ranges (approx. 1-1000 MHz) for measuring all nuclei with not too high relaxation times, T_1. They are therefore by far the most widely used type. To measure the NQR absorption, the sample is placed in the high-frequency magnetic field of the oscillator inductance. The high-frequency oscillation is interrupted periodically by low-frequency (10-50 kHz) quench modulation of the oscillator circuit amplification. At the same time, the high-frequency amplitude that occurs in the oscillating circuit periodically traverses exponentially rising and falling regions. In the case of resonance absorption, the amplitude response is changed by the preceding nuclear magnetization in a characteristic manner (Fig. 4.8), which manifests itself in an increase in the oscillation amplitude in the decaying and starting phases. To mark this very slight change in the signal caused by resonance absorption, either the absorption effect is modulated by low-frequency Zeeman field modulation (10-200 Hz) of the NQR transition or

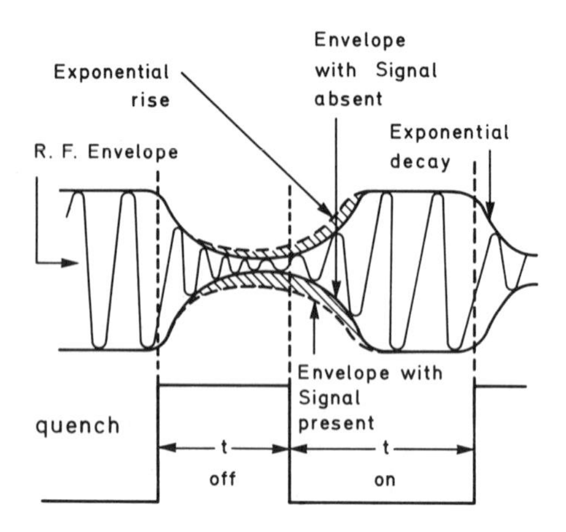

Fig. 4.8 Quench cycle of a superregenerative oscillator.

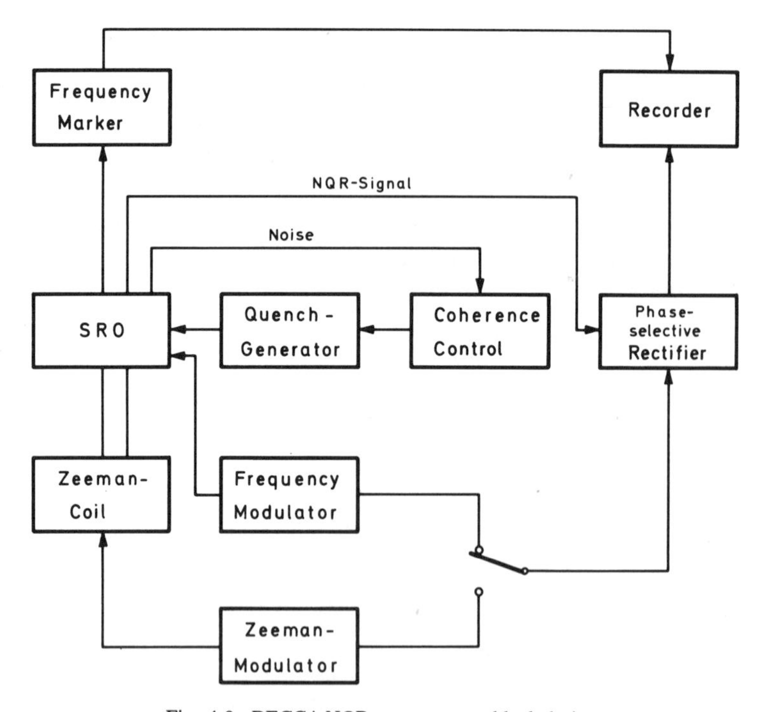

Fig. 4.9. DECCA NQR spectrometer, block design.

the oscillator is frequency modulated with a frequency deviation of one-half the linewidth. To maintain constant detection sensitivity of the SRO spectrometer when scanning larger frequency ranges, so-called coherence control is used. This controls the extent of the decay of the high-frequency oscillation, mostly by controlling the pulsewidth t_{off} of the quench modulation (Fig. 4.8), taking the oscillator noise in the range below the quench frequency as a parameter. Fig. 4.9 shows the block design of

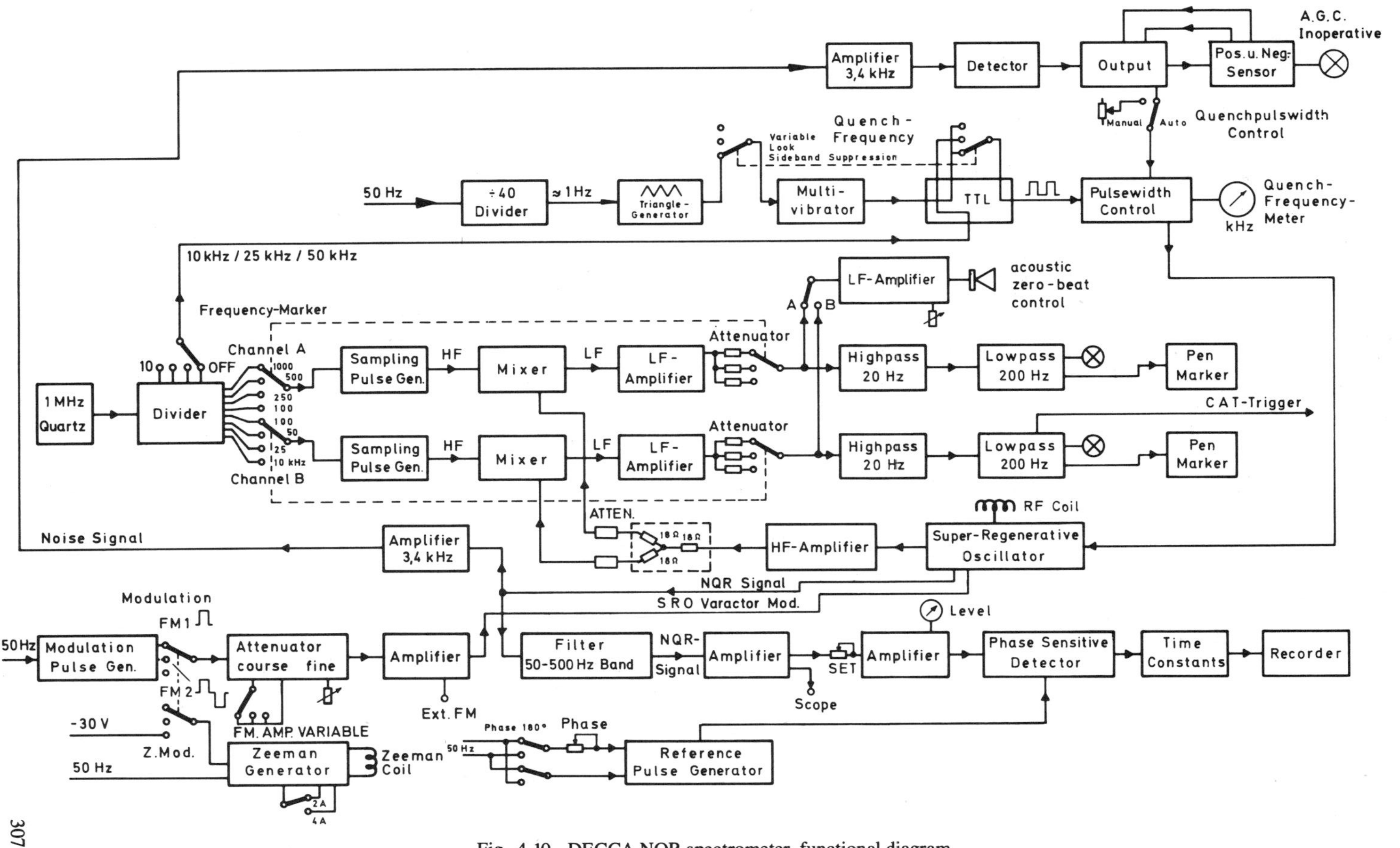

Fig. 4.10. DECCA NQR spectrometer, functional diagram.

a modern NQR spectrometer which works with an SRO. Details of the electronic function are illustrated in Fig. 4.10, using as an example an instrument manufactured by DECCA RADAR (19). This spectrometer is fitted with a number of additional devices which are indispensable for analytical applications. It provides not only for automatic sensitivity (coherence) control (18,19,62) by control of the quench pulse width t_{off}, but also for accurate frequency marking of the spectra, with the possibility of selecting a marker spacing of 10–25–50 or 100 kHz in the one channel and 100–250–500 or 1000 kHz in the second channel. The accuracy of the generation of markers is guaranteed by a derivation of the quench frequency (10–25 or 50 kHz) from the same 1-MHz quartz oscillator. The apparatus is also equipped with a device for suppressing the sidebands caused by the quench modulation of the oscillator. For this purpose the quench frequency is varied in 1-Hz rhythm over a certain range (approx. 10–40 kHz), so that the sidebands are spread over a corresponding range. The oscillator frequency is generally preselected through exchangeable coils for the frequency range of 2 MHz to approximately 60 MHz. With other oscillator units, frequencies of up to about 400 MHz are reached. The fine tuning within the bands is mostly carried out capacitively with a geared motor with variable reduction.

The NQR signal after the first demodulation is modulated up to a carrier frequency of the value of the Zeeman field modulation or the frequency modulation. For recording the NQR spectra with a line recorder, there is a further phase-sensitive rectification, the time constant of which can be varied in the range of approximately 0.3–30 s to adapt to the sensitivity. For cooling the sample, either a Dewar flask can be provided within the oscillator coil or the whole complex of coil and sample is cooled.

The special advantages of the SRO spectrometer lie in the fact that it is simple to operate and works fully automatically with fixed adjustment over a range of roughly half an octave. The detection sensitivity fluctuates by only about ±20%. With a careful design of the oscillator circuit, no "false" resonances occur in the range up to at least 100 MHz (19,31,32). The high-frequency performance can be easily adapted to practically all nuclei, with the exception of ^{14}N and possibly also ^{11}B. These nuclei require rf powers that are so low that it is impossible to attain them with a conventional SRO and at the same time maintain sufficient detection sensitivity. However, by using a pulse-operated high-frequency field of absolute coherence and a more highly developed SRO technology, sensitive spectrometers can be constructed for these nuclei as well (58,59).

b. Marginal Oscillator Spectrometer

For recording the NQR spectra of nuclei such as ^{14}N and ^{11}B, which have a long spin–lattice relaxation time, so-called marginal oscillators are used to prevent radiation saturation (15,64,67,89,91). The characteristic feature of the marginal oscillator is that it still has relatively high detection sensitivity even when the intensity of the high frequency magnetic field is very low. The principle of this oscillator detector system is that, contrary to the SRO, the amplitude of the oscillation remains constant, with the result that no sidebands occur. By means of a special device in the oscillator circuit, the oscillator is kept constantly in the initial oscillation phase. This

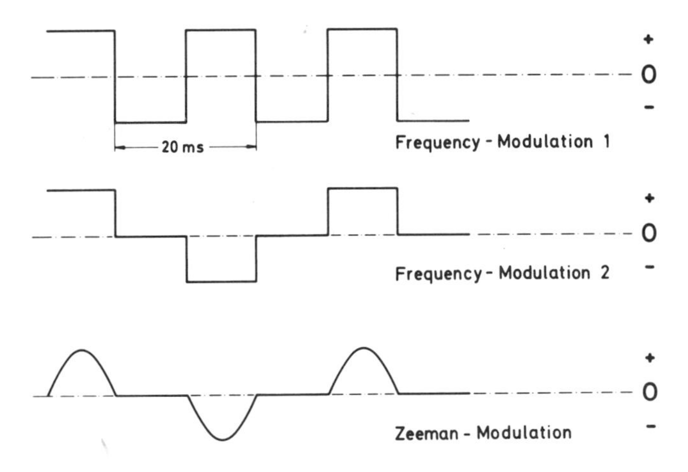

Modulation-Signals

Fig. 4.11. Typical modulation envelopes.

means that due to NQR absorption the amplitude is extremely sensitive to the damping of the frequency-determining oscillation circuit. For better marking of the resonance absorption, the marginal oscillator is usually frequency-modulated with a bisymmetrical square-wave voltage (FM 2 in Fig. 4.11), a frequency deviation of approximately one-half the line width being selected. In the case of phase-sensitive rectification of the nuclear resonance signal, the first derivative of the spectral line results, while if the lock-in amplifier is adjusted at double the modulation frequency the second derivative occurs. The Zeeman modulation is not particularly efficient for nitrogen compounds with a higher asymmetry parameter η and requires very high modulation-field intensities for material in powder form (13). In contrast to the SRO, the marginal oscillator cannot operate automatically over larger-frequency ranges and usually the marginality and intensity are readjusted manually at frequency intervals of 100–300 kHz to maintain sufficient detection sensitivity. Consequently, in spite of some improvements (38,66,71,89), the marginal oscillator is not widely used, in particular because of its difficult adjustment.

c. PULSE FT SPECTROMETER

Pulse spectrometers, which operate in a manner similar to that known from nuclear magnetic resonance spectroscopy (^{13}C) (28), can be used for recording NQR spectra that are very weak, either because of greater linewidths due to incomplete crystallization of the sample, as is the case with polymers, for example (3,70), or because of a high content of impurities. They are also suitable where the amount of substance is very small and where the absorption is easily saturated. In contrast to spectrometers with SRO or marginal oscillators, the sample is subjected to one or several short high-frequency pulses (90° pulse) and the spin–echo signal which then occurs is recorded in the form of free-induction decay (12,42,49,63). This signal must then undergo Fourier transformation to transform it from the time to the frequency domain (computer). The advantages of the pulse FT spectrometer lie in its comparatively high detection sensitivity, the simple measurement of the relaxation times T_2^*, T_2, and T_1, and the avoidance of saturation problems. The disadvantages are that it is expensive and above all that it does not permit automatic recording of larger spectral ranges.

3. Spectrometer Systems

For better control of the electronic function, detection sensitivity, and resolving power, and for monitoring the calibration of the spectra, the spectrometer should be equipped with a spectrum analyzer and a frequency counter (33). The spectrum analyzer makes it possible to measure accurately the frequency deviation in the case of frequency modulation which with regard to the NQR linewidth must not rise above or fall below certain limits (Section III.B). With this device, it is also possible to check the symmetry of the sidebands of the SRO (quench modulation) and the coherence adjustment and to minimize the frequency deviation of the carrier in the case of sideband suppression. A digital frequency counter is mainly required for checking the frequency calibration of the spectra (frequency markers). It is particularly necessary when noncalibrated measuring coils are used and when individual frequencies have to be adjusted quickly and precisely to record certain regions of the spectrum. For measuring frequency, the quench modulation must be switched off if the gate control of the counter is not synchronized. Consequently it is impossible to measure the frequency during recording.

If, on the other hand, a counter is used on which the gate is controlled by the quench modulation (e.g., Hewlett Packard (5360A), digitial frequency measurements can be carried out during recording of the spectra even when the quench modulation is switched on. Of course, it is also possible to make frequency measurements with an uncontrolled counter during recording of the spectra when using a marginal oscillator which is subjected to a frequency modulation with only a small deviation (0.2–1 kHz). In this case, the degree of accuracy depends on the extent of the frequency deviation. The most effective, undirectional interconnection of SRO spectrometer, spectrum analyzer, and counter is illustrated in Fig. 4.12. Details of possible applications of the spectrum analyzer are given in the following Section III.B.

B. SIGNAL SEARCHING

1. Methods of Using a Spectrometer

a. General

There are basically two different methods of using an NQR spectrometer, the one being characterized by maximum detection sensitivity, the other by maximum resolving power. To detect the absorption lines, the first step is to prepare a survey spectrum that will definitely cover the expected frequency range. For this purpose, the spectrum is recorded without regard to distortions and resolution and at the highest amplification, high rf level of the oscillator, and lowest bandwidth of the receiver (time constant 10–30 sec). It may take up to 12 h to record a survey

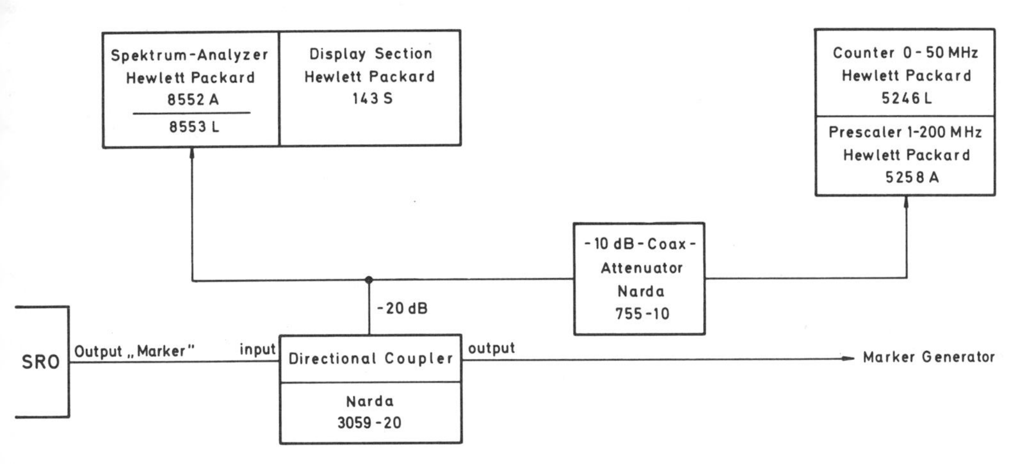

Control Circuit for Oscillator Sideband-Suppression, Symmetry and Frequency

Fig. 4.12. Connection of spectrum analyzer, frequency counter, and spectrometer: control circuit for oscillator side band, symmetry, suppression, and frequency.

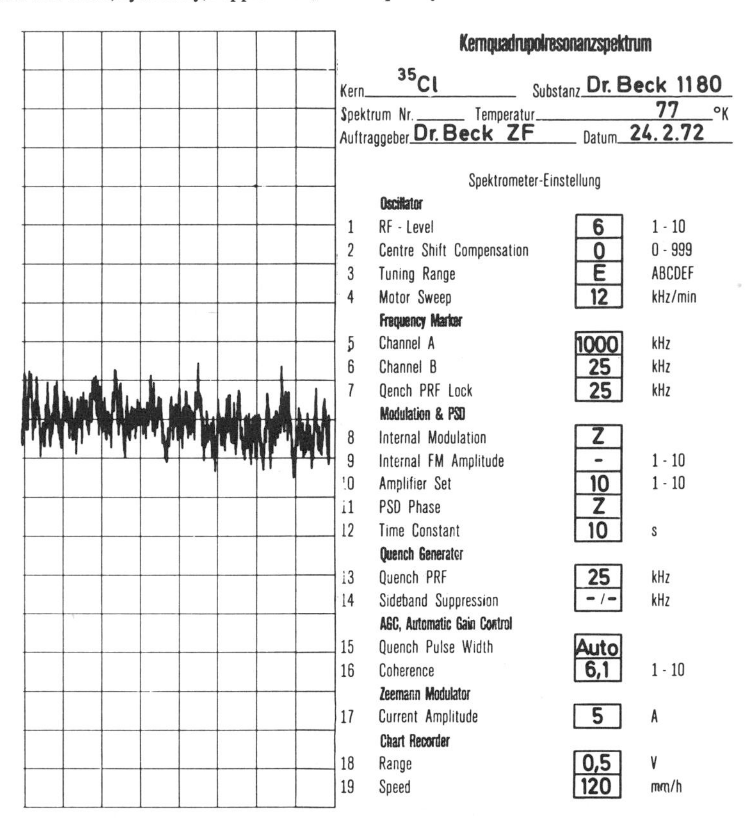

Fig. 4.13. Typical SRO spectrometer settings.

spectrum. The modulation normally used for this purpose is of the Zeeman type, with a high amplitude of ± 50 to ± 200 G. Once the line frequencies are roughly known, extended spectra are run with optimized instrument settings to determine the line frequencies more precisely. The table at the spectrum head (Fig. 4.13) provides a survey of the required instrument settings. The setting of the instrument includes the selection of the measuring coil which is required for the frequency range and into which the sample tube is to be inserted, and the setting of the oscillator level, taking into consideration the saturation properties of the nuclei concerned. When recording survey spectra, it is in most cases feasible to use such a high rf power that saturation results. The intensity-graduated sidebands yield in every case good coverage of the absorption line free from saturation effects.

At 77K, saturation generally occurs more easily than at 300K. This means that the

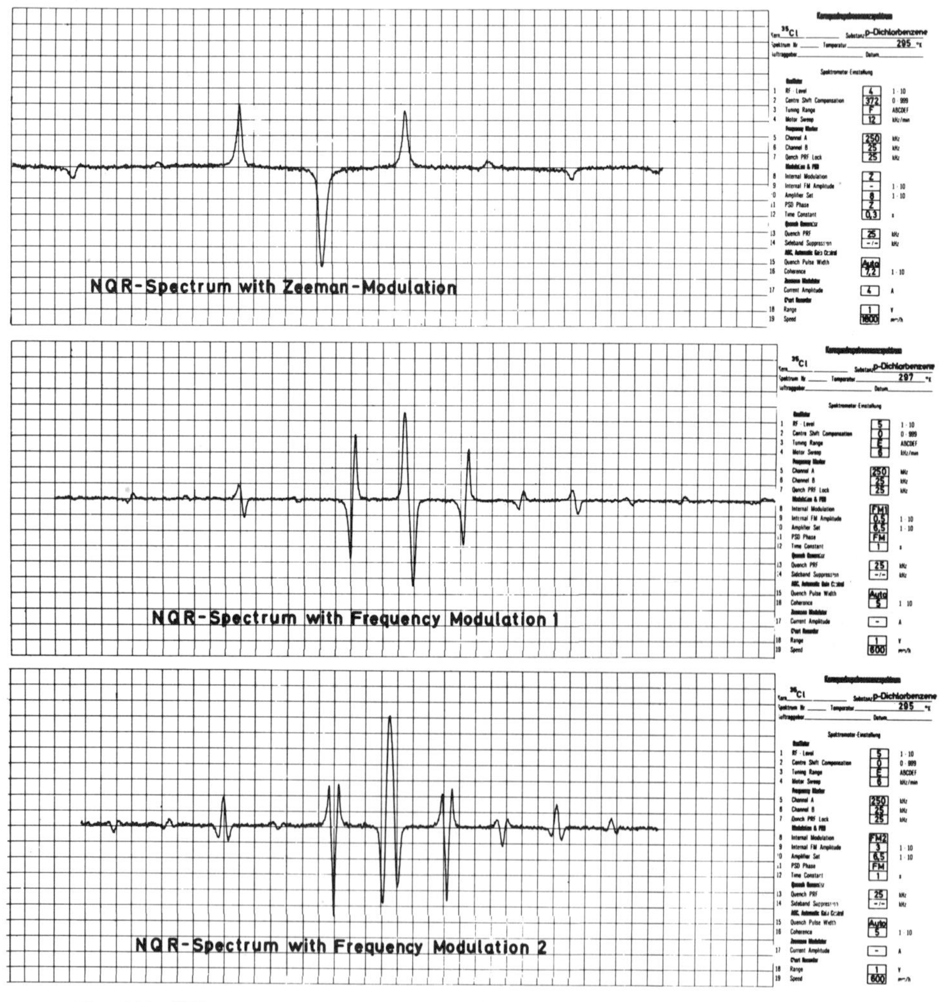

Fig. 4.14. ^{35}Cl-NQR spectrum of *p*-dichlorobenzene taken with different modulation signals.

advantage of the intensity increase in the NQR signal by cooling to 77K is partly canceled by the lower oscillator rf level required due to easier saturation.

b. TYPES OF MODULATION AND LINE SHAPE

There are three possible ways of modulating the absorption signal. For recording over large frequency ranges, Zeeman modulation (50 Hz sine half-waves, bisymmetric, see Fig. 4.11) gives a representation of the natural line profile (Fig. 4.14), high detection sensitivity, and good stability of the zero line. In addition, it is not necessary to adapt the modulation amplitude to the expected line width, as is the case in frequency modulation (79). Frequency modulation (50 Hz) with different wave shapes (Fig. 4.11) has no advantages over Zeeman modulation. In general, it can only be used for recording over frequency ranges up to 0.5 MHz, usually leads to drift of the zero line, and may cause interferences with parasitics of the oscillator (false signals). When the lines are not too wide, but closely arranged, bisymmetric square-wave frequency modulation (Fig. 4.11, middle) (double-reference frequency of the lock-in rectifier) may give somewhat better separation, as the NQR lines can then be written in the form of the second derivative (Fig. 4.14). The spectrum of CCl_4 is used here to demonstrate the resolving power attainable in the range of chlorine resonance. Lines with 2 kHz spacing can still be distinguished, with 4 kHz spacing they can be clearly separated (Fig. 4.15).

c. QUENCH MODULATION AND SIDE-BAND SUPPRESSION

For quench modulation of the oscillator, quench frequencies of about 25 kHz are mostly used. In the case of extremely wide lines and high NQR frequencies, it may be advisable to increase the quench frequency to 50 kHz to achieve better separation of the sidebands, although this does not improve the detection sensitivity. Due to the quench modulation of the oscillator, a sideband spectrum is produced in which the

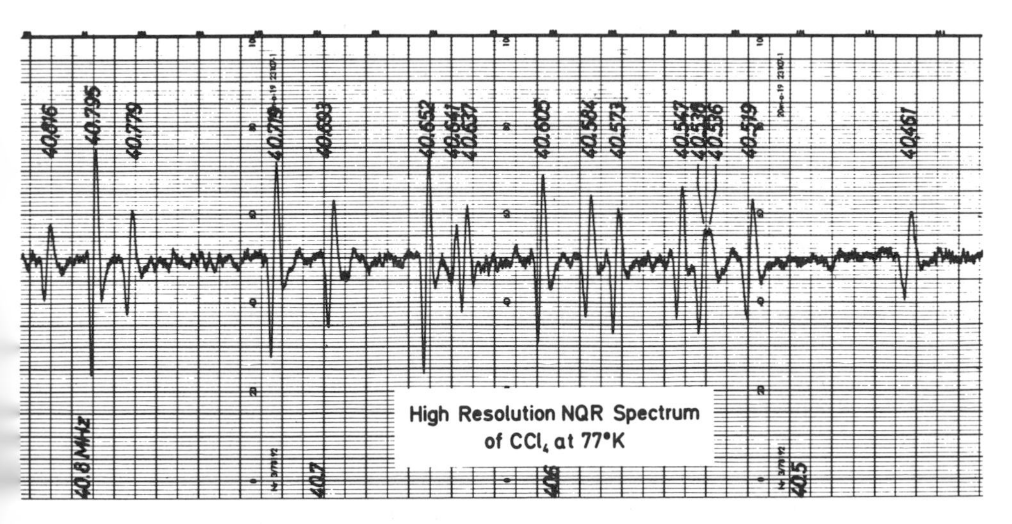

Fig. 4.15. ^{35}Cl-NQR spectrum of CCl_4 with maximum resolution.

signals are spaced according to the quench frequency and which gives an analogous representation of the NQR line (Fig. 4.16).

The intensity of the sidebands should decrease rapidly as the distance from the carrier increases to guarantee the best possible separation and detection of closely arranged lines (Fig. 4.17). For the same reason, the amplitude of the sidebands on both sides of the carrier signal should also decrease symmetrically (check with spectrum analyzer). This symmetrization (instrument setting center shift compensation) makes it easier to determine the exact frequency position of the NQR line, in that apart from the main signal only small sideband lines occur, provided saturation is avoided.

Spectra with several closely arranged lines should be recorded with sideband suppression for better distinction between line and sideband signal (Fig. 4.18). Sideband suppression is based on the periodic variation of the quench frequency (cf. also Fig. 4.10) and consequently of the distance between sideband and carrier (17,19). This operation can be most simply observed and adjusted with the aid of a spectrum analyzer with oscillographic representation. For this purpose, the center shift compensation must be set so that the frequency modulation of the carrier due to the 1-Hz frequency modulation of the quench frequency disappears. After some practice, it is possible for the operator to set the instrument on a purely acoustic basis, although not very plastically, by using the interference of the carrier with a signal of the frequency marker generator. The 1-Hz modulation leads to an inevitable amplitude-modulation of the carrier and consequently to a reduction in the detection sensitivity. When recording over larger frequency ranges, the setting of the center shift compensation must be corrected about every 1–2 MHz, but this can be done using the spectrum analyzer without disturbing the recording of the spectra.

2. Spectra Calibration and Measurement of Line Frequencies

a. Frequency Measurements

To determine the exact frequency of an NQR line, automatic frequency calibration of the spectra by means of fringe marks is absolutely essential. It is recommended to use frequency markers arranged on either side of the recorder chart, with marker spacings of 10, 25, 50, or 100 kHz on the one channel and 100, 250, 500, or 1000 kHz on the other channel. The markers are produced by interference of the oscillator signal with harmonic frequency markers of the corresponding spacings. Unambiguous frequency markers are obtained due to the fact that both the harmonic marker frequencies and the quench frequency are derived from one common frequency standard (19) (Fig. 4.10). If the symmetrization of the sidebands of the SRO is sufficient, the frequency marker generator picks up only the carrier signal, and possibly also the first-order sidebands. The fringe marks appear in the case of Zeeman modulation as single marks, in the case of frequency modulation 1 as double marks and in the case of frequency modulation 2 as treble marks (Fig. 4.14). The spacings of the double and treble marks indicate the amount of frequency deviation.

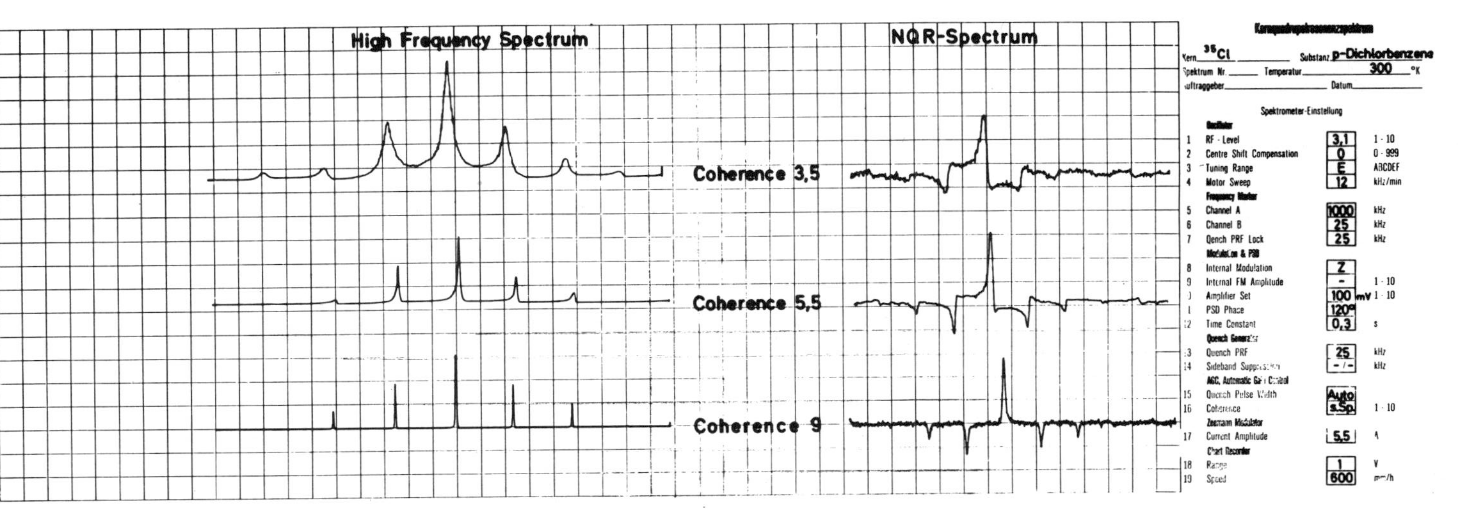

Fig. 4.16. Relation between the high-frequency spectrum of the oscillator and the NQR spectrum.

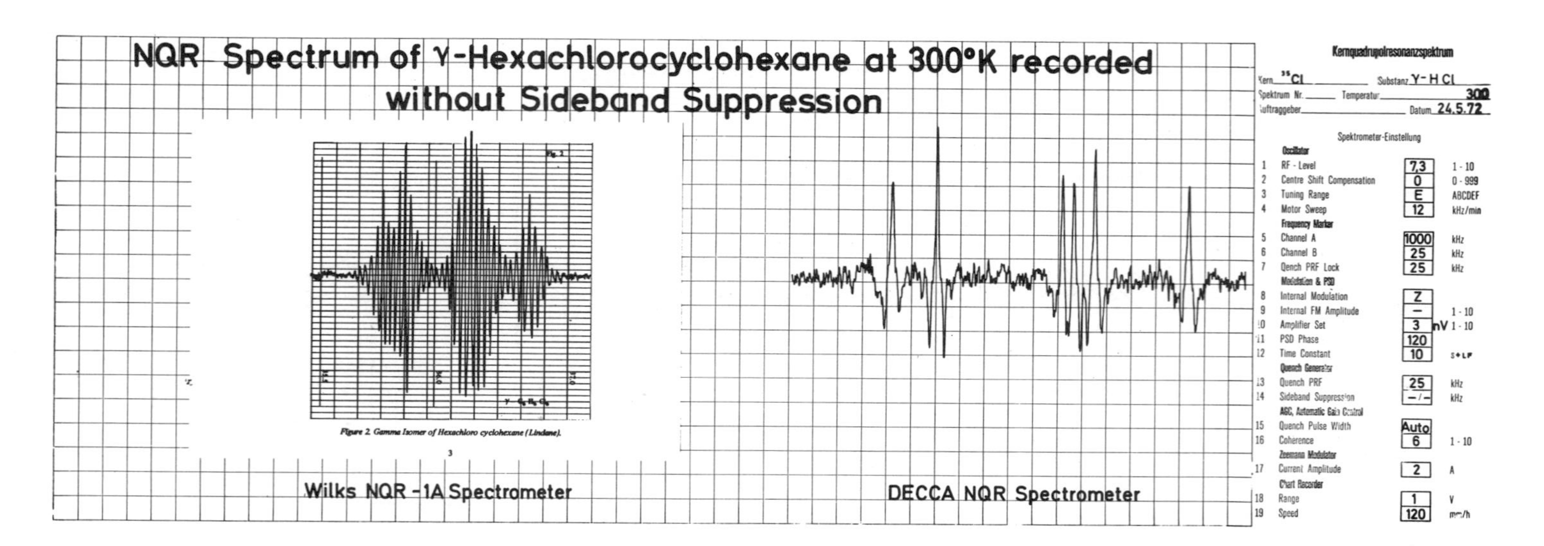

Fig. 4.17. ^{35}Cl-NQR spectrum of γ-hexachlorocyclohexane at 300K, taken without side-band suppression. Comparison of two commercial spectrometers.

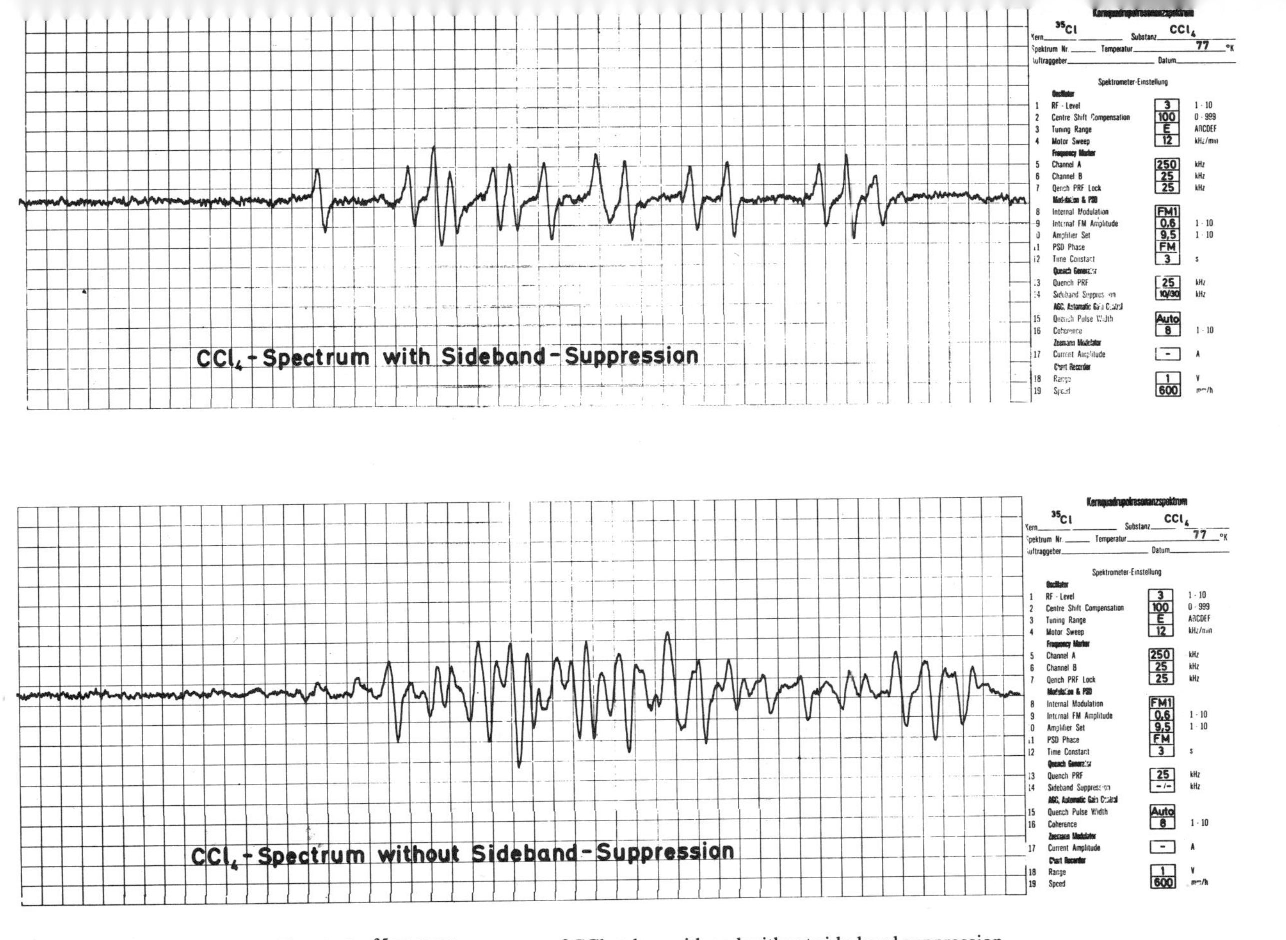

Fig. 4.18. ^{35}Cl-NQR spectrum of CCl_4 taken with and without side-band suppression.

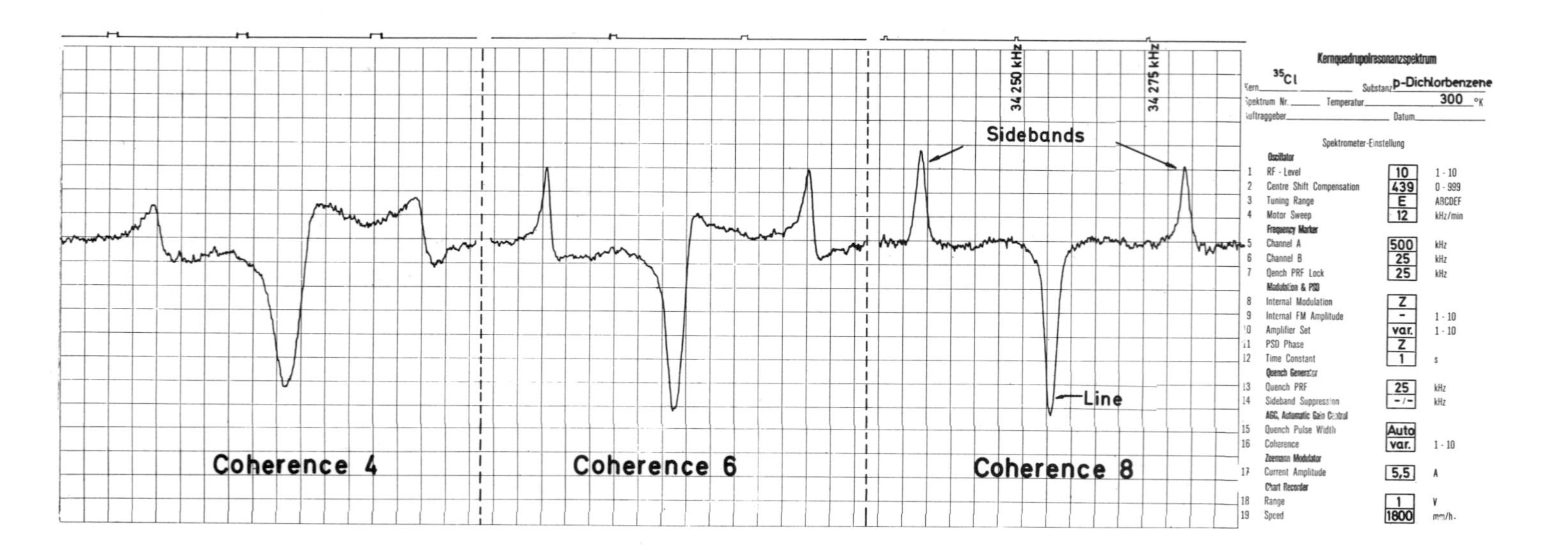

Fig. 4.19. Disturbance of line symmetry by false setting of coherence and rf level.

To determine the line frequency exactly, a completely symmetrical line form is desirable. After phase-sensitive rectification, if the instrument is correctly set, the Zeeman modulation yields more or less the natural line form (19,78), frequency modulation the first and second derivative (cf. also Fig. 4.14). In practice, admixtures of higher derivatives frequently occur that result in an asymmetrical line form and prevent precise location of the line center. In particular, when the lines are relatively narrow, it is possible to restore the line symmetry (Fig. 4.19) and determine the exact frequency of the line center by systematic variation of the rf power and coherence.

b. COHERENCE ADJUSTMENT

As shown in Fig. 4.16, the coherence adjustment also affects the resulting NQR linewidth due to the fact that it changes the carrier linewidth. Low coherence means marked damping of the oscillator amplitude in the quench phase t_{off} (Fig. 4.8) and a considerable degree of incoherence in the starting-up phase. This leads to a widening of the oscillator signal due to phase noise. High coherence means little damping of the oscillator amplitude, good coherence between consecutive high frequency impulses and relatively small carrier line width (Fig. 4.16). The resulting line width in the NQR spectrum depends to a considerable extent on the spectral width of the oscillator signal, provided the width of the oscillator signal is greater than the natural line width. The highest possible detection sensitivity is achieved roughly when the width of the oscillator signal determined by the coherence adjustment coincides with the expected line width (at a given high-frequency power). As can be seen in Fig. 4.16, the signal-to-noise ratio (S/N) of the NQR signal decreases from this optimum value towards both higher and lower coherence values.

3. Sample Handling

a. SAMPLE PREPARATION

In view of the low S/N of NQR lines, sample preparation is of particular importance (19,31,32,88). To obtain strong signals, the substance must be as pure as possible (>99%) so that well-crystallized substances are formed. As the crystal field is one of the factors determining the line frequency, compounds that do not crystallize satisfactorily give either very widened, weak signals or none at all. It is thus impossible to obtain spectra from liquids that have solidified in a glassy form. Substances crystallizing in the form of fibers are also unsuitable. The best results are obtained with clearly crystalline powders with particle sizes of around 0.2 - 1 mm. Good results—and a high filling factor—can also be achieved by slow solidification of the melt in the sample tube.

Impure substances can in most cases be sufficiently cleaned by recrystallization, distillation, or sublimation. For example, after purification by the zone melting method (10 passages), the signal obtained from technically pure *p*-dichlorobenzene improved by a factor of 3. When purifying by the zone-melting method, however, it must be borne in mind that a number of compounds have a tendency to thermal decomposition. The effect of impurities on the signal intensity (19,29,45) depends on the nature of the impurity, whether it crystallizes separately or whether it causes distortion of the crystal field by incorporation in the matrix lattice. Solid substances should always be disintegrated using as little mechanical force as possible to prevent major crystal defects. Treatment in a mortar, for instance, may reduce the signal intensity by a factor of 2 to 4. When cooling down from 300°K to 77°K, too, larger crystals in particular may show considerable mechanical strains, causing noticeable line broadening (cf. also Chapter IV.B.2). Modification changes likewise contribute to line broadening. Liquids that have a tendency to solidify in the form of glasses should be treated by repeated solidification by immersing in liquid nitrogen and by partial thawing until complete crystallization has been achieved.

b. SAMPLE CONTAINERS

Compared with other spectroscopic methods, the amount of pure crystallized substance required for NQR measurement is very large. For example, for ^{14}N-measurement (range 1–3 MHz) approximately 5–10 g are required, for Cl measurements (Fig. 4.20) (range 5–50 MHz) approximately 0.5–2 g and for Br measurements (50–200 MHz) approximately 0.2 g. The substance is not changed for the purpose of measuring. The line intensity depends to a considerable extent on the amount and quality of the substance used (33).

With well-crystallized, pure compounds, the amounts required are as a rule less than those given above. Suitable sample containers should at any rate guarantee a high filling factor of the measuring coil (19,33) and be able to withstand cooling to 77°K. Suitable sample containers for liquids are thin-walled glass ampules with a level bottom, such as are commonly used for pharmaceutical purposes. As these ampules have good dimensional stability, they fit exactly into the frequently self-

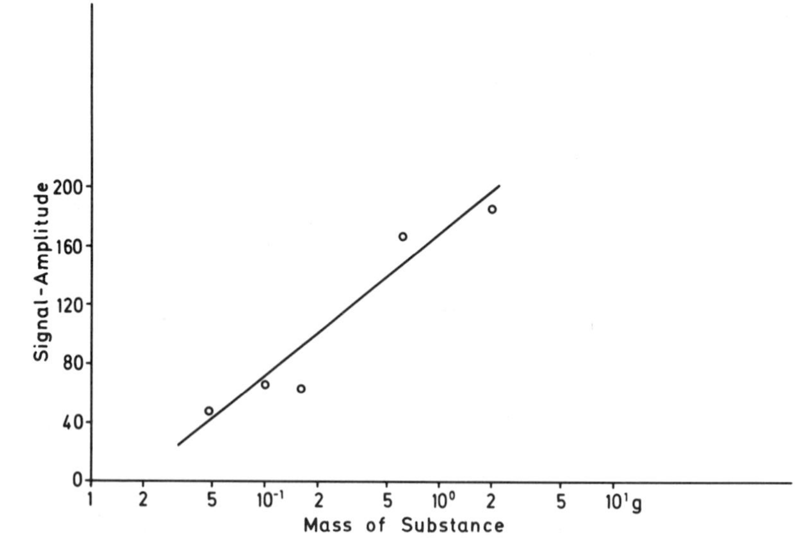

Fig. 4.20. NQR signal as a function of the quantity of substance for optimized dimensions of the oscillator coil.

Fig. 4.21. NQR sample containers (0.1 and 1 g) with oscillator coils and stainless steel coil shield.

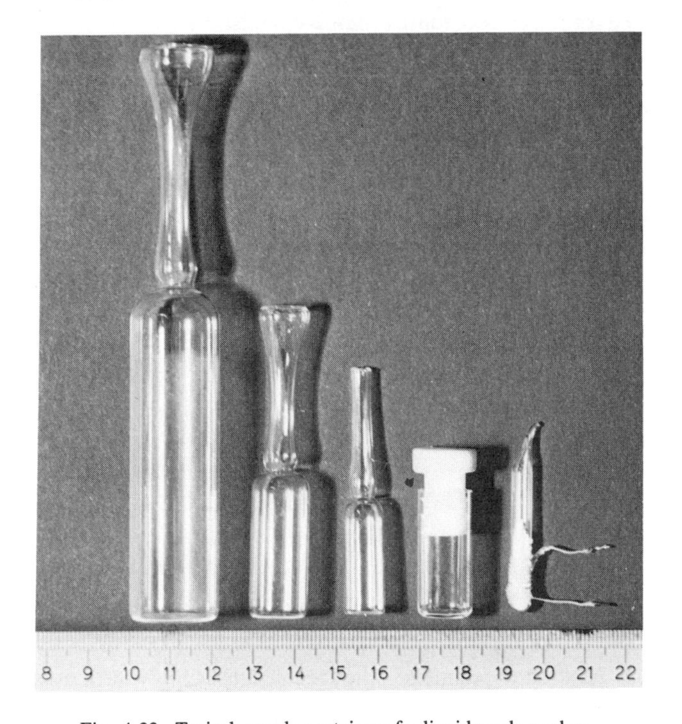

Fig. 4.22. Typical sample containers for liquids and powders.

supporting measuring coil (Fig. 4.21) and they can also be melted down without any problems. For powders, cylindrical quartz glass containers with a level bottom are more suitable. A tightly sealing Teflon stopper (Fig. 4.22) allows the material to be gently compressed (N.B. filling factor) and prevents moisture from entering during freezing. To ensure a high filling factor, the sample tube should completely fill the interior of the coil. The arrangement in a self-supporting coil (19) is therefore favorable, the coil preferably being made of silk-insulated copper wire (Fig. 4.21). To obtain good detection sensitivity, the coil must be linked to the tuning condenser of the oscillator circuit by a low-impendance coaxial line which should be as short as possible (33). Plug and socket connections should therefore not be used.

c. TEMPERATURE

The frequency of NQR lines depends to a considerable extent on the temperature of the sample (4,45,52). As the intensity of the lines increases as the temperature drops, measurements are usually made at 77°K, the temperature of liquid nitrogen. If the sample along with the measuring coil is immersed in liquid nitrogen, electrical disturbances may occur in the oscillator signal due to the formation of gas bubbles. For this reason it is better to seal off the sample and coil tightly (Fig. 4.21) or to introduce a Dewar flask into the coil in which the sample is located, although this leads to a reduction of the filling factor (65). Conventional Dewar flasks permit measurement for 8 h at 77°K with one LN_2 filling. Automatic replenishing devices for LN_2 can be used for longer recording times. Replenishing the LN_2 does not disturb the recording of the spectra. The evaporation rate can be noticeably lowered by using special devices that reduce the eddy current losses (in the Zeeman modulation field) in the Dewar flask and the coil shield (33). For analytical purposes, the frequency values at 77°K in particular are of interest, as these are compiled in reference tables (6,33). For identifying overlapping and in particular trichloromethyl groups, it may sometimes be of advantage to carry out measurements at a higher temperature, e.g. 300°K, as well. Air bath thermostats are in most cases adequate for this purpose. Temperature gradients over the cross section of the sample should be less than 0.1° to prevent line broadening. Consequently, an adaptation time of approximately 20–30 min is necessary when cooling to 77°K, amongst other things to even out mechanical stresses. In the range of ^{14}N resonance, piezoelectric resonances frequently lead to false signals at a low temperature (15,19,56).

IV. APPLICATIONS

A. METHODS OF STRUCTURE ANALYSIS

1. Structure-Specific Elements of the Spectra

To elucidate the structure of nuclei, the spectra must supply three essential items of information (31,32).

1. The frequency position (shift) of the line, which depends on the nature of the chemical bond of the quadrupole nucleus.
2. The total number of lines in the spectrum, which must be equal to the number of quadrupole nuclei per molecule, or, in the case of there being several molecules per unit cell, the corresponding multiple (calculated on a given NQR transition).
3. The multiplet grouping of the lines resulting from chemically equivalent bonding in certain functional groupings of the molecule, such as, for example, $-CCl_3$, $-PCl_4$, or $-NCl_2$.

For the purpose of structure analysis, therefore, spectrometers must permit unambiguous and sufficiently finely divided frequency calibration of the spectra. In addition, the measuring sensitivity and amplification must be constant when operating over larger frequency ranges, so that in the case of overlappings the number of equivalent nuclei can be determined from the line intensity (Fig. 4.23). It is also advantageous to have a number—limited in the case of SRO spectrometers—of sidebands for each line to make the spectrum easier to read by avoiding overlapping of sidebands from closely arranged lines (Fig. 4.17).

2. Shift, Multiplet Grouping, and Number of Lines per Molecule

These three elements form the basis for elucidating the structure of monomers by NQR spectroscopy and can be determined directly from the spectra. Using organic chlorine compounds as an example, Fig. 4.24 shows which frequencies can be assigned to the individual, differently bound Cl nuclei. The nature of the functional group can be determined directly by the number of lines within a multiplet, but in particular the frequency position (shift) also supplies important information. Thus, for example, the ^{35}Cl shift ranges of organic chlorine compounds, which are particularly interesting for NQR spectroscopy, lie within certain frequency limits, depending on the bond partner Si, P, S, C, or N (Fig. 4.25), so that in most cases it is quite simple to work out roughly which frequency can be assigned to which nucleus (cf. also Fig. 4.24). Within these shift ranges, more detailed allocations can be carried out, as is shown in Fig. 4.6, using the N–Cl bond as an example. In this case, too, the general rule applies that the proximity of highly electronegative nuclei and groupings increases the resonance frequency of the nucleus being considered. Fig. 4.7 shows that these inductive effects are easily active over several bonds. Shift tables such as this can easily be drawn up to solve special analytical problems, using spectra tables which contain all the line frequencies of compounds measured so far (6) (cf. also Chapter IV.3). Analogous shift tables can also be compiled on the basis of ^{14}N resonance, for which there is a larger amount of experimental material available (Fig. 4.26). It must be remembered that the shift range of ^{14}N resonance is considerably smaller than that of chlorine resonance, but because of the smaller line width (0.3–1 kHz in the case of ^{14}N as compared with approximately 2–20 kHz for ^{35}Cl), the ratio of shift range to linewidth is roughly the same as for chlorine. As with chlorine, the electronegativity of the bond partners has the largest effect on the position of the ^{14}N resonance, e.g., in the series NF_3, $N(CH_3)_3$, NH_3. Another typical feature is the monotonic increase in the shift in the series C≡N, C=N, and C–N. A factor of analytical importance is that, in contrast to IR spectroscopy, it is possible to identify C=N in the presence of C=C. The differences between primary amines of the $C{-}NH_2$ type and those of the $N{-}NH_2$ type are also worth noting. Additional information on the symmetry of the molecule is provided by the spectra in the form of two transitions, the frequency spacing of which is proportional to the asymmetry of the field gradient. Symmetrically triple-coordinated nitrogen, as in hexamethylenetetramine, gives only one transition. The same also applies for singly coordinated C≡N compounds with a disappearing π component (11). When there are two molecules per unit cell, the number of lines per molecule may double as a

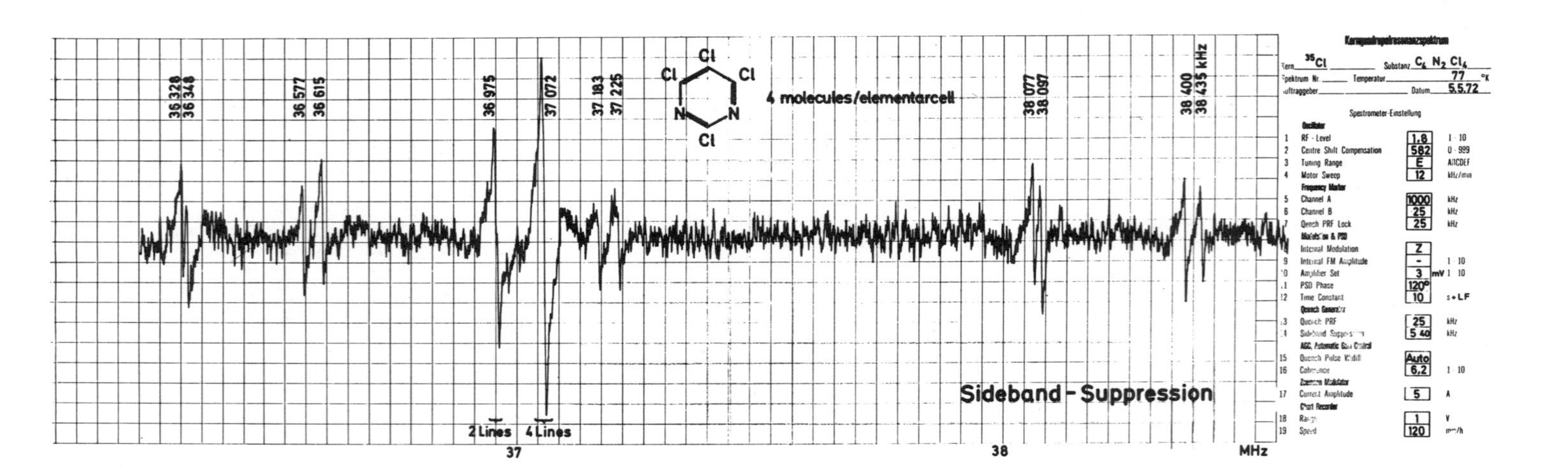

Fig. 4.23. ^{35}Cl-NQR spectrum of tetrachloropyrimidine. Note the good recognizability of all lines of the compound, even when partly superimposed.

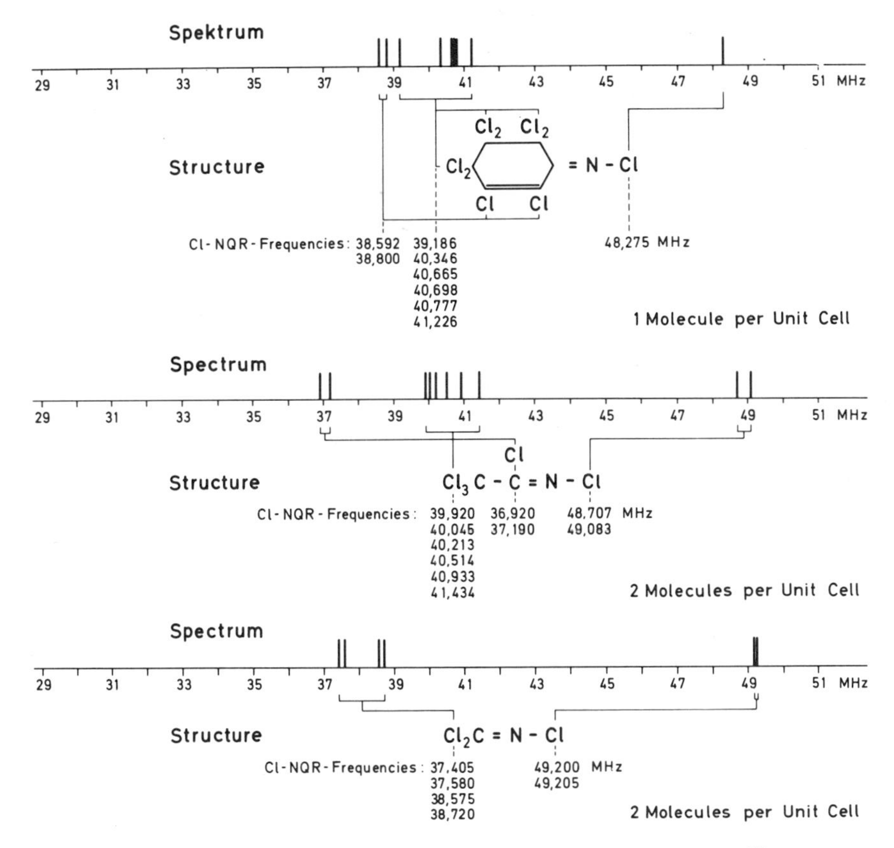

Fig. 4.24. Structure assignment by number, multiplet grouping, and shift of the ^{35}Cl-NQR lines.

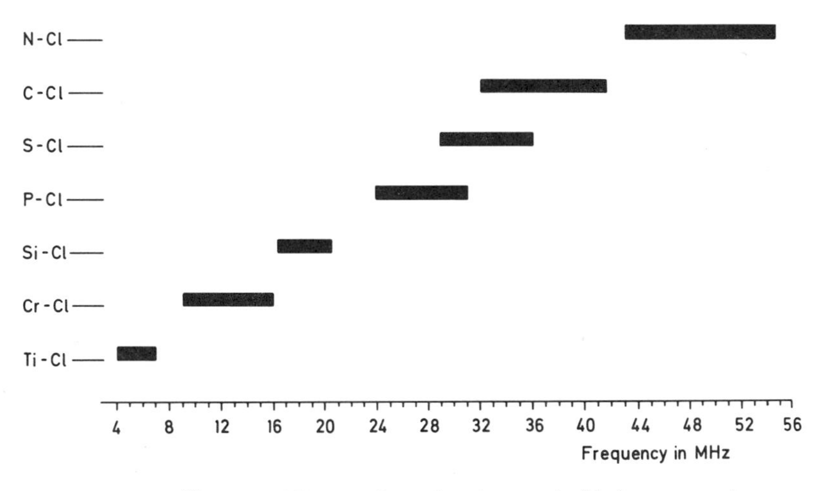

Fig. 4.25. ^{35}Cl-NQR shift range of organic and anorganic chlorine compounds.

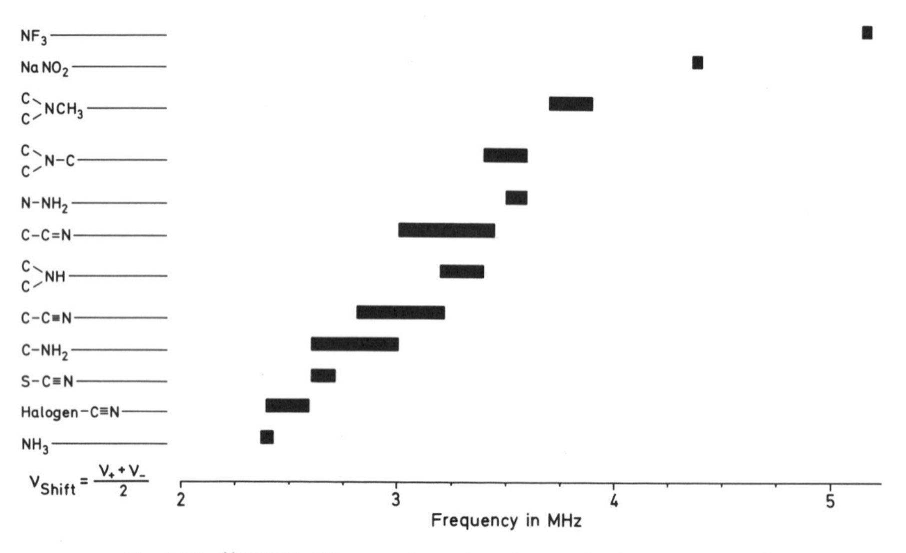

Fig. 4.26. ^{14}N-NQR shift range of organic and anorganic nitrogen compounds.

result of crystal field effects. To ensure accurate analysis, it is necessary in each case to find at least the $\nu\pm$ transitions. Where there are several molecules per unit cell, it is advisable, in order to pair the $\nu\pm$ transitions accurately, to determine the difference frequencies ν_0 as well, although these have only very weak intensities at very low frequencies.

3. Examples of Solving Structure Problems

So far, NQR spectroscopy has not been widely used for the routine analysis of structures. For this reason, three examples have been included here to illustrate the procedure, which shows a certain similarity to that for NMR spectroscopy, in somewhat greater detail.

In the elucidation of unknown structures, an especially frequent task is the decision between several possible arrangements. Fig. 4.27 shows a typical example solved by NQR. It was possible to distinguish between structures I and II on the basis

Cl$_a$ Cl$_b$ Cl$_c$

I NC - CCl$_2$ - CCl$_2$ - N = PCl$_3$
II NC - CCl$_2$ - C(Cl) = N - PCl$_4$

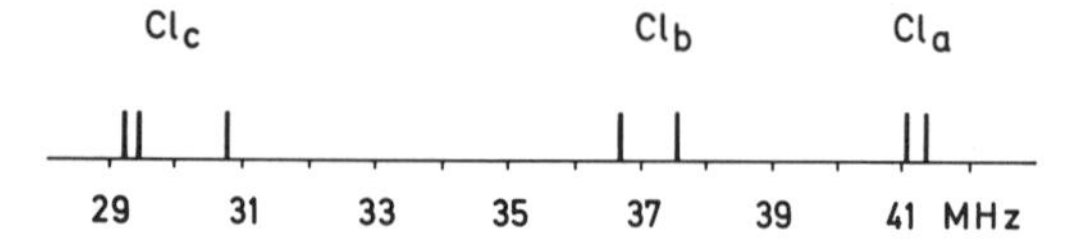

Fig. 4.27. Molecular structure and [35]Cl-NQR spectrum.

of the number and arrangement of the lines and also on the basis of the shift values. Structure I results from the three lines of the PCl_3 grouping at 30 MHz. PCl_4 would have required four lines. The Cl_b lines lie at medium shift values, while the Cl_a lines lie relatively high, above 41 MHz, because of the adjacent, strongly electronegative CN group. The nonequivalence of the Cl_a, Cl_b, and Cl_c frequencies can be attributed to crystal field effects and hindered rotation of the chain.

In another case, the structure of a compound with the molecular formula $C_6Cl_6O_3$, for which three structures, I, II, and III, were suggested, was to be determined (Fig. 4.28). The spectrum gave six lines (1 molecule per unit cell) with the frequencies:

40.875 MHz	mean value 40.6	$-CCl_3$
40.850		
40.150	39.4	$-CCl$
39.425		
38.995	mean value 38.8	$=CCl_2$
38.570		

On the basis of the multiplet grouping and the shift values, suggestion I appears to be the most probable structure and II and III are practically eliminated. In addition to the shift values, the saturation behavior (spin–lattice relaxation) may also be used for assignment. This shows that the three lines with the highest frequencies have equal saturation behavior, as do the two lines with the lowest frequencies. From this it can be concluded in accordance with the above assignment that there are groups of three and two equal chlorine nuclei and one single Cl (68).

For further verification of the assignment, the spectrum to be expected for I is now estimated on the basis of comparison substances.

Position of the $-CCl_3$ Frequency: Because of the proximity of $-CCl$ and C=O, the $-CCl_3$ frequencies are very high. The frequencies for trichloromethyl groups in an electronegative surrounding (8,9,31,32) can be

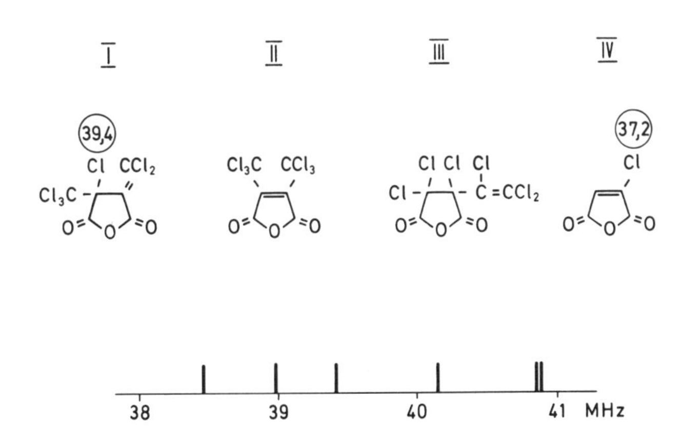

Fig. 4.28. Structure assignment by multiplet grouping and shift comparison.

estimated by means of a shift table (Fig. 4.29). The result is a typical position between 40.5 and 40.7 MHz. Structure II can be excluded because of the wide scattering of the line frequencies over 2.3 MHz. In comparison, the lines of Cl_3-CCl_3 lie between 40.6 and 40.7 MHz and the lines of *p*-bis(trichloromethyl)benzene between 38.3 and 39.6 MHz, so that here the effect of the crystal field (including the rotation of the $-CCl_3$ group) is noticeably less, and in the compound $C_6Cl_6O_3$, different chemical linkage is the cause of the wide scattering of the frequencies.

Position of the C–Cl Frequency (Structure I): The frequency of the single chlorine is at 39.425 MHz very high. The main factor responsible for the high shift value is the proximity of the CCl_3 group, the electronegativity of which is equal to that of a directly (geminally) bound chlorine. For example, on transition from $H-CH_2Cl$ to Cl_3C-CH_2Cl, the Cl frequency increases from 34 to 36.4 MHz ($\Delta = 2.4$). Also on transition from H_3C-CCl_3 (38 MHz) to $Cl-CCl_3$ (40.6 MHz), there is an approximately equal change of 2.6 MHz. On transition to I, when $\Delta = 2.4$ MHz, a frequency of 39.6 MHz is to be expected for the single Cl, in which case the frequency value for the comparison substance will probably be somewhat too high because of the double bond of the carbon involved, so that the agreement with the measured value of 39.425 is to be considered good.

Position of the $C{=}CCl_2$ Frequency: $Cl_2C{=}CH_2$ yields 36.5 MHz. The spatial proximity of CCl_3 and the proximity of the carbonyl group give a strong shift to higher frequencies of approximately 2 MHz (measured mean value 38.8 MHz), as is shown in Table 4.XI using the comparison substances.

In the last example, a structure arrangement for γ-chlordene (36), a compound with an insecticidal effect, is described, using a number of similar and other compounds for comparison. As regards the structures I and II already discussed (Fig. 4.30) (60), it was possible, not least on the basis of the NQR

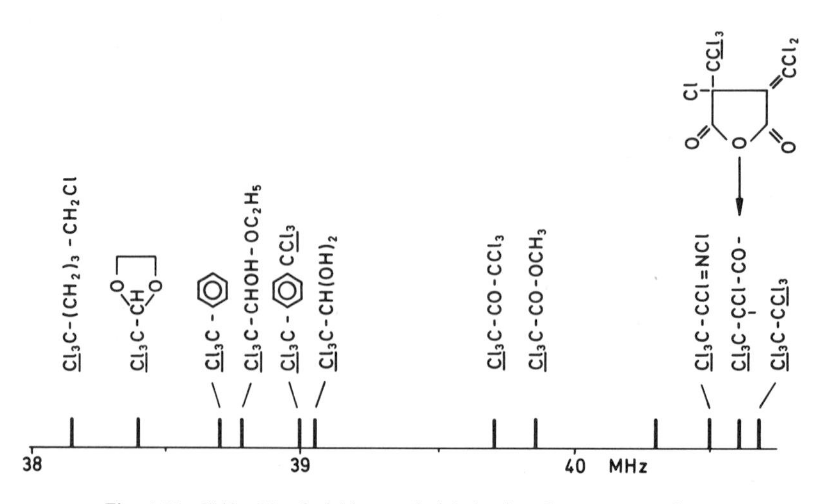

Fig. 4.29. Shift table of trichloromethyl derivatives for structure assignment.

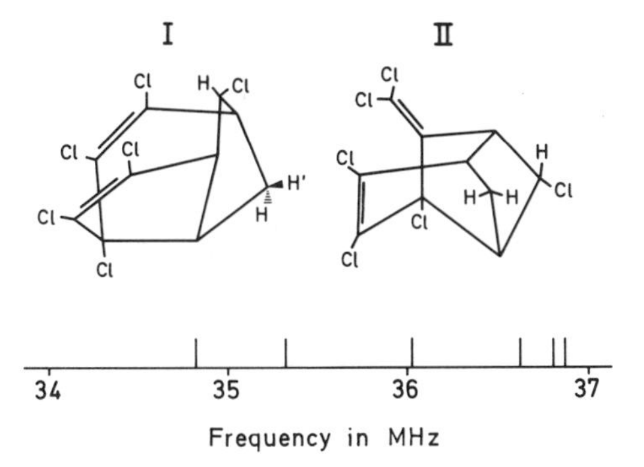

Fig. 4.30. γ-Chlordene, structure assignment by NQR.

TABLE 4.XI

^{35}Cl NQR Frequencies of the Group C=CCl$_2$ at 77°K

Compound	Frequency (MHz)	Reference
$CH_2{=}CCl_2$	36.837	6
	36,524	
	36.524	
	36.268	
$=CCl_2$ $=CCl_2$	38.839	[a]
	38.639	
	38.072	
	37.901	
$=CCl_2$	38.314	[a]
	38.190	
$=CCl_2$ $=CCl_2$	38.994	[a]
	38.570	

[a] Hashimoto, H., K. Mano, *Bull. Chem. Soc. Japan*, 45, 706 (1972).

TABLE 4.XII

^{35}Cl NQR Frequencies of γ-Chlordene at 77°K and 300°K and Correlation to the Chlorine Nuclei of the Molecule (see Fig. 4.30)

Frequency (MHz) 77°K	Frequency (MHz) 300°K	Cl bonded to
34.815	34.480	C-8
35.315	34.910	C-1
36.030	35.600	vinyl. C
36.625	36.330[a]	vinyl. C
36.805	36.360[a]	vinyl. C
36.870	36.505	vinyl. C

[a] Superposition.

spectrum, to establish that γ-chlordene has form I and not form II (36). The assignment of the chlorine signals of γ-chlordene is given in Table 4.XII. A C═CCl_2 configuration such as occurs in II could be excluded (cf. also Table 4.XI). The frequency position of vinylic Cl in a similarly electronegative environment is shown in Table 4.XIII and in the spectra representation in Fig. 4.31, which also contains shift values for chlorine incorporated in a similar manner to Cl (8) and Cl (3a) (Fig. 4.32).

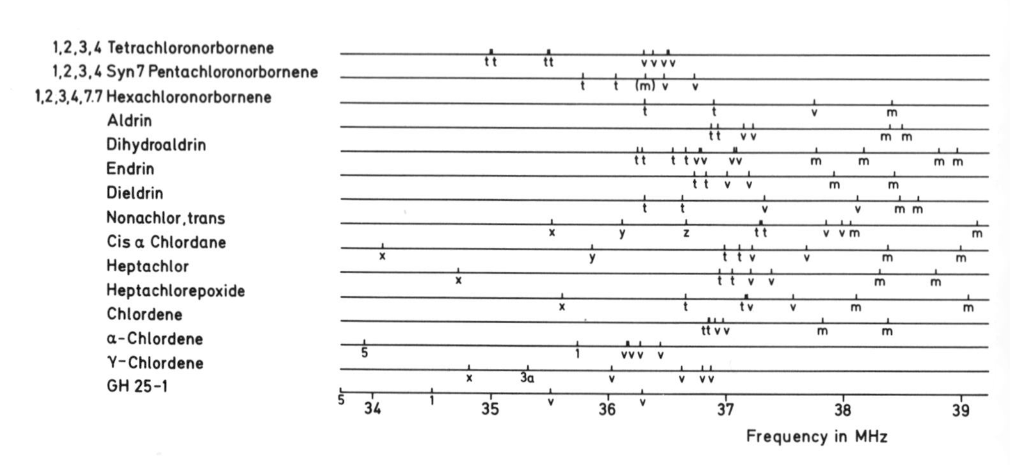

Fig. 4.31. ^{35}Cl-NQR shift values of vinylic Cl (v), tertiary Cl (t), and bridgehead Cl (m) from some insecticide compounds, (1,3a, 5,8 : Cl bonded to C (1), C (3a), . . .).

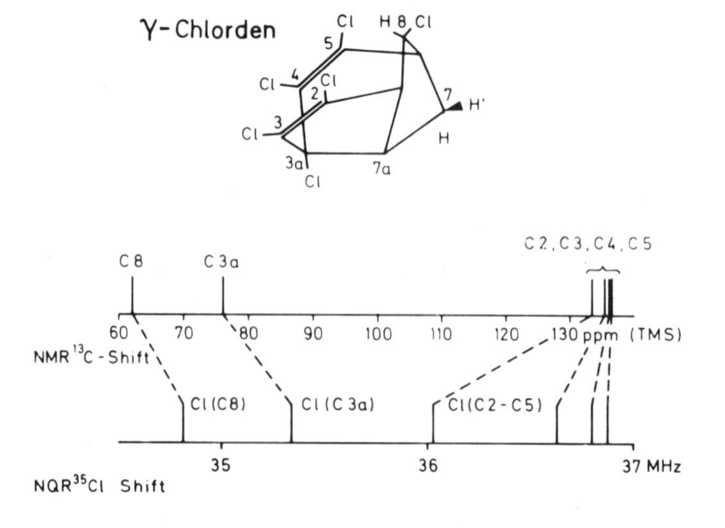

Fig. 4.32. Additional verification of structure by comparison of ^{35}Cl-NQR shift with ^{13}C-NMR shift.

TABLE 4.XIII

^{35}Cl NQR Frequencies of the Group $-ClC{=}CCl-$ (Vinylic Cl) at 77°K

Compound		Frequency (MHz)	Reference
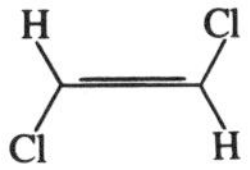		34.497	36
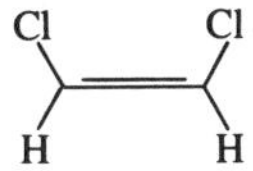		34.837 34.894	36
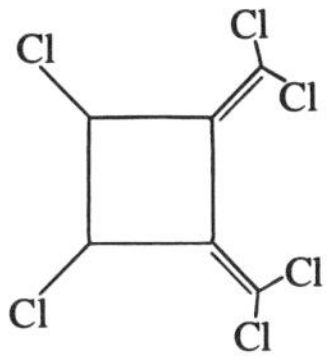		36.371 36.520	36
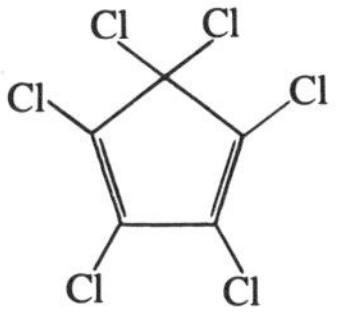		36.953 37.279 37.457 37.457	36
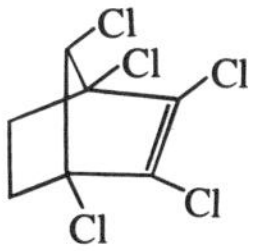		36.724	25,36
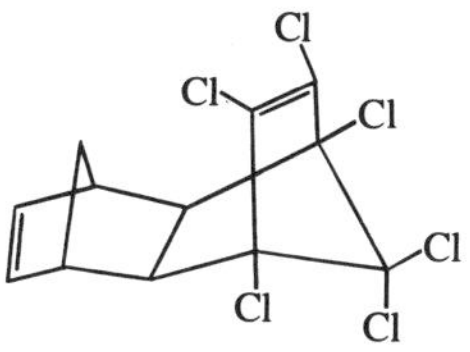	Aldrin	37.140 37.225	36,37
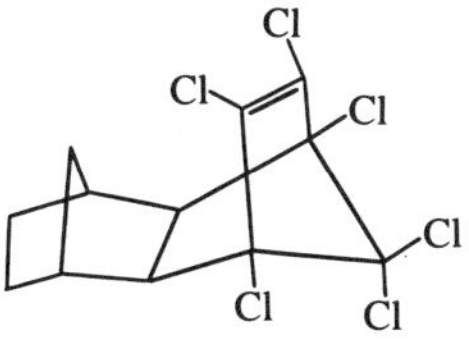	Dihydroaldrin	36.773 36.773 37.058 37.080	[a], 37

TABLE 4.XIII *(Continued)*

Compound	Frequency (MHz)	Reference
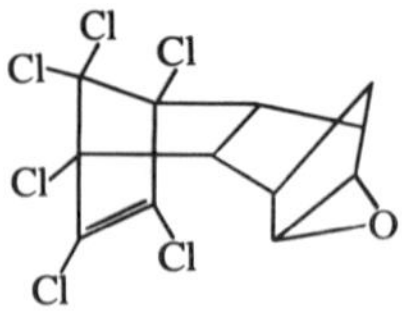 Endrin	37.000 37.188	[a], 69
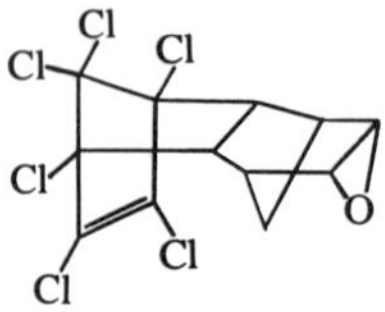 Dieldrin	37.140 37.225	36,69
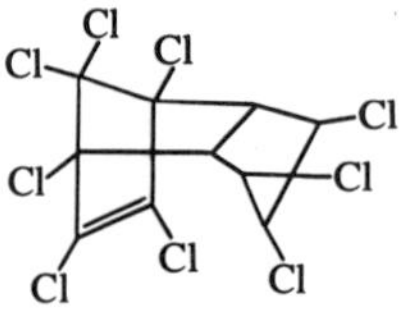 Nonachlor, *trans*	37.846 37.980	[a]
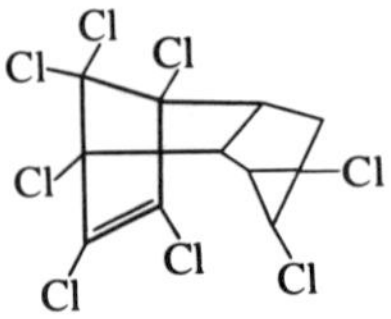 α-Chlordane, *cis*	37.221 37.680	[a], 69
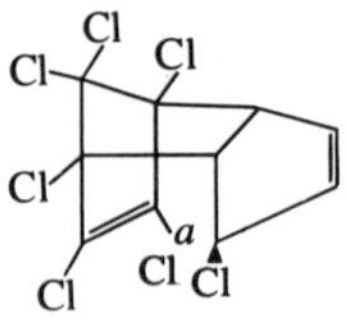 Heptachlor	37.210 37.385	[a], 69
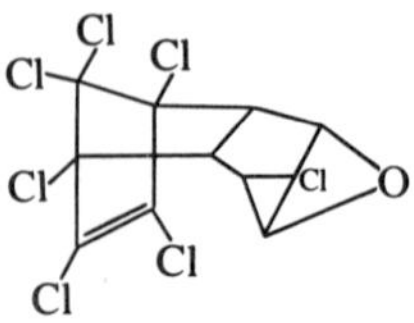 Heptachlorepoxide	37.164 37.571	[a], 69

TABLE 4.XIII *(Continued)*

Compound	Frequency (MHz)	Reference
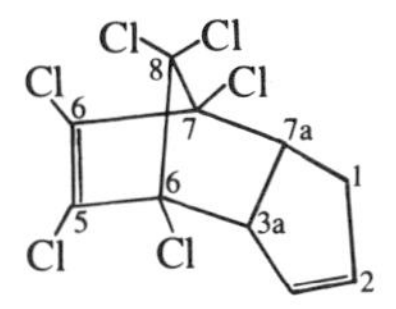 Chlordene	36.905 36.975	36
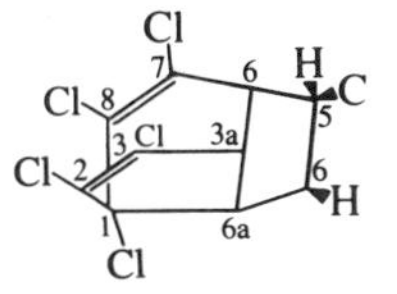 α-Chlordene	36.160 36.160 36.270 36.445	36
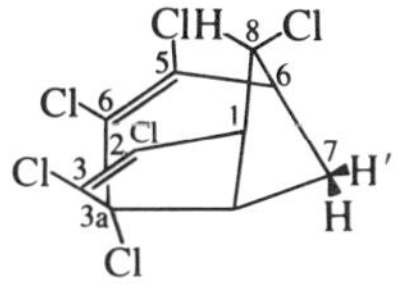 γ-Chlordene	36.030 36.625 36.805 36.870	36

[a] S. Gäb, H.G. Fitzky, unpublished results.

4. Combination of Methods for Spectroscopic Analysis

For successful analysis by the NQR method, it is useful to employ other spectroscopic methods first to reduce the number of potential structures. First of all the molecular formula of the compound must be established by means of low-resolving mass spectroscopy. Thereafter the various structure possibilities are worked out using IR spectroscopy before finally NQR and NMR methods—in the case of highly chlorinated organic compounds preferably ^{35}Cl NQR and ^{13}C NMR—are applied to determine which structure appears to be the correct one. The last two methods mentioned above are more or less equally informative, so that in many cases the results obtained by each method independently can be used to verify the results obtained by the other method. In the investigation of certain types of material in particular, there are parallelisms in the shift values, as is explained in Fig. 4.32 at the example γ-chlordene. In some other cases, e.g., in elucidating the structure of nitrogen compounds, it may be advantageous to employ photoelectron and Auger electron spectroscopy (ESCA) as well (34).

B. OTHER APPLICATIONS

1. Purity Measurements

Foreign atoms or molecules in the crystal lattice lead to wide stress ranges in the crystal with corresponding variations in the crystal field. This broadens the NQR lines and reduces the intensity of the peaks. Measurements of the effect of the molecule size and the concentration of the impurity have been carried out mainly on organic chlorine compounds. According to these, one foreign molecule can affect up to 600 neighboring chlorine nuclei. Only readily soluble compounds have a considerable effect on the line intensity, the concentration range extending from about 0.02–2 mole-%. Impurity concentrations of 0.5 mole-% generally change the intensity by 50% (29,45,88).

2. Temperature and Pressure Measurements

The frequency of the nuclear quadrupole transition is linked to the thermal expansion or elastic deformation of the crystal through the portion of the crystal field in the field gradient (29,45,88). Thus, the temperature and pressure can be determined directly by measuring the frequency. The advantage of the method lies in the fact that for the measuring, the properties of an atomical system which guarantees good constancy of the calibration are utilized. For example, in the range from 50–297°K it is possible to achieve a degree of accuracy above ±0.001°K, in the range up to 20K ± 0.01°K. For measuring the temperature, an NQR spectrometer in which the oscillator frequency (approx. 28 MHz) is controlled by the resonance line of the ^{35}Cl in the $NaClO_3$ is used (84,85,86,87).

A similar arrangement can be used for measuring pressure (27,35,86). Typical measuring ranges go up to 2 kbar, the maximum being 6 kbar with $NaClO_3$ as a measuring substance. The accuracy of the pressure measurement depends on how good the temperature stabilization is and reaches ±2.5 bar at ±0.015°K.

The position of the resonance frequency which depends on the pressure (48) can also be used to determine elastic stress in solid high polymers (43), introducing as sensors inert fillers in powder form, the NQR frequency of which is measured during compression or strain. At strains of up to 1.2%, epoxy resins with Cu_2O as a sensor gave frequency shifts of the Cu resonance line of 8 kHz (measuring accuracy ±0.02 kHz).

REFERENCES

1. Ardalan, Z., and E.A.C. Lucken, *Helv. Chim. Acta*, **56**, 1715 (1973).
2. Ardalan, Z., and E.A.C. Lucken, *Helv. Chim. Acta*, **56**, 1724 (1973).
3. Babushkina, T.A., V.I. Robas, and G.K. Semin, *Dokl. Akad. Nauk SSSR*, **164** (1964).
4. Bayer, H., *Z. Physik*, **130**, 227 (1951).
5. Bersohn, R., *J. Chem. Phys.*, **22**, 2078 (1954).
6. Biryukov, I.P., M.G. Voronkov, and I.A. Safin, *Tables of Nuclear Quadrupole Resonance Frequencies*, Israel Program for Scientific Translations, IPST Press, Jerusalem, 1969.
7. Boyd, R.J., and M.A. Whitehead, *J. Chem. Soc.* (*Dalton*) **73**, 78, 81 (1972).
8. Brame, E.G. Jr., *Anal. Chem.*, **39**, 918 (1967).
9. Brame, E.G. Jr., *Anal. Chem.*, **43**, 35 (1971).
10. Bray, P.J., *J. Chem. Phys.*, **23**, 703 (1955).
11. Bronswyk, W. van, "The Application of Nuclear Quadrupole Resonance Spectroscopy to the Study of Transition Metal Compounds," in *Structure and Bonding*, Springer-Verlag, Berlin, Heidelberg, New York, 1970, Vol. 7, p. 87.
12. BRUKER-PHYSIK AG, *Transient Techniques in Nuclear Quadrupole Resonance Spectroscopy*, Application note, D-7501 Karlsruhe-Forschheim, W.Germany.

13. Casabella, P.A., and P.J. Bray, *J. Chem. Phys.*, **46**, 1186 (1958).
14. Chihara, H., N. Nakamura, and T. Irie, *Bull. Chem. Soc. Jpn.*, **42**, 3034 (1969).
15. Colligiani, A., *Rev. Sci. Instrum.*, **38**, 1331 (1967), *J. Chem. Phys.*, **52**, 5022 (1970).
16. David, S., and L. Guibe, *Carbohydr. Res.*, **20**, 440 (1971).
17. Dean, C., and M. Pollak, *Rev. Sci. Instrum.*, **29**, 630 (1958).
18. Dean, C., *Rev. Sci. Instrum.*, **29**, 1047 (1958).
19. DECCA RADAR Ltd., *N.Q.R. Spectrometer Handbook*, Instrument Division, Lyon Road, Walton-on-Thames, Surrey, England, Aug.70.
20. Dehmelt, H.G., and H. Krueger, *Naturwiss.*, **37**, 111, (1950).
21. Dehmelt, H.G., and H. Krueger, *Naturwiss.*, **37**, 398 (1950).
22. Dehmelt, H.G., *Z. Physik*, **129**, 401 (1951).
23. Dehmelt, H.G., *Z. Physik*, **130**, 356 (1951).
24. Delay, F., M. Geoffroy, E.A.C. Lucken, and P. Mueller, *J. Chem. Soc. Faraday Trans.* **II**, *71*, 463 (1975).
25. Dewar, M.J.S., M.L. Herr, and A.P. Marchand, *Tetrahedron*, **27**, 2371 (1971).
26. Dewar, M.J.S., and M.L. Herr, *Tetrahedron*, **27**, 2377 (1971).
27. Early, D.D., R.F. Tipsword, and C.D. Williams, *J. Chem. Phys.*, **55**, 460 (1971).
28. Farrar, T.C., and E.D. Becker, *Pulse and Fourier Transform NMR*, Academic, New York and London, 1971.
29. Fedin, E.I., and G.K. Semin, "Use of Nuclear Quadrupole Resonance in Chemical Crystallography," in *The Mössbauer Effect and its Applications in Chemistry*, Consultants Bureau, New York, 1964, p. 67.
30. Fitzky, H.G., Die Kernquadrupolresonanz als neues Hilfsmittel zur Strukturaufklärung, dargestellt am Beispiel der Stickstoff und Chlor-Resonanz, in "*Colloquium Spectroscopicum Internationale XVI*, Adam Hilger, London 1971, Vol. I, p. 64.
31. Fitzky, H.G., *G-I-T Fachz. Lab.*, **15**, 992 (1971) and **15**, 1100 (1971).
32. Fitzky, H.G., *G-I-T Fachz. Lab.*, **17**, 10 (1973).
33. Fitzky, H.G., Nuclear Quadrupole Resonance Spectroscopy for the Analysis of chlorinated Organic Compounds in Industry, in "*Advances in Nuclear Quadrupole Resonance*," Heyden & Son Ltd., London, New York, Rheine, 1974, Vol. 1, p. 79.
34. Fitzky, H.G., D. Wendisch, and R. Holm, *Ange w. Chem., Int. Edit.*, **11**, 979 (1972).
35. Frisch, R.C., and D.C. van der Hart, *J. Res. Nat. Bur. Stand*, **74C**, 3 (1970).
36. Gaeb, S., H. Parlar, W.P. Cochrane, D. Wendisch, H.G. Fitzky, and F. Korte, *Liebigs Ann. Chem.*, **1** (1976).
37. Gegiou, D., *Anal. Chem.*, **46**, 742 (1974).
38. Gill, D., M. Hayek, Y. Alon, and A. Simievic, *Rev. Sci. Instrum.*, **38**, 1588 (1967).
39. Gordy, W., W.V. Smith, R.F. Trambarulo, *Microwave Spectroscopy*, J. Wiley, New York, 1953.
40. Gordy, W., and R.L. Cook, *Microwave Molecular Spectra*, Chemical Applications of Spectroscopy, Interscience Publishers, New York, London, Sydney, Toronto 1970, Part II.
41. Grechishkin, W.S., and G.B. Soifer, *Pribory Tekh. Eksper. USSR*, **1**, 1 (1964).
42. Guibe, L., "Application of Pulse Methods to Nuclear Quadrupole Resonance Spectroscopy," in *Proceedings of the Second International Symposium on Nuclear Quadrupole Resonance Spectroscopy*, A. Vallerini, Pisa, 1975, p. 95.
43. Hewitt, R.R., and B. Mazelski, *J. Appl. Phys.*, **43**, 3386 (1972).
44. Hooper, H.O., and P.J. Bray, *J. Chem. Phys.*, **33**, 334 (1960).
45. Jeffry, G.A., and T. Sakurai, *Progr. Solid State Chem.*, **1**, 380 (1964).
46. Kiichi, T., N. Nakamura, and H. Chihara, *J. Magnet. Res.*, **6**, 516 (1972).

47. Kopfermann, H., *Kernmomente*, Akademische Verlagsgesellschaft, Frankfurt/M., 1956, p. 314.
48. Kushida, T., G.B. Benedek, and N. Bloembergen, *Phys. Rev.*, **104**, 1364 (1956).
49. Lenk, R., "Fourier-Transform NQR (FT-NQR) Spectra and Spin Relaxation Decays in $AsCl_3$ and PCl_3," in *Proceedings of the Second International Symposium on Nuclear Quadrupole Spectroscopy*, A. Vallerini, Pisa, 1975, p. 111.
50. Linscheid, P., and E.A.C. Lucken, *Chem. Commun.*, 425 (1970).
51. Livingston, R., *J. Phys. Chem.*, **57**, 496 (1953).
52. Lucken, E.A.C., *Nuclear Quadrupole Coupling Constants*, Academic, London, 1969.
53. Lucken, E.A.C, "Stereochemistry and Nuclear Quadrupole Resonance," in *Advances in Nuclear Quadrupole Resonance*, Heyden & Son Ltd., London, New York, Rheine, 1974, Vol. 1, p. 235.
54. Lucken, E.A.C., and M.A. Whitehead, *J. Chem. Soc.*, 2459 (1961).
55. Lucken, E.A.C., *Z. Anal. Chem.*, **273**, 337 (1975).
56. Matzkanin, G.A., T.N. O'Neal, and T.A. Scott, *J. Chem. Phys.*, **44**, 4171 (1966).
57. Moores, R., "The NR2 Nuclear Quadrupole Spectrometer System," in *Proceedings of the Second International Symposium on Nuclear Quadrupole Resonance Spectroscopy*, A. Vallerini, Pisa, 1975, p. 79.
58. Murgich, J., T. Coor, and R. Dallenbach, "A versatile NQR Spectrometer using Time Shared Modulation," in *Proceedings of the Second International Symposium on Nuclear Quadrupole Resonance Spectroscopy*, A. Vallerini, Pisa, 1975, p. 85.
59. PRINCETON APPLIED RESEARCH, *NQR-Spectrometer Model 320*, Specifications T-323-2M-8/73.
60. Parlar, H., and F. Korte, *Chemosphere*, **1**, 125 (1972).
61. Petersen, G.E., Bridenbaugh, *Rev. Sci. Instrum.*, **35**, 698 (1964).
62. Petersen, G.E., Bridenbaugh, *Rev. Sci. Instrum.*, **36**, 702 (1965).
63. Petersen, G., and T. Oja, "A Pulsed Nuclear Quadrupole Resonance Spectrometer," in *Advances in Nuclear Quadrupole Resonance*, Heyden & Son Ltd., London, New York, Rheine, 1974, Vol. 1, p. 179.
64. Pound, R., and W. Knight, *Rev. Sci. Instrum.*, **21**, 800 (1950).
65. Riedel, E.F., and R.D. Willett, *J. Am. Chem. Soc.*, **97**, 701 (1975).
66. Robinson, F.N.H., *J. Sci. Instrum.*, **36**, 481 (1959).
67. Robinson, F.N.H., *Rev. Sci. Instrum.*, **34**, 1260 (1963).
68. Roedig, A., H.H. Bauer, B. Heinrich, and D. Kubin, *Chem. Ber.*, **104**, 3525 (1971).
69. Roll, D.B., and F.J. Biros, *Anal. Chem.*, **41**, 407 (1969).
70. Rubenstein, M., and P.C. Taylor, *Phys. Rev. Lett.*, **29**, 119 (1972).
71. Schemp, E., and P.J. Bray, *J. Chem. Phys.*, **46**, 1186 (1967).
72. Schemp, E., and P.J. Bray, *J. Chem. Phys.*, **49**, 3450 (1968).
73. Schemp, E., "Nuclear Quadrupole Resonance Spectroscopy," in *Physical Chemistry*, Academic, New York, 1970, Vol. IV.
74. Scrocco, E., *Adv. Chem. Phys.*, **5**, 319 (1963).
75. Semin, G.K., T.A. Babushkina, and G.G. Jacobson, *Applications of Nuclear Quadrupole Resonance to Chemistry*, Chimia, Leningrad, 1972, russ.
76. Sugden, T.M., and C.N. Kenney, *Microwave Spectroscopy of Gases*, van Nostrand, London, 1965.
77. Todd, J.E., M.A. Whitehead, and K.E. Weber, *J. Chem. Phys.*, **39**, 404 (1964).
78. Tong, D.A., *J. Sci. Instrum.*, **1**, Ser. 2, 1153 (1968).
79. Tong, D.A., *J. Sci. Instrum.*, **1**, Ser. 2, 1162 (1968).
80. Townes, C.H., and B.P. Dailey, *J. Chem. Phys.*, **17**, 782 (1949).
81. Townes, C.H., and B.P. Dailey, *Phys. Rev.*, **78**, 346 (1950).

82. Townes, C.H., and B.P. Dailey, *J. Chem. Phys.*, **20**, 35 (1952).

83. Townes, C.H., and B.P. Dailey, *J. Chem. Phys.*, **23**, 118 (1955).

84. Utton, D.B., *Metrologia*, **3**, 98 (1967).

85. Utton, D.B., *NBS Techn. News Bull.*, 28 (1968).

86. Utton, D.B., "Non Chemical Applications of NQR," in *Proceedings of the Second International Symposium on Nuclear Quadrupole Resonance Spectroscopy*, A. Vallerini, Pisa, 1975, p. 341.

87. Vanier, J., *Metrologia*, **1**, 135 (1965).

88. Vargas, H., J. Pelzl, A. Chappe, and D. Dautreppe, "Impurity Effects on the Chlorine-35 Quadrupole Resonance in Chlorates," in *Proceedings of the Second International Symposium on Nuclear Quadrupole Resonance Spectroscopy*, Vallerini, Pisa, 1975, p. 313.

89. Viswanathan, T.L., T.R. Viswanathan, and K.V. Saue, *Rev. Sci. Instrum.*, **41**, 477 (1970).

90. Voronko, M.G., V.P. Feshin, P.A. Nikitin, and N.I. Berestennikov, "NQR-Studies of Electronic Effects in Geminal Systems," in *Proceedings of the Second International Symposium on Nuclear Quadrupole Resonance Spectroscopy*, Vallerini, Pisa, 1975, p. 207.

91. Wang, T., *Phys. Rev.*, **99**, 566 (1955).

92. WILKS SCIENTIFIC CORPORATION, *Model NQR-1A Nuclear Quadrupole Resonance Spectrometer, Principles, Applications, Specifications*.

Part I
Section I

Chapter 5

SECONDARY-ION MASS SPECTROMETRY

BY H.W. WERNER* AND A.E. MORGAN‡, *Philips Research Laboratories, Eindhoven, The Netherlands*

Contents

*Department Head, "Structure Analysis."
‡Philips Research Laboratories, Sunnyvale (c/o Signetics Co., Sunnyvale, California.)

I. INTRODUCTION

A. CONTEXT OF SECONDARY-ION MASS SPECTROMETRY

A well-equipped analytical laboratory has at its disposal several instrumental methods such as spectrochemistry, spark-source mass spectrometry, X-ray fluorescence etc., to complement the classical, wet chemical approach. These methods permit multielement detection, but often with very different sensitivities for the individual elements, and therefore calibration standards are generally required for a quantitative analysis. A new generation of microanalytical instrumental techniques has arisen to meet the demands of surface and thin-film characterization. Basic to all of these methods is the capability of small area examination (local or microspot analysis) and of obtaining the analytical information from a restricted depth.

A narrow beam of electrons, ions, neutral particles, or electromagnetic radiation is used for excitation. The beam can be made to strike a well-defined and predetermined area on the sample surface. Rastering of this beam is often employed to investigate sample homogeneity across its surface. The degree of sample perturbation is controlled by the nature, dose, and energy of the exciting species. The information depth is governed by the escape depth of the emitted particles or radiation utilized for the analysis. The signal obtained from different positions within the sample is used for an in-depth analysis.

The broad term, layer characterization, encompasses such diversities as crystalline perfection, stoichiometry, trace element detection and determination, inorganic and organic compound identification, electronic properties, lattice location of impurity atoms, etc. Therefore, in addition to the nature and dimensions of the layer involved, a particular technique is selected on the basis of the essential information required. However, the capacibilities and limitations of most modern surface and thin-film analytical techniques have yet to be fully realized.

In the following review, we shall attempt to present the current status regarding one of these methods, viz. secondary-ion mass spectrometry (SIMS). Later on, in Section IV, we briefly compare and contrast SIMS with some other common methods.

For principles and construction of mass analyzers, detectors, and basic spectra identification, peak-stripping etc., the reader is referred to textbooks on mass spectrometry (7A,22A,43A,43B,50C,54A,90,112B,191C,248A,255A,303A).

A useful introduction to the fundamental physics behind the method will be found in the books by Carter and Colligon (62), and Kaminsky (186). General reviews concerning the analytical capabilities of SIMS have been given by many authors (12,32,34,72,108,109,118,164,221,248,251,356,360,405,406,407,409,413).

B. PRINCIPLE OF SIMS

The solid sample, maintained in a vacuum chamber, is bombarded by a beam of primary ions (Fig. 5.1), having a fixed energy of (usually) several keV. As a result, particles are removed from the solid surface, some of which are in the form of positive and negative ions. These are extracted from the target into a mass analyzer and separated therein on the basis of their mass-to-charge ratio. These ions are then detected by suitable means and the resulting signal is fed to a recorder or a computer.

A scan through the mass spectrum reveals the elements present in the sample or, more strictly speaking, since these ions originate from the first few Å of the target, the scan reveals the surface composition of the sample. By continuing the bombardment, however, one digs deeper and deeper into the solid, thereby revealing the bulk composition. By monitoring one or more of the mass peaks as a function of

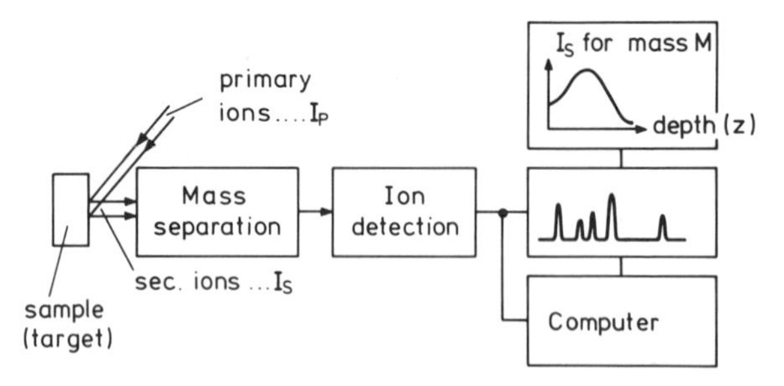

Fig. 5.1. Principle of SIMS.

bombardment time, i.e., the eroded depth, one obtains an in-depth concentration profile.

Fairly simple spectra (since only signals > 10^{-14} A are included) resulting from Ar^+ impingement onto an aluminum specimen, are shown in Figs. 5.2 and 5.3. Several features are worthy of note:

1. The atomic ions of electropositive elements appear with greater intensity in the positive spectrum (Fig. 5.2), while the electronegative elements favor negative ion formation (Fig. 5.3).

2. Molecular ions exist in either spectrum, and therefore mass spectral coincidences are to be expected.

3. Multiply-charged positive ions are also found.

4. Impurities such as Na, K, F, and Cl are readily detected. Contrary to appearances, these elements are in fact present in extremely minute amounts and the first lesson that a budding analyst must learn is that a quantitative assessment directly from a SIMS spectrum is foolhardy.

5. Isotopes (e.g., $^{24,25,26}Mg^+$, $^{35,37}Cl^-$) are easily separated, which is of potential use for self-diffusion studies and geological dating.

6. The ability to detect H is apparent.

7. The signals extend over a large dynamic range, which is in fact far greater than that indicated since ion currents can be measured down to the 10^{-19} A range; one singly-charged ion detected per second $\triangleq 1.6 \times 10^{-19}$ A.

8. The spectra change with bombardment time due to the gradual removal of the surface oxide layer, i.e., the inherent capacity of the technique for both surface and bulk analysis is revealed.

9. The Al^+ signal decreases with time despite more Al being exposed, i.e., matrix effects are to be expected.

By focussing the impinging ion beam to μm dimensions, microspot analyses can be accomplished. The erosion rate chosen will depend on the specimen thickness; for analyses in the monolayer range, an erosion rate of around 1 Å/hr is desirable (static SIMS). The maximum uniform rate generally attainable of some μm/hr is the deciding factor for an in-depth concentration profile of a thick film. The minimum

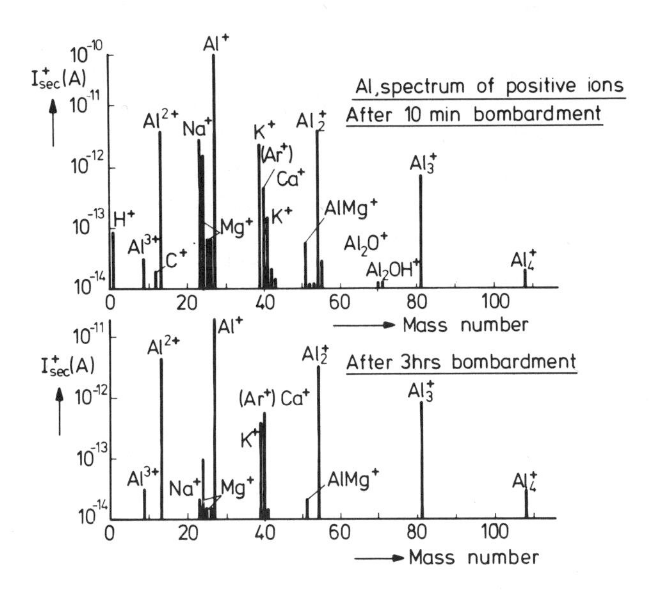

Fig. 5.2. Positive secondary-ion mass spectrum from an aluminium target obtained by Ar^{+} bombardment.

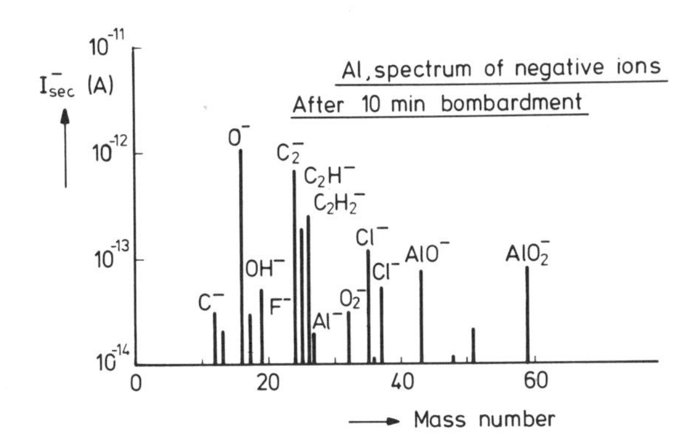

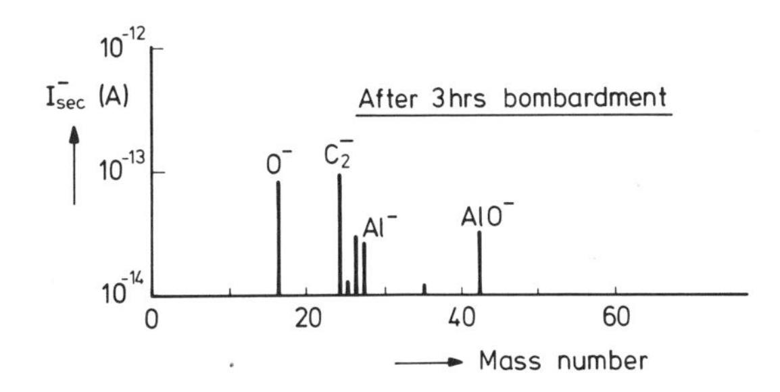

Fig. 5.3. Negative secondary-ion spectrum under the same conditions as in Fig. 5.2.

detectable concentration, besides varying from element to element and from matrix to matrix by orders of magnitude, will also depend upon the amount of sample that can permissibly be sputtered away. (Bombardment of a 7×10^{-4} cm^2 area of Si with a 5.5 keV, 60° incident O_2^+ beam of current density 4×10^{-4} A/cm^2, leads to an erosion rate of 8 Å/sec, an eroded volume of 6×10^{-11} cm^3/sec and a sample consumption of 1.3×10^{-10} g/sec.) Detection limits lying in the ppm to ppb range normally be expected.

Thus, in principle, the SIMS technique provided a method of obtaining:

1. Identification of all elements with low detection limits in samples having thicknesses ranging between one monolayer and several μm.
2. Microspot analyses in the μm range.
3. Lateral distributions of elements across the sample surface with μm resolution.
4. In-depth concentration profiles.

II. BASIS OF SIMS

A. PHYSICAL BASIS

1. Mechanism of Sputtering

The most profitable initial kinetic energy of secondary ions is some eV (25,102,147,183,199,259,368). They are emitted as a result of momentum transfer through the so-called collision cascade, initiated by penetration of the energetic primary particle into the solid (Fig. 5.4). The primary particle loses its kinetic energy

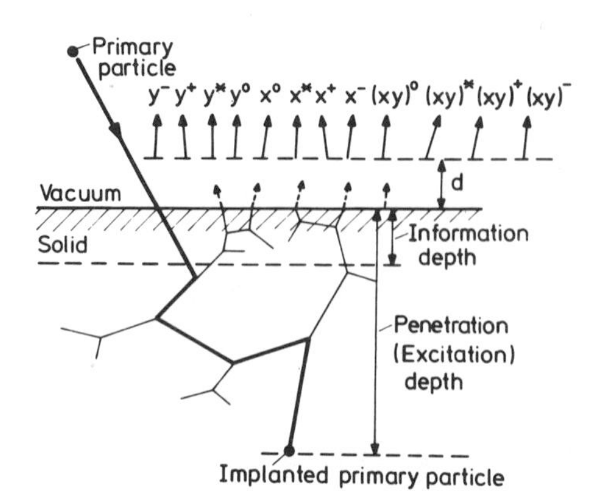

Fig. 5.4. Schematic representation of an energetic projectile–solid interaction, leading to the emission of neutral, excited, and ionized target atoms (x, y) and molecules (xy). Within a distance $d(\lesssim 50$ Å) from the surface, processes may occur that change the original state of an ejected particle. Particles of lower energy are more susceptible to change of state because of their lower velocity.

through a series of binary collisions with the target atoms, which are set into motion and displaced in turn from their lattice positions and thereby giving rise to many collision sequences. The primary eventually comes to rest at a mean penetration depth [a typical value being 250 Å for a 20 keV normal incident primary (333)] and is thus implanted in the solid. When a collision sequence intersects the surface region, an atom or a group of atoms may receive enough momentum in a suitable direction to be ejected from the solid. This phenomenon is called sputtering. Ejection by direct impact between a primary and a target atom occurs fairly infrequently.

The primary comes to rest in about 10^{-13} sec and the average cascade lifetime is $< 10^{-12}$ sec (351). This is much shorter than the time between the impingement of successive primaries onto that surface region influenced by the cascade (typically about 100 Å in diameter), even at the maximum current densities of several mA/cm^2 commonly employed. Therefore, on the average, no overlap occurs between the individual cascades generated by each primary particle (except in the case of molecular bombardment, when closely spaced cascades may result due to dissociation of the molecules upon impact at the surface). The statistical properties of the collision cascades can be expressed using transport theory in terms of a linear Boltzmann equation (246A,322A,322B). Solutions are available for amongst other things the longitudinal and transverse extention (226A,226C,332A) and the spatial distribution of energy as a result of elastic and inelastic collisions (422G,226B). For further details, a review (350) and tabulated results (422F,422H,128A) may be consulted.

Sigmund has presented (349A,351A) a general theory for the sputtering of amorphous targets based on the above considerations. The sputter yield, S, defined as the number of ejected atoms (irrespective of their charge) per impinging ion, is given by

$$S(\cos\Theta, E) = \frac{3}{4\pi^2}\frac{F_D(\cos\Theta, E, 0)}{NU_0C_0} \qquad (1)$$

for projectiles with energy E incident at an angle Θ. The average density of the energy deposited at the target surface located at $x = 0$ by a projectile as a result of nuclear stopping is denoted by F_D, the density of the target atoms by N and the surface binding energy by U_0. The constant C_0 may be set equal to 1.808×10^{-16}. For convenience, a dimensionless factor α is introduced so that

$$F_D = (\cos\Theta, E, 0) = \alpha\left(\frac{dE}{dx}\right)_n = \alpha N S_n(E) \qquad (2)$$

where $(dE/dx)_n$ is the stopping power for elastic atomic collisions and $S_n(E)$ the nuclear stopping cross-section. The factor α depends on the ratio M_2/M_1 of target to projectile mass, the screening constant used in the calculation of the nuclear stopping power, the projectile energy, and the incident angle. Values have been given for instance by Winterbon (422H).

For perpendicular incidence, the sputter yield reduces to

$$S(E) = 0.042\,\alpha\,\frac{S_n(E)}{U_o\,10^{-16}} \qquad (3)$$

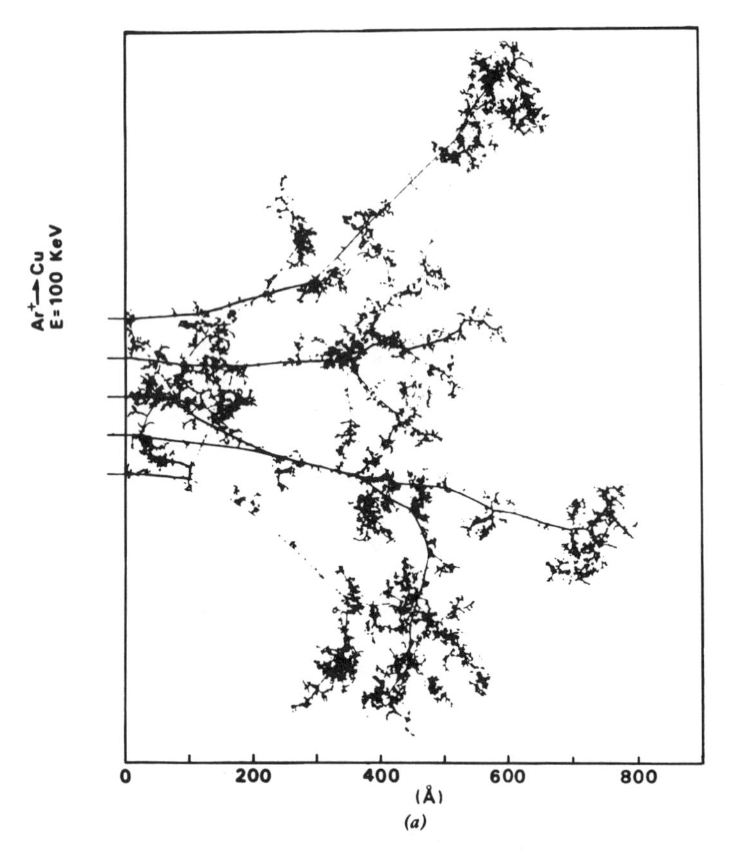

Fig. 5.5. Trajectories of projectiles (broad lines) and recoils (thin lines) for (a) 100-keV Ar^+ bombardment of Cu and (b) 100-keV Au^+ onto Au (427E). (Courtesy of *Radiation Effects*.) (c) Calculated distribution of the original depth of sputtered atoms versus depth (174). (Courtesy of *Applied Physics*.)

with U_o given in ergs, Andersen's article (15C) contains an extensive list of surface-binding energies, together with calculations of S using the above equations for Ne^+, Ar^+, and Xe^+ in the 10 keV range.

The formulae given above refer to a linear collision cascade, i.e., for a situation in which the lattice structure is essentially preserved and only a small fraction of the atoms is actually set in motion. At the other extreme is the thermal spike situation in which the structure is destroyed locally and all atoms within the spike volume are essentially in motion (351,15B). This may apply when a heavy projectile encounters a heavy target. In many instances, the actual situation will lie between these two extremes (7B,65A,278B,376B,376C,373A,373B,373C). As in the case of the spike, the sputter yield cannot then be derived from first principles.

Computer simulations of the sputtering process have also been presented employing the same collision cross-sections and mean free paths used in the basic theory (422E), (145D). One of the most comprehensive programs ("Marlowe") has been written by Robinson and Torrens (311A). Littmark et. al. (228B) demonstrated complete agreement between analytical and computer calculations for 100 eV–

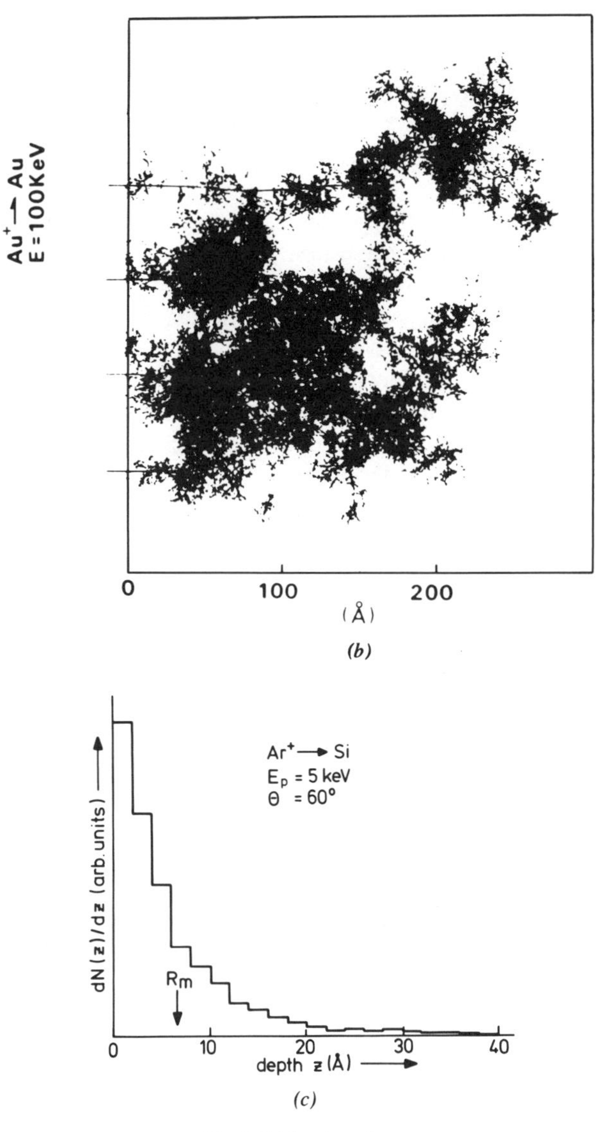

Fig. 5.5 *(Continued)*

20 keV helium bombardment of niobium. A large number of computer simulations have been performed by the Japanese school (173B,173C,174,175,347A,427E); Fig. 5.5*a* depicts the development of linear cascades whereas spikelike cascades are shown in Fig. 5.5*b*. (For a review of experimental and theoretical aspects see reference 283).

The *information depth* of the SIMS method is the depth within the solid from which those sputtered particles monitored as ions have originated. Fig. 5.5*c* indicates that most sputtered atoms originate from the near-surface region with a mean escape depth of 6 Å. Giber (130) has calculated for Cu crystals bombarded with 0.5–

5 keV Ar^+ that the number of atoms sputtered from the different atomic layers decreases approximately exponentially with depth. The information depth, defined here as the depth of the layer from which e-times less atoms are sputtered than from the surface layer, was given as 3 ± 1 atomic layers. The escape depth for particles detected as ions is probably identical but might even be less than that pertaining to sputtered atoms.

Not every cascade leads to the ejection of one or more sputtered neutrals (e.g., if no collision sequence reaches the surface), and even fewer lead to eventual ionic emission. At present, it is by no means well established exactly how secondary ions are formed, although many suggestions have been put forward which are reviewed in references 322 and 427. In particular, no theory has yet emerged enabling one to calculate the ionization probability in a general situation from first principles. Later, in Section II.D.3c, we shall describe an empirical approach that has had some success in this respect. Here, we shall now summarize current thinking concerning the mechanism of secondary ion formation.

2. Models and Theories of Secondary-Ion Emission

Ionization processes are conveniently classified into two categories, intrinsic ion emission for the case of noble gas ion bombardment of clean elemental surfaces and chemical emission applying to the dramatic increases brought about by the presence of reactive species, either initially present (compounds, alloys) or introduced during the bombardment. Most first-principle theories and calculations apply to the former category and will be considered first.

a. Kinetic Model

This has been developed (181B,182) to describe the emission of singly- and multiply-charged ions with relatively large kinetic energies ($\geqslant$ 25 eV). Some of the displaced atoms in the collision cascade may escape from the solid in a neutral state but carrying a bound-level excitation. De-excitation outside the target will lead to a secondary ion and an Auger electron. More energetic collisions could create more than one inner-level vacancy and then Auger de-excitation would lead to a multiply-charged ion. Calculations of the probability of vacancy production have been performed using the electron promotion model developed for inelastic atomic collisions in gases (145D). Wittmaack (427B) has proposed that H^+ sputtered from hydrogenated silicon is formed by the dissociation in vacuum of $(SiH)^{2+}$ subsequent to Auger de-excitation of an ejected $(Si_{2p}H)^+$ molecule containing an excited silicon atom with a $2p$ hole.

b. Surface-Effect Models

These autoionization model (45,44B) considers that the perturbation experienced by the outer electronic shells of the atom as it crosses the solid-vacuum interface can lead to the formation of autoionizing excited states. Auger de-excitation in vacuum is again held responsible for ionic formation. Calculations for some metals and dilute alloys have produced encouraging results.

Other proposals consider the probability of ionization of a ground-state neutral atom as it crosses the solid surface by quantum-mechanicas transitions of its valence electron to the top of the conduction band of the metallic solid. Calculations have been made using various approximations (68,181E,334,365,365A). For instance, quantum (18) considered an atom moving through the metal target as an interstitial whose local excess of positive charge is compensated by an electron cloud moving together with the atom. This electron cloud was represented as a plane de Broglie wave which is partially reflected at the potential barrier of the metal surface. The ion yield was then equated to the quantum-mechanical reflection coefficent, yielding good agreement with experiment for several metals but unfortunately with orders of magnitude discrepancies for the noble metals.

The most sophisticated quantum model has been forwarded by Blandin et al. (50A) They derived an exact solution for the case in which the time-dependent Andersen Hamiltonian can be written in one-body form (i.e., the atomic structure is reduced to a single orbital). The ionization probability P^+ for slow ions was given by

$$P^+ = \frac{2}{\pi} \exp\left(\frac{-\pi/\Phi - E_d}{hva}\right) \tag{4}$$

where Φ is the work function, E_d the d-level energy with respect to the vacuum level, h Planck's constant, and v the velocity of the departing particle normal to the surface. An exponential time decrease of the width Δ of the d level was assumed, and a^{-1} characterizes this decrease with respect to distance from the surface; i.e., $\Delta = \Delta_o \exp(-avt)$. Joyes (182A) obtained the same result by another method.

c. Thermodynamic Models

In the LTE model (13,15), the sputtering region is assumed to resemble a dense plasma in local thermal equilibrium. The generation of positive ions is assumed to follow the equilibrium reaction.

$$M^o \rightleftharpoons M^+ + e \tag{5}$$

enabling the ionization probability to be calculated from the Saha-Eqqert equation;

$$\frac{n_{M^+}}{n_{M^o}} = \frac{2Z_{M^+}}{Z_{M^o}} \cdot \frac{(2\pi mkT)^{1/2}}{n_e h^3} \exp\left[-(E_M - \Delta E_M)(kT)\right] \tag{6}$$

where n_{M^+}, n_{M^o} are respectively the numbers of M^+ ions and M^o neutrals emitted per second from a sample containing the element M; Z represents an electronic partition function; k and h are respectively the Boltzmann and Planck constants; m is the electron mass, and n_e the electron density in the plasma; E_M is the first ionization potential of M and ΔE_M the ionization potential depression to the plasma; T is the plasma temperature. Similarly (53A), for negative ions one may write that

$$\frac{n_{M^-}}{n_{M^o}} = \frac{g_{M^-}}{2g_{M^o}} \cdot \frac{n_e h^3}{(2\pi mkT)^{\frac{3}{2}}} \cdot \exp\left(\frac{A_M}{kT}\right) \tag{7}$$

where g represents a ground state statistical weight and A_M is the electronic affinity of

M. Similar equations apply to the formation of multiply-charged and molecular ions.

Although experimental results appear to support an exponential dependence upon ionization potential or electron affinity, the plasma model is not generally accepted and hence there has been much discussion in an attempt to rationalize this dependence and to give a meaning to the parameter T (62A,133A).

Coles (75A) suggested a secondary electron model. The plasma would be formed outside of the solid by all departing species and T would be associated with secondary electron emission. Williams (422B) assumed thermal equilibrium amongst the exated electrons at the sputtering site and interaction with the departing atom via resonant exchange processes. Morgan and Werner (263,406,415A) suggested that kT should be regarded as the energy available for ionization in an individual cascade. It has also been suggested (185,202A) that the Dobretsov equation, developed to describe nonequilibrium surface ionization, should provide a more satisfactory expression for secondary ion emission:

$$P^{+} = \frac{Z_{M^{+}}}{Z_{M^{0}}} \cdot \exp\left[\frac{\Phi - E_{M} + (e/4x_{c})}{kT}\right] \tag{8}$$

and

$$P^{-} = \frac{g_{M^{-}}}{g_{M^{0}}} \cdot \exp\left[\frac{A_{M} - \Phi + (e/4x_{c})}{kT}\right] \tag{9}$$

where x_c represents the critical distance for charge exchange with the surface.

d. Ion Neutralization Models

The sputtered particle is assumed to leave the surface in the form of an ion with a constant velocity v normal to the surface and may undergo neutralization as a result of resonance tunnelling in the vicinity of the surface (28,398). The probability to escape as an ion is given by

$$P^{+} = \exp(-A/va) \tag{10}$$

where A is the transition rate at the surface and a is a characteristic distance determined by the overlap of the electron wave function. Recently, Prival (303B) has derived an alternative expression for this neutralization probability.

e. Chemical Emission Models

Several proposals have been forwarded to rationalize the dramatic enhancements in positive ion yields brought about by the presence of an electronegative element such as oxygen and in negative ion yields by an electropositive element such as cesium. In the bond-breaking model (354A,401,27D,63A,28) the ionic or partial ionic character of the bonding between a metal atom M and the reactive surface atom (O or Cs) is supposed to lead to a greatly increased probability that the atom M is detached from the solid as M^{+} or M^{-} respectively. The ejection of M^{+} has been described pictorially (46A) in terms of curve-crossing in the potential energy diagram of the M-O gas phase system. In fact, if it is postulated that the final collision responsible for the sputtering event leads to the momentary ejection of an MO quasimolecule (269,376A) (a) gas-phase diagrams may then realistically be employed to discuss the probability of molecular dissociation into $M^+ + O^-$ instead of into $M^\circ + O^\circ$, and (b) ionic species are formed at a short distance from the surface thereby escaping Auger or resonance neutralization at the surface.

In the work-function model (11), oxygen was supposed to increase the work function thereby impeding neutralization by thermionic electrons excited over the surface potential barrier. In fact, measurements have shown (45A) that the work function may decrease during positive ion yield enhancement by oxygen. Yu (430) found that the negative ion yield increased and the work function decreased with increasing cesium coverage and used a tunnelling model to rationalize his results.

In the band structure model (189A), oxygen is supposed to deplete the conduction band of electrons (leading ultimately to a band gap in the case of oxide formation) leaving none available at the energy required for neutralization of the departing ion at the surface. However, the concept and application of bulk structure and band gap values to the surface region of a solid, highly disordered through sputtering, is questionable. Further, the model does tie in completely with experimental observations (191B).

A surface polarization model (422A) has been involved to explain the observation that oxygen also enhances negative secondary ion yields somewhat. Differently orientated surface dipoles in the oxygenated surface are supposed to lead to localized electron emissive or electron retentive sites which dominate the emission of negative and positive ions, respectively.

f. OTHER MODELS

Sparrow (362A) has proposed a formula for the determination of the relative ion yield of compounds. His considerations in good agreement with the work of Morgan and Werner (263), are based on the following.

For an atom to be emitted from the sample, ionized, and then detected, chemical bonds must be broken, the atom M must be ionized (ionization energy E_M) and emitted. He assumed an exponential dependence of the ion yield on the ionization potential; the ionic bond strength is assumed to be represented by X, the electronegativity of the element; the number of bonds n involved is taken equal to the valence state of the atom:

$$S^+_{\text{rel}} = S^+_M / S^+_{\text{ref}} \exp(-E_M/kT) nMX^2 \tag{11}$$

Lodding (229) was one of the first to use SIMS for investigation of biological hard tissues. He studied the F^+/Ca^{++} content of apatites familiar as constituents of, among other things, human teeth. He found that the measured F^+/Ca^{++} ratio varied, depending on the prevailing experimental conditions. To overcome this experimental influence, he plotted log (F^+/Ca^{++}) against log (P^+/Ca^{++}) obtained from a number of samples with known concentrations. The linear curves obtained were used as a sort of working curve for the determination of the absolute concentration of F^+ in apatite.

Lodding interpreted these linear curves as an exponential dependence of the secondary ion current on the ratio of ionization energy divided by kT, similar to the model of Andersen.

g. Cluster Ion and Molecular Ion Emission

Cluster ions (150,163C,163D,211A) have been observed with metals (cf. Fig. 5.2) and metal alloys (cf. Figs. 5.20*a* and 5.20*b*). Molecular ions (cf. Fig. 5.21) have been observed with inorganic compounds (163A,375A) including crystals (57) and with organic compounds (cf. Fig. 5.23 and section V.A and V.D). Clusters of molecular ions have been observed with salts (163A,375A) and in crystals (57).

The mechanisms which lead to the emission of cluster and molecular *particles*—i.e., charged or uncharged—are far from being completely understood. In particular, controversial opinions have been published regarding the question if a cluster (e.g., a dimer) irrespective of its charge state is a emitted *as such*, i.e., contains atoms from nearest or next nearest sites in the sample (*molecular and direct-cluster emission*) or is generated by recombination of atoms not originating from nearest or next nearest sites (*recombination mechanism*) (422C,422D,125A,145D,422E,427A).

Moreover, diverging ideas exist whether the recombination takes place closely above the surface (a few Å) or further away. (211A, 181C, 208A, 181D, 29A, 32A, 32B, 272A, 272B, 422C, 422D, 422E, 125A, 145D, 427A, 44A, 375A, 163A, 163B).

The ejection of clusters or molecules (as ions or neutrals) is thought to take place via one or a combination of the two mechanisms described above.

The recombination mechanism has been elucidated somewhat by means of computer calculations of particles (irrespective of their charge) in single crystals (422C,125A,422D,145D,44A,145A) ejected due to displacement of particles via a series of biparticle collisions following ion bombardment. Recombination was considered as a statistical bombardment. Recombination was considered as a statistical process of two particles, originating from distant sites (however within a circle of 70 Å^2 area), which may come close enough to each other within 4 Å above the surface. These two (or more) particles may then recombine into a dimer (multimer) if $T + V < O$, i.e., when their relative kinetic energy T in the center of mass system is smaller then their potential energy $V(< O)$, the lattice being responsible for the attraction between the two particles (422E,126A). This requirement can often be met with, because the excess energy is delivered to the surface atoms. If $T + V > O$, no dimer (multimer) generation can take place.

Following the results of Harrison (145D) that 90% of the ejected particles come from the topmost layer, most of the computer calculations for the recombination mechanism only take into account the few top layers. While dimers can originate further apart, trimers are forced by geometrical considerations to have a nearest neighboring pair (422C) either in a low index face or in the first and second (100) face [of Cu; (422C)].

Direct-cluster emission, i.e., emission of a cluster as such (in contrast to the statistical considerations in the recombination model) has been treated for dimers of metals by Joyes et al. (211A,181C,181D) and Bitenskii et al. (44A).

According to these authors, only 10–30% of the observed dimer yield is due to direct-cluster emission, while the statistical mechanism is responsible for the remaining 90–70%.

Note that these figures may change with the number of atoms in a cluster (422D); cf. also cluster ion abundance in Figs. 5.20*a* and 5.20*b*).

Molecular (particle) ejection occurs with organic and inorganic compounds, because the excess energy, which would lead to dissociation is soaked in the molecule system (422E).

Significant rearrangement of the atoms in the ejected molecule during the emission has been mentioned (375A,163A). Obviously the statistical model would fail when trying to explain the emission of large organic molecules (of Fig. 5.23 and Table 5.IX). Instead, one must think in terms of the emission of large molecules from the top surface layer as a whole (409).

In the emission of ions from salts (375A), up to 13 molecules were found in one emitted cluster ion. The emission of molecules as a whole is probably due to the large binding energy, which disfavors dissociation (cf. Fig. 5.19).

The direct emission of Si_n-cluster ions from Si, as discussed by Wittmaack (427A) has been questioned by Winograd (422D).

So far, no distinct mechanism leading to the emission of charged clusters or molecular ions has been reported. In general, one tacitly assumes an ionization mechanism in addition to the process leading to neutral particle emission. In this respect, the experimental work of Oechsner (283,284,126B,283A,298A,298B, 283B) is most promising.

The experimental results of Staudenmaier (368A,368B) on *W*-cluster ion emission agree well with the calculated ones of Können et al. (196A).

Independent of the above considerations, Plog and Benninghoven (298A) have derived a model that predicts abundances of molecular oxide ions in different oxidation states. The basic idea is that the valence of a given atom in the original lattice (lattice valence) and the "nominal" valence of a fragment ion (fragment valence) are related to each other.

The authors have shown that ions with a given atom/oxygen ratio are emitted in great abundance when the fragment valence is equal to the lattice valence, in good agreement with the experimental results of Werner (404) and the considerations of Morgan and Werner (263).

B. BASIC FORMULA FOR A SIMS ANALYSIS

The ion current measured with the mass spectrometer and corrected for isotopic abundance, I_{M^+}, of positive, monatomic, secondary ions of a certain element M present in the sample can be considered a product of five parameters:

$$I_{M^+} = I_p \cdot S \cdot c_M \cdot \beta_{M^+} \cdot \eta \qquad (12)$$

An analogous expression can be written for negative secondary ions. I_p is the primary ion current, S the sample sputter yield, and c_M the fractional atomic concentration of the element M. β_{M^+} is called the degree of ionization, and is equal to the number of M^+ ions emitted from the sample per second, n_{M^+}, divided by the total number of M-

TABLE 5.I

Typical Values for Primary Beam Current I_p, Beam Diameter d_p, Primary Ion Current Density j_p (assuming stationary square beam), and Erosion Rate $\dot{z}$ (assuming $S = 3$ and $M/\rho = 10$)

Mode	I_p	d_p	j_p	$\dot{z}$ $(S = 3)$
Static	10^{-11} A	1 mm	1 nA/cm²	1 Å/hr
	10^{-9} A	1 cm	(= 0.01 nA/mm²)	
Dynamic	3×10^{-8} A	1 mm	3 μA/cm²	300 Å/hr
			(= 30 nA/mm²)	
	10^{-11} A	1 μm	1 mA/cm²	10 μm/hr
			(= 10 μA/mm²)	
	10^{-5} A	1 mm		

containing species emitted per second, n_M. The product of S, c_M, β_{M^+}, which is thus equal to the number of M^+ ions emitted per primary ion. The instrumental factor for the particular isotope monitored, η, is the ratio of the measured ion current to that emitted from the target. We shall now consider each of the parameters contained in equation (12) in turn.

1. Primary Ion Current

Values of the primary ion current, I_p, between 10^{-11} and 10^{-5} A are generally used (see Table 5.I). Another related parameter of importance is the primary ion current density, j_p, which is given by I_p/A_b where A_b is the effective bombarded area. With a

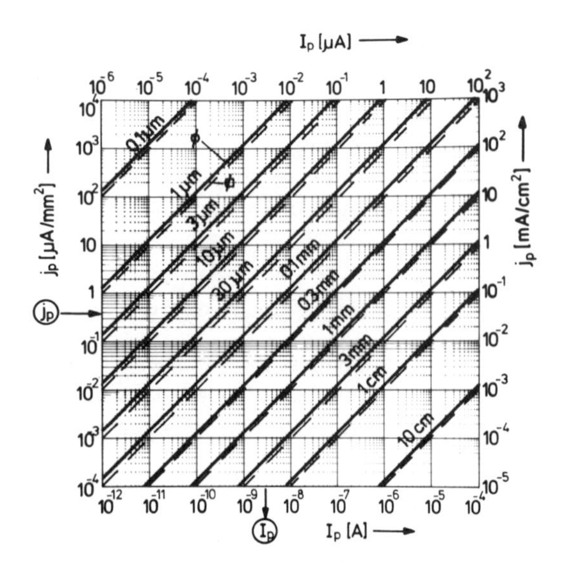

Fig. 5.6. Relationships between the primary ion current density j_p and the primary ion current I_p for different beam dimensions; circular beam diameter indicated by ϕ, square-beam diameter indicated by ◻.

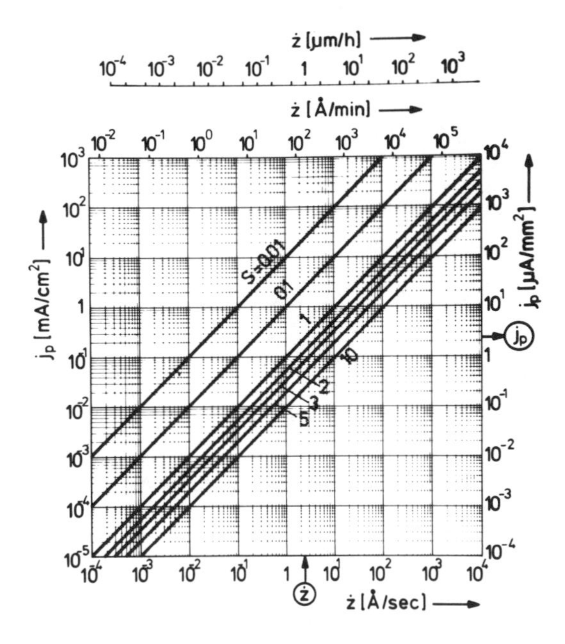

Fig. 5.7. Erosion rate $\dot{z}$ as a function of j_p for different sputter yields S; $M/\rho = 10$.

stationary beam, A_b is determined by the beam dimensions and with scanning beam, by the rastered area; it may also be necessary to bear in mind the extent of the area from which secondary ions can be or are accepted into the mass analyzer. Fig. 5.6 shows the connection between j_p and I_p; the upper line in each pair refers to a stationary circular beam of diameter d_b, while the lower line refers to a square rastered area of side d_b. The value of j_p is typically between 10^{-6} and 100 mA/cm², the upper limit being dictated by the brightness of the available primary ion source.

The erosion rate, $\dot{z}$, depends on j_p and is given by:

$$\dot{z}(\mu\text{m/hr}) = 3.6 \times 10^{-4}\, [M(\text{amu})/\rho(\text{gm/cm}^3)] j_p(\mu\text{A/cm}^2) S, \tag{13}$$

the atomic weight and density of the target being denoted by M and ρ, respectively. Fig. 5.7 permits a rapid estimate of $\dot{z}$ to be obtained for different combinations of j_p and S, taking $M/\rho = 10$.

2. Sputter Yield

The sputter yield, S, depends on many parameters (62,186,283,350,399) such as mass, energy, and angle of incidence of the primary ions; mass and surface-binding energy of the target atoms; crystal structure, lattice orientation, surface roughness, etc. For instance, Fig. 5.8 shows the variation of S for Cu as a function of bombardment energy for normal incident Ar^+ primaries. In general, S first increases linearly with primary energy up to about 10 keV. At higher incident ion energies, S increases less rapidly and eventually a slight decrease sets in since the primaries

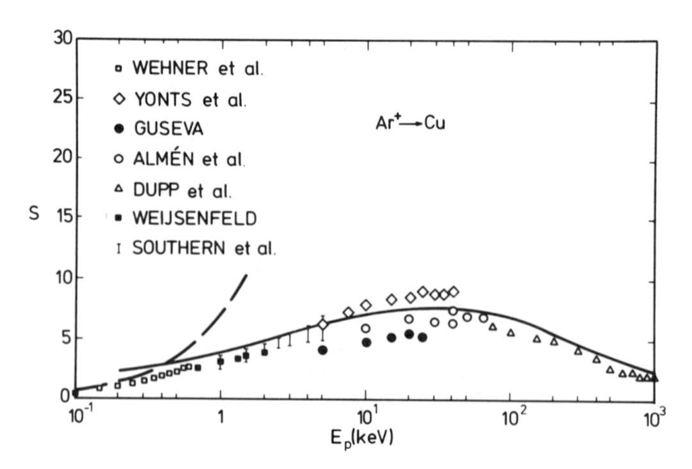

Fig. 5.8. Sputter yield S as a function of the primary energy E_p for Ar^+ on Cu; lines denote calculated values (350). (Courtesy of *Review Roumanian Physics*.)

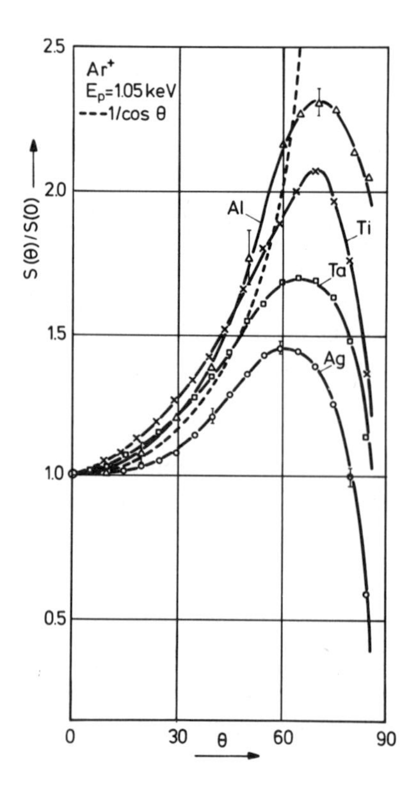

Fig. 5.9. Variation of sputter yield S with angle of incidence (against the normal) θ to the target normal for various metals (283). (Courtesy of *Applied Physics*.)

penetrate deeper into the lattice, leaving less energy available for atomic displacements at the surface. An exact proof of this behaviour (349A,351A) can be derived from the fact, that $S \propto S_n$ (see above) and the similar dependance of S_n with energy. Yields increase (23A) with angle of incidence (Fig. 5.9), because of (15B) decreased primary particle penetration, then reach a maximum and finally decrease at larger incident angles. This decrease reflects the increasing likelihood that the projectile will emerge from the solid before completely depositing its energy in a cascade or that it will be reflected at the surface. Heavier primaries (Fig. 5.10) sputter more efficiently than lighter ones. Values of S for many elements from 400 eV Ar^+ bombardment at normal incidence are shown in Fig. 5.11; S is proportional to the inverse of the surface binding energy (≈heat of sublimation).

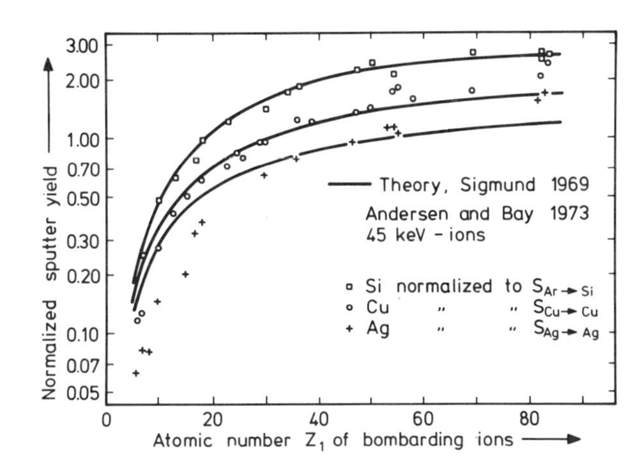

Fig. 5.10. Sputter yield as a function of the atomic number of the bombarding ions for three elemental targets (283). (Courtesy of *Applied Physics*.)

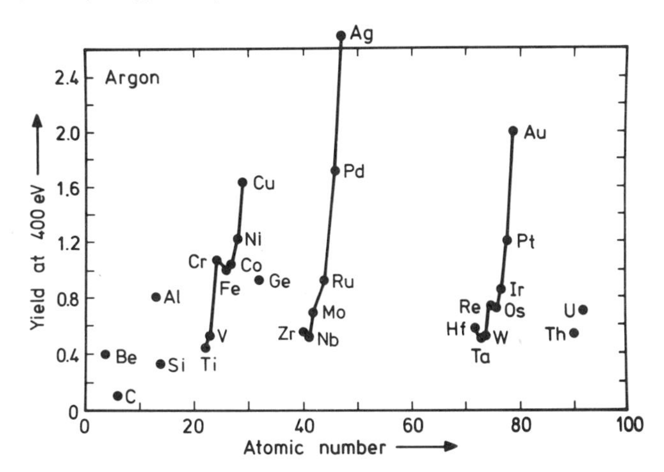

Fig. 5.11. Sputter yield as a function of the atomic number of the target (205). (Courtesy of *Journal of Applied Physics*.)

There are three general approaches to the experimental determination of the sputter yield:

1. Volume loss. The dimensions of the sputtered crater, generated by bombardment with an ion beam of known current for a specified time, are measured. For the sake of accuracy, a flat-bottomed crater is obviously desirable. This entails either scanning a focused ion beam over a large enough area or using a large beam of homogeneous current density (59A). Also, since an appreciable depth is generally required for accurate measurement, sputtering for a considerable time is required. This limits the determination to that of the stationary sputter yield which may differ from the initial (momentary) yield due to the influence of projectiles implanted into the target. Depths may be obtained using a mechanical stylus (416B) or by interferometry; the latter may be used for in situ measurements (191A) as may Rutherford backscattering and energy-dispersive X-ray measurements. With these last two techniques, the decrease in thickness of a thin film may be followed during the sputtering.

2. Weight loss. This more sensitive method involves determining the change of weight of the target with a microbalance (361A,398A). This however entails either the cumbersome demounting of the target or an in situ measurement, which can restrict the scope of the experiment unduly. It should be realized that the target experiences not only weight loss through the sputtering but also weight gain through the implantation of the projectiles. To take this into account, and also to study the variation in the sputter yield as a function of projectile fluence, one requires a fast and sensitive weight determination by means of a vibrating quartz crystal (236A,15A) coupled with a technique such as Rutherford backscattering (50B) to determine the implanted projectile concentration. Kirschner and Etzkorn (194A) used Auger electron spectroscopy to measure this concentration along with thickness and hence sputter yield determination by X-ray analysis.

3. Material deposition. The sputtered material is deposited onto a collector placed in the vicinity of the target and the actual amount determined by any of the following methods: (a) weighing, (b) Rutherford backscattering or X-ray analysis, (c) radioactive tracer or activation analysis, (d) atomic absorption or optical transmission spectroscopy, or (e) resistance change. This last method has the advantages of simplicity, sensitivity, and high spatial resolution, and has been used (399A) to study variation of the sputter yield with ejection angle. The material deposition technique requires knowledge of the sticking coefficient at the collector.

In the case of multielement samples, initially the elements of higher sputter yield are preferentially removed, thereby leading to an enrichment at the surface of the lower sputter yield elements. When steady-state conditions are reached, (a) the overall erosion rate is determined by the element with the lowest sputter yield, and (b) the elements are sputtered at rates proportional to their *bulk* concentrations.

Bombardment, as already mentioned, leads to implantation of the primary species in the target. After erosion of a certain depth (roughly equal to the most probable range of the primaries), a time-independent surface concentration of implanted atoms is achieved (381). Their relative surface density is given in the simplest case by the ratio of the trapping probability to the sputter yield (61A,199A,381,50B,194A). Use of a chemically reactive primary, such as oxygen, can initially cause S to change dramatically as a result of bombardment implantation. S for an oxide can be greater or less (189) than that for the corresponding element.

3. Secondary-Ion Yield

The degree of ionization, β_{M^+}, depends essentially upon (a) the element, (b) the matrix, (c) the chemical nature of the bombarding species and of any other element deliberately introduced in the sample (e.g., by residual gas adsorption or by continuous metal film deposition), and (d) residual gas pressure p (or deposition rate) in relation to the erosion rate. The secondary-ion yield, S_{M^+}, depends in addition upon the parameters that influence the sputter yield S.

Relatively few *absolute* secondary ion yields have been reported (34,43,107,148). Some results from Ar^+ bombardment of elemental samples are shown in Tables 5.II and 5.III; note the increase in the *positive* ion yield by orders of magnitude

TABLE 5.II

Absolute Positive Secondary Ion Yields for Clean and Oxygen-covered Elements; 3 keV, Ar^+, $\theta = 70°$, $p = 10^{-10}$ torr (34)

M (Element)	S_{M^+} (Clean surface)	S_{M^+} (Oxygen-covered surface)
Mg	0.01	0.9
Al	0.007	0.7
Ti	0.0013	0.4
V	0.001	0.3
Cr	0.0012	1.2
Mn	0.0006	0.3
Fe	0.0015	0.35
Ni	0.0006	0.045
Cu	0.0003	0.007
Sr	0.0002	0.16
Nb	0.0006	0.05
Mo	0.00065	0.4
Ba	0.0002	0.03
Ta	0.00007	0.02
W	0.00009	0.035
Si	0.0084	0.58
Ge	0.0044	0.02

TABLE 5.III

Absolute Positive Secondary Ion Yields from 8 keV Ar^+ Bombardment at Normal Incidence in a Residual Vacuum of 2×10^{-7} torr (148)

M	Mg	Al	Si	Ti	Fe	Ni	Cu	Ag	Ta
S_{M^+}	0.004	0.025	0.001	0.0016	0.001	0.0008	0.0006	< 0.00005	0.0005

(Table 5.II), brought about by the presence of the *electronegative* element, oxygen. This pinpoints the absolute necessity of determining these yields under well-defined experimental conditions (e.g., either complete exclusion of oxygen or guarantee of enough oxygen to ensure saturated emission) in order to obtain any meaningful results. *Relative* secondary ion yields of the elements from Ar^+ bombardment in high or ultrahigh vacuum have also been reported (184,401,425).

Storms et al. (371) have measured relative positive ion yields for O^- bombardment of (mainly) elemental samples. Their results are shown in Fig. 5.12. In the case of compound samples, the measured ion currents have been divided by c_M. *Negative* ion yields are considerably enhanced by the presence of an *electropositive* element. Figure 5.13 shows the relative negative yields (371) for the elements from Cs^+ bombardment. Andersen and Hinthorne (12) have also measured positive and negative elemental yields for O^- and Cs^+ bombardment, respectively. It should be noted that S_{M^+} and S_{M^-} may also be increased somewhat by the presence of electropositive and electronegative elements respectively (129,422).

For practical quantitative analyses, however, knowledge is required of S_{M^+} (or more strictly speaking, β_{M^+}) for the various elements under the chosen experimental conditions, in the particular *matrix* under investigation. From all the foregoing, we

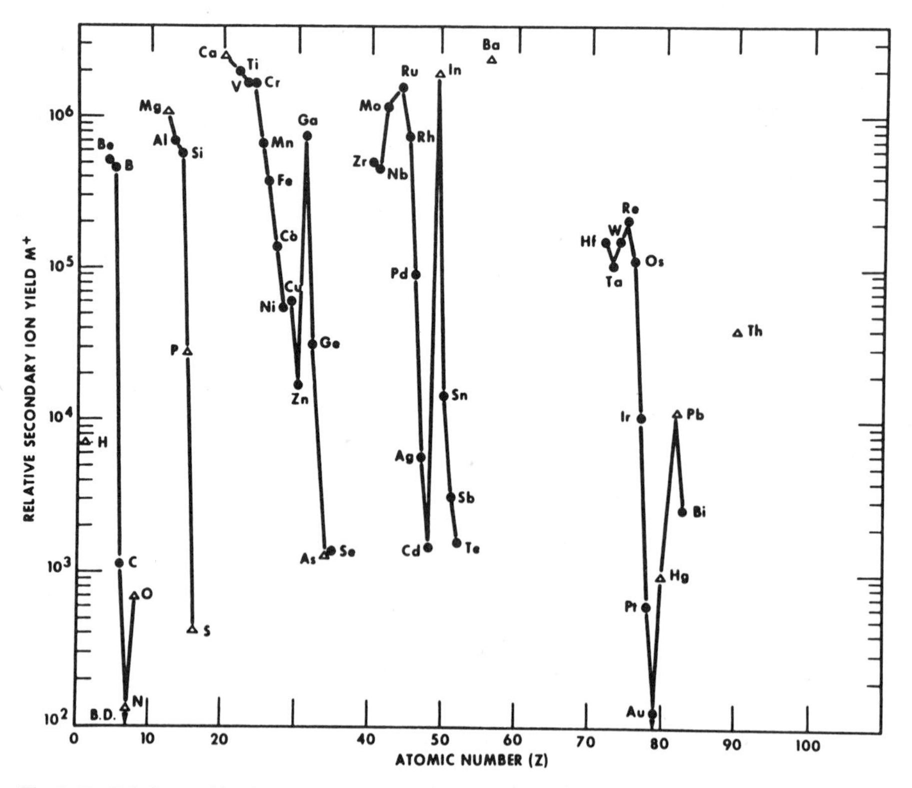

Fig. 5.12. Relative positive ion yields for 13.5-keV O^- bombardment at normal incidence; Δ signifies that a compound sample was used; B.D., barely detected (371). (Courtesy of *Analytical Chemistry*.)

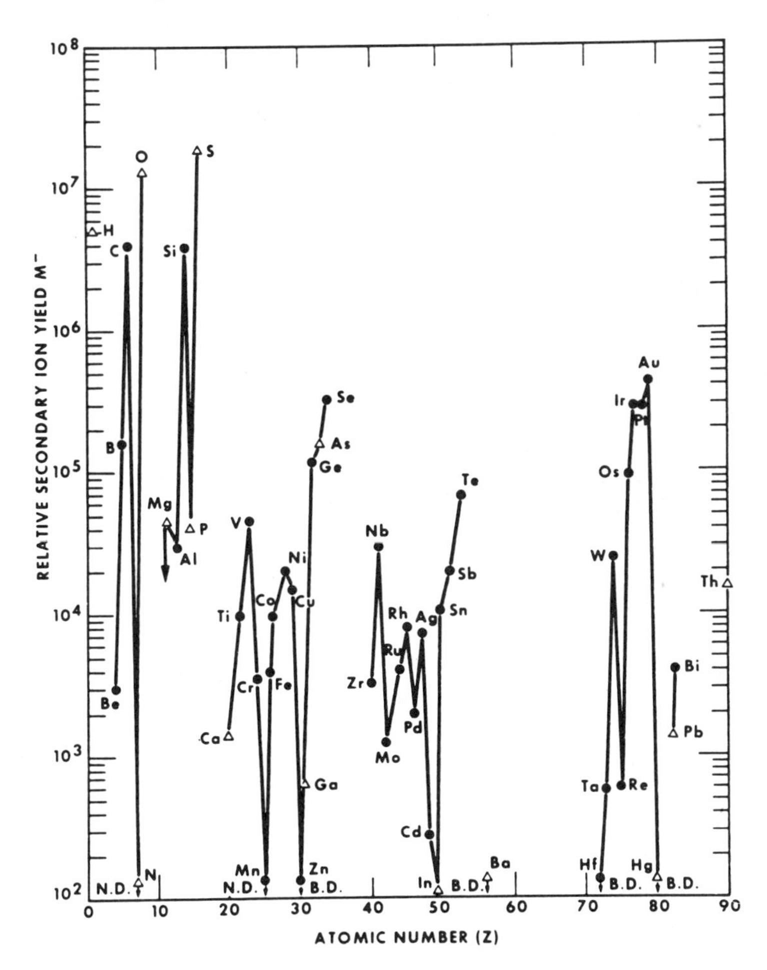

Fig. 5.13. Relative negative ion yields for 16.5-keV Cs^+ bombardment at normal incidence; N.D., not detected (371). (Courtesy of *Analytical Chemistry*.)

can expect a value for β_{M^+} lying anywhere in the range from about 10^{-5} to 1.

Deline et al. (92A) measured $\beta_{M^\pm}\eta$ for several elements M each contained in several matrices, oxygen bombardment being used for positive ion emission and cesium for the negative ions. For a fixed bombardment area and current density, they found that for each element

$$\beta_{M^\pm}\eta \propto [1/\dot{z}]^{x^\pm} \tag{14}$$

with $x^+ \approx 3$ and $x^- \approx 2.5$. To a first approximation, the matrix erosion rate $\dot{z}$ is inversely proportional to the volume concentration of the implanted primary ion species. They thus claimed that the SIMS matrix effect is understandable merely in terms of how much of the enhancing species can be implanted into the matrix or is already contained therein. They then compared the ion yields for the major elements in the different matrices (92B). A correction was made for the matrix effect via equation 14 through measurement of $\dot{z}$. Using these corrected values, they found a linear relationship between log β_{M^+} and the ionization potential of M, and a fairly linear plot of log β_{M^-} versus the electron affinity.

4. Instrumental Factor

Values for the instrumental factor, η, have rarely been given. Benninghoven and Mueller (30) report that $\eta = 10^{-5}$ for their single-focusing 60° sector field instrument. Vallerand and Baril (391) have calculated a value of 10^{-3} for their unique double-focusing instrument. By assuming that $\beta_{M^+} = 1$ for potassium on nickel, Šroubek (365) calculated $\eta = 3.2 \times 10^{-4}$ for his quadrupole instrument. The instrumental factor comprises the efficiency of (a) collection into the mass spectrometer, (b) transmission through the mass analyzer, and (c) ion detection. The transmission will obviously decrease with increasing mass resolution. Any dependence of η upon the actual ion monitored will come about because of differences in the initial energies and emission angles of secondary ions, and because of mass-dependent transmission and/or detection in the spectrometer.

A convenient way of comparing SIMS instruments is to report the number of ions (corrected for isotopic abundance) detected from a given sample and bombarded area per primary ion of a given species, when using a mass resolution of 200–300. Count rates of 10^9 counts/sec per μA of primary current for I_{Fe^+} and I_{Si^-} from O_2^+ and Cs^+ bombardment, respectively, of the pure elements have been reported. These values are considered adequate for trace analyses.

5. Limit of Detection

The *minimum detectable concentration*, $(c_M)_{min}$, is from equation (12)

$$(c_M)_{min} = (I_{M^+})_{min}/I_p S \beta_{M^+} \eta \qquad (15)$$

where $(I_{M^+})_{min}$ is the minimum detectable ion current after correction for isotopic abundance. A signal is considered to be detected so long as its magnitude is at least three times the standard deviation of the background. With a properly designed ion detector, a background of less than 1 count/sec can be achieved. The sample consumption is determined by I_p and S. If no restrictions are imposed, the maximum I_p available can be utilized, and S can then be increased somewhat by judicious selection of the mass, energy, and incident angle of the primary ions. β_{M^+} may be increased substantially by a correct choice of bombarding species and/or the adsorption or deposition of a suitable species onto the sample surface during the analysis. η is optimized by adequate instrumental design and by ensuring that the secondary ions are collected from the entire bombarded area. In addition, the detection limit may be improved by an appropriate choice of the secondary ion to be monitored, i.e., positive or negative, atomic or molecular.

Table 5.IV shows some calculated values of $(c_M)_{min}$, in the ideal situation of the absence of spectral interference, when using a primary ion current of 1 μA. It should be remembered that reduction of I_p will increase these values proportionately.

TABLE 5.IV

Calculated Values for the Lowest Detectable Concentration from 1 μA Ar^+ Bombardment (assuming $(I_{M^+})_{min} = 2 \times 10^{-18}$ A ($\triangleq$ 12 counts/sec) and $\eta = 10^{-3}$, and taking $S\beta_{M^+}$ from Table 5.II.

Elements	$S\beta_{M^+}$	ppma
Cu	0.0003	7
Ni	0.0006	13
Al	0.007	0.3
Oxygen-covered	$S\beta_{M^+}$	ppba
Cu	0.007	300
Ni	0.045	44
Al	0.7	3

In practice, material limitations may well have to be taking into account, due either to a restricted area (microspot analysis) or to a restricted sample thickness (surface analysis). Therefore, considerations of *sample consumptions* during the analysis become important. We rewrite equation (12) in an alternative form to give the rate of sample volume removal, $\dot{V}$ (cm³/sec);

$$\dot{V} = (i_{M_i^+}/f)/(Nc_M\beta_{M^+}\eta) \quad (16)$$

where, $i_{M_i^+}$ = number of ions of the *isotope* M_i of the element M registered at the detector per second (counts/sec); f = fractional isotopic abundance; N = sample atomic density (atoms/cm³), which is equivalent to $\rho\, N_o/M$ where ρ is the sample density (gram/cm³), N_o is Avogadro's number, and M is the atomic weight. c_M, β_{M^+}, and η have already been defined.

For *uniform* erosion, we may write

$$\dot{V} = \dot{z}\, A_b \quad (17)$$

where $\dot{z}$ = erosion rate (cm/sec) and A_b = effective bombarded area (cm²). Further, the total volume consumed during the analysis, V(cm³), is

$$V = \dot{V}\, t_a \quad (18)$$

where t_a is the time for the analysis (sec). For instance, t_a can correspond to the count time at each individual mass number or the time to scan through the complete mass range (typically 200 sec).

We use Eqs. (16)-(18) to discuss the following points:

1. To collect a certain number of ions, we must remove a certain amount of material. For instance, suppose we require 100 ions/sec from a monoisotopic element (f = 1), present at a concentration c_M of 1 ppm in a sample of density $N = 6 \times 10^{22}$ atoms/cm³, i.e., ρ/M = 10. If $\beta_{M^+} = 10^{-2}$ and $\eta = 10^{-3}$, then with a bombarded area of 100 μm × 100 μm, we must erode the sample at a rate $\dot{z}$ of

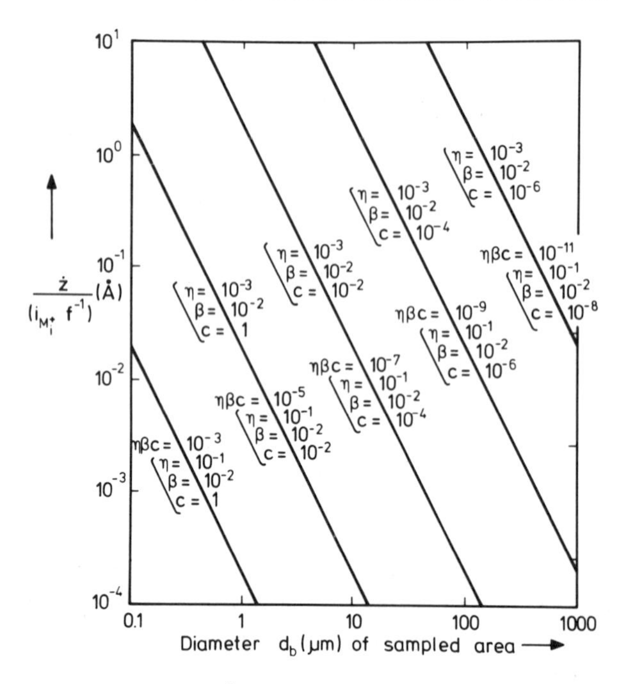

Fig. 5.14. Plot of uniform erosion rate $\dot{z}$ (Å/sec) divided by count rate ($i_{M_i^+}/f$) (counts/sec) for circular beam of diameter d_b, for various combinations of ($c_M \beta_{M^+} \eta$); f is the fractional isotopic abundance, c_M is the fractional atomic concentration, β_{M^+} is the degree of ionization, and η is the instrumental factor. Sample atomic density $N = 6 \times 10^{22}$ atoms/cm^3 (221). (Courtesy of *Journal of Physics*.)

≈170 Å/sec. Fig. 5.14, taken from Liebl (221), we can estimate the sample consumption rate, in relation to the count rate desired, for various combinations of (c_M, β_{M^+}, η).

2. The precision of measurement (339,354B,262A) depends on the number of ions counted. The relative standard deviation, σ_r, is given by $\pm\ (i_{M_i^+} t_a)^{-1/2}$. For example, a precision of ±10% requires 100 ions, whereas 2500 ions are needed for a ±2% relative standard deviation. For isotopic ratio measurements, the precision required is typically ±0.1%, which means that 10^6 ions are required. To obtain a specific precision, therefore, a certain minimum volume of sample must be sputtered away. For the above example, this volume equals 1.7×10^{-10} and 4.3×10^{-9} cm^3 for a σ_r of ±10% and ±2%, respectively.

3. Table 5.V shows how much of the sample requires to be sputtered away in a fairly typical situation to attain a desired detection limit. It can be seen just how incompatible are the desires for minimum sample destruction and low detection limits. In this context, we must try to ensure *efficient use of sputtered material*:

Wastage results from a mass scan in which only one collector for the ion currents is employed, thus permitting only one mass at a time to be registered. Detection either by photoplate or by electrical multicollection should remedy this. The decision

TABLE 5.V

Calculated Interrelations (for uniform sputtering parallel to the surface) between Primary Current I_p, Primary Current Density j_p, Erosion Rate $\dot{z}$, Total Amount W of Material Sputtered in 200 sec, Total Layer Thickness Removed $\dot{z}t_a$, and Minimum Detectable Concentration $(c_M)_{min}$

$I_p(A)$	$j_p(\mu A/mm^2)$	$\dot{z}$(Å/sec)	W (gram)	$\dot{z}t_a(\mu m)$	$(c_M)_{min}$ (ppma)
10^{-8}	1	2	2×10^{-9}	0.04	2
10^{-7}	10	20	2×10^{-8}	0.4	0.2
10^{-6}	100	200	2×10^{-7}	4	0.02

Assumed values of parameters: $A_b = 100\ \mu m \times 100\ \mu m$ (effective bombarded area); $S = 2$ (sputter yield); $\rho = 5$ gram/cm^3 (sample density); $N = 6 \times 10^{22}$ atoms/cm^3 (sample atomic density); $t_a = 200$ sec (analysis time); $\beta_{M^+} = 5 \times 10^{-2}$ (degree of ionization); $\eta = 10^{-3}$ (instrumental factor); $(I_{M^+})_{min} = 2 \times 10^{-18}$ A [minimum detectable ion current, i.e., $(i_{M_i^+})_{min} = 12$ counts/sec, $f = 1$].

TABLE 5.VI

Comparison of Limits of Detection with Single-collector Electrical Detection (multiplier) and Photoplate Detection for Different Amounts of Material Sputtered

Volume V (cm^3) sputtered in $t_a = 1$ sec	Number of particles in V	Number of ions collected per sputtered atom	Limit of detection: Multiplier (1 line)	Limit of detection: Photoplate (ca. 100 lines)
2×10^{-12}	10^{11}	10^{-5}	10 ppm	10^4 ppm
2×10^{-9}	10^{14}	10^{-5}	10 ppb	10 ppm

between photographic and electrical detection in this context depends on the minimum sample consumption allowable, the number of elements to be analyzed and the required detection limit (see Table 5.VI). Single-collector electrical detection has in its favor the lower limit since only about 10 ions/sec are needed, but it is restricted to one element at a time. The photoplate, needing about 10^4/mm^2 for a barely detectable line with typical values of secondary ion mass and energy, has the higher detection limit but measures all ions throughout the spectrum at the same time. The use of multicollector electrical detection (e.g., oblong channel plates with a sufficiently small diameter) may shift the balance in favor of electrical detection (307)(see Section III.D.1.).

C. BASIC EXPERIMENTAL APPROACH

1. Samples

Sample cleaning may be performed in situ by ion etching. Hygroscopic samples are first dried in a dessicator and then quickly transferred to the vacuum chamber so as to avoid a lengthy pump-down period. A typical procedure for biological tissues

(27C) involves fixation, dehydration, imbedment in paraffin, sectioning, deposition on gold discs, removal of paraffin with solvents, and drying. High-vapor-pressure solids might prove amenable to analysis when the sample stage is cooled to liquid nitrogen temperatures. Samples in the form of chips can be pressed onto a metal foil for mounting purposes, and then examined directly. Powder samples are usually pressed into pellets, sometimes mixed with spectroscopically pure silver or graphite powder for adhesion. The latter procedure provides one possible method for avoiding charge build-up in insulator specimens due to ionic bombardment; such charging leads to instability, reduction or even complete suppression of the secondary ion currents. Other methods used for thick insulators and fairly narrow primary beams include:

1. Employment of neutral primary beams (48,96), which can only circumvent charging if the ejection of secondary electrons is prevented. However, neutral beam bombardment may help to reduce any distortion of existing concentration profiles, caused by the field-enhanced migration of ions released inside the solid insulator by irradiation with charged-particle beams (247).

2. Bombardment with metal ions for continuous generation of a conducting layer (388); a wise choice of the metal will enhance the negative ion yield considerably.

3. Deposition of a conducting film or grid onto the insulator surface (354) which, in the case of positive ion bombardment, limits the analysis to the sputter lifetime of the overcoat.

4. Heating the insulator to increase its conductivity (7). With photoconductors, direct light onto the sample has proved beneficial (415). The most widely-adopted methods are:

5. Compensation of positive charging through the use of electrons (49,149,276). Critical adjustment is required to obtain the negative secondary-ion spectrum (273).

6. Use of negative primary ions (10), the negative charging being compensated by the departing secondary electrons. An auxiliary conducting electrode in the sample region, such as a metal diaphragm lying on the sample surface (411), is necessary to draw the secondary electrons away from the bombarded region.

2. Vacuum Requirements

a. Surface Contamination

To reduce surface contamination, the erosion rate $\dot{z}$ and hence the primary ion current density, j_p, must at least exceed the residual gas adsorption rate. In a pressure of 10^{-6} torr of an active gas, approximately one monolayer is adsorbed per second. Therefore, we deduce from Fig. 5.7 that the above wish is approximately fulfilled if

$$j_p(\mu\text{A/cm}^2/p(\text{torr}) \geqslant 10^8 \qquad (19)$$

where p is the background pressure in the sample surface region. To analyze very thin layers, $\dot{z}$ must be maintained small and therefore surface contamination has to be reduced by minimizing p. A bakeable UHV system is capable of reaching a background pressure of, at least, 10^{-10} torr.

However, irrespective of the desire to perform a true surface analysis or a thin-film analysis, the necessity of having a clean vacuum environment cannot be overemphasized. Many SIMS systems are wanting in this respect. Even a slight amount of surface contamination can obscure trace element detection due to spectral coincidences. In particular, the system base pressure should be free from hydrocarbons (oil-free vacuum pumping) and from water (bakeable apparatus). Hydrocarbon adsorption leads to the appearance of a large number of mass peaks (of greater intensity in the negative spectrum) containing hydrogen and carbon. The presence of hydrogen-containing ions in the spectrum, presumably from the adsorption of water vapor, is particularly disturbing; an MH^+ ion can prevent the detection of the next highest mass number element, and the already numerous types of molecular ions are increased substantially by the incorporation of one or more hydrogen atoms.

When the primary gas is introduced into a differentially pumped ion source (at typically $10^{-4}-10^{-3}$ torr), the base pressure in the target chamber nevertheless still rises due to leakage from the source. A further increase may occur when the primary beam is utilized, the pressure then increasing with increasing primary ion current. Thus, due to the fairly high pressure in the sample region *during* the measurement, some residual gas adsorption is likely. Surface contamination is minimized by having a bakeable source and by using a source gas of high purity. The above considerations impose stringent experimental demands upon a trace analysis for H, C, N, O, etc.

We have already noted how the presence of a chemically reactive gas can affect ion yields drastically. If such a gas is introduced deliberately into the vacuum chamber to increase sensitivities, it should be pure and dry. Exchange with adsorbed gases on the chamber walls can be minimized by continuous pumping (no release of previously pumped gases from the pump is then required), and by squirting the gas directly at the target. This latter trick enables one to use higher tolerable working pressures in the sample region.

b. VARIATION OF SECONDARY-ION CURRENTS WITH PRESSURE

For Ar^+ bombardment of metallic specimens in vacuum, the secondary ion current per primary ion is often dictated by the residual gas adsorption rate which, in turn, is dependent on the erosion rate. Also, nonuniform erosion leads to increased adsorption at the periphery of the bombardment area and thereby to enhanced secondary ion yields from this region. Fig. 5.15 shows the Al^+ ion image obtained by Ar^+ bombardment of an Al sample. At the circumference of the focused beam, enhancement of the Al^+ current through active residual gas adsorption is clearly apparent. Therefore, the relatively small peripheral region can contribute disproportionately to the total signal. It is thus often difficult to obtain meaningful, reproducible and/or stable quantitative measurements from Ar^+ bombardment of metals in vacuum (9).

Yields are increased by oxygen bombardment in vacuum; the signal per primary ion is dependent upon the oxygen sample concentration in the near-surface region, which is governed both by implantation and by residual gas adsorption. (The residual gas content is determined essentially by leakage from the ion source). This may not be

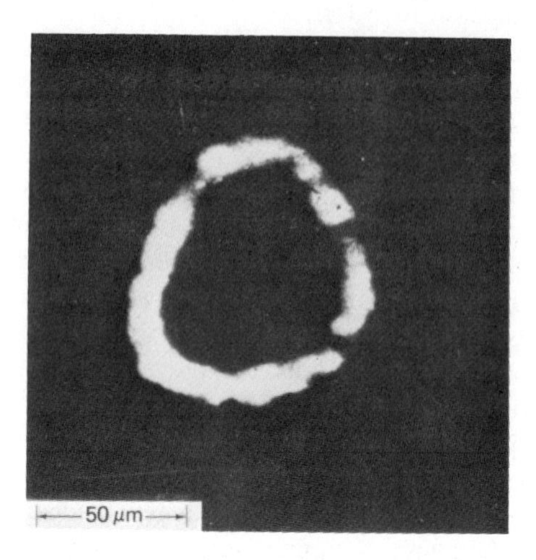

Fig. 5.15. Al^+ image from Ar^+ bombardment of an Al sample, obtained in an ion microscope, showing the enhanced Al^+ yield at the periphery of the primary ion beam due to residual gas adsorption (251). (Courtesy of C.A. Evans.)

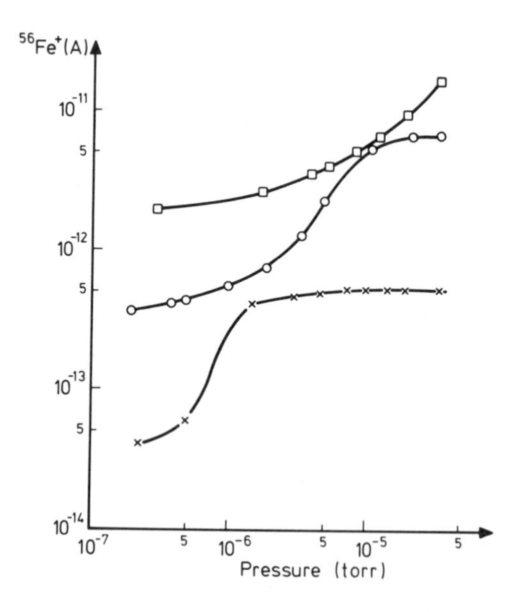

Fig. 5.16. Intensity of Fe^+ secondary ions from 6-keV O_2^+ bombardment of iron versus nominal residual oxygen pressure at different primary current densities: (□) 21 μA, spot of diameter ≈1 mm; (○) 2.1 μA, spot of 300-μm diameter rastered over an area ≈1 mm^2; (X) 1 μA, spot of 150-μm diameter rastered over ≈6 mm^2. Secondary-ion currents measured from central 300-μm diameter region of the eroded area (264). (Courtesy of *Applied Physics*.)

sufficient to maximize the secondary ion currents (Fig. 5.16); when O_2 is admitted into the target chamber, the $^{56}Fe^+$ current rises until a saturation value is reached at a certain oxygen pressure which depends on the erosion rate. Further, for a given erosion rate, the oxygen pressure dependence can vary drastically for different elements in the same metal (212,263,268); this applies both to the magnitude and to the rate of enhancement, as well as to the oxygen pressure at which a maximum value is achieved. Therefore, quantitative measurements should be carried out in a sufficient oxygen pressure to saturate *all* secondary ion currents (204,212). Saturated yields from oxide samples are generally achieved by oxygen bombardment in vacuum, but this is not necessarily the case with Ar^+ primaries (265,270).

To maintain a sufficiently long mean free path for the ions, the maximum O_2 pressure is limited to about $10^{-5}-10^{-4}$ torr. As can be seen from Fig. 5.16, this places a restriction on the erosion rate, and hence on the detection limit, if saturation conditions are to be achieved. In this context, it is preferable to use oxygen rather than Ar^+ bombardment so that a higher erosion rate may be used when working in the saturation region.

Oxygen flooding, besides increasing the positive ion yield, has other beneficial effects (see Section II.C.3a).

3. Primary Beam

a. SPECIES

The use of neutral primaries has been reported (97). These are obtained from a keV ion beam by cross-collision with a molecular beam in a charge exchange chamber. The kinetic energy of the original ions and therefore the sample sputter rate are essentially unchanged. Bombardment with neutral particles is useful for insulator specimens. Disadvantages are that (a) focusing into a very fine spot and (b) beam rastering for uniform erosion (and for ion imaging) are not easily achieved.

Ar^+ primaries were the original choice since fairly high sputter rates are obtainable and it is possible to obtain reasonable currents in a small spot. Drawbacks, particularly low ion yields, have been mentioned before. However, Ar^+ is still utilized (a) when an inert gas is required for the study of oxidation processes, to obtain chemical information etc., and (b) for ion imaging (see Fig. 5.26) to highlight topographical and phase differences and, through matrix effects, different chemical environments.

O_2^+, O^-, or N_2^+ primaries are commonly used to increase positive secondary ion yields. Negative primary ions are useful for insulators (10,411) but the maximum primary ion current obtainable is about 10 times less with O^- than with O_2^+. Cs^+ primaries are used to enhance negative secondary ion yields; Ar^+ bombardment with simultaneous Cs deposition (41) is another possibility. Cs (200) or Na (4) deposition may further increase the negative yield over that obtained by Cs^+ bombardment alone. This is similar to the situation with O_2^+ bombardment, when further O_2 admission is often necessary to maximize the positive ion yield.

After commencement of bombardment, an ion signal often increases due to the removal of surface contamination, before decreasing to an eventual constant level as the surface oxide layer is sputtered away. However, with active gas bombardment, the signal starts to increase again until a constant concentration of implanted projectiles is reached. This time-independent concentration is reached after an approximate depth of $R_p + 2\Delta R_p$ has been removed, where R_p is the mean projected range and ΔR_p its standard deviation (381). (For ions at normal incidence, the mean projected range equals the mean penetration depth). Therefore, the first 100 Å or so must be sputter-removed before constant signals are attained. This depth can be reduced by simultaneous O_2 flooding (214) to about 10 Å, or by minimizing R_p.

Nonuniform sputter erosion degrades the depth resolution in concentration profiling. For even erosion, it is helpful to amorphize the bombarded sample region. This can be accomplished by ensuring the presence of enough active gas to form oxide or nitride surface layers. N_2^+ (156) or O_2^+ (384) bombardment, or O_2 flooding (39,40) have been used successfully to obtain even erosion.

Heavier projectiles can increase the sample sputter yield (i.e., shorten analysis times) and decrease the mean penetration depth. A judicious choice of primary projectiles can reduce interference problems from molecular ions. For instance, if oxide ions prove troublesome, an analysis in an oxygen-free atmosphere may prove worth while. On the other hand, detection limits for a given element can sometimes be improved considerably by ensuring an abundant emission of one of its molecular ions.

b. Purity

Ionized impurities in the primary ion beam arise from the source gas, from source electrodes, and insulators. These impurities can be removed by primary ion mass separation, using either a sector-type mass spectrometer or a Wien filter (crossed electric and magnetic fields).

c. Incident and Azimuthal Angles

Both oblique and normal incidence is adopted. Oblique incidence increases the sputter yield and reduces the penetration depth. However, the likelihood of obtaining sufficient active gas implantation may be lessened with oblique incidence, so that additional gas adsorption may become more necessary.

The secondary ion yield varies with crystallographic orientation due to the orientation dependence of the sputter yield. Also, a given secondary ion signal originating from a monocrystalline sample not rendered amorphous by the bombardment, periodically increases and decreases with sample rotation (40,431). This reflects the variation of the sputter yield due to channelling.

d. Energy

A primary ion energy between a few hundred eV and 20 keV is normally selected. The ability to vary this energy permits the optimum choice for a particular analysis. The maximum I_p attainable generally increases with increasing primary energy, as does the sputter yield. Very low energies (< 500 eV) may lead to the lateral migration of surface atoms (375). The use of O_2^+ or N_2^+ is essentially equivalent to sputtering with atomic projectiles possessing half the initial kinetic energy.

Lattice damage from the primary particles and the ensuing collision cascades is increased when higher primary energies are used. A lower energy is particularly desirable to minimize knock-in effects (i.e., the pushing of the atoms of a certain element deeper into the sample by energetic primaries and recoils), which cause a degradation in depth resolution and a broadening of concentration profiles. Low energies are also desirable if lattice or structural information is required (57).

e. CURRENT DENSITY

A stable primary ion current, I_p, is vital for quantitative measurements. Improved detection limits are obtained through the use of a higher I_p. For a given I_p, a higher current density, $j_p = I_p/A_b$, produces a faster erosion rate, $\dot{z}$. For a surface analysis, $\dot{z} \approx 1$ Å/hr is desirable and j_p must correspondingly be chosen small. This is best accomplished by increasing the effective bombarded area, A_b, rather than by minimizing I_p so that favorable detection limits can be maintained. Depth profiling of thick films, on the other hand, necessitates the use of a high j_p for a fast erosion rate. Therefore, a smaller A_b is used. The requirements of maximum sensitivity along with both a low erosion rate *and* and a small A_b are incompatible.

f. DIAMETER

A focused beam is required for microspot analyses. The beam diameter also determines the lateral resolution in ion imaging obtained by the beam-rastering technique. Minimum beam diameters of 1–2 μm are in use. However, the maximum available primary current decreases fairly rapidly with decreasing beam diameter; with a spot diameter of a few μm, it is already in the nA range, resulting in poorer detection limits.

A focused ion beam is generally considered to have a Gaussian current density profile; see Fig. 5.17;

$$j = j_o \exp{(-r/r_s)^2} \tag{20}$$

The maximum current at the center is j_o and r_s can be regarded as the nominal radius of the beam. The total current is $j_o \pi r_s^2$. (An ion beam with this current but of uniform current density, j_o, would have a radius r_s.) Such a focused beam lead to nonuniform erosion with the generation of a bell-shaped crater. Rastering the beam over a square area with a side approximately five times the beam diameter generates an acceptably flat-bottomed crater for profiling measurements.

For microspot location, microscopic viewing of the sample in its bombardment position is helpful. The desired analytical area can be positioned under the bombarding beam by fine mechanical adjustment of the sample holder (backlash $\approx$1 μm). Another possibility is to try and direct the primary beam onto the desired area by rastering and simultaneous imaging (either the total secondary ion or electron currents could be used for the latter). This however is a destructive process. An auxiliary focused and rastered electron beam with its attendant secondary electron micrograph is also useful for fine area location.

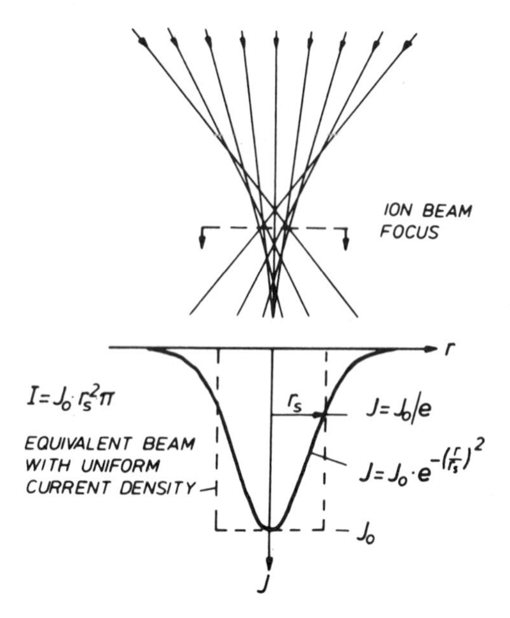

Fig. 5.17. Gaussian current density profile across a focused ion beam (222). (Courtesy of *Journal of Vacuum Science and Technology.*)

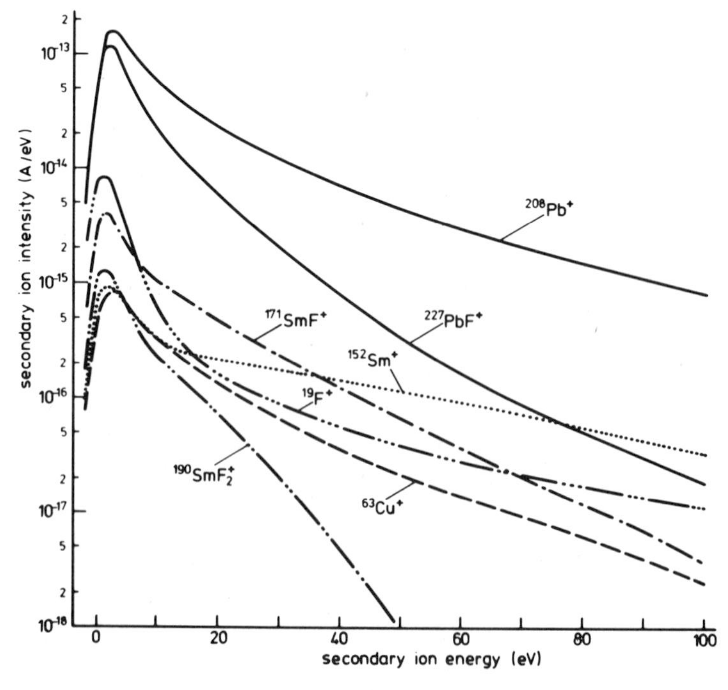

Fig. 5.18. Secondary-ion currents in A/eV from 6-keV Ar^+ bombardment of PbF_2 containing 0.3 at. % of both Sm and Cu as a function of initial kinetic energy (269). (Courtesy of *Journal of Chemical Physics*.)

4. Secondary Ions

The angular distribution of secondary ions (42,93,148) resembles that of sputtered neutrals; maximum emission occurs in a direction normal to the surface and drops off approximately with a cosine distribution. The energy distribution varies according to the type of secondary ion (Fig. 5.18) and can also be different for atomic ions in the same sample. Nevertheless, nearly all distributions peak sharply at a few eV; the high energy tails of atomic ions, however, can extend into the keV region (150,183). Efficient collection entails allowance for this angular and energy spread; this and contact potential differences can make precise measurements of secondary ion energy distributions rather exacting (26,91,199,359).

Broadening of the spectral peaks and a resultant loss in mass resolution would occur if secondary ions of all energies were to be transmitted through the spectrometer. For this reason, it is necessary to select an energy band with a width from a few eV up to about 100 eV, depending on apparatus design and required resolution. This band is usually located in the peak region of the energy distribution for maximum sensitivity. However, due to differences in these distributions, the positioning of the band may have to be varied for each individual mass peak to achieve maximum detection. This might also apply to the collection voltages, since light ions, for instance, leave the sample with relatively high velocities.

The mass spectrometer may be of the magnetic-sector, quadrupole, or time-of-flight type; mass-independent transmission is preferable. The mass range required can extend from unity up to perhaps 1000 amu if studies of molecular ions or high-molecular-weight organics are needed. The resolution should at least permit separation of all integral mass peaks. Automatic and rapid switching between preselected mass peaks is a very useful facility.

Signals of up to at least, say, 10^9 counts/sec need to be detected for both positive and negative secondary ions, with a detector background of < 1 count/sec. Peak shapes should be such that neighboring masses having a signal ratio of, say, 10^6, can be distinguished. Positive and negative secondary ions should preferably be generated under the same bombardment conditions, with comparable collection and detection efficiencies for either type of ion; fast interchange between the two spectral modes is useful.

D. INFORMATION AVAILABLE

1. Elemental Identification

In principle, all elements are detectable with SIMS to low detection limits; this limit in a given matrix may differ by as much as 10^6 for the different elements if positive ions only are monitored. However, by using the negative spectrum as well, this range can be reduced to about 10^3. In practice, mass spectral coincidences often prevent detection for atomic concentrations $\lesssim 10^{-3}$. It is not uncommon to find peaks at practically every mass number. Many elements have more than one isotope. Doubly charged and triply charged positive atomic ions are present; these generally pose no problems since they often occur at nonintegral mass numbers and in any case have relatively low intensities. Some molecular ions on the other hand are even more

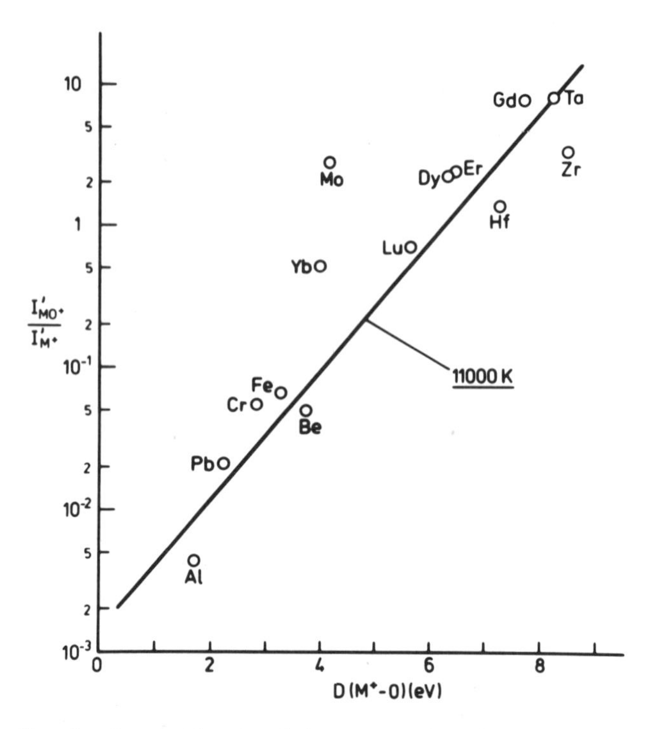

Fig. 5.19. Semilog plot of monoxide to atomic ion current ratio against the bond dissociation energy $D(M^+-O)$ for various elements in PbO and Bi_2O_3 matrices; 6-keV Ar^+ bombardment, 0–20 eV secondary ions monitored (269). (Courtesy of *Journal of Chemical Physics*.)

prevalent than the corresponding atomic ion, and thus may be utilized to improve detection limits (403A). Which molecular ions occur and their intensities depend in the first place upon the nature and the amounts of the sample constituents, the bombarding ions, and the residual gas. Fortunately, their intensities tend to decrease with increasing atomic complexity. Figure 5.19 illustrates the important role played by stability of the molecular ion. The exponential dependence upon the bond strength of the molecular ion is responsible for the large differences observed in molecular ion intensities (269).

Methods to circumvent spectral interferences include:

1. The use of natural isotopic abundances. For a multiisotopic element, one can attempt to find an interference-free mass peak. For instance, $^{60}Ni^+$ would be used to monitor Ni traces in iron due to the presence of $^{58}Fe^+$. When more than one type of ion contributes to a mass peak, the isotope stripping technique may prove useful; in the 56–60 mass range, for instance, one can expect major contributions in a mineral sample from Fe^+, Si_2^+, CaO^+, and $CaOH^+$. Thus, five equations containing four unknowns are obtained by assuming that the intensity of each mass peak is made up of the sum of the contributions from each type of secondary ion; these contributions are

equivalent to the total intensity of the particular secondary ion multipled by the relative abundance of its isotope at that particular mass number. The identification of an element can only be considered definite if its correct isotopic abundance has been found (either directly or indirectly after peak stripping).

2. Restriction of the analysis to secondary ions with higher initial kinetic energies. It can be seen from Fig. 5.18 that the spectrum would become progressively more dominated by atomic ions if measurements were to be restricted to, say, 60–80 eV ions rather than to 0–20 eV ions. This would be accompanied however by an overall loss in sensitivity. Mass spectra from an Al–Mg alloy obtained by using low- and high-energy secondary ions respectively are shown in Fig. 5.20*a* and *b*). The first spectrum contains a large number of polyatomic ions. However, when the energy acceptance of the instrument is shifted so that only higher energy secondary ions are

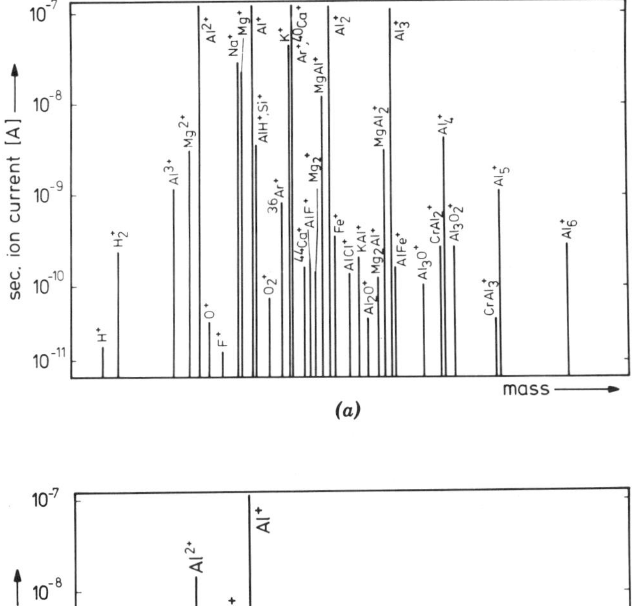

(*a*)

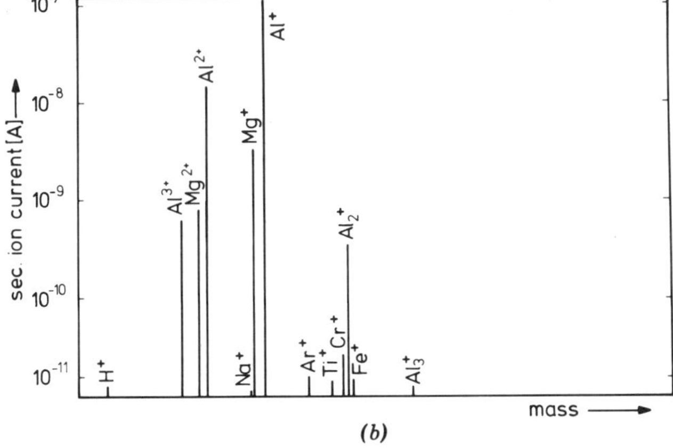

(*b*)

Fig. 5.20. Mass spectrum of Al–Mg alloy bombarded with 12 keV Ar^+ (150); analyzer adjusted to accept only (*a*) low-energy secondary ions, (*b*) high-energy secondary ions. (Courtesy of *Radiation Effects*.)

TABLE 5.VII

Mass Resolution Required to Separate Desired Analytical Ion from an Interfering Ion (20)

Interference type	Interfering ion	Analytical ion	Required resolution
Multiply-charged matrix ions	$^{28}Si^{2+}$	$^{14}N^{+}$	950
	$^{62}Ni^{2+}$	$^{31}P^{+}$	3,200
Matrix polymer ions	$^{16}O_2^{+}$	$^{32}S^{+}$	1,800
	$^{28}Si_2^{+}$	$^{56}Fe^{+}$	2,950
Matrix plus projectile ions	$^{63}Cu_2^{65}Cu^{16}O^{+}$	$^{207}Pb^{+}$	1,050
	$^{29}Si^{30}Si^{16}O^{+}$	$^{75}As^{+}$	3,250
	$^{27}Al^{16}O_2^{+}$	$^{59}Co^{+}$	1,500
Hydride ions	$^{30}SiH^{+}$	$^{31}P^{+}$	4,000
	$^{54}FeH^{+}$	$^{55}Mn^{+}$	3,300
	$^{120}SnH^{+}$	$^{121}Sb^{+}$	19,500
Hydrocarbon ions	$^{12}C_2H_3^{+}$	$^{27}Al^{+}$	650
	$^{12}C_5H_3^{+}$	$^{63}Cu^{+}$	650
	$^{12}C_2H_2^{+}$	$^{12}C^{14}N^{+}$	2,000

monitored, almost all polyatomic ions disappear and a number of trace elements may now clearly be distinguished.

3. The use of high mass resolution; the true isotope mass differes slightly ("mass defect") from its nominal integral value. If the mass resolving power of the spectrometer is greater than the ratio of the integral mass to the mass defect, then two isotopes at the same integral mass can be separated. Table 5.VII shows that a resolution of 4000–5000 should be adequate in most cases. Increasing resolution, however, also results in a loss in sensitivity (309,418).

5. Skill and experience of the analyst; careful choice of the bombarding species and of the secondary ion monitored, and control of the vacuum environment can reduce spectral interferences. For instance, N_2^+ bombardment can be used to achieve high positive atomic ion yields if the occurrence of oxide ions would lead to interferences. On the other hand, O_2 flooding (264,302) can appreciably reduce the intensities of molecular ions such as $M_3O_n^+$, M_n^+ and $(M_nM'_m)^+$, where M' signifies a second element. Utilization of both the positive and negative spectrum often enables more elements to be detected; in the negative spectrum, metallic elements can often be monitored via their oxide ions. The detection limit for As in Si (see Table 5.VII) has recently been lowered to ~50 ppb (421) by recording the $AsSi^-$ species under Cs^+ bombardment; the use of Cs considerably enhances the As^- and $AsSi^-$ yields while at the same time minimizing interferences from molecular ions containing silicon and oxygen.

2. Chemical and Structural

a. Fingerprint Spectra

Under a given set of bombardment and measuring conditions, the mass spectrum of each element and compound is characterized by the relative abundances of its atomic and molecular ions. Fig. 5.21 shows the fingerprint spectra of Si, CoSi, and Co_2Si resulting from 5.5 KeV Ar^+ bombardment of these samples in vacuum. Also

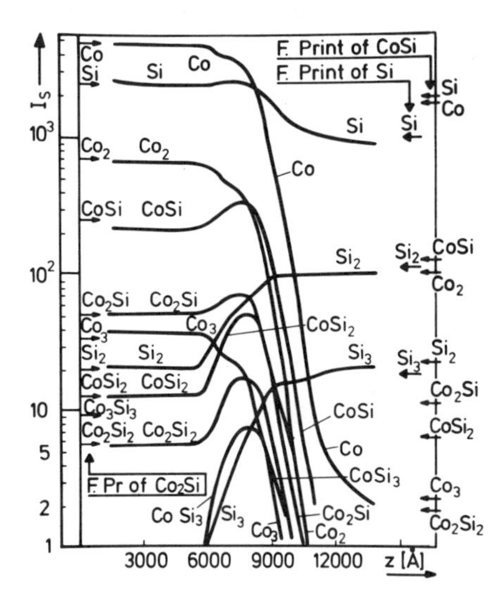

Fig. 5.21. Positive secondary ion currents as a function of depth from a cobalt-silicon layer on a Si substrate. Arrows on the left- and right-hand sides indicate the fingerprint spectra of Co_2Si, CoSi, Si, obtained from pure samples of these materials.

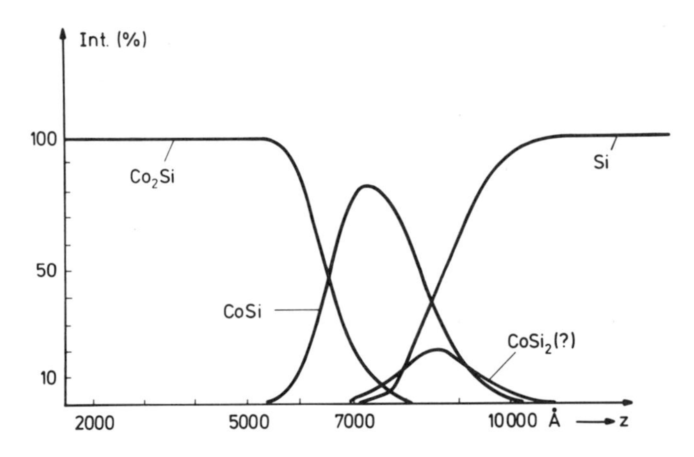

Fig. 5.22. Data of Fig. 5.21 reduced by application of the fingerprint spectra supposition theory. As a result, the concentration of various cobalt–silicon layers as a function of depth have been obtained.

shown is the variation in secondary ion intensities during an in-depth concentration profile of a Co_xSi_y layer on a Si substrate. From a comparison with the previously determined fingerprint spectra, one can deduce that the outermost region of the layer comprises Co_2Si. By assuming that if more than one phase is present, the resulting spectral intensities are made up of contributions from each phase weighted in proportion to the concentration of that phase (403,404), one can then derive the phase distribution shown in Fig. 5.22.

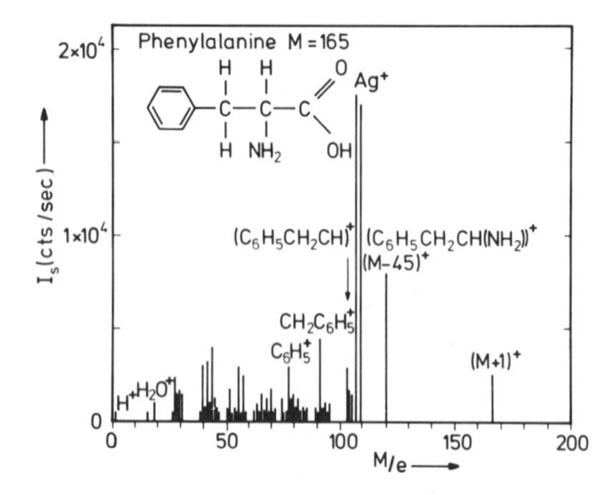

Fig. 5.23. Positive secondary ion spectrum from phenylalanine on an Ag foil (33); 2.25-keV Ar^+, $I_p = 2 \times 10^{-10}$ A, $A_b = 0.1$ cm^2, $p = 10^{-8}$ torr. (Courtesy of *Applied Physics*.)

b. SURFACE ATOM ARRANGEMENTS

Molecular ion emission is used to deduce the short-range structural order of atoms on, in, or just below the sample surface (32,34,57,105). A molecular ion is assumed to contain only those atoms that had occupied adjacent sites at the surface. Static SIMS is necessary to minimize the accumulative effects of atomic mixing; the use of a very low primary ion current density helps to ensure that each primary ion encounters an unperturbed surface region (34). Each collision cascade in SIMS extends over a projected surface area of typically 100 Å in diameter. Thus, the place of origin of a molecular ion is generally far removed from the point of disturbance at which the energetic projectile enters the lattice. However, the extent to which the composition and structure of the emission area are altered before the secondary ion is eventually ejected from it remains uncertain. Also, it should be remembered that the presence or absence of molecular ions is not merely a result of stability considerations (see Fig. 5.19).

c. STRUCTURE AND MOLECULAR WEIGHTS OF ORGANIC OVERLAYS

Fig. 5.23 shows the positive static SIMS spectrum from an Ag foil which had been dipped into an aqueous solution of phenylalanine. The molecular weight can be determined from the $(M + 1)^+$ protonated parent peak $(M - 1)^-$ in the negative spectrum. Further, some information on the molecular structure may be obtained from the fragment ions. For instance, the $(M - 45)^+$ peak derives from removal of the COOH group, the latter in fact being prominent at mass 45 in the negative spectrum (33).

3. Quantitative Analysis

To obtain c_M from the measured I_{M^+}, I_p, and S, Eq. (12), one needs to determine η absolutely and to know β_{M^+}. However, more often than not, relative concentrations suffice. From equation 12, we may write

$$c_M/c_R = (I_{M^+}/I_{R^+})\,(\beta_{R^+}/\beta_{M^+}) \qquad (21)$$

assuming that I_p, S, and η are the same for the monitored isotope of the reference element R. By summing the concentrations to 100%, the sample composition can be obtained, providing that all constituents have been determined. Qualitative analysis then boils down to a knowledge of the relative degrees of ionization, β_{M^+}/β_{R^+}. Measurements with pure elements are unlikely to be of much use because of strong matrix effects.

a. USE OF EXTERNAL STANDARDS

Figure 5.24 shows some calibration curves for the determination of Si, Cr, Ni and Mn in steel alloys (342). Note the linearity extending over several decades of concentration. Calibration plots always turn out to be linear (providing that the standards are homogeneous!) up to concentrations of some %, after which curvature may result due to the changing matrix. This makes SIMS analyses more difficult in the higher concentration range; however, in this instance, the use of alternative analytical techniques that are less influenced by matrix effects is advisable. The prime advantage of SIMS after all lies in its trace element capabilities. The calibration plots may safely be extrapolated into the ppb region, thus circumventing the difficulty of

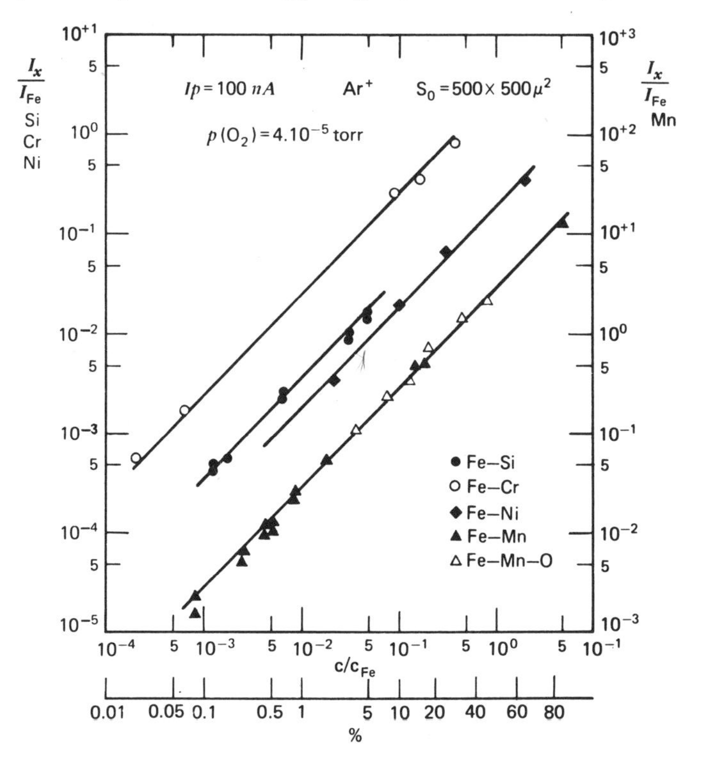

Fig. 5.24. Calibration curves from the analysis of various elements in steel samples (342). (Courtesy of *Le Vide*.)

obtaining standards at these concentrations. In fact, because of the linearity of the plots, measurements with just one standard may suffice, providing that this standard is well characterized and homogeneous. Standards commonly used have been characterized by bulk analytical methods which at best can certify homogeneity only down to a volume corresponding to the milligram region. Homogeneity testing with SIMS is limited by statistical fluctuations in ion counting (Section II.B.5); for instance, to obtain a precision of $\pm 2\%$ for an element present at 1 ppm, we have shown that a typical sample consumption of 2×10^{-8} grams is necessary. The maximum consumption rate achievable in a SIMS instrument is about 10^{-9} grams/sec. The quantity of bulk standard that needs to be sampled in a SIMS analysis to establish homogeneity on a microscale has been discussed (339).

b. USE OF RELATIVE ELEMENTAL SENSITIVITY FACTORS

McHugh (252) and others (358) have measured values of $(I_{M^+}/c_M)/(I_{R^+}/c_R)$ for an element M contained in a number of standards of widely different matrix compositions (but possessing a common reference element), under a given set of experimental conditions. From this set of results, an attempt is made to select the sensitivity factors most appropriate to the matrix under investigation. Pivin et al. (298) have shown how the relevant factors for an alloy may be estimated from those pertaining to the pure metals.

c. USE OF INTERNAL STANDARDS

Andersen and Hinthorne (13,15) proposed the utilization of a Saha-Eggert type equation to describe secondary ion emission;

$$n_{M^+}/n_{M^0} = K\ [Z_{M^+}(T_i)/Z_{M^0}(T_i)] \exp(-E_M/kT_i) \qquad (22)$$

where n_{M^+}, n_{M^0} are respectively the numbers of M^+ ions and M^0 neutrals emitted per second from a sample containing the element M. K is possibly of the form $A\exp(B/kT_i)$, and is assumed to be the same for all elements in the sample. Z represents an electronic partition function whose magnitude is temperature dependent. E_M is the first ionization potential of M, k is Boltzmann's constant, and T_i is a matrix-dependent fitting parameter called the ionization temperature (263) to distinguish it from the macroscopic temperature. The precise form and origin of equation 22 remain under discussion.

Now the total number of M-containing species sputtered per second, n_M, is given by

$$n_M = n_{M^0} + n_{M^+} + n_{M^-} + n_{MM_1^0} + n_{MM_1^+} + n_{MM_1^-} + \ldots\ldots \qquad (23)$$

How many of these terms are obtainable or are in fact necessary for analytical purposes? Andersen and Hinthorne (13) tried to take into account, through the use of equation 22 and its like, all sputtered atomic, monoxide, and dioxide species and thereby obtained a semiquantitative analysis (concentrations specifiable to within a factor of 2 or 3) from the positive secondary ion spectrum. However, lengthy computations are involved. Others (320,345,346,352) demonstrated that no loss in analytical accuracy resulted if only the first two terms on the right-hand side of equation 23 are taken into account. Therefore, combining this approximation with equations 21 and 22, relative concentrations are obtainable from

$$\frac{c_M}{c_R} = \frac{I_{M^+}}{I_{R^+}} \left[\frac{1 + Z_{M0}(Z_{M^+}K)^{-1} \exp{(E_M/kT_i)}}{1 + Z_{R0}(Z_{R^+}K)^{-1} \exp{(E_R/kT_i)}} \right] \tag{24}$$

As in reference 13, a computer search is made for the K and T_i values leading to the best match of calculated to known concentrations for at least two constituents, i.e., a minimum of two internal standards is still required.

The simplest approximation is to consider only the first term in equation 23, leading to

$$\frac{c_M}{c_R} = \frac{I_{M^+}}{I_{R^+}} \cdot \frac{Z_{M0}Z_{R^+}}{Z_{M^+}Z_{R0}} \cdot \exp\left(\frac{E_M - E_R}{kT_i}\right) \tag{25}$$

Since only one fitting parameter T_i is now involved, only one internal standard is required. Semiquantitative analyses are also possible with this approach (231,263,265,267,270). Figure 5.25 shows some T_i values derived from equation 25

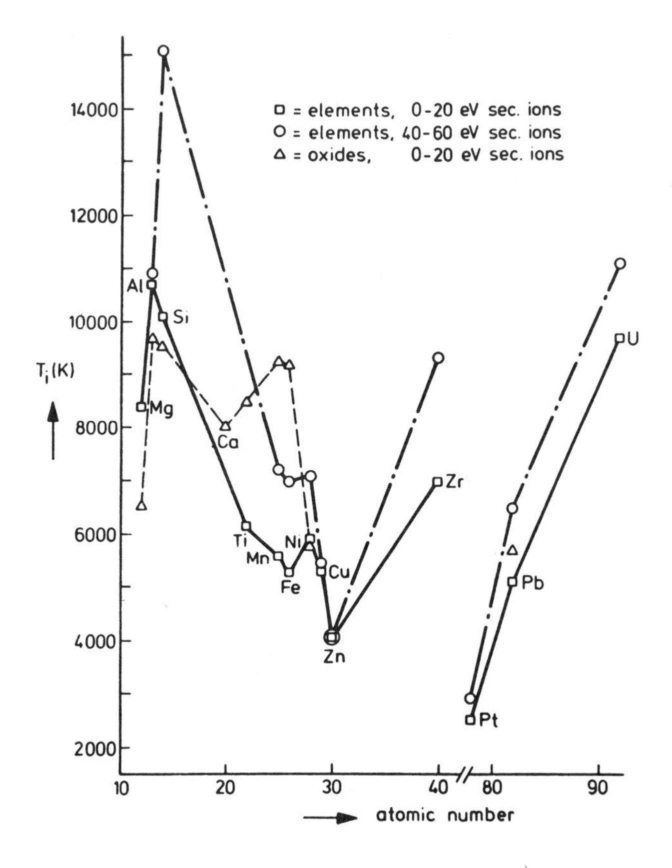

Fig. 5.25. T_i values with enough oxygen present to ensure saturated positive secondary-ion emission (270). Element—the sample matrix before bombardment comprises essentially a single element. Oxide—the matrix contains predominantly one other element besides oxygen. (Courtesy of *Mikrochimica Acta*.)

by using standard samples (270). Notice that the T_i value is dependent upon which energy range of the secondary ions is monitored, and also that values from mineral samples are not necessarily the same as those found with elemental samples in the presence of oxygen. Prior knowledge of T_i permits an analysis without the necessity for internal standards.

All these analytical methods based on equation 22 lead to poor results (a) for the very electronegative elements and (b) in those instances when a monoxide ion current, I_{MO^+}, is large in relation to I_{M^+}. The latter difficulty can be circumvented (a) by using empirical correction factors based on the measured I_{MO^+} to I_{M^+} ratio (15,263,268,270) or (b) by restricting measurements to higher-energy secondary ions (263,265,268,270) so that $I_{MO^+} \ll I_{M^+}$ for all elements (cf. Fig. 5.18).

4. Elemental Mapping

With the mass spectrometer tuned to a particular mass number, an elemental distribution across the surface can be obtained either by using the (scanning) ion microprobe or the direct-imaging approach (221). In the *ion microprobe*, a small diameter primary ion beam (typically 1–2 μm) is scanned across the surface. The secondary ion current measured at the detector is displayed as a modulation on a cathode ray tube, scanned synchronously with the primary beam. The lateral resolution in the resulting ion image is determined by the beam diameter. In the *ion microscope*, a relatively large sample area is simultaneously bombarded and the point of origin of the given secondary ion is viewed via the ion optics, giving an image obtained in a comparable fashion to that with a conventional optical microscope. The image is made visible on a fluorescent screen or on a photoplate. The lateral resolution is now controlled by aberrations in the ion optics and is typically ≤ 1 μm (316).

An ion image of $^{40}Ca^+$ from a ferrite sample is shown in Fig. 5.26. One must use

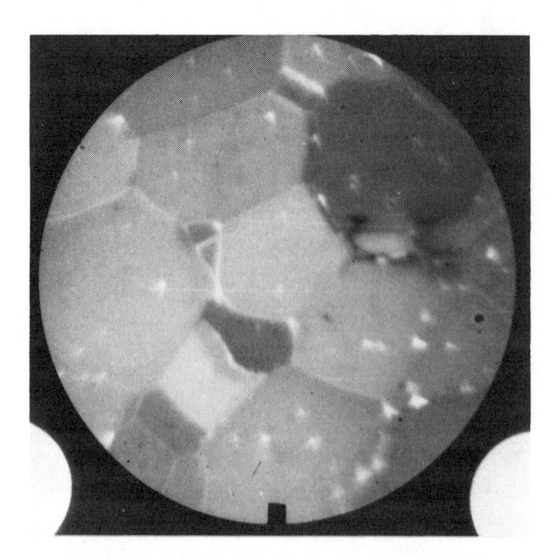

Fig. 5.26. $^{40}Ca^+$ image from a ferrite obtained from an ion microscope using 5.5-keV Ar^+ bombardment.

caution in interpreting such images since brightness contrast is not necessarily synonomous with differences in concentration across the field of view; as is evident from equation 12, variations in the secondary ion current, I_{M^+}, can arise from other causes than changes in the concentration, c_M. Some of these effects are visible in Fig. 5.26; contrast due to the dependence of the sputter yield S upon crystallographic orientation can clearly be seen. Also, the bright spots arise from a higher degree of ionization, β_{M^+}, owing to the matrix effect. Comparison with the ion images of other constituents should reveal the element or elements responsible for this increased ion yield. With rough surfaces, the mass spectrometer acceptance for a given secondary ion may not be uniform across the surface; further, the extent of this nonuniformity may differ for each element on account of differences in secondary ion energy distributions. With glancing incidence, nonuniform bombardment of an uneven surface due to shadowing effects will lead to topographic contrasts. Other problems associated with samples containing nonconducting areas have been reported (332).

Most of these artefacts can be eliminated from the ion images (a) by oxygen bombardment and flooding, or (b) by normalizing the secondary ion current either to that of the matrix element or to the total secondary ion current.

Ion imaging at various depths permits a complete three-dimensional characterization of the sample.

5. In-Depth Concentration Profiles

The depth distribution of a particular element is obtained by monitoring a given secondary ion signal as a function of sputter time. With a constant erosion rate, the time scale may be converted into depth by simply measuring the final crater depth, either interferometrically or by using a mechanical stylus. The change of sputter yield until steady-state conditions are reached may be important for very thin layers. Also, for sandwich layers, the depth scale is not simply proportional to sputter time.

We have already remarked that variation of a secondary ion current is not always a reliable indication of concentration changes. This applies especially when profiling through different matrices and through the near surface region, owing to removal of the surface contamination layer and to implantation artefacts. Oxygen flooding may help to minimize intensity changes resulting from the latter effect (214); it may also help to promote uniform sputter erosion (39,40). It is always advisable to monitor several mass peaks during a depth profile, especially those of oxygen (or cesium) and the matrix elements.

We have mentioned that a focused beam of Gaussian profile erodes a bell-shaped crater, which means that the secondary ions originate from different depths (Fig. 5.17). Efforts are therefore made to limit acceptance to those ions that originate from the central flat portion of a larger sputtered area, formed by defocusing or by rastering the primary beam. When raster-scanning a Gaussian beam, the uniformity of the summed overlapping current density will be nearly 0.1% if the line spacing is chosen to be r_s (equation 20). The limited acceptance is accomplished (222,251) with a mechanical and or an electronic aperture—the detector electronics are gated on and off synchronously with the rastered beam such that the counts are obtained only when the beam is in the central portion of the rastered area. The latter scheme may not prove entirely satisfactory because of (a) the nonfocused component of the primary beam (251) made up of energetic neutrals and scattered ions, and (b) the many high-energy sputtered neutrals and back-scattered primaries in the bombardment vicinity (250). All of these particles may cause secondary ions to be released from surface areas outside of that being covered by the raster scan, and these ions will be detected when the electronic gate is open. This may create problems when profiling from a high surface to a low-bulk concentration. Energetic neutral particles formed by charge-exchange collisions in the ion gun may be separated from the primary beam by deflecting the beam off-axis. However, the other sources of potential trouble remain. Thus, it may be necessary when employing an electronic aperture to limit the secondary ion acceptance area by using a mechanical aperture or by suitable design of the extraction optics (120,196).

The *depth resolution* is the sample thickness Δz, which must be sputtered away before the secondary ion signal from an element, distributed in depth according to a step function, changes by a specified amount; a possible definition is the distance (on the depth scale) between 84% and 16% of the experimentally determined step height (± 1 standard deviation from the half maximum value). The ultimate depth resolution limit is about 10 Å, controlled by the secondary ion escape depth and by the unavoidable development of surface topography on an atomic scale due to the random nature of the sputter process (160). In practice, the depth resolution is usually limited (155,161,162,163,243) by any of the following effects:

1. Instrumental: Uneven current distribution over the sputtered area will lead to a nonflat crater bottom. It is easy to show (406) in this case that the relative depth resolution, $\Delta z/z$, should be constant. Another instrumental effect is the resputtering of material from crater edges.

2. Initial surface roughness: A clean, flat and optically polished specimen is obviously desirable.

3. Included surface roughness: Sputtering of certain materials by certain projectiles can lead to changes in surface microtopography, particularly to the appearance of sharp cones (278A) with crystalline materials. Cone formation may be reduced by sputtering with a reactive gas (156).

4. Preferential sputtering: This will not influence the depth resolution per se. However it will induce a transitional period needed to establish a surface composition different from that of the bulk. Thus, transients in surface composition may be encountered when profiling through layers of substantially differing bulk compositions.

5. Atomic mixing (15C,144,337A,215A,158A); this may be classified into three categories, viz., cascade mixing, recoil implantation (or knock-on), and radiation-enhanced diffusion brought about by radiation-generated vacancies. Isotopic atomic displacements in the collision cascade cause the smearing of originally sharp features over a depth roughly equal to the mean projected range of the primary ions (144,175,250). The second category refers to those few atoms that receive large

momenta in the forward direction through direct impact with incoming projectiles. Their ranges are thus much larger than most of the displaced atoms. The degradation in depth resolution due to atomic mixing may be lessened (15C) by decreasing the primary ion range, i.e., by selection of a heavier projectile, by increasing the angle of incidence, and by decreasing the projectile energy. By measuring the profile at different primary energies, the profile without atomic mixing can be obtained by extrapolating the results to zero energy. In this fashion, Honig (164) calculated a depth resolution of 25 Å at a depth of 230 Å. This increased with projectile energy to 32 Å at 1.75 keV, 45 Å at 4.25 keV, and 105 Å at 7.75 keV. Similar results were reported by Andersen (15C).

6. Field-induced migration of atomic species in insulating matrices (McCaughan).

Four profiles from a garnet epilayer, measured (266) in the same run by automatic switching between preselected mass peaks, are shown in Fig. 5.27. Charge build-up was avoided by using negative primary ions and by having a 300-μm orifice Ta diaphragm in contact with the sample. A 10-μm diameter aperture limited the analyzer acceptance area. The sputter rate was 2.8 μm/hr. The finite signal at M/e = 158 in the epitaxial layer arose from molecular ions; the relative depth

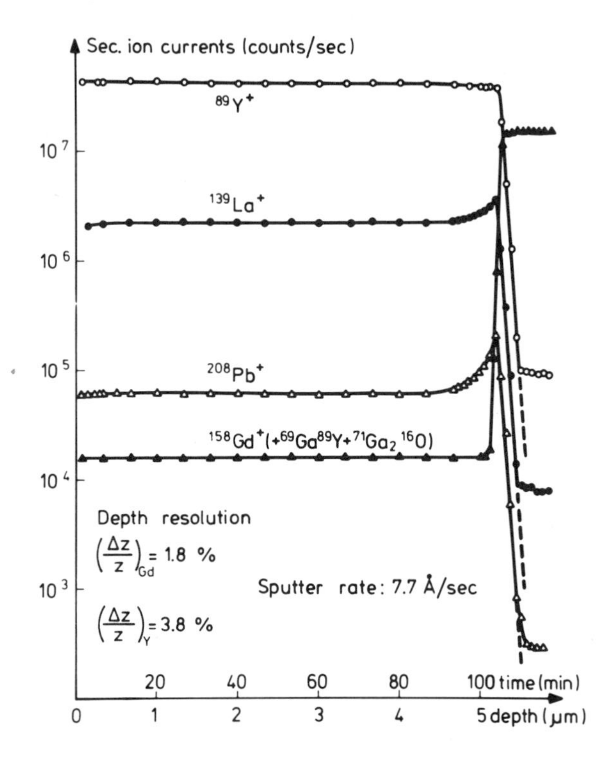

Fig. 5.27. Concentration profiles from a $Y_{2.8}La_{0.2}Pb_{0.009}Fe_{3.92}Ga_{1.08}O_{12}$ epitaxial layer on a $Gd_3Ga_5O_{12}$ substrate. Primary beam parameters; 15 keV O^-, I_p = 0.8 μA, θ = 40°, d_b = 100 μm, j_p = 0.3 mA cm^{-2} (266). (Courtesy of *Applied Physics*.)

resolution in this profile is 1.8%. Besides the depth limitation of the SIMS method, this figure can also be influenced by thickness nonuniformity of the film and by an unsharp interface. The high background levels (tails) in the Y^+, La^+, and Pb^+ profiles restrict the *dynamic range* in this run to between 2.5 and 3 decades; the dotted profiles result from subtraction of these background signals. The tailing also degrades the depth resolution to 3.8%. The Δz values pertaining to Fig. 5.27 have been calculated as the thickness over which the secondary ion intensity changes between 1% and 80% of the total intensity change at the interface. Adoption of the 16–84% definition would improve these values to 0.7% and 1.5% for the *Gd* and *Y* profiles respectively. However, perhaps the former definition leads to a more realistic appraisal since one is interested in measuring genuine secondary ion signals over several decades in order to exploit the trace element capabilities of the SIMS method.

With care, a dynamic range (without background subtraction) of at least 4 decades can be achieved when profiling from a region of high to zero concentration; this is of course provided that the signal does not first reach either the background count level or a constant level due to the presence of another secondary ion at the same mass number. This range is ultimately limited:

1. By deposition of material sputtered from the crater edges and walls into its central region. Sharp rather than slanting crater walls are beneficial here, as are also normal incidence and a large area crater. A small-area crater eroded by a scanning microbeam might prove troublesome in this context.

2. By material transferred onto a nearby surface (e.g., an extraction lens) and resputtered back onto the sampled region by reflected primary ions or by secondary ions accelerated through a strong extraction field. This so-called memory effect may be lessened by extended sputtering of samples not containing the element of interest before the commencement of measurements or by removing nearly collecting surfaces. An alternative approach to depth profiling, especially suitable for thicker films, is to make a line scan with a finely focussed primary ion beam across an angle-lapped surface (100).

E. RELATED ANALYTICAL TECHNIQUES

1. Ionized Neutral Mass Spectrometry

We have already seen that it is difficult to derive atomic concentrations directly from secondary ion intensities due to the large spread in the degrees of ionization and to matrix effects. It seems logical therefore to attempt to analyse the sputtered neutrals by means of postionization.

In *glow-discharge mass spectrometry* (GDMS), a rare gas glow discharge plasma ($p \approx 0.1$ torr) is utilized both for sample bombardment and for ionization of the sputtered neutrals, which are then mass analyzed (71,304,349). Ionization proceeds by the Penning mechanism via metastable gas atoms produced in the discharge:

$$Ar^* + M \rightarrow Ar + M^+ + e$$

In *sputtered neutral mass spectrometry* (SNMS), a substantially lower Ar pressure ($\approx 10^{-3}$ torr) is employed, and postionization in the rf-excited plasma is due to electron impact (126A,126B,282,283A).

Using either method, essentially all sputtered species are ionized in the plasma, thus permitting a semiquantitative analysis to be obtained directly from the measured ion currents. Disadvantages compared with SIMS are (a) all gas phase species are ionized and therefore residual gas contaminants will always show up in the mass spectrum, (b) less sensitivity in absolute terms due to the lower extraction efficiency of ionized neutrals into the mass spectrometer, (c) lack of lateral resolution, (d) careful shielding is necessary to prevent sputtering of the sample holder, and (e) crater edge and wall problems in depth profiling.

In the approach introduced by Blaise and Castaing (47), the sputtered flux impinging on the walls of an oven is ionized for subsequent analysis in a mass spectrometer. In this way, matrix effects and spectral interferences from cluster ions are avoided. Drawbacks include contamination from the inner walls of the oven, low sensitivity, and difficulties with high-ionization-potential elements. The *gas ion probe* (192) has been developed for the detection of rare gases in solids. Gases sputter-released from the target are first thermalized and then ionized by electron impact for detection in a quadrupole mass spectrometer. This technique has been successfully applied to the measurement of rare gas depth profiles in lunar samples.

2. Bombardment-induced Light Emission

In addition to secondary ions, excited particles are sputter-ejected from the solid. These can de-excite in the gas phase with the emission of light of a characteristic wavelength. A spectral analysis of this emitted light can provide information on the sample composition. This method has much the same advantages and disadvantages as SIMS (377,416).

In *glow-discharge optical spectroscopy* (GDOS), the characteristic emission is monitored from sputtered neutrals which are collisionally excited in a dc glow discharge (138).

3. Ion-Induced Auger Electron Spectroscopy (AES) (115,136,190,393).

As opposed to the conventional electron excitation (60,181), ion-induced AES seems to suffer from (a) complicated spectra due to the Auger electrons originating from sample atoms present both in the gas and in the solid phase, (b) sensitivity decreases, and (c) matrix effects.

III. SIMS INSTRUMENTATION

An instrument consists essentially of a primary ion source, a target chamber, a secondary-ion energy selector and mass separator, and an ion detector. Usually it is coupled with equipment for automatic data acquisition and processing, and, probably in the future, for instrument control also. The main distinguishing feature between

TABLE 5.VIII

Some Terminology Used to Describe Various Types of Secondary Ion Mass Spectrometers

Primary ions	
Current density	high—dynamic SIMS low—static SIMS
Beam diameter	~mm—macroprobe SIMS ~μm—microprobe SIMS
Secondary ions	
Mass separation	quadrupole, magnetic sector (time-of-flight occasionally)
Energy selection	electrostatic designs
Mass resolution	low (300) with single-focusing instruments high (up to 10^4) with double-focusing instruments
Element mapping	by scanning beam (μm diameters) —ion microprobe by ion optical imaging —ion microscope
Vacuum system	UHV (bakeable), $p = 10^{-10}$ torr HV, $p = 10^{-7}-10^{-8}$ torr
Modular design	compatible with other thin-film analytical techniques

instruments is the use of a magnetic-sector or a quadrupole mass analyzer. Table 5.VIII contains some terms commonly applied to SIMS instruments.

A. PRIMARY ION COLUMN

Usually a gas (commonly Ar, O_2, or N_2) at a typical pressure of 10^{-3} torr is ionized in the source chamber; see Fig. 5.28. Alternatively, the source may be utilized to generate ions from solid materials (e.g., Cs). The extraction optics, comprising two or more electrostatic cylindrical or aperture lenses, are used for beam extraction, preliminary focusing, and acceleration. The accelerating potential, U, applied to these optics controls the primary beam energy. The ions are extracted from the ionization region by the electric field, dU/dx, penetrating into the source. The maximum energy spread in the beam, ΔU, is equal to $\Delta x(dU/dx)$, where Δx is the

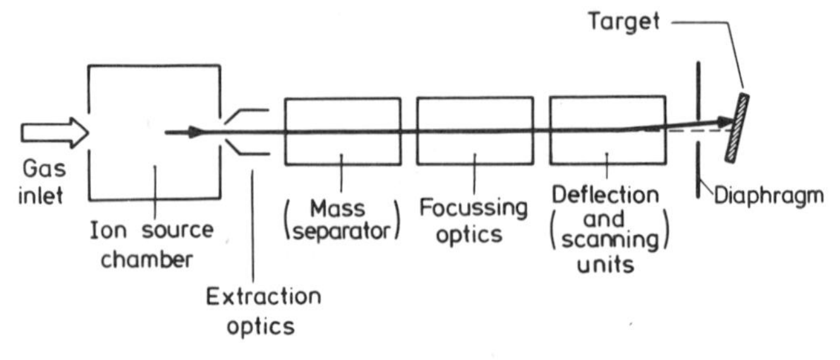

Fig. 5.28. Diagram of primary ion gun. Options are given in parentheses.

extent of the extraction region. A stronger field will extract a higher ion current but will also lead to a larger energy spread. In general, positive ions are extracted but sometimes negative ions (e.g., O^-, I^-) are used instead.

After optional mass separation, beam focusing to the desired diameter on the target is performed. An Einzel lens, in which the two outer elements of a three-electrode lens are at the same potential, is commonly employed. Two pairs of deflection plates are used to define the bombardment sample area and the incident angle, and also for raster scanning of the beam. The diaphragm, whose orifice is not much larger than the beam diameter at that point, restricts gas flow into the target chamber; the source column is differentially pumped to remove nonionized gas. A vacuum valve isolating the column from the target chamber is a useful addition. Also, energetic neutrals formed by charge exchange in the column may be removed from the beam by introducing a slight bend in its path.

Desirable properties of a primary ion source include:

1. a high total current output for favourable detection limits;
2. generation of low-energy ions to minimize sputter erosion of source components;
3. a small inherent energy spread of these ions so that fine beam focusing can be achieved;
4. a high brightness (i.e., a high current density within a small angular aperture) to obtain reasonable currents with focused beams;
5. a high ionization efficiency so that low source pressures may be utilized on account of vacuum requirements;
6. compatibility with reactive gases (i.e., the avoidance of heated filaments);
7. beam purity, particularly if no mass separation is available;
8. uniform current density profiles over an appreciable area for uniform sputter erosion.

Electron impact sources (146,279) are particularly suitable for static SIMS experiments since a low source pressure (10^{-7}–10^{-4} torr) can be utilized, and the low current output is not disadvantageous. Beam intensities can be increased somewhat by lengthening the travel path of the ionizing electrons. This is accomplished by inducing oscillatory motion (e.g., Finkelstein source (29,116)) and/or by increasing the volume of the ionization region (modified Bayard-Alpert or extractor ionization gauge (Pittaway). Another possibility is to use an inhomogeneous magnetic field to increase the electron density in the ionizing region.

With a *radio frequency source*, a gas (plasma) discharge is initiated by high electric rf fields, which are also used to promote electron oscillations. Extra electrons needed to initiate and/or to sustain the discharge are provided by thermionic emission, by field emission, or by ion impact upon a cold cathode (secondary electrons). A hollow cathode is often employed in ion sources for the latter purpose; solids deposited on the inside of this cathode may then be directly analyzed via their secondary ions (54,353). Long electron paths are obtained in a *Penning discharge* source (43,400) through cycloidal trajectories in superimposed magnetic and electric fields. However, unstable discharges are experienced with both Penning and rf sources, along with short source lifetimes due to sputter erosion.

A *duoplasmatron source* (19,217) is commonly employed in SIMS for gaseous elements; see Fig. 5.29. A plasma discharge of a few hundred volts is maintained between the cathode and anode. Electrons from the cathode are accelerated towards the anode through an intermediate electrode (held at a potential between those of the cathode and anode) into the discharge region. A double constriction of the plasma is achieved by having a cone-shaped intermediate electrode and by using an inhomogeneous magnetic field along the source axis. A high concentration of ions (and electrons) is thereby obtained in the region of the extractor aperture. The source is characterized by a high brightness with a relatively small energy spread. An adjustable magnetic field can be used to optimize conditions for positive or negative ion extraction. Use of a duoplasmatron as a source for negative atomic iodine ions has been described (226). An *unoplasmatron* source (19) utilizes geometric constriction alone. A *colutron* source (256) also dispenses with the auxiliary magnet but relies upon the magnetic field from a coiled heated filament for additional constriction.

The ultimate limit of lateral resolution for SIMS is about 100 Å (typical collision cascade diameter). The useful limit of beam diameters that should be achievable with a duoplasmatron is about 0.1 μm (223,224). (Note that this would be associated with a current of about 10^{-11} A, resulting in poorer detection limits and in fast erosion rates due to the large current density of 100 mA/cm^2). This limit is set by the brightness of the source and the chromatic aberration of the objective lens. *Field ionization* sources

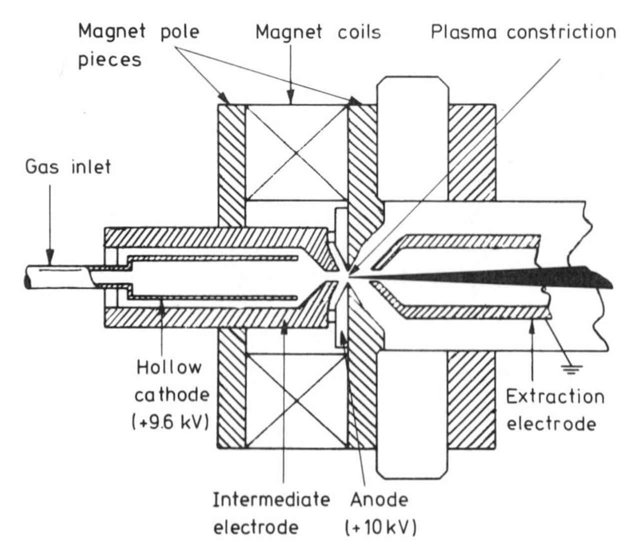

Fig. 5.29. Diagram of duoplasmatron ion source with a hollow cathode (for positive ion extraction); the plasma is constricted twice, (a) by the shape of the intermediate electrode, and (b) by the inhomogeneous magnetic field in the vicinity of the intermediate electrode.

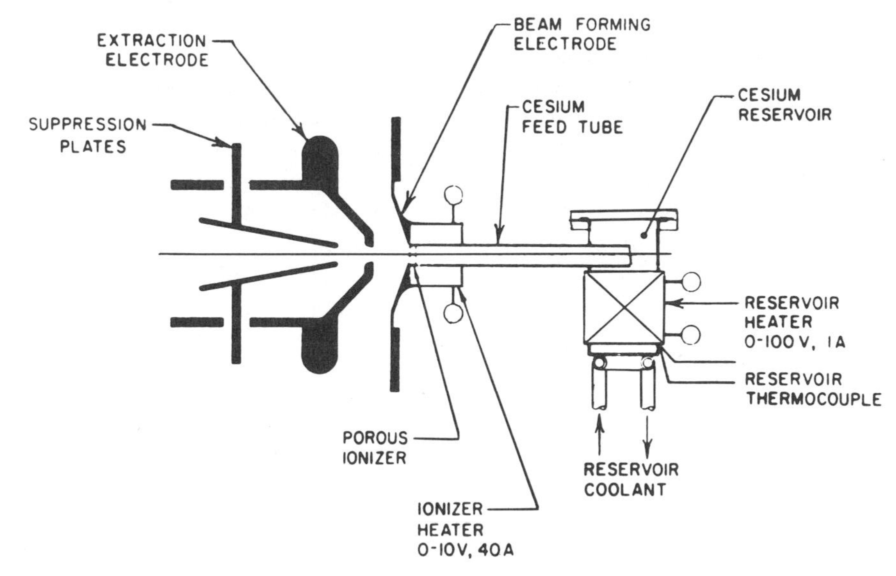

Fig. 5.30. Diagram of cesium ion gun, fitted to an AEI IM-20 instrument (420). (Courtesy of *Analytical Chemistry.*)

are advantageous in this respect since they have brightness values several orders of magnitude higher than duoplasmatrons yet have a similar energy spread. A high electric field between a very sharp, positively charged needle and an extraction electrode is employed. This source can be used for gases (338), but the total current output is rather low. This is not the case for "field ionization" (69,202) of a liquid metal at the tip of the needle; these sources can be employed for a wide range of elements.

However, up to now, only *surface ionization* sources have been used in SIMS for Cs^+ primaries. Operation at low pressures is possible, and a large current-density ion beam of high purity is achievable. The ions produced by thermal evaporation have also a very small energy spread so that, for a given primary ion current, smaller beam diameters should be obtainable than with a duoplasmatron source. A Cs salt on a heated filament (11,388), out-diffusion from a Knudsen cell (22,390), or Cs diffusion through porous tungsten (200,371,420) as illustrated in Fig. 5.30, have been utilized. These sources are limited to low ionization potential elements, but this is not a severe restriction for SIMS applications. By using porous lanthanum hexaboride as a substrate, an I^- surface ionization source has recently been developed (306).

B. TARGET CHAMBER

Without baking, a background pressure in the 10^{-7}–10^{-8} torr range can be achieved; indium, viton, or Teflon seals can be used and rapid sample change is possible. A background pressure of 10^{-10} torr can be achieved by baking overnight at 250°C; Cu or Au gaskets are then employed.

Nowadays, the main chamber is usually pumped either by an ion pump or by a turbomolecular pump; a Ti-getter pump is a useful addition for rapid pumpdown after sample introduction. An ion pump does not handle heavy gas loads well, has a slow pumping speed for rare gases, and suffers from memory effects. A turbomolecular pump must be mounted on bellows to prevent the transfer of vibrations and has a small but still satisfactory pumping speed for low-mass elements such as hydrogen. The primary ion source is best pumped by a small turbomolecular pump.

A well-designed apparatus should have the following features: it should permit rapid sample changeover, precise sample positioning, sample observation in the bombardment position, and the possibility of in situ heating, cooling, and fracture of specimens. It should also allow electrons to be directed at the sample and metal films to be deposited, and permit the incorporation of other analytical techniques for sample examination, either simultaneously or sequentially.

C. MASS ANALYZERS

1. Magnetic Sector

Ions subjected to an accelerating potential U undergo momentum analysis upon entering a homogeneous, magnetic sector field (normal to the plane of the paper in Fig. 5.31), and are separated according to their mass-to-charge ratio, M/e. Only ions of a given ratio can pass through to the collector, the others striking the sides of the collector slit or the analyzer tube. Given a correct choice of geometry (angle of magnetic sector, positioning of slits), all ions of the same mass and the same energy within a small angle of emittance are focused at the collector slit; focusing only in angle is referred to as single or directional focusing.

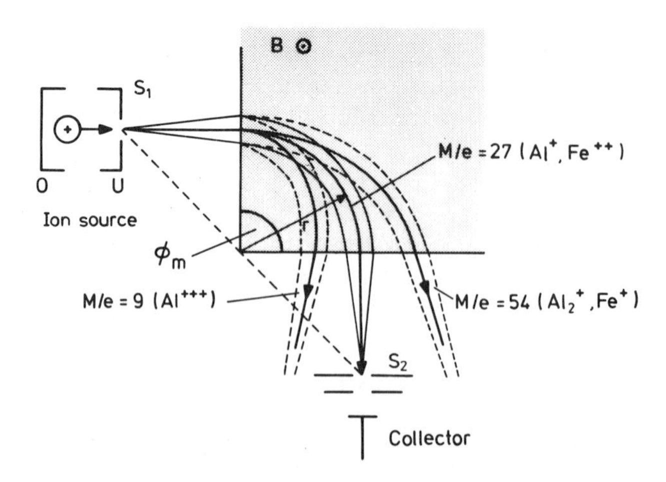

Fig. 5.31. Illustration of mass separation and directional focusing in a magnetic sector field; angle of magnetic sector $\phi_m = 90°$, radius of curvature of central beam $= r$, entrance and exit (collector) slits S_1 and S_2 respectively.

The path radius r in cm is given by

$$r = \frac{144\sqrt{(M/n)U}}{B} \tag{26}$$

where B is the magnetic field strength in Gauss (1 Gauss = 10^{-4} V · sec/m^2), M is the mass in amu, n is the number of charges on the ion, and U is given in volts. Variation of B with time for a fixed sector radius permits each mass number in turn to pass through the collector slit, and a mass spectrum is obtained by recording the temporal variation of the collector current. Since the mass number is proportional to B^2, a quadratic mass scale is obtained. A simultaneous record of the entire spectrum can be taken if a photoplate or a multiple electrical detector is placed in the collector region and the analyzer tube is large enough to accomodate all the separated beams.

The mass dispersion D is the separation at the collector slit between two masses M and $M + \Delta M$ (with ΔM usually taken as unity) which turns out to be equal to $D = r\Delta M/M$. Thus, for a given ΔM, the dispersion decreases with increasing mass. The peak width W for single-focusing instruments (homogeneous magnetic field, normal beam incidence) is approximately equal to $r(\Delta U/U) + S_1 + S_2$ for an incident beam of small angular divergence and small relative energy spread, $\Delta U/U$, and is thus mass-independent; S_1 and S_2 represent the slit widths.

The mass resolution R may be formally defined as $M/\Delta M$, where M is the mass number at which two peaks of mass M and $M + \Delta M$ are completely separated or resolved; ΔM is conventionally taken as 1 amu. For resolution, the separation D between the centers of the two peaks must at least equal their width W, i.e., $D = W$. Thus for peaks of uniform width,

$$R = r/(r(\Delta U/U) + S_1 + S_2) \tag{27}$$

For a fixed r, good resolution therefore entails a beam of small relative energy spread and the use of narrow slits. Further the resolution is the same for all masses, as is the transmission of the mass analyzer (which, to a first approximation, is inversely proportional to the resolution). The occurrence of a transmission that is not independent of mass is referred to as mass discrimination.

There is no universally accepted criterion of resolution. Fairly common nowadays is the 10% valley definition—two symmetrical peaks of equal intensity at mass positions M and $M + 1$ are said to be resolved if the valley between them is 10% of the peak height. An operational definition that is technically equivalent to this (Fig. 5.32) for a wide range of situations is the 5% peak-width definition. The resolution achieved at any isolated peak in the mass spectrum may be calculated from the ratio of the mass corresponding to that peak divided by the width (in mass units) at 5% of the peak height. Perhaps more useful for SIMS is a specification in terms of contribution resolution (abundance sensitivity), i.e., the intensity of the mass peak M at the mass position ($M + 1$).

Due to the large energy spread of secondary ions, an energy analyzer is needed to cut out a narrow energy band in order to be able to achieve good mass resolution (equation 27). The resolution of any energy is defined as the ratio of the width of the energy spread after analysis, ΔU, to the energy U of the beam, i.e., $\Delta U/U$. It is usual

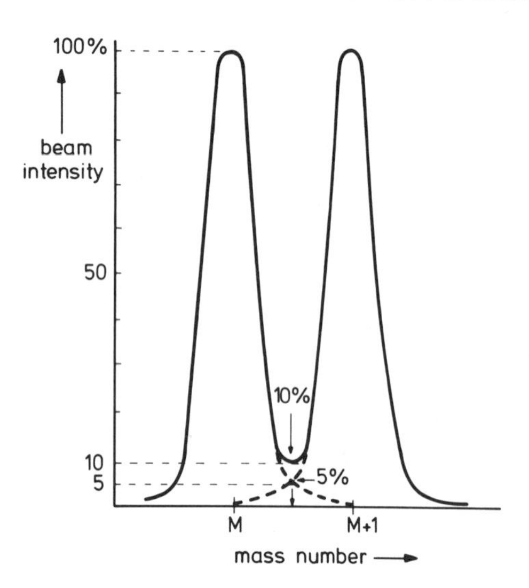

Fig. 5.32. 10% valley definition of mass resolution.

to take ΔU as the full width at half maximum (FWHM) intensity. Fig. 5.33*a* illustrates the use of an electrostatic deflection-type analyzer, consisting of two concentric cylinders at different potentials. The electrostatic field separates the ions with respect to energy only. Focusing of a narrowly divergent incident beam at the entrance slit to the magnetic sector S_{e2} is achieved by correct geometrical design. Energy analyzers used for SIMS in conjunction with magnetic mass spectrometers include cylindrical (21,97,107,374) and spherical (218,219) sectors, and single (63) and parallel plate (22,52) mirrors.

Higher mass resolution is achieved using the principle of double focusing, i.e., focusing in both angle and energy, (Fig. 5.33*b*), by which all ions having a fairly small energy spread may be brought to a focus at the collector slit by arranging that the energy dispersion in the electric field is equal, but of opposite sign, to the energy dispersion in the magnetic field. In contrast to Fig. 5.33*a*, an intermediate slit A_e of relatively large opening can be used, resulting in little intensity loss.

2. Quadrupole (89,90)

An oscillating field is established between four hyperbolically shaped long parallel rods (electrodes); see Fig. 5.34. Opposite electrodes are connected together; to one pair is applied a potential of $\phi = U + V\cos(2\pi \text{ ft})$ and to the other, the same potential but of opposite sign. U is a dc voltage and V is the zero-to-peak amplitude of an rf voltage at the frequency f. Ions injected in the z direction move along oscillatory trajectories through the rod system. At any specified frequency, only ions of a given mass number will undergo stable oscillations, which are necessary for them to pass through the length of the field without being collected by the electrodes. This mass number is determined by the choice of both U and V. Mass spectra are recorded by continuously varying both voltages while maintaining their ratio constant. The

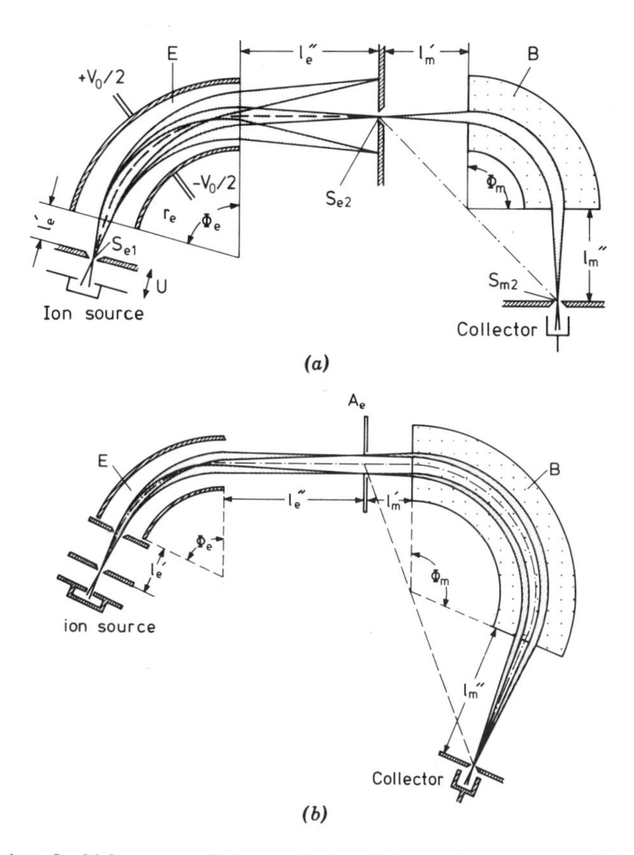

Fig. 5.33. Designs for high-mass resolution, sector-type instruments using sequential electric E and magnetic B fields; for simplicity, only ion trajectories of a particular mass are shown. (a By use of an energy analyzer with slits S_{e1} and S_{e2}. Electric sector separates ion beam with respect to energy while mass separation is achieved in the magnetic sector. (b) By use of double focusing. A beam of larger energy spread (wider slit opening A_e) can be mass-separated, giving a relatively large intensity at high mass resolution.

Fig. 5.34. Quadrupole mass filter: r_r = rod radius = 1.148 r_0 where r_0 is field radius; $\phi = U + V \cos (2\pi ft)$.

transmitted mass (amu) is given by

$$M = \frac{0.825U \text{ (volts)}}{f^2(\text{MHz}) r_0^2(\text{cm})} \tag{28}$$

The mass resolution is controlled simply by selecting the ratio of U to V. For a mass filter of unlimited cross-section and length, an infinite resolution $M/\Delta M$ is theoretically achieved if $U/V = 0.16784$.

Equation 28 shows that operation of a quadrupole in the $U(t)$ mode leads to a linear mass scale. Further, the mass dispersion D is mass-independent so that, unlike the situation in magnetic sector-type instruments, the masses are equidistantly spaced along the mass scale.

For a filter of *finite length*, the mass resolution obtainable depends upon how long the ions remain in the rf field. This means that their maximum axial injection energy must be limited to typically 5 – 10 eV. For a filter of finite *aperture*, ions have to enter the quadrupole within a specific zone around its axis and within a specific angle of divergence to be transmitted. Since this acceptance volume decreases with increased resolution, the transmission of the filter is thus a function of the resolution.

Theoretically, a quadrupole operates in the constant resolution ($M/\Delta M$) mode if U/V is kept constant. This mass-independent resolution means that the actual peak width will be directly proportional to M. Detailed considerations (287) of the mass filter acceptance indicate that, in this mode, the transmission is in fact *mass-dependent*, with discrimination being made against the lighter masses.

On the other hand, if in theory $U = 0.16784\ V \pm \delta$, where δ is an added fixed dc potential, the quadrupole can be made to operate such that a constant peak width is obtained throughout the mass range, i.e., the resolution ($M/\Delta M$) will now be directly proportional to M. In this mode, the transmission is expected to decrease as M^{-1}.

In practice, a quadrupole operates somewhere between these two modes. By letting $U = \gamma\ V \pm \delta$, where γ is the constant controlling the theoretical resolution, the desired resolution mode can be reinforced by selection of γ and δ. Notice that because of transmission considerations, the former mode is preferable for heavy-ion detection and the latter for the light ions. A hybrid resolution mode is chosen when mass-independent transmission over a large mass range is desired.

a. Energy Selection

An electrostatic energy filter (369) is necessary to prevent high-energy secondary ions and backscattered primaries from being injected into the quadrupole since this degrades the mass resolution. Another function of this energy filter is to prevent direct line of sight between the sample and the ion detector of the quadrupole, to reduce signal background arising from photons and from sputtered neutrals and metastables. More offset of the detector from the quadrupole axis (168) is insufficient; however, even when using an energy filter, this offset leads to some improvement.

Because of fringing fields at the end of the rods, the resolution of a quadrupole mass filter depends on the axial injection velocity of low-energy ions. Therefore, it is also prudent to select a very narrow energy band of secondary ions to improve the mass resolution; the position of this band should be variable so that the mean-pass energy corresponding to the optimum ion velocity at that particular resolution setting can be chosen. Notice that this band must always be relatively narrow, since double focusing in the sense used when describing magnetic mass spectrometers is not possible; i.e., the energy dispersion cannot be compensated in the quadrupole since this analyzer is not based on ion optical principles.

Electrostatic analyzers used in conjunction with quadrupoles for SIMS have included (a) a deflection-type electrode (364), (b) parallel plate capacitors (123,423), (c) straight axially symmetric types with axial-beam stop (Bessel box) (103,117), (d) cylindrical mirrors (325,335), (e) gridded spherical retarding/accelerating potential analyzers with central stop (91,199), (f) an electrostatic prism (135), (g) cylindrical sectors (288,295), and (h) spherical sectors (26,117,173,240,312). Focusing of the secondary ion beam in one direction is possible with type (g) and in two directions with types (d), (e), and (h). A converging ion beam at the quadrupole entrance that matches the mass filter acceptance is obviously desirable.

3. Secondary-Ion Extraction

Secondary ions are ejected with considerable angular and energy spread, and often from a fairly large sample area (particularly in static SIMS). Efficient collecting optics are therefore required to transfer as many ions as possible from the sample into the energy analyzer. In magnetic-sector-type instruments, the ions are usually accelerated from the target by a strong extraction field of say 1 kV/mm and then focused onto the entrance slit of the analyzer. An optical system that gathers about 10% of the secondary ions into a high-resolution mass spectrometer has been described (201).

With quadrupoles however, only very weak extraction fields may be applied since the secondary-ion injection energy must not be too high. Acceleration of the secondary ions from the target followed by deceleration before injection leads to some improvement (8,424), but is restricted to fairly small accelerating potentials, since defocusing of the ion beam may occur in the deceleration stage. Collecting optics have been described that limit secondary ion extraction to the central region of the bombarded sample area (120,240). Use of a spherical retarding grid energy analyzer such as the type described by Staib (366) permits ion acceptance from a large target surface area and emittance angle, and so is particularly useful for static SIMS studies.

4. Magnetic Sector or Quadrupole?

A mass resolution of up to 10^4 is possible with a double-focusing sector-type instrument, as opposed to the quadrupole limit of about 10^3. Also, the former analyzer offers the possibility of simultaneous multielement detection. Apart from these two features, all other advantages seem to be in favor of the quadrupole.

However, for SIMS analyses, the question of instrument sensitivity must claim first priority. It would appear that a well-designed quadrupole instrument might match the trace element detection capabilities of the more conventional double-focusing magnetic instruments. The overall transmission of the quadrupole mass filter should be more favorable, since narrow defining slits are not necessary to obtain satisfactory resolution. On the other hand, the magnetic sector can accept a wider energy band of secondary ions, and more efficient collection can be realized through the use of strong extraction fields.

5. Time of Flight

Mass analysis is achieved by measuring the time taken for a pulsed beam of ions of a given energy to travel a specified distance. Over a flight path of 1 m, separation between $M/e = 100$ amu and 101 amu can be achieved for 10 keV ions by means of a time discrimination of 150 nsec. This type of analyzer offers several advantages: a transmission of about 50%, simplicity of construction, and speed of analysis, since the whole spectrum can be scanned in a few microseconds with a 100-nsec ion pulse. Such analyzers have occasionally been used (259,359) for SIMS experiments. Liebl (220) advocated the use of a pulsed primary beam and analysis of each burst of secondary ions with a time-of-flight spectrometer for simultaneous element detection with minimum sample consumption.

D. ION DETECTION

Ion currents between $10^{-8}-10^{-14}$ A are detected with a *Faraday cup* (i.e., a hollow metal container) connected to a dc amplifier. The current collected on the cup is measured as a voltage drop across the input resistor of the dc amplifier. The minimum detectable voltage is limited by noise due to thermal motion of free electrons in the input resistor.

The principle of operation of a (focusing-type) *electron multiplier* is sketched in Fig. 5.35. It consists of a number of metal electrodes (dynodes), curved to give a focusing effect and held at successively higher positive potentials. Each ion striking the conversion dynode C releases γ secondary electrons ($\gamma \approx 2-3$ electrons/ion). These electrons are attracted and electrostatically focused onto dynode 1 by the positive potential difference between the two dynodes. Each of the original electrons causes δ (≈ 2) additional electrons to be emitted from dynode 1 and so on up to the nth dynode (10–20 stages being common). Thus, the overall gain of the multiplier is $\gamma\delta^n(10^4-10^8)$. The gain, however, deteriorates with time and with exposure to the atmosphere. The multiplier can be used in two modes:

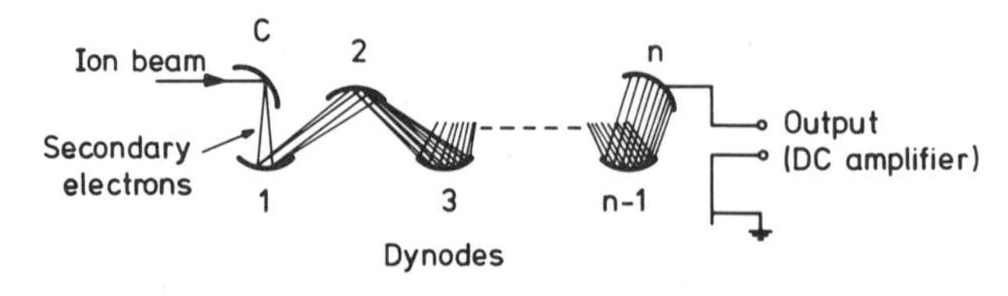

Fig. 5.35. Principle of secondary electron multiplier. Multiplier dynode voltage $V_C = -5000$ V (typically), $V_n = 0$V. Total voltage across the multiplier is divided equally among the dynodes. Counting efficiency at first dynode increases with V_C.

1. DC mode: The observed dc current is a product of the intensity of the impinging ion beam times the multiplier gain. The minimum detectable current, limited by noise in the output circuit, corresponds to around 10 ions/sec. This mode is not suited for precise quantitative measurements since the efficiency γ of the converter dynode depends upon the mass (and the energy) of the impinging ion. Also, detection of negative ions is not simple since they will be repelled by the high negative potential (see Fig. 5.35). It is now necessary for $V_C = 0\ V$ and consequently for V_n to be a high positive voltage. Maintaining the dc amplifier at such a high potential creates problems. Ways around this difficulty have been described (357).

2. Pulse-counting mode: The impinging ions are counted individually via voltage pulses at the multiplier output. With pulse-height discrimination to distinguish pulses from detector noise (402), a background of < 0.1 count/sec can be achieved. The signal is now independent of multiplier gain, and of the dependence of γ upon the incident ion, provided that the pulse height sufficiently exceeds that of the minimum detectable pulse, i.e., the setting of the discriminator level must be such as to permit a counting efficiency of close to 100%. Also, negative ions may easily be counted since the counting electronics can be dc isolated from the last dynode (held at a high positive potential) by means of a blocking capacitor through which the pulses can pass.

At high count rates, counting losses must be considered. These arise since the counting circuitry requires a definite time between events to recover and to accept another pulse (dead time). A time constant τ (sec) means that the detection electronics will accurately respond to τ^{-1} evenly spaced ions/sec. However, if this number of ions actually reaches the detector, many ions will be missed due to coincidences caused by the random nature of the signal, e.g., only one ion will be counted if the arrival time between two ions is less than τ. It is easy to show that the actual number of ions N counted per sec from an impinging beam of N' ions per sec is given by $N = N'(1 - N\tau)$, from which the true count rate may be obtained if τ is known. Today, τ values between 10^{-6} and 10^{-9} sec can be achieved.

The ultimate limit to the measurement of a flux of particules is the statistical fluctuation caused by the random arrival of individual particles. If N is the count rate and t is the measurement time, the relative error in the current measured in either mode will be $(Nt)^{-1/2}$ from this effect (see Section II.B.5.).

The *channel electron multiplier* (channeltron) is an electrostatic-focusing electron multiplier of high gain, made up of a thin glass tube (typically of 1-mm diameter and 1-cm length) coated internally with a secondary electron-emitting metal to generate a continuous dynode. A two-dimensional array of even thinner tubes ($\approx$ 20-μm diameter) constitutes the CEMA (channel electron multiplier array), sometimes referred to as a channel plate. The gain of a channeltron can be as much as 10^9 at an applied voltage of 7 kV. Although the gain is practically unaltered after atmospheric exposure, its actual value is unfortunately dependent upon the residual gas pressure.

In a *Daly-type detector* (87), the ion beam is deflected onto a metal converter electrode held at a high negative potential ($>$ 20 kV) and releases typically 3–5 electrons per ion. The secondary electrons are accelerated towards an aluminized scintillator at ground potential, creating flashes of light upon impact. These photons in turn release secondary electrons from the photocathode of a photomultiplier (either pressed against the rear of the scintillator or mounted external to the vacuum system). The (stable) electron gain inside a good multiplier can be as high as 10^8. Furthermore, γ of the converter is unaffected by gas adsorption. Careful shielding from stray light enables a background of < 0.1 count/sec to be achieved when employing pulse-height discrimination. In comparison with the electron multiplier, the high ion-impact energy and the acute angle of incidence ensure a high value of γ; this leads to a reduction in counting losses and hence to more sensitive and reliable counting (i.e., no mass discrimination) since, in a well-designed arrangement, better pulse height discrimination is possible. Dark currents lead to a fairly high background in the dc mode, which can be suppressed somewhat with the aid of coincidence circuits. Negative secondary ions can now be detected in this mode too, because the converter is dc-isolated from the photocathode.

The device can be modified (88) to form an energy-selective detector as follows. By directing the ion beam towards the scintillator now held at a potential of $+V_s$, only ions that have an energy of $<eV_s$ are reflected towards the converter maintained at earth potential. Ions of higher energy are buried in the thin metal coating of the scintillator. The secondary electrons from the converter travel towards the positive scintillator and are detected. By placing a retarding potential grid in the incident ion beam, a low-energy cutoff can also be achieved. In this way, a magnetic sector can be used (372) for SIMS measurements without the necessity of an electrostatic stage.

In ion microscopes, the ion image is converted into a secondary electron image for viewing either directly on a phosphor screen or indirectly on a photoplate (64,66).

1. Multielement Detection

Photoplate detection was for many years the only available possibility for simultaneous mass measurements, but this method is tedious, has a limited dynamic signal range, and is rather impractical for quantitative purposes. However, the desire to monitor many elements with minimum sample consumption has led to the development of multiple electric detection methods utilizing many channel multipliers in parallel or a channel plate.

In one design (307), an ion-to-electron converter is placed along the focal plane of the mass spectrometer and the emerging secondary electrons are focused (241) by the stray magnetic field onto a CEMA. Sixty wires feed the magnified electron signals into a commercial 60-channel on-line data collection unit. The mass spectrum is electrically scanned to cover the portions between these collecting wires. A spectrometer with the equivalent of 60 exit slits is thus obtained.

Electro-optical devices have been reported (131,344,387). A channel plate is used to amplify the ion currents striking different positions on the spectrometer focal plane. These currents may represent different masses or, in the case of an ion microscope, the ion image of one mass. The electrons at the output of the channel plate array are accelerated onto a phosphor screen. The generated light is transferred via fiber optics to a Vidicon camera, followed by an optical multichannel analyzer

(OMA) and displayed on a cathode ray tube. The use of a conversion grid in front of the channel plate array might prove useful to prevent the ions from directly striking the CEMA and thereby causing rapid detioration by sputtering.

A similar device for image display in an ion microscope (324) uses an ion-to-electron converter, with direct acceleration of the secondary electrons to the phosphor screen, Vidicon, OMA set-up.

E. COMPUTERIZATION

1. Data Acquisition

This consists of data handling and processing. Raw data must be offered to the input of the computer as voltages—$V_1(t)$ will correspond to the mass spectrum $I(t)$; $V_2(t)$ to the magnetic field $B(t)$ in a sector-type instrument; $V_3(t)$ to the dc voltage $U(t)$ in a quadrupole. Further, the information must be supplied in digital as opposed to analog (i.e., continuous) form. A mass peak, for instance, must be subdivided into a succession of currents, each pertaining to a very small time interval (sampling period). The current integral over a sampling period is stored (physical smoothing). Further smoothing is carried out by the computer, i.e., mathematically, and the resulting peak shape (intensity in digits per sampling period) is used to find the peak maximum. The intensity of the peak can either be taken as the peak height or the integral area under the peak. With a suitable computer program, this digital information can be *reduced* to another form, e.g., V_1 together with the corresponding V_2 or V_3 can be converted into a readout of I as a function of (M/e), if the mass scale calibration is also provided to the computer. As we have already seen, Section III.C.1, V_2 (the Hall voltage) is proportional to $(M/e)^2$, but hysteresis can modify this ideal relationship. With a quadrupole, V_3 is directly proportional to (M/e) (see Section III.C.2.).

In *off-line data processing*, the signals V_1, V_2, and V_3 are transformed and stored in a digital magnetic tape recorder for subsequent transfer to a computer. Nowadays, *on-line processing* is mainly used (187) whereby the information flows directly into a dedicated computer of small memory capacity. The reduced data thus obtained can either be displayed or stored for further treatment with a larger computer. As indicated in Fig. 5.36, the analog signals from the mass spectrometer are first converted into digital form in an ADC (analog-to-digital converter). The latter is dc-isolated from the computer by opto-insulators consisting of light-emitting diodes (LED), which enable digital information (pulses) to be transferred to the computer via light flashes. Digital signals from the mass spectrometer (i.e., ion currents measured in the counting mode, V_1') are transmitted directly to the computer, also by an opto-insulator; such transfer is less sensitive to disturbance than analog data transfer.

Data acquisition and reduction in a computer is performed in real time, i.e., simultaneously with the incoming information flow. The display of the reduced data occurs with some slight delay because the (mechanical) printing equipment cannot follow the information flow fast enough.

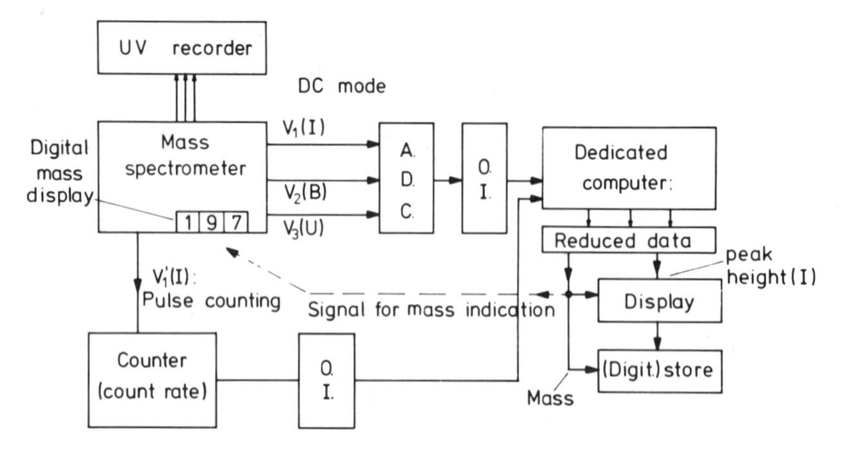

Fig. 5.36. Flow scheme for on-line data processing. Analog signals, V_1, V_2, V_3, indirectly transferred to a computer. ADC, analog-to-digital converter; OI, opto-insulator; digital signal, V_1', directly fed into the computer.

Data handling procedures can be used for:

1. Digital display of peak intensity at each mass (Fig. 5.36).

2. Digital display of the mass number (Fig. 5.36).

3. Display mass peak intensities as a function of sputter time in depth profiling; also rations of secondary ion currents to that of oxygen or of the matrix element, automatic background subtraction, conversion of time scale into depth and intensity scale into concentration (using a calibration factor obtained from standards) (410).

4. Calculation of absolute concentrations from secondary ion signals for quantitative analysis by means of a suitable correction procedure (13,265,320).

5. Isotope stripping and peak identification (303,319,370).

6. Treatment of elemental mapping, i.e., recording of contours of equal concentration (55,113,321,370A).

7. Detection of preselected mass peaks; the magnetic field or the dc voltage are switched successively to different values corresponding to the desired mass peaks. Switching of the magnetic field is difficult due to hysteresis, but good results are obtained in practice by sequential switching. Faster switching is possible with a quadrupole and also less dwell-time since it is easier to locate the peak maximum. The different ion currents can either be handled by a multichannel UV recorder or, better, by the computer. Possibilities for switching include (a) an array of solid-state switches, closed in turn by the computer so that all signals can be handled on one ADC; (b) a digital solid state switch array (multiplexer) in which each signal is converted by its own ADC, and the computer with one command line activates the digital switches; (c) every ADC has its own command line to the computer.

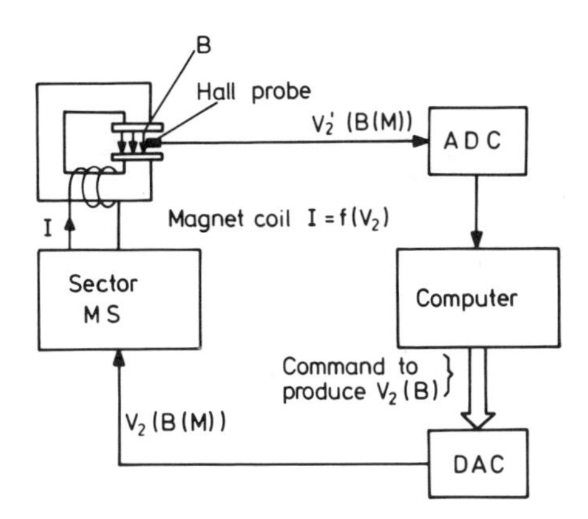

Fig. 5.37. Hardware for computer control of sector-type mass spectrometer.

2. Instrument Control

In modern instrumentation, the mass spectrometer parameters can be controlled by the computer, e.g., to scan a given mass interval of the spectrum. Now we require a digital-to-analog converter (DAC); see Fig. 5.37. Computer software is also needed to ensure that the computer command has been correctly executed, e.g., no disturbances from hysteresis effects. The software will control the current through the magnet field-coil so that V_2' always equals the desired value V_2. Long (233) reported computer control of the magnetic field to a few parts in 10^6 in his high-mass resolution instrument.

Parameters such as beam diameter or beam position can be computer-controlled, providing that the relation between the primary ion optics and the applied voltages has previously been established. Also, the computer can check the constancy of operating parameters (121,139) such as primary ion acceleration voltage, ion current, gas pressure, etc. On-line computer control and data processing for SIMS combined with other thin-film techniques have been reported (329,394). The ease of operation and time-saving achievable with computer control will enable the analyst to devote himself better to the real problems at hand.

F. EXISTING INSTRUMENTS

1. Commercial

Table 5.IX is designed to give a quick survey of those instruments known to the authors that are currently commercially available or, at least, have been until fairly recently. Further details can be found in the references cited or obtained from the manufacturer directly. Lack of space prevents a detailed description of all of these instruments, many of which have been well described in other reviews (108,220,221,223,225,248,251,254,360). Merely for the sake of illustration, we

TABLE 5.IX

Commercially Available SIMS Instruments

Firm	AEI	Appl.Res.Labs.		Atomika	Balzers[c]	Cameca		Hitachi
Model	IM 20	IMMA	QMAS	A-DIDA	101	IMS 300	IMS 3F	IMA-2
Primary Ions								
Ion source	dp[a]	dp	dp	gas discharge	ei	dp	dp	dp
Reactive gases	yes	yes	yes	yes	no	yes	yes	yes
Negative ions	yes	yes	yes	no	no	yes	yes	no
Differential pumping	yes	yes	no	yes	no	yes	yes	no
Acceleration voltages (kV)	3–25	10–22.5	1–10	0.5–15	0.3–3	10	5–20	1–15
Mass separation	yes	yes	no	yes	no	no	no	yes
Minimum beam diameter	2 μm	2 μm	80 μm	5 μm	0.2 mm	15 μm	2 μm	2 μm
Beam raster	yes	yes	yes	yes	no	yes	yes	yes
Bakeable apparatus	yes	no	no	yes	yes	no	yes	no
Secondary Ions								
Ion image display	yes	yes	no	yes	no	with lateral resolution ≤1 μm		yes
Energy analyzer			160°	// plate	no	(a) mirror		
Mass analyzer	(es ms) df	(es ms) df	spher. q	defl. q	q	(b) (es ms) df	(es ms) df	(es ms) df
Detector	em photopl.	Daly	Daly	cem	em	(a) Daly (b) em + Faraday cup	Faraday cup em channel plate	Daly
Reference	20,21	12,218	121,312	423	58,168	272,316	317	374

Firm	Leybold	3M	PHI	Riber		VG
Model	LH-SIMS[b]	610[b]	2500[b]	[b]	MIQ156	[b]
Primary Ions						
Ion source	ei	ei	ei	ei	dp	ei
Reactive gases	no	no	no	no	yes	no
Negative ions	no	no	no	no	yes	no
Differential pumping	no	no	no	yes	yes	yes
Acceleration voltages (kV)	0.1 – 5	0.05 – 5	0.5 – 5	0.2 – 5	3 – 15	0.2 – 5
Mass separation	no	no	no	no	yes	yes
Minimum beam diameter	2 mm	100 μm	2.5 mm	0.1 mm	2 μm	1 mm
Beam raster	no	yes	yes	yes	yes	no
Bakeable apparatus	yes	yes	yes	yes	yes	yes
Secondary Ions						
Ion image display	no	yes	no	no	yes	no
Energy analyzer	Bessel box	Bessel box	double-pass axial	45° cylinder	45° cylinder	Bessel box
Mass analyzer	q	q	q	q	q	q
Detector	em	cem	em	em	cem	cem
Reference	331	362	127			

[a] *Abbreviations*: dp-duoplasmatron; ei-electron impact; (c)em-(channel) electron multiplier; q-quadrupole; ms-magnetic sector; df- double focusing; es-electric sector.

[b] Intended primarily as a modular addition to existing surface analytical equipment manufactured by the same company.

[c] Obsolete

have selected two of the newer instruments—one using a double-focusing magnetic sector and the other a quadrupole.

The latest ion microscope from Cameca is shown in Fig. 5.38. Microspot analysis is accomplished by using a fine-focused beam and by accurate positioning of the sample under this beam. Negative primary ions are available for insulator specimens. Primary-beam raster and secondary-ion detection limited to a 25-μm, 150-μm, or 400-μm diameter sample area by means of a diaphragm are available for depth profiling. A high-extraction field (target held at 5 kV) and a 150 eV energy bandpass through the double-focusing spectrometer ensure adequate sensitivities for trace element analysis. Flat-topped mass peaks permit isotope ratio measurements. A mass spectrum is recorded either with the Faraday cup plus a dc amplifier (dc mode) or with the electron multiplier (pulse counting), and a mass resolution (10% valley definition) of up to 5000 is possible. This can be increased to 10^4 when the instrument is used as a mass spectrograph, the channel plate ensuring high sensitivity in this mode also. Direct simultaneous viewing of a large field of view, but with a spatial resolution of <1 μm (determined by image aberrations), is possible since the ion image formed by the immersion (objective) lens is maintained in the final mass-resolved beam. The projection lens magnifies the image onto the channel plate; the secondary electrons so released strike a fluorescent screen for viewing purposes. The gain of the channel plate permits imaging with weak secondary ion currents. An incorporated minicomputer is provided for control of optical and electronic instrumental parameters, and for calculation and display of analytical results.

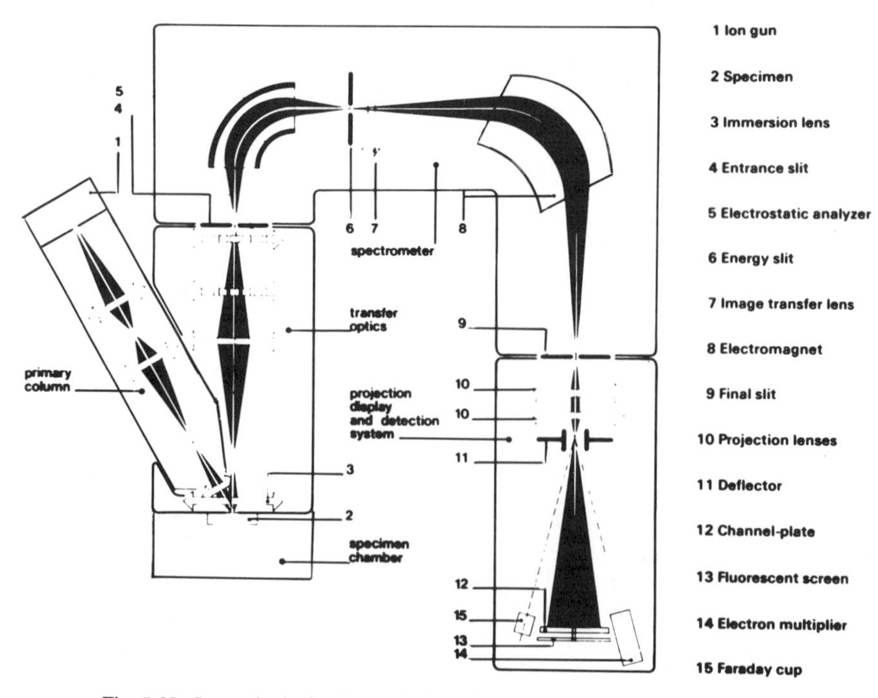

Fig. 5.38. Ion optics in the Cameca IMS 3F instrument. (Courtesy of Cameca.)

The QMAS instrument from ARL is shown in Fig. 5.39, which is fairly self-explanatory. The liquid nitrogen cold plate is used for vacuum improvement in the sample region. Beam raster and electronic gating are used for depth profiling. In addition to providing efficient extraction, the multielectrode extraction lens system limits sampling to the central portion of the bombarded area and forms a real but magnified image at the entrance to the spherical electrostatic energy analyzer. Since the latter is a stigmatic device, the bombarded area is re-imaged at the quadrupole entrance. An energy band of a few eV is selected. The ion collimator lens ensures maximum transmission to the Daly detector. The apparatus has an abundance sensitivity of 2×10^6 (peak at mass 27 to valley at $26\frac{1}{2}$).

An intending purchaser should, of course, test an apparatus in the light of his own specific analytical requirements. However, other points to consider but not included in Table 5.IX are: (a) maximum mass resolution obtainable and how rapidly a given signal decreases with increase of mass resolution; (b) mass range; (c) maximum primary ion current obtainable; (d) variation of this current with the nature and charge of the primary ion, with beam diameter, and with beam energy; (e) stability of primary and secondary ion currents; (f) detection sensitivity (e.g., Fe^+ count rate per μA of O_2^+ at a reasonable mass resolution); (g) arrangement for insulator specimens, particularly in obtaining the negative secondary ion spectrum; (h) abundance sensitivity, particularly on the low-mass side of a high peak; (i) peak shape; (j) ability to monitor H^+ and H^-; (k) background count in both the positive and negative spectra, particularly in a situation with a high overall count rate; (l) direct viewing of the sample in the analysis position; (m) rapid sample introduction and turnover; (n) residual vacuum during intense ion bombardment; (o) spectral cleanliness with

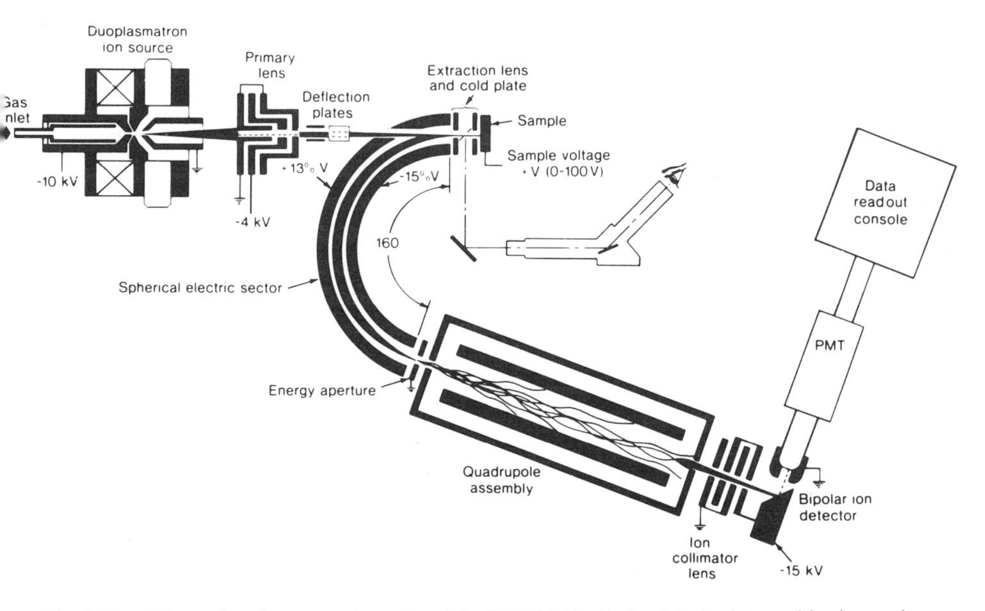

Fig. 5.39. ARL quadrupole mass analyzer for solids (QMAS); bipolar ion detector is in positive ion mode (121). (Courtesy of *Research/Development*.)

respect to H, C, N, O, etc.; (p) the detection of large negative secondary-ion currents; (q) depth profiling performance, i.e., dynamic range, maximum area of raster without change of primary ion current, and typical sputter rate; (r) imaging capabilities, i.e., lateral resolution, trace elements, and insulators; (s) arrangement for O_2 flooding; microspot analysis capabilities; (t) secondary-ion energy bandpass, i.e., width and positioning; (u) accuracy of isotope ratio measurements; (v) compatibility with other analytical techniques; (w) capability for automatic data acquisition and instrument control. One desirable feature not yet automatically included is a Cs primary-ion source. As far as price is concerned, the quadrupole-equipped instruments tend to be considerably cheaper.

2. Homemade

We mention here a few instruments possessing some unique features not already mentioned. Liebl (219) has designed a combined ion-electron microprobe housed in a UHV system. Ion and electron bombardment are performed either separately or simultaneously. The primary ion beam is mass-separated in a 180° magnet without beam defocusing occurring. The secondary particles are energy-analyzed, and then either detected with an energy-dispersive X-ray analyzer or passed through a 180° magnet for mass analysis. A prototype has been constructed by Varian MAT in Bremen.

Rüdenauer and Steiger (318) built a UHV, 1-μm focus, SIMS instrument containing a tandem mass spectrometer, i.e., two double-focusing mass spectrometers in series. The exit slit of the first spectrometer inside a 180° uniform-field magnet serves as the entrance slit to the second. High abundance sensitivities are thus achieved by the elimination of the spectral background due to the elastic scattering of matrix secondary ions in the gas phase.

Baril and Vallerand (22) have constructed a unique double-focusing instrument by combining a parallel plate mirror with a magnetic sector. Clement et al. (70) are at present constructing a 1–2-μm microprobe incorporating a magnet radius of 1 m, designed for high sensitivity when working at a mass resolution of 10^4. Degreve et al. (92) have improved the secondary ion extraction optics on the AEI-IM20 apparatus for better transmission when working at high mass resolution.

McHugh et al. (253) have combined the primary-ion bombardment and secondary-ion extraction systems of the ion microscope with the mass analyzer and detection systems of the scanning ion microprobe (Fig. 5.40). A primary-ion beam of relatively large diameter (100 μm–500 μm) is used, and a global secondary-ion image is formed by the immersion lens at a focal plane in front of the mass spectrometer. This ion image can be raster-scanned across a variable-diameter aperture in the focal plane. The secondary ions that pass through this aperture enter a double-focusing mass spectrometer. A mass-resolved secondary-ion image is obtained by displaying the amplified detector output on a cathode ray tube rastered in synchronism with the raster of the global ion image. Cherepin and Vasilev (67) built an instrument on similar principles but used a quadrupole mass filter instead, the secondary ions being extracted at 10 kV and decelerated before entering the quadrupole. A mass resolution of only 60 was reported.

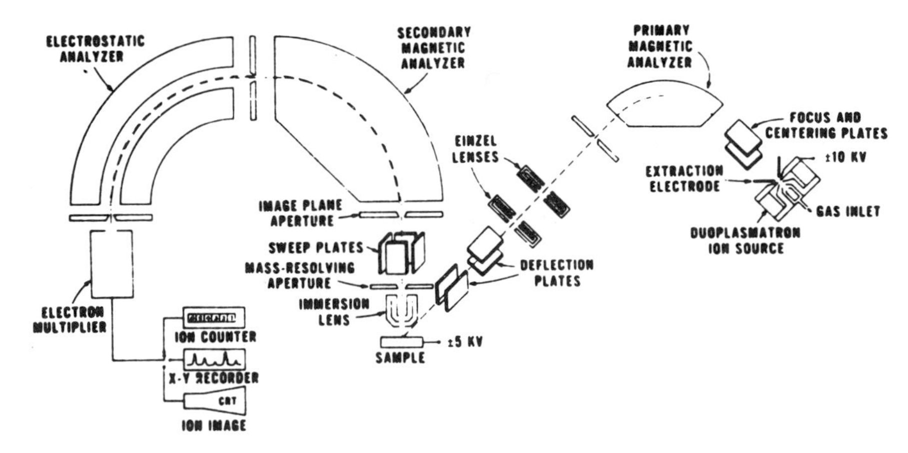

Fig. 5.40. Direct-imaging scanning ion microprobe (253). (Courtesy of American Society for Mass Spectrometry.)

By using a mirror-type ion-electron converter (157) for secondary-ion detection, which is insensitive to high-energy secondary ions, sputtered neutrals and photons, Hofer and Thum (158) obtained reasonable SIMS spectra from an on-axis alignment of sample, quadrupole mass filter and detector without using an energy filter. This type of detector, however, is unsuitable for negative ions. Cherepin and Maifet (66) built an instrument utilising a monopole mass filter.

3. Combination with Other Techniques

The easiest technique to combine with SIMS (237,347) is bombardment-induced light emission (see section II.E.2), since the same ion gun can be used for primary excitation. Thus, all the analytical SIMS modes can be duplicated, for example elemental images by scanning (135,257), and with about the same sensitivity. It is hoped that in this way spectral interference problems in the one technique can be avoided by utilizing the other. A window in the vacuum chamber, a few mm free space in front of the target, and equipment for monitoring light are required.

Apart from the SIMS electron microprobe example mentioned above, all SIMS combination instruments have involved a quadrupole since this is compact, easily bakeable, and has no stray magnetic fields to interfere with the other measuring techniques. Combinations with AES, XPS, and LEIS (see Section IV) are fairly common; the last example is particularly simple since this can be accomplished (142) with a quadrupole preceded by a sector-field energy analyzer having an energy resolution of a few percent. For LEIS, the quadrupole is simply tuned to the mass of the noble primary ion and an energy spectrum is taken from which a mass spectrum of the sample can be obtained. [A combined instrument has been described (378) utilizing only a magnetic-sector momentum analyzer.]

Depth profiling using SIMS and electron-induced AES signals obtained simultaneously from the same sputtered crater have been reported (65,191,198,278). Ploog and Fischer (299) used a combination of SIMS, AES, and RHEED (reflected high-energy electron diffraction) for in situ characterization of semiconductor films grown by molecular beam epitaxy. An ion gun and a quadrupole mass filter have been added to scanning electron microscopes (206,215).

IV. COMPARISON WITH OTHER THIN-FILM ANALYTICAL METHODS

All thin-film analytical techniques in current use have advantages and disadvantages. Fortunately, the weak points of one technique are often the strong points of another. A modern analytical laboratory therefore requires many of these techniques, to be employed either sequentially or simultaneously, for thorough characterization of a sample. Combination of several techniques in one apparatus leads to faster sample throughput, facilitates microspot analysis, and, in particular, avoids sample contamination during transfer between instruments. Many manufacturers provide equipment for a particular method which is suitable for attachment to an existing analytical apparatus. For sequential examination, the sample position can be altered (usually by rotation in vacuum). On the other hand, by using a fixed sample position, it is easier to ensure that the same area is always analyzed and simultaneous examinations; however, a fixed sample position can mean that the set-up for one or more of the measuring techniques is not optimal. Also, having a single apparatus for combined analyses often restricts the time available to develop each of the individual methods to its full.

In the following, we present a brief comparison of SIMS with other surface and thin-film analytical methods in an attempt to demonstrate their distinctive features [for more information, see review articles (31, 34, 35, 110, 111, 112, 165, 246, 271, 412,414)]. We use Table 5.X, which lists the characteristics of the most popular and versatile methods. Note that some of these properties represent extremes which, although having been reached, are not necessarily obtainable in every analysis and with every instrument.

In AES (Auger electron spectroscopy) and XPS (X-ray photoelectron spectroscopy, sometimes referred to as ESCA, electron spectroscopy for chemical analysis), the energy spectrum of secondary electrons emitted by electron and X-ray beam excitation, respectively, is recorded. Elements are identified by the energy of their

TABLE 5.X
General Comparison of Thin-Film Analytical Methods

	AES	XPS	SIMS	LEIS	HEIS
Elemental range	⩾Li	⩾Li	⩾H	⩾Li	⩾He
Elemental Resolution	$\Delta Z=1$	$\Delta Z=1$	isotopes	poor for high Z	
Sensitivity variation	10	20–30	10^3	10	100
Detection limit (at.%)	0.1–1	1	10^{-4}	0.1–1	10^{-2}–10
Quantitative analysis			standards required		direct
Lateral resolution	0.1 μm	1 mm	1 μm	100 μm	few μm
Depth resolution (Å)	5–50	5–50	3–10	3–6	20–200
Atom location	no	no	no	yes	yes
Chemical information	some	yes	some	no	no

Auger and photoelectrons. LEIS (low-energy ($\leqslant$ few keV) ion scattering, also called ISS, ion-scattering spectrometry) and HEIS (high-energy (0.5–3 MeV) ion scattering, or RBS, Rutherford backscattering) utilize energy losses experienced by ions backscattered from the topmost one or two atomic layers of the sample in the former case, and simultaneously from various depths in the sample in the latter case. Measurements of the backscattered energy makes it possible to identify the mass of the scattering atom.

The elemental range with HEIS is essentially conditional upon a heavier element being present in a matrix of a light element, unless channelling (with single-crystal specimens) or resonance scattering is utilized. However, nuclear reactions may also be used to detect light elements with the HEIS apparatus. These reactions are specific for each isotope and permit detection of hydrogen, but usually only elements lighter than argon are accessible. The capability of SIMS for isotope identification and for monitoring of hydrogen should be noted. The mass resolution obtainable with the ion-scattering methods decreases with increasing atomic number Z, so that it is impossible, for instance, to distinguish between Pb and Bi. However, in a HEIS experiment, the elemental constituents in the target can be identified from their characteristic X-radiation induced by the incident beam. With this method of *particle-induced X-ray emission* (PIXE), all elements from Be onwards can be detected and resolved. Nevertheless, HEIS is best suited for samples containing only a few elements.

The sensitivity variation gives a rough indication of the maximum amount that is possible between different elements in the same sample. The figure quoted for SIMS assumes that both positive and negative secondary ions are monitored. The sensitivity is not directly dependent upon Z with AES, XPS, and SIMS. The backscattering yield with LEIS, however, increases by about a factor of 10 when going from light to heavy elements. With HEIS, the scattering cross-section is directly proportional to Z^2 and thus the sensitivity increases in a predictable fashion. The limit of detection represents a fairly average practical value taking all elements into consideration; the lower limit given for HEIS pertains to a very high Z element on a very low Z substrate, while the upper limit corresponds to a light element on a heavy substrate and is only achievable if channelling is employed. With HEIS, a direct quantitative analysis with an accuracy of a few percent is possible, including stoichiometry determinations of thin films. All the other methods require standards of a more or less similar matrix for calibration purposes.

The ultimate lateral resolution in AES is controlled not so much by the beam diameter but by the backscattering effect, i.e., the spatial extent of backscattered primaries and other electrons within the target having sufficient energy to release Auger electrons from surface areas not directly irradiated by the primary beam. The width of the X-ray beam limits the lateral resolution with XPS, although Hovland's approach (166) may improve the situation in certain cases. If a too-narrow beam is employed for LEIS, the increased current density leads to a fast erosion rate, which negates the surface sensitivity advantage of this method. HEIS is generally a broad beam method (few mm^2), but beam diameters of a few μm have been achieved (53,79) with highly specialized and complicated equipment.

The information depth with AES and EPS is determined by the inelastic mean free path of the energetic electrons in the solid; this quantity depends upon the actual energy and also somewhat upon the type of matrix under investigation. The HEIS signal may be restricted to the first monolayer or so by utilizing the surface peak in the channelled spectra of monocrystalline specimens. Generally, however, all depths (up to about 1 μm or more, depending on the mass, energy, and incident angle of the projectile) are sampled simultaneously and an in-depth concentration profile may be obtained directly without the necessity for sputtering. Further, the depth scale may also be derived directly from the known average energy loss per unit distance travelled by the projectile in the target. The depth resolution at the surface is typically 200 Å (which may be improved by about a factor of 10 by using glancing incidence or a detector of higher energy resolution) but degrades in depth due to energy straggling. With the other methods, the depth resolution in sputter profiling should be similar to that achievable with SIMS.

Information on the location of impurity atoms within the lattice (i.e., substitutional, interstitial sites), on the extent of damage or amorphization induced in single crystals by, say, ion implantation, and on the structure of single-crystal surfaces may be obtained with HEIS by using channelling. Surface structure analysis is also possible with LEIS. XPS is the preferred technique for chemical information since the peak position of a given element is dependent upon its chemical environment.

V. ANALYTICAL APPLICATIONS

The number of applications has been increasing at an accelerating pace during the past few years so that it is now impractical to quote every example. In this brief survey, we have tried to include the more recent work and some of the more novel earlier applications. Many examples can also be obtained from earlier reviews (12,72,108,109,164,221,248,251,356,360,405,406,407,413).

A. BIOLOGICAL

A comprehensive review has been given by Burns-Bellhorn (27A,27B). We consider first analysis made on hard tissues. Lodding and co-workers (122,208,232,290,291) have studied the in-depth distribution of elements (particularly fluorine) in biological hard tissues such as human and shark teeth. Fig. 5.41 shows the localized distribution of fluorine, which acts as a protective agent against caries, in the outermost surface layers. The F concentration was derived from calibration curves. It was found that even with a homogeneous standard, the measured F^+/Ca^{++} ratio could vary widely, depending on the prevailing experimental conditions (207,229). To overcome this, the calibration curve was constructed by plotting $\log(F^+/Ca^{++})$ versus $\log(Ca^{++}/Ca^+)$; a set of parallel lines was obtained with the spacings between directly proportional to the F/Ca concentration ratios. Lefèvre and co-workers (85,86,210,211,209A,209B,84A) have studied calcium biomineralizations in plants and animals, and Tousimis (378A) studied bone formation.

Several studies on biological soft tissues have been reported

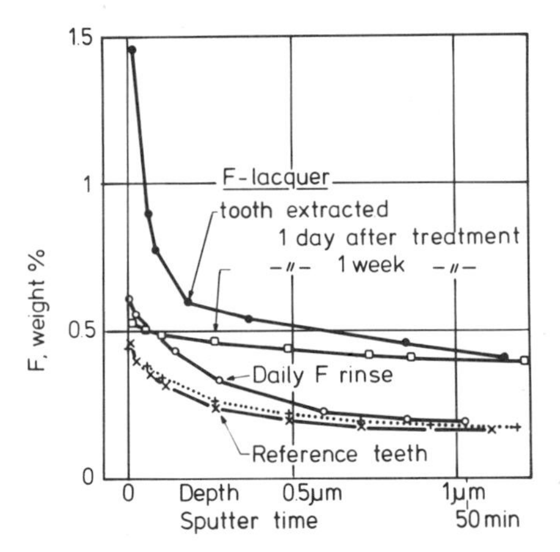

Fig. 5.41. Variation of F concentration with depth in dental enamel which had been treated with a fluoride-containing varnish (232). (Courtesy of A. Lodding.)

(124,178,272,379,380,165A,242A,242B,378C,379A). Sample preparation difficulties are considerable. Fig. 5.42 shows the distribution of Ca in the transversal section of an insect abdomen. The element is present in all tissues; cuticle (c), fat body (f), pericardial cells (p) where it is accumulated, and ovocytes (o) whose nuclei are clearly visible. Tissue samples from known and suspected cases of beryllium disease in human lungs have been analyzed (5,6). The use of deuterium as a tracer in biological samples has been investigated (195). The production of standards for quantitative analyses of biological specimens has been reported (27,27C,12). Con-

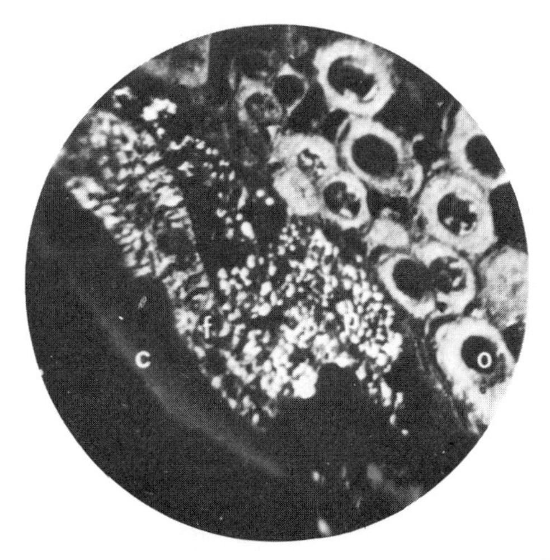

Fig. 5.42. $^{40}Ca^{+}$ ion micrograph of transversal section of insect abdomen, showing distribution in cutile (c), fat body (f), pericardial cells (p) and ovocytes (o); viewing field 250 μm (272). (Courtesy of *Analytical Chemistry.*)

centration profiles in fir tree needles have been measured (125), and the retina of cats studied (26A). Galle (124A) made ion images from rabbit kidneys and from dried frog red blood cells.

B. ELECTRONIC MATERIALS AND DEVICES

One of the most popular and rewarding applications of SIMS has been the determination of ion-implantation profiles in semiconductors. Compared to the more classical approaches (radiotracers, electrical activity, X-ray generation by heavy-ion bombardment), the SIMS method offers good detection limits, depth resolution, simplicity, rapidity, and versatility. Werner (410) has reviewed activities in this context, in particular for B implants in Si where a detection limit of 10^{15} atoms/cm^3 may be realized. Colby (72) and others (261) provide several examples illustrating how knowledge of these profiles is helpful in transistor technology. Typical studies involve measurements of the range distributions of the implanted ions as a function of primary flux and energy, target orientation, implanting species, annealing conditions, etc. Some recent applications include P in Si (59,173A), Cs in Si and in SiO_2/Si double layers (123,173), As and P in Si (426,383B) and in SiO_2/Si (383A,154B), Be in Si (172), As and Ga in Si_3N_4 (84), B, P, As, and N in Si after implantation in Si_3N_4/Si systems (154,154A,B) in diamond (50), N in GaP and N, O, F in Si (382), Al in Si (98), Na in SiO_2 (240A) Si (209), B (125B), Be (170,77A), and Zn and Se (113A) in GaAs, and Zn and N in $GaAs_{1-x}P_x$, SiO_2, and Si_3N_4 (274,275A,275B). Depth profiling of Cr in GaAs is currently attracting much interest (240B,240C,112A,167B) as is pulsed-laser annealing of ion-implanted silicon (427C,128B,416A). The surface segregation of alkaline elements implanted into GaAs has been examined in relation to photoemissive properties (7C). Impressive detection limits have been achieved (216,383,420,421,422) in the profiling of electronegative elements (e.g., P, As, Se, Au, etc.) in Si and in GaAs by using Cs^+ bombardment and by monitoring negative secondary ions. The solid solubility limit of Se in GaAs has been determined (216A). Magee et al. (238,239) have measured H depth profiles in electronic materials. Fig. 5.43 shows such an implantation profile obtained by using 5 keV Ar^+ bombardment. By taking adequate precautions (low hydrogen background pressure, primary-beam mass separation, and neutral component elimination, restricted-area secondary-ion acceptance), a detection limit of around 5×10^{17} atoms/cm^3 (10 ppm) has been achieved.

Many further examples can be found pertaining to integrated circuit fabrication (101). Migration of Na (171,293) and K (94) in SiO_2, and segregation of B (73) and other implants (337,22B) at the SiO_2/Si interface have been investigated in connection with metal-oxide-semiconductor (MOS) structures; also SiO_2-Ge MIS structures have been analyzed (397). Composition, chemical bonding, and contamination of SiO_xN_y films were examined with several techniques including SIMS (15D). The SiO_2/Si interface itself has been studied (228,228A,276A), as has the mechanism of Si (83) and of GaAs (74,351B,397A) anodization. Epitaxial Si films grown on sapphire have been characterized (145,203,294,378B). Other topics include the origin of high ohmic resistance in Al-Si contacts (126), contaminants in Pt films

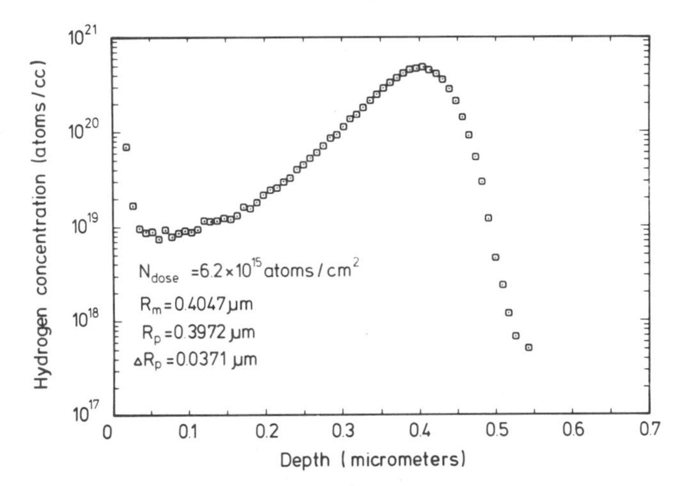

Fig. 5.43. Hydrogen depth profile of 40-keV proton implant into silicon. R_m is the modal range, R_p is the mean projected range, and ΔR_p is the standard deviation in the projected range (239). (Courtesy of *Nucl. Instr. and Methods*.)

sputter-deposited on Si (17,43C) and in vacuum-deposited Si (114) and B (51) films, carbon residues on surfaces of Si integrated circuits (119), oxygen accumulation at the Au-Si interface (301), gold migration in Au-Y_2O_3-Y junctions (281A), gold-zinc contacts on *n*-type InP (386A), oxidation of sputtered tantalum nitride layers (196B), the degradation characteristics of heat-treated metal-GaAs Schottky barrier diodes (193,285,394A), and evaluations of several elements as barrier metals for preventing Ga and As out-diffusion from sintered Pt-GaAs contacts (37,275C). Purity control of InP (104) and GaAs (300) substrates used in molecular beam epitaxy has been performed along with characterization and evaluation of ZnTe (428), CdTe (144A), GaN (16), GaP and $GaAs_{1-x}P_x$ (245), and GaAs and $Al_xGa_{1-x}As$ (300) films and heterostructures. Annealing studies of Be-doped GaAs grown by molecular beam epitaxy demonstrated a diffusion coefficient two orders of magnitude lower than that for implanted Be at the same concentration (254A). The oxygen content and its depth distribution in thin GaAs epitaxial layers have been analyzed (167). Direct nonassignment of oxygen to the mean electron traps in GaAs was concluded (167A). The surface contaminant on hybrid microcircuit capacitors was identified as AgCl (143).

The importance of an alkali-rich surface layer for high-area detection efficiency of channel electron multiplier arrays was demonstrated (286). Impurities in the ZnO phosphor layer in vacuum fluorescent display tubes were shown (348) to cause low brightness. High temperature decalibration of Pt/Rh thermocouples was found to be due to reaction with the Al_2O_3 insulating sheath (82,67A). Si (311,417) and Cu_2S/CdS (327) solar cell samples have been analyzed, as have HgI_2 crystals, which are used for nuclear radiation detectors (308). Concentration profiles in Nb-Ge sputtered films have been determined in an investigation of the origin of their high superconducting critical temperature (323). The diffusion coefficients of Ga in Si_3N_4 (230), ^{30}Si in intrinsic silicon (185A) Cd in $CuInSe_2$ (187A) and deuterium in hydrogenated silicon (60A) have been derived, as was reported earlier for In in Ge (401).

C. METALLURGICAL

Many studies have been reported concerning irons and steels. The composition of stainless steel surfaces after various heating and cleaning treatments has been investigated (133,169,213), including Cr segregation and depletion (336). The oxide layers on steels and stainless steels have been analyzed (212,343). Fig. 5.44 shows the ion images of Cr^+ and Si^+ at various depths in the oxide film. The Cr^+ image at 1100 Å indicates that the element is homogeneously distributed within the grains, but that it is absent from some of the grain boundaries. At 3000-Å depth, however, the Cr is situated almost exclusively along the grain boundaries (as the oxide). At 7000-Å depth, it has almost been sputtered away, with some Cr residues remaining along the boundaries. Si displays a somewhat different behavior. At 1600-Å depth, Si is present (white) within the *grains* and at 7200 Å, it is present only along the grain *boundaries*. Phillips (292) has published similar images of the C distribution in stainless steel, revealing a carbon-containing precipitate at the grain boundaries. The morphology of graphite precipitates in cast iron (56), chemical segregation in the fracture surface and grain boundaries of irons and steels (244,328), and Al segregation in Fe–0.1 Al % alloys (386) have provided further topics. The concentration profiles of Cr, Ni, and Cu implanted into mild steel have been measured (176,177). Quantitative analyses of B (61) and C (314,427D) in steels have been made, and the effect of carbide formation on the quantitative analysis of constituents in low-alloy steels have been investigated (385).

Secondary-ion imaging was used (55) to identify the chemical nature of precipitates in stainless steel that lead to cracking and embrittlement. Fig. 5.45 shows that the precipitates responsible for the embrittlement contain Ti, S, and C. The capability

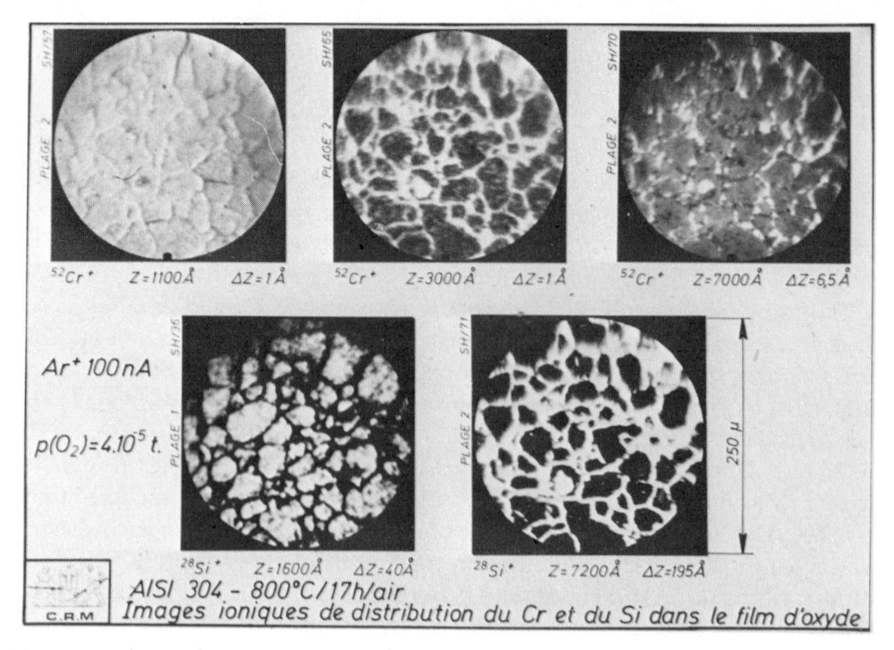

Fig. 5.44. Cr^+ and Si^+ images at different depths in oxidized stainless steel (343). (Courtesy of *C.R.M.*)

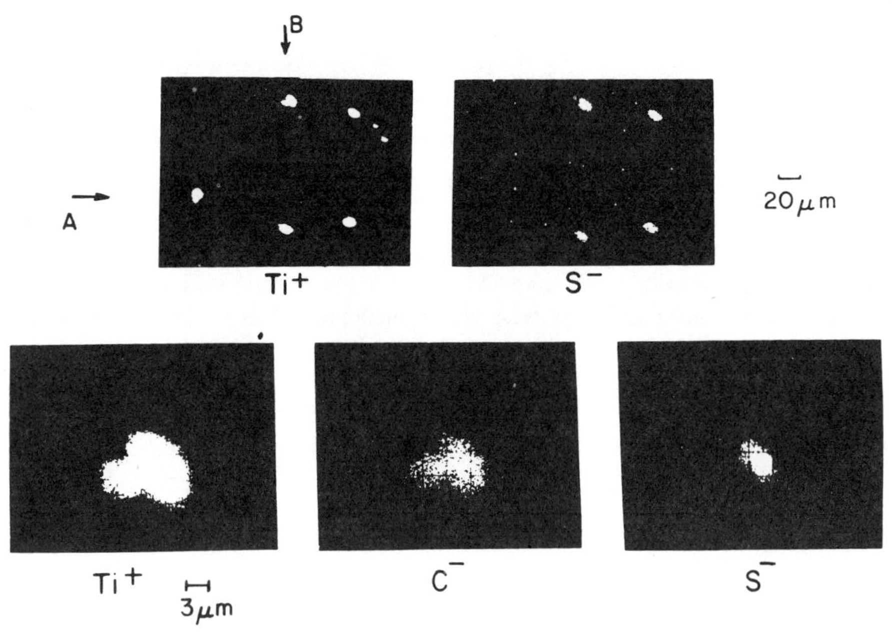

Fig. 5.45. Elemental images of precipitates in a ferritic stainless steel matrix. Precipitate A was used for a quantitative analysis and was thereby sputtered away, which explains its absence from the sulphur image. Precipitate B is shown with higher magnification (55). (Courtesy of R. Dobrozemsky et al.)

of the SIMS method for local analysis was utilized to obtain a quantitative analysis of the core of a precipitate.

Hydrogen embrittlement in Ti (137,284) and in Nb (140,141,419) has been investigated, as has the hydrogen distribution in proton-irradiated steel and Ta (188), and the hydrogen permeability of V (95). The concentration of deuterium in Ti, V, Zr, Nb, and Ta has been determined (361). A microdetermination of dissolved oxygen and nitrogen in a Ti alloy has been made (395). The diffusion coefficients of O in Zr (80,305) and of Ni in Cu (341) have been measured. The protective coatings on superalloys (376), Nb-coated Mo and Ti (260), and the anodic barrier oxide films on Al (1,2,3,254) and Ta (254) have been examined. The oxide phases on an Al alloy (196) and on Fe (277) have been identified using fingerprint spectra in combination with depth profiling as was done earlier for Cu (403). The passive layer on Ni has been characterized by using fingerprint spectra, ion imaging, and depth profiling (408). Imaging revealed internal oxidation along the grain boundaries of a Cu–Be alloy (313). The nature and size of small precipitates formed during the internal oxidation of Cu–Al alloys have also been studied (236), as has the composition of oxide layers formed on Ni–Cr, Fe–Ni–Cr alloys (296,297).

Surface changes resulting from various treatments in Al and Ti alloy adherents (23) and the removal of C from Mo (197) have been studied. Heating was found to cause Rh surface segregation in Pt–Rh alloys (46). Ion microprobe studies provided direct evidence (396) of boron segregation at the grain boundaries in a nickel-base alloy, which has been believed responsible for the substantial improvement in creep behavior. The embrittling agent responsible for the tensile failure of a Pt–Rh–W alloy was identified as Si (81). SIMS analyses of doped-tungsten lamp filaments (180,262,289,340,373) have been reported, as has the quantitative analysis of tungsten base composite alloys (231).

D. ORGANIC

Static SIMS has been applied to thin films of organic materials on metal substrates. Benninghoven and Sichtermann (36) investigated the secondary-ion emission of many biologically important compounds and found relatively high yields of "parent-like" ions such as $(M + H)^+$ and $(M - H)^-$ (Table 5.XI), thus permitting the determination of molecular weights. Also, fragment ions corresponding to functional groups such as the phenyl $C_6H_5^+$ and the pyridyl $C_5H_4N^-$ ions were emitted with high yields. The fragmentation patterns from polymers (99), fatty acid overlayers (77) and, in the area of tribology, carbonaceous overlayers on ball-bearing surfaces (76) have been examined.

E. GEOLOGICAL

Lovering has considered the possibilities of SIMS in this area (234) and has provided a comprehensive review of actual applications in geochemistry and cosmo-

TABLE 5.XI

Absolute Yields $S^{\pm}(X)$ of "Parent-like" Secondary Ions of Organic Compounds on Silver. Primary ions: 2.5 keV Ar^+, current density of 4×10^{-6} A/cm² (static SIMS). [Courtesy of Organic Mass Spectrometry (36)]

		Molecular weight	100x Yield $S^{\pm}(X)$ (no. of secondary ions per primary ion)		
Compound	Formula	M	$(M + H)^+$	$(M - H)^-$	$(M - COOH)^+$
Amino acids					
Glycine	$C_2H_5NO_2$	75	120.0	3.2	52.0
α-Alanine	$C_3H_7NO_2$	89	21.0	40.0	53.0
β-Alanine	$C_3H_7NO_2$	89	88.0	19.5	7.2
Phenylalanine	$C_9H_{11}NO_2$	165	4.0	0.3	13.0
Serine	$C_3H_7NO_3$	105	61.0	18.0	61.0
Threonine	$C_4H_9NO_3$	119	8.3	2.6	13.8
Proline	$C_5H_9NO_2$	115	19.2	8.8	72.0
Valine	$C_5H_{11}NO_2$	117	8.0	8.3	32.0
Leucine	$C_6H_{13}NO_2$	131	0.8	26.4	40.0
Norleucine	$C_6H_{13}NO_2$	131	24.8	6.5	76.0
Arginine	$C_6H_{14}N_4O_2$	174	7.2	2.4	2.1
Tyrosine	$C_9H_{11}NO_3$	181	7.4	—	13.6
Tryptophan	$C_{11}H_{12}N_2O_2$	204	3.5	0.8	3.5
Cysteine	$C_3H_7NO_2S$	121	12.0	11.0	15.0
Cystine	$C_6H_{12}N_2O_4S_2$	240	4.0	1.6	1.8
Methionine	$C_5H_{11}NO_2S$	149	13.1	5.4	9.4
Ethionine	$C_6H_{13}NO_2S$	163	13.6	5.6	12.0
Glutamine	$C_5H_{10}N_2O_3$	146	7.2	8.3	4.3

TABLE 5.XI *(Continued)*

Compound	Formula	Molecular weight M	100x Yield $S^{\pm}(X)$ (no. of secondary ions per primary ion) $(M+H)^+$	$(M-H)^-$	$(M-COOH)^+$
Derivatives of amino acids			$(M+H)^+$	$(M-H)^-$	$(M-Cl)^+$
Glycine ethyl ester HCl	$C_4H_{10}ClNO_2$	139	—	1.6	180.0
Alanine ethyl ester HCl	$C_5H_{12}ClNO_2$	153	—	—	48.0
Cysteinium	$C_3H_8ClNO_2S$	157	—	4.0	19.7[a]
Taurine	$C_2H_7NO_3S$	125	4.8	—	—
Peptides			$(M+H)^+$	$(M-H)^-$	$(M-COOH)^+$
Glycylglycine	$C_4H_8N_2O_3$	132	41.6	4.8	—
Glycylglycylglycine	$C_6H_{11}N_3O_4$	189	4.0	0.4	2.0
Glycylleucine	$C_8H_{16}N_2O_3$	188	1.6	4.2	3.0
Phenylalanylglycine	$C_{11}H_{14}N_2O_3$	222	8.0	1.6	—
Drugs			$(M+H)^+$	$(M-H)^-$	$(M-OH)^+$
Barbital	$C_8H_{12}N_2O_3$	184	—	44.0	—
Ephedrine	$C_{10}H_{15}NO$	165	16.0	—	40.0
Atropine	$C_{17}H_{23}NO_3$	289	84.8[b]	—	—
Epinephrine	$C_9H_{13}NO_3$	183	—	6.4	—
Vitamins			$(M+H)^+$	$(M-H)^-$	
Ascorbic acid (C)	$C_6H_8O_6$	176	3.7	17.6	
Biotin (H)	$C_{10}H_{16}N_2O_3S$	244	0.3	4.2	
Nicotinic acid (PP)	$C_6H_5NO_2$	123	—	46.4	
Nicotinamide	$C_6H_6N_2O$	122	2.1	15.2[c]	
Sulfonamides			$(M+H)^+$	$(M-H)^-$	
Sulfanilic acid	$C_6H_7NO_3S$	173	—	16.3	
Sulfanilamide	$C_6H_8N_2O_2S$	172	0.6	17.6	
Sulfacetamide	$C_8H_{10}N_2O_3S$	214	—	20.8	
Other compounds			$(M+H)^+$	$(M-H)^-$	
Thymidine	$C_{10}H_{14}N_2O_5$	242	1.9	1.3	
Acriflavine	$C_{14}H_{14}ClN_3$	259	—	—	96.0[d]
Creatine	$C_4H_9N_3O_2$	131	2.9	—	3.4[e]
Creatinine	$C_4H_7N_3O$	113	16.0	6.0	6.0[f]

[a] M' = mass of related amino acid, identical to $(M+H)^+$.
[b] $(M)^+$.
[c] $(M)^-$.
[d] $(M-Cl)^+$.
[e] $(M-OH)^+$.
[f] $(M+H_3O)^+$.

chemistry (235). Terrestrial and lunar materials (rocks, soils) and meteorites (399C,399B) have been examined. Typical information obtained (355) includes the distribution and identification of mineral phases, trace element distribution within these phases and along their boundaries, and quantitative determinations of trace elements such as B (152) and H and F (153). Radiometric age dating of individual grains and phases has been accomplished by isotope abundance analysis ($^{207}Pb/^{206}Pb$ or Rb/Sr), e.g., of individual mineral grains in lunar rock sections (14). This capability of the SIMS method for in situ localized isotope analysis has been further exploited in self-diffusion measurements [e.g., K in biotite (159); 0 in silicates (132), and Li in LiF (75)] and in studies of Ni diffusion in iron meteorites (280). Thin sections of (terrestrial) soil containing root material have been examined (44). The composition and age of surface films on lunar glass particles has been established (258). Solar wind activity has been investigated by depth profile measurements on lunar rock crystal and soil grains (432,433). Spatial variations in cathodoluminescence emission was correlated with the spatial distribution of impurities within the mineral samples (310).

F. MISCELLANEOUS

The use of SIMS for studying surface kinetic processes such as adsorption, oxidation, and catalysis has been reviewed by Benninghoven (34) and by Fogel (118). The surface composition of graphite fibers has been investigated (106) after various treatments designed to enhance their surface energetics. Also, the coatings on TiO_2 pigments have been characterized (275). Airborne particulates (227,249,326) including Pb-bearing species from automobile exhausts (24) have been analyzed in connection with pollution control. The isotope ratios and absolute isotope production cross-sections for Li, Be, and B generated in ^{16}O by high-energy protons have been measured (429) in connection with astrophysical studies.

In the field of nuclear technology, Na corrosion of metals has been examined (38). The oxygen self-diffusion coefficient in UO_2 has been derived (78,389). Plasma-wall interactions such as impurity deposition during the discharge (367) and the erosion and changes in the surface composition of stainless steel after low-energy light ion bombardment (315) have been investigated. Surface effects of materials related to fission and fusion reactors have been studied (179) including 14-MeV neutron sputtering of Nb (255). Several studies (134,151,242,281,330,363,392,134A) have been directed towards an analysis of the surface composition and a determination of elemental depth distributions (particularly of Na) in glass samples. The surface segregation of Al (spinel precipitation) in MgO crystals has also been investigated (194). An analysis has been made (128) of the interface between various silane-based primers and a mild steel substrate in connection with adhesion studies.

VI. CONCLUSIONS

It is to be hoped that his article has provided the reader with a helpful insight into the principle, methodology, and applicability of SIMS. This versatile analytical method, distinguished by its favorable detection limits, is already finding widespread use in solving many practical problems associated with a wide range of scientific disciplines. It is a safe prediction to assert that the method will grow rapidly in popularity and that the range of applications is by no means exhausted. However, more work including apparatus improvements remains to be done before SIMS can be regarded as a routine analytical tool.

More basic studies into the process of secondary-ion ejection are needed, including reliable measurements of escape depth and of absolute ion yields under well-defined experimental conditions. An understanding of matrix influences and chemical effects is required. At present, a unifying theory of secondary-ion emission is lacking, preventing quantitative analysis from first principles. Further work is required to establish a reliable empirical approach to quantification without direct recourse to standards.

Factors influencing the depth resolution obtainable in concentration profiling should be further investigated so that improvements can be brought about and so that reliable estimates of, say, interface sharpnesses can be made. It is also necessary to develop methods of avoiding matrix artefacts that complicate the interpretation of depth profiles and of ion images. The actual potentiality of the static SIMS method for gas adsorption studies and for deducing short-range atomic order needs to be established. Further it is likely that SIMS will be used to determine structural arrangements in organic and polymeric thin films. It may be foreseen that increased use will be made of the negative secondary-ion spectrum to improve detection limits for the electronegative elements, particularly when employing cesium (or other electropositive metals).

More attention should be given in instrument design to optimizing the dynamic signal range obtainable when depth profiling from a high surface to a low-bulk concentration. Better secondary-ion collection and detection efficiencies will not only ensure improved detection limits, but will also permit the utilization of smaller-beam diameters, since the primary-ion current density (and hence the erosion rate) can then be minimized by reducing the primary ion current. Contamination of the analyzed area needs to be minimized to permit a better analysis for H, C, N, O, etc. Flexibility in SIMS instrumentation is also necessary to allow the incorporation of other analytical methods.

The next few years should witness a further understanding of the secondary-ion emission process and of the analytical potentiality of the SIMS method, together with an extension of analytical applications and improved instrument performance.

ACKNOWLEDGMENT

The authors wish to thank Dr. N. Warmoltz for many helpful discussions and contributions to this paper, Mrs. P.H.J.C. Vermeulen-Pellen for arranging the references, and Mr. H.M. Wielink for assistance in organizing the lay-out of this paper.

REFERENCES

1. Abd Rabbo, M. F., J. A. Richardson, G. C. Wood, and C. K. Jackson, *Corrosion Science*, **16**, 677 (1976).
2. Abd Rabbo, M. F., J. A. Richardson, and G. C. Wood, *Corrosion Science*, **16**, 689 (1976).
3. Abd Rabbo, M. F., J. A. Richardson, and G. C. Wood, *Electrochim. Acta.*, **22**, 1375 (1977).
4. Abdullayeva, M. K., A. K. Ayukhanov, and U. B. Shamsiyev, *Radiat. Eff.*, **19**, 225 (1974).
5. Abraham, J. L., R. Rossi, N. Marquez, and R. M. Wagner, in O. Johari and I. Corvin, Eds., *Scanning Electron Microscopy/1976*, IIT Res. Inst., Chicago, 1976, p. 501.
6. Abraham, J. L., and T. A. Whatley, *Fed. Proc.*, **36**, 1090 (1977).
7. Abroya, I. A., V. P. Leivrov, and I. G. Fedorova, *Sov. Phys. Solid State*, **7**, 2954 (1966).

7a. Ahearn, A. J., *Trace Analysis by Mass Spectrometry*, Academic, New York and London, 1972.

7b. Ahmed, N. A. G., C. E. Christodoulides, and G. Carter, *Radiat. Eff.*, **38**, 221 (1978).

7c. Alexandre, F., *J. Physique*, **39**, 701 (1978).

8. Alpat'ev, Yu. S., I. N. Dubinskii, V. L. Ol'khovskii, A. P. Pilipenko, and V. T. Cherepin, *Instrum. Exp. Techn.*, **15**, 798 (1972).
9. Andersen, C. A., *Int. J. Mass Spectrom. Ion Phys.*, **2**, 61 (1969).
10. Andersen, C. A., H. J. Roden, and C. F. Robinson, *J. Appl. Phys.*, **40**, 3419 (1969).
11. Andersen, C. A., *Int. J. Mass Spectrom. Ion Phys.*, **3**, 413 (1970).
12. Andersen, C. A., and J. R. Hinthorne, *Science*, **175**, 853 (1972).
13. Andersen, C. A., and J. R. Hinthorne, *Anal. Chem.*, **45**, 1421 (1973).
14. Andersen, C. A., and J. R. Hinthorne, *Geochim. Cosmochim. Acta.*, **37**, 745 (1973).
15. Andersen, C. A., *Natl. Bur. Std. (U.S.) Spec. Publ.*, **427**, 79 (1975).

15a. Andersen, H. H., and H. L. Bay, *Radiat. Eff.*, **19**, 139 (1973); *J. Appl. Phys.*, **46**, 1919 (1975).

15b. Andersen, H. H., and H. L. Bay, *J. Appl. Phys.*, **46**, 2416 (1975).

15c. Andersen, H. H., *Appl. Phys.*, **18**, 131 (1979).

15d. Anderson, G. W., W. A. Schmidt, and J. Comas, *J. Electrochem. Soc.*, **125**, 424 (1978).

16. Andrews, J. E., A. P. Duhamel, and M. A. Littlejohn, *Anal. Chem.*, **49**, 1536 (1977).
17. Andrews, J. M., and J. M. Morabito, *Thin Solid Films*, **37**, 357 (1976).
18. Antal, J., *Phys. Lett. A*, **55**, 493 (1976).

18a. Antal, J., *Phys. Lett. A*, **55**, 281 (1976).

19. Ardenne, M. von, *Tabellen der Elektronenphysik, Ionenphysik und Uebermikroskopie*, Vol. 1, Deutscher Verlag der Wissenschaften, Berlin, 1956, p. 544.
20. Bakale, D. R., B. N. Colby, and C. A. Evans, *Anal. Chem.*, **47**, 1532 (1975).
21. Banner, A. E., and B. P. Stimpson, *Vacuum*, **24**, 511 (1974).
22. Baril, M., and P. Vallerand, *Can. J. Phys.*, **52**, 482 (1974).

22a. Barnard, G. P., *Modern Mass Spectrometry*, The Institute of Physics, London, 1953.

22b. Barsony, I., D. Marton, and J. Giber, *Thin Solid Films*, **51**, 275 (1978).

22c. Barsony, I., and J. Giber, *Appl. Surf. Sci.*, **4**, 1 (1980).

23. Baun, W. L., N. T. McDevitt, and J. S. Solomon, *ASTM STP 596* (1976), p. 86.

23a. Bay, H. L., and J. Bodhansky, *Appl. Phys.*, **19**, 426 (1979).

24. Bayard, M. A., *Proc. 7th Natl. Conf. Electron Probe Anal.*, (1972), p. 37a.
25. Bayly, A. R., P. J. Martin, and R. J. MacDonald, *Nucl. Instrum. Methods*, **132**, 459 (1976).
26. Bayly, A. R., and R. J. MacDonald, *J. Phys. E*, **10**, 79 (1977).

26a. Bellhorn, M. B., and Lewis R. K., *Exp. Eye Res.*, **22**, 505 (1976).

27. Bellhorn, M. B., and D. M. File, *8th Internl. Conf. on X-Ray Optics and Microanalysis*, Boston, U.S.A., 1977, paper No. 137.

27a. Bellhorn, M. B., *Proceedings of the Mass Analysis Society* (1978), p. 28A.

27b. Bellhorn, M. B., in J. Gennaro, Ed., *Advanced Techniques in Biomedica Electron Microscopy*, Masson et Cie., 1979.

27c. Bellhorn, M. B., and D. M. File, *Anal. Biochem.*, **92**, 213 (1979).

27d. Benninghoven, A., *Z. Naturforsch.*, **22A**, 841 (1967).

28. Benninghoven, A., *Z. Phys.*, **220**, 159 (1969).

29. Benninghoven, A., and E. Loebach, *Rev. Sci. Instrum.*, **42**, 49 (1971).

29a. Benninghoven, A., *Surf. Sci.*, **28**, 541 (1971).

30. Benninghoven, A., and A. Mueller, *Phys. Lett. A*, **40**, 169 (1972).

31. Benninghoven, A., *Appl. Phys.*, **1**, 1 (1973).

32. Benninghoven, A., *Surf. Sci.*, **35**, 427 (1973).

32a. Benninghoven, A., and A. Mueller, *Surf. Sci.*, **39**, 416 (1973).

32b. Benninghoven, A., and L. Wiedmann, *Surf. Sci.*, **41**, 483 (1974).

33. Benninghoven, A., D. Jaspers, and W. Sichtermann, *Appl. Phys.*, **11**, 35 (1976).

34. Benninghoven, A., *Crit. Rev. Solid State Sci.*, **6**, 291 (1976).

35. Benninghoven, A., *Thin Solid Films*, **39**, 3 (1976).

36. Benninghoven, A., and W. Sichtermann, *Organic Mass Spectrom.*, **12**, 595 (1977).

37. Berenz, J. J., G. J. Scilla, V. L. Wrick, L. F. Eastman, and G. H. Morrison, *J. Vac. Sci. Technol.*, **13**, 1152 (1976).

38. Berkey, E., G. G. Sweeney, and W. M. Hickam, *Nucl. Technol.*, **16**, 263 (1972).

39. Bernheim, M., and G. Slodzian, *Int. J. Mass Spectrom. Ion Phys.*, **12**, 93 (1973).

40. Bernheim, M., *Radiat. Eff.*, **18**, 231 (1973).

41. Bernheim, M., and G. Slodzian, *J. Phys. Lett.*, **38**, L-325 (1977).

42. Bernheim, M., and G. Slodzian, Presented at the Internl. SIMS Conf., Muenster, Germany, 1977.

43. Beske, H. E., *Z. Naturforsch.*, **22A**, 459 (1967).

43a. Beynon, J. H., *Mass Spectrometry and its Applications to Organic Chemistry*, Elsevier Publishing Company, Amsterdam, 1960.

43b. Biemann, K., *Mass Spectrometry, Organic Chemical Applications*, McGraw-Hill Book Company, Inc., New York, 1962.

43c. Bindell, J. B., J. W. Colby, R. Wonsidler, J. M. Poate, D. K. Conley, and T. C. Tisone, *Thin Solid Films*, **37**, 441 (1976).

44. Bisdom, E. B. A., S. Henstra, A. Jongerius, J. D. Brown, A. P. von Rosenstiel, and D. J. Gras, *Neth. J. Agric. Sci.*, **25**, 1 (1977).

44a. Bitenskii, I. S., and E. S. Parilis, *Sov. Phys. Tech. Phys.*, **23**, 1104 (1978).

44b. Blaise, G., and G. Slodzian, *J. Physique*, **31**, 93 (1970); **35**, 237 (1974); **35**, 243 (1974).

45. Blaise, G., *Radiat. Eff.*, **18**, 235 (1973).

45a. Blaise, G., and G. Slodzian, *Surf. Sci.*, **40**, 708 (1973).

46. Blaise, G., J. P. Contour, and C. Leclerq, *J. Microsc. Spectrosc. Electron.*, **1**, 247 (1976).

46a. Blaise, G., *Surf. Sci.*, **40**, 65 (1976).

47. Blaise, G., and R. Castaing, *C.R. Acad. Sc. Paris, Ser. B*, **284**, 449 (1977).

48. Blanchard, B., J-C. Brun, and N. Hilleret, *Analusis*, **3**, 312 (1975).

49. Blanchard, B., P. Carrier, N. Hilleret, J. L. Marguerite, and J. C. Rocco, *Analusis*, **4**, 180 (1976).

50. Blanchard, B., J. L. Combasson, and J. C. Bourgoin, *Appl. Phys. Lett.*, **28**, 7 (1976).

50a. Blandin, A., A. Nourtier, D. W. Hone, *J. Physique*, **37**, 369 (1976).

50b. Blank, P., and K. Wittmaack, *J. Appl. Phys.*, **50**, 1519 (1979).

50c. Blauth, E. W., *Dynamische Massenspektrometer*, Friedr. Vieweg & Sohn, Braunschweig, 1965.

51. Blum, N. A., C. Feldman, and F. G. Satkiewicz, *Phys. Status Solidi A*, **41**, 481 (1977).

52. Bolduc, L., and M. Baril, *J. Appl. Phys.*, **44**, 657 (1973).

53. Bosch, F., A. El Goresy, B. Martin, B. Povh, R. Nobiling, D. Schwalm, and K. Traxel, *Nucl. Instrum. Methods*, **149**, 665 (1978).

53a. Brown, J. B., and J. M. Short, paper Nr. 138, Proc. 12th Ann. Conf. Microbeam Analysis Society, Boston, August 18–24, 1977, MAS, Ed.

54. Bruhn, C. G., B. L. Bentz, and W. W. Harrison, *Anal. Chem.*, **50**, 373 (1978).

54a. Brunnee, C., and H. Voshage, *Massenspektrometrie*, Verlag Karl Thiemig, KG, Munich, Germany, 1965.

55. Bueger, P. A., and J. H. Schilling, *Proc. of the 7th Internl. Vac. Congr. and the 3rd Internl. Conf. on Solid Surfaces*, Vienna, 1977, p. 2589.

56. Bueger, P. A., and J. H. Schilling, Presented at the Internl. SIMS Conf., Muenster, Germany, 1977.

57. Buhl, R., and A. Preisinger, *Surf. Sci.*, **47**, 344 (1975).

58. Buhl, R., W. K. Huber, and E. Loebach, *Vakuum-Technik*, **24**, 189 (1975).

59. Burkhardt, F., and C. Wagner, *Phys. Status Solidi A*, **39**, K63 (1977).

59a. Cantagrel, M., and M. Marchal, *J. Materials Sci.*, **8**, 1711 (1973).

60. Carlson, T. A., *Photoelectron and Auger Spectroscopy*, Plenum, New York, 1975.

60a. Carlson, D. E., and C. W. Magee, *Appl. Phys. Lett.*, **33**, 81 (1978).

61. Carpenter, B. S, and R. L. Myklebust, *Anal. Chim. Acta.*, **81**, 409 (1976).

61a. Carter, G., J. S. Colligon, and J. H. Leck, *Proc. Phys. Soc.*, **79**, 299 (1962).

62. Carter, G., and J. S. Colligon, *Ion Bombardment of Solids*, Heinemann, London, 1968.

62a. Carter, G., D. G. Armour, K. J. Snowdon, *Radiat. Eff.*, **35**, 175 (1978).

63. Castaing, R., and G. Slodzian, *J. Microscopie (Paris)*, **1**, 395 (1962).

63a. Castaing, R., and G. Slodzian, *C.R. Acad. Sc. Paris*, **255**, 1893 (1962).

64. Castaing, R., and G. Slodzian, *Proc. 1st Interl. Conf. Electron and Ion Beam Sci. and Technol.*, Wiley, New York, 1965, p. 780.

65. Chamberlain, M. B., and R. J. Lederich, *Thin Solid Films*, **41**, 167 (1977).

65a. Chapman, G. E., B. W. Farmery, M. W. Thompson and I. H. Wilson, *Radiat. Eff.*, **13**, 121 (1972).

66. Cherepin, V. T., and Yu. P. Maifet, *Sov. Phys. Tech. Phys.*, **17**, 766 (1972).

67. Cherepin, V. T., and M. A. Vasilev, *Vtorikhanaya Ionno-Ionnaya Emissiya Metallovii Splavov*, Naukova Dumka, Kiev, 1976, p. 198.

67a. Christie, W. H., R. E. Ely, R. L. Anderson, and T. G. Kollie, *Appl. Surf. Sci.*, **3**, 329 (1979).

68. Cini, M., *Surf. Sci.*, **54**, 71 (1976).

69. Clampitt, R., and D. K. Jefferies, *Nucl. Instrum. Methods*, **149**, 739 (1978).

70. Clement, S., W. Compston, and G. Newstead, Presented at the Internl. SIMS Conf., Muenster, Germany, 1977.

71. Coburn, J. W., and E. Kay, *Appl. Phys. Lett.*, **18**, 435 (1971); **19**, 350 (1971); *Proc. 6th Internl. Vac. Congr. 1974*, Japan. *J. Appl. Phys.*, Suppl. 2, Pt. 1, 501 (1974).

72. Colby, J. W., in J. I. Goldstein and H. Yakowitz, Eds., *Practical Scanning Electron Microscopy*, Plenum, New York, 1975, p. 529.

73. Colby, J. W., and L. E. Katz, *J. Electrochem. Soc.*, **123**, 409 (1976).

74. Coleman, D. J., D. W. Shaw, and R. D. Dobrott, *J. Electrochem. Soc.*, **124**, 239 (1977).

75. Coles, J. N., and J. V. P. Long, *Philos. Mag.*, **29**, 457 (1974).

75a. Coles, J. N., *Surf. Sci.*, **55**, 721 (1976); **79**, 549 (1979).

76. Colton, R. J., J. S. Murday, J. R. Wyatt, and J. J. Decorpo, *Proc. of the 25th Annual Conf. on Mass Spectrom. and Allied Topics*, Washington, D.C., May 29–June 3, 1977, p. 322.

77. Colton, R. J., J. S. Murday, J. R. Wyatt, and J. J. Decorpo, Presented at the Internl. SIMS Conf., Muenster, Germany, 1977.

77a. Comas, J., P. K. Chatterjee, W. V. Levige, K. V. Vaidyanathan, and B. G. Streetman, *Ion Implantation in Semiconductors*, J. A. Borders, Ed., Plenum, 1977.

78. Contamin, P., and G. Slodzian, *Appl. Phys. Lett.*, **13**, 416 (1968).

79. Cookson, J. A., and F. D. Pilling, *Thin Solid Films*, **19**, 381 (1973).

80. Cox, B., and J. P. Pemsler, *J. Nucl. Mater.*, **28**, 73 (1968).

81. Christie, W. H., D. H. Smith, and H. Inouye, *J. Radioanal. Chem.*, **32**, 85 (1976).

82. Christie, W. H., and T. G. Kollie, Proc. *25th Annual Conf. on Mass Spectrom. and Allied Topics*, Washington, D.C., May 29-June 3, 1977, p. 435.

83. Croset, M., and D. Dieumegard, *Corrosion Science*, **16**, 703 (1976).

84. Croset, M., and D. Dieumegard, *J. Microsc. Electron.*, **2**, 329 (1977).

84a. Cuif, J. P., and R. Lefevre, *C.R. Acad. Sc. Paris*, **278**, 2263 (1975).

85. Cuif, J. P., Y. Dauphin, R. Lefevre, M. T. Venec-Peyre, *J. Microsc. Spectrosc. Electron.*, **2**, 313 (1977).

86. Cuif, J. P., Y. Dauphin, and R. Lefevre, *C.R. Sc. Paris, Ser. D*, **285**, 81 (1977).

87. Daly, N. R., *Rev. Sci. Instrum.*, **31**, 264 (1960).

88. Daly, N. R., A. McCormick, and R. E. Powell, *Rev. Sci. Instrum.*, **39**, 1163 (1968).

89. Dawson, P. H., *Int. J. Mass Spectrom. Ion Phys.*, **17**, 447 (1975).

90. Dawson, P. H., Ed., *Quadrupole Mass Spectrometry and its Applications*, Elsevier, Amsterdam, 1976.

91. Dawson, P. H., and P. A. Redhead, *Rev. Sci. Instrum.*, **48**, 159 (1977).

92. Degreve, F., R. Figaret, and P. Laty, Presented at the Internl. SIMS Conf., Muenster, Germany, 1977.

92a. Deline, V. R., W. Katz, C. A. Evans Jr., and P. Williams, *Appl. Phys. Lett.*, **33**, 832 (1978).

92b. Deline, V. R., C. A. Evans Jr., and P. Williams, *Appl. Phys. Lett.*, **33**, 578 (1978).

93. Dennis, E., and R. MacDonald, *Radiat. Eff.*, **13**, 243 (1972).

94. Derbenwick, G. F., *J. Appl. Phys.*, **48**, 1127 (1977).

95. Deventer, E. H. van, T. A. Renner, R. H. Pelto, and V. A. Maroni, *J. Nucl. Mater.*, **64**, 241 (1977).

96. Devienne, F. M., *Le Vide*, **28**, 193 (1973).

97. Devienne, F. M., and J-C. Roustan, *C.R. Acad. Sc. Paris, Ser. B*, **283**, 397 (1976).

98. Dietrich, H. B., W. H. Weisenberger, and J. Comas, *Appl. Phys. Lett.*, **28**, 182 (1976).

99. Dillon, A. F., R. S. Lehrle, J. C. Robb, and D. W. Thomas, *Adv. Mass Spectrom.*, **4**, 477 (1968).

100. Dobrott, R. D., *Electrochem. Soc. Ext. Abstr.*, **100**, 250 (1975).

101. Doi, H., I. Kanomato, and N. Sakudo, Proc. of the 7th Conf. on Solid State Devices, Tokyo, 1975, p. 71. Suppl. to *Jpn. J. Appl. Phys.*, **15**, (1976).

102. Doucas, G., *Int. J. Mass Spectrom. Ion Phys.*, **25**, 71 (1977).

103. Dowsett, M. G., R. M. King, and E. H. C. Parker, *J. Phys. E*, **8**, 704 (1975).

104. Dowsett, M. G., R. M. King, and E. H. C. Parker, *Appl. Phys. Lett.*, **31**, 529 (1977).

105. Dowsett, M. G., R. M. King, and E. H. C. Parker, *Surf. Sci.*, **71**, 541 (1978).

106. Drzal, L. T., *Carbon*, **15**, 129 (1977).

107. Duesterhoeft, H., R. Manns, and S. Rogaschewski, *Exp. Techn. Phys.*, **25**, 117 (1977).

108. Evans, C. A., *Anal. Chem.*, **44**, 67A (1972).

109. Evans, C. A., *Thin Solid Films*, **19**, 11 (1973).

110. Evans, C. A., *Anal. Chem.*, **47**, 818A, (1975); **47**, 855A (1975).

111. Evans, C. A., *J. Vac. Sci. Technol.*, **12**, 144 (1975).

112. Evans, C. A., *Natl. Bur. Std. (U.S.) Spec. Publ.*, **400-23**, 219 (1976).

112a. Evans, C. A., Jr., V. R. Deline, T. W. Sigmon, and A. Lidow, *Appl. Phys. Lett.*, **35**, 291 (1979).

112b. Ewald, H., and H. Hintenberger, *Methoden und Anwendungen der Massenspektroskopie*, Verlag Chemie, GmbH, Weinheim/Bergstrasse, 1953.

113. Fassett, J. D., J. R. Roth, and G. H. Morrison, *Anal. Chem.*, **49**, 2322 (1977).

113a. Favennec, P. N., and H. L'Haridon, *Appl. Phys. Lett.*, **35**, 699 (1979).

114. Feldman, C., and F. G. Satkiewicz, *J. Electrochem. Soc.*, **120**, 1111 (1973).

115. Ferrante, J., and S. V. Pepper, *Surf. Sci.*, **57**, 420 (1976).

116. Finkelstein, A. T., *Rev. Sci. Instrum.*, **11**, 94 (1940).

117. Fite, W. L., and M. W. Siegel, *Proc. of the 25th Annual Conf. on Mass Spectrom. and Allied Topics*, Washington, D.C., May 29–June 3, 1977, p. 197.

118. Fogel, Ya. M., *Int. J. Mass Spectrom. Ion Phys.*, **9**, 109 (1972).

119. Fontana, P. V., J. P. Decosterd, and L. Wegmann, *J. Electrochem. Soc.*, **121**, 146 (1974).

120. Fralick, R. D., *J. Vac. Sci. Technol.*, **13**, 388 (1976).

121. Fralick, R. D., and R. L. Conrad, *Research/Development*, **28**, 32 (1977).

122. Frostell, G., S. J. Larsson, A. Lodding, H. Odelius, and L. G. Petersson, *Scand. J. Dent. Res.*, **85**, 18 (1977).

123. Fuller, D., J. S. Colligon, and J. S. Williams, *Surf. Sci.*, **54**, 647 (1976).

124. Galle, P., and J. P. Dumery, in J. F. Gross, R. Kaufmann, and E. Wetterer, Eds., *Modern Techniques in Physiological Sciences*, Academic Press, London, 1973, p. 385.

124a. Galle, P., in T. Hall, P. Echlin, and R. Kaufmann, Eds., *Microprobe Analusis as Applied to Cells and Tissues*, Academic, New York (1974).

125. Garrec, J. P., B. Blanchard, J. C. Brun, A. M. Bisch, R. Bligny, and A. Fourcy, *C.R. Acad. Sc. Paris, Ser. D*, **277**, 805 (1973).

125a. Garrison, B. J., N. Winograd, and D. E. Harrison, Jr., *Surf. Sci.*, **87**, 101 (1979).

125b. Gauneau, M., *Analusis*, **5**, 357 (1977).

126. Gerber, R. M., and J. W. Dzimianshi, *J. Vac. Sci. Technol.*, **10**, 1072 (1973).

126a. Gerhard, W., *Z. Physik*, **B22**, 31 (1975).

126b. Gerhard, W., and H. Oechsner, *Z. Phys.*, **B22**, 41 (1975).

127. Gerlach, R. E., and L. E. Davis, *J. Vac. Sci. Technol.*, **14**, 339 (1977).

128. Gettings, M., and A. J. Kinloch, *J. Mater. Sci.*, **12**, 2511 (1977).

128a Gibbons, D. F., W. S. Johnson, and S. W. Mylroie, *Projected Range Stastistics, Semiconductors and Related Materials*, 2nd Ed., (1975), Dowden, Hutchinson, and Ross, Stroudsberg, Penn.

128b. Gibbons, J. F., J. Peng, J. D. Hong, W. Katz, and C. A. Evans Jr., *J. Appl. Phys.*, **50**, 4388 (1979).

129. Giber, J., and V. K. Josepovits, *Proc. of the 7th Internl. Vac. Congr. and the 3rd Internl. Conf. on Solid Surfaces*, Vienna, 1977, p. 2585.

130. Giber, J., Presented at the Internl. SIMS Conf., Muenster, Germany, 1977.

131. Giffin, C. E., H. G. Boettger, and D. D. Norris, *Int. J. Mass Spectrom. Ion Phys.*, **15**, 437 (1974).

132. Giletti, B. J., M. P. Semet, and R. A. Yund, *EOS, Trans. Am. Geophys. Union*, **57**, 350 (1976).

133. Golowacz, H., and J. Marks, *Proc. of the 7th Internl. Vac. Congr. and the 3rd Internl. Conf. on Solid Surfaces*, Vienna, 1977, p. 2543.

133a. Good-Zamin, C. J., M. T. Shehata, D. B. Squires, and R. Kelly, *Radiat. Eff.*, **35**, 139 (1978).

134. Gossink, R. G., H. A. M. de Grefte, and H. W. Werner, *Silicates Industriels*, **44**, 35 (1979).

134a. Gossink, R. G., and T. P. A. Lommen, *J. Am. Cer. Soc.*, **61**, 539 (1978); *Appl. Phys. Lett.*, **34**, 444 (1979).

135. Goutte, R., C. Guilland, R. Javelas, and J-P. Meriaux, *Optik*, **26**, 574 (1967).

136. Grant, J. T., and M. P. Hooker, *Phys. Lett. B.*, **58**, 167 (1975).

137. Gray, H. R., *Corrosion-NACE*, **28**, 47 (1972).

138. Greene, J. E., F. Sequeda-Osorio, and B. R. Natarajan, *J. Appl. Phys.*, **46**, 2701 (1975); *J. Vac. Sci. Technol.*, **12**, 366 (1975).

139. Gregory, M. C., J. D. Stein, and H. A. Storms, *Proc. of the 24th Annual Conf. on Mass Spectrom. and Allied Topics*, San Diego, California, May 9 - 13, 1976, p. 651.

139a. Gries, W. H., and F. G. Ruedenauer, *Int. J. Mass Spectrom. Ion Phys.*, **18**, 111 (1975).

140. Grossbeck, M. L., H. K. Birnbaum, P. Williams, and C. A. Evans, *Phys. Status Solidi A*, **34**, K97 (1976).

141. Grossbeck, M. L., and H. K. Birnbaum, *Acta. Metall.*, **25**, 135 (1977).

142. Grundner, M., W. Heiland, and E. Taglauer, *Appl. Phys.*, **4**, 243 (1974).

143. Guthrie, J. W., *J. Electrochem. Soc.*, **121**, 1617 (1974).

144. Haff, P. K., and Z. E. Switkowski, *J. Appl. Phys.*, **48**, 3383 (1977).

144a. Hage-Ali, M., R. Stuck, A. N. Saxena, and P. S. Herf, *Appl. Phys.*, **19**, 25 (1979).

145. Harrington, W. L., C. W. Magee, G. W. Cullen, and J. F. Corboy, *Proc. of the 24th Annual Conf. on Mass Spectrom. and Allied Topics*, San Diego, California, May 9 - 13, 1976, p. 310.

145a. Harrison, D. E., Jr., W. L. Moore, Jr., and H. T. Holcombe, *Radiat. Eff.*, **17**, 167 (1973).

145b. Harrison, D. E., and C. B. Delaplain, *J. Appl. Phys.*, **47**, 2252 (1976).

145c. Harrison, D. E., Jr., P. W. Kelly, B. J. Garrison, and N. Winograd, *Surf. Sci.*, **76**, 311 (1978).

145d. Harrison, D. E. Jr., B. J. Garrison, and N. Winograd, in A. Benninghoven, C. A. Evans Jr., R. A. Powell, R. Shimizu, and H. A. Storms, Eds., *Secondary Ion Mass Spectrometry (SIMS-II)*, Proc. of the Secondary Internl. Conf. on Secondary Ion Mass Spectrometry (SIMS-II), Stanford University, Stanford, California, August 27 - 31, 1979, Springer-Verlag, Berlin - Heidelberg - New York, 1979, p. 12.

146. Heil, H., *Z. Phys.*, **120**, 212 (1942).

147. Hennequin, J. F., *J. Phys.*, **29**, 655 (1968).

148. Hennequin, J. F., *J. Phys.*, **29**, 957 (1968).

149. Herzog, R. F. K., W. P. Poschenrieder, and F. G. Satkiewicz, NASA Contract No. NAS5-9254, Final Report, GCA-TR-67-3N, 1967.

150. Herzog, R. F. K., W. P. Poschenrieder, and F. G. Satkiewicz, *Radiat. Eff.*, **18**, 199 (1973).

151. Heyndryckx, P., *Glastechn. Ber.*, **44**, 543 (1976).

152. Hinthorne, J. R., and P. H. Ribbe, *Am. Mineral.*, **59**, 1123 (1974).

153. Hinthorne, J. R., and C. A. Andersen, *Am. Mineral.*, **60**, 143 (1975).

154. Hirao, T., K. Inoue, S. Takayanagi, and Y. Yaegashi, *Appl. Phys. Lett.*, **31**, 505 (1977).

154a. Hirao, T., K. Inoue, S. Takayanagi, and Y. Yaegashi, *J. Appl. Phys.*, **50**, 193 (1979).

154b. Hirao, T., K. Inoue, Y. Yaegashi, and T. Shigetoshi, *Jpn. J. Appl. Phys.*, **18**, 647 (1979).

155. Ho, P. S., and J. E. Lewis, *Surf. Sci.*, **55**, 335 (1976).

156. Hofer, W. O., and H. Liebl, *Appl. Phys.*, **8**, 359 (1975).

157. Hofer, W. O., and U. Littmark, *Nucl. Instrum. Methods*, **138**, 67 (1976).

158. Hofer, W. O., and F. Thum, *Nucl. Instrum. Methods*, **149**, 535 (1978).

158a. Hofer, W. O., and U. Littmark, *Phys. Lett. A*, **71**, 457 (1979).

159. Hofman, A. W., B. J. Giletti, J. R. Hinthorne, C. A. Andersen, and D. Comaford, *Earth Planet Sci. Lett.*, **24**, 48 (1974).

160. Hofmann, S., *Appl. Phys.*, **9**, 59 (1976).

161. Hofmann, S., *Appl. Phys.*, **13**, 205 (1977).

162. Hofmann, S., J. Erlewein, and A. Zalar, *Thin Solid Films*, **43**, 275 (1977).

163. Hofmann, S., *Proc. of the 7th Internl. Vac. Congr. and the 3rd Internl. Conf. on Solid Surfaces*, Vienna, 1977, p. 2613.

163a. Honda, F., G. M. Lancaster, Y. Fukuda, and J. W. Rabalais, *J. Chem. Phys.*, **69**, 4931 (1978).

163b. Honda, F., Y. Fukuda, and J. W. Rabalais, *J. Chem. Phys.*, **70**, 4834 (1979).

163c. Honig, R. E., *J. Chem. Phys.*, **22**, 126 (1954).

163d. Honig, R. E., *J. Appl. Phys.*, **29**, 549 (1958).

164. Honig, R. E., *Adv. Mass Spectrom.*, **6**, 337 (1974).

165. Honig, R. E., *Thin Solid Films*, **31**, 89 (1976).

165a. Hourdry, J., and M. Truchet, *Tissue and Cell*, **8**, 175 (1976).

166. Hovland, C. T., *Appl. Phys. Lett.*, **30**, 274 (1977).

167. Huber, A. M., G. Morillot, N. T. Linh, J. L. Debrun, and M. Valladon, *Nucl. Instrum. Methods*, **149**, 543 (1978).

167a. Huber, A. M., N. T. Linh, M. Valladon, J. L. Debrun, G. M. Martin, A. Mitonneau, and A. Mircea, *J. Appl. Phys.*, **50**, 4022 (1979).

167b. Huber, A. M., G. Morillot, N. T. Linh, P. N. Favennec, B. Deveaud, and B. Toulouse, *Appl. Phys. Lett.*, **34**, 858 (1979).

168. Huber, W. K., H. Selhofer, and A. Benninghoven, *J. Vac. Sci. Technol.*, **9**, 482 (1972).

169. Huber, W. K., and E. Loebach, *Vacuum*, **22**, 605 (1972).

170. Hubler, G. K., J. Gomas, and L. Plew, *Nucl. Instrum. Methods*, **149**, 635 (1978).

171. Hughes, H. L., R. D. Baxter, and D. Phillips, *IEEE Trans. Nucl. Sci.*, **NS-19**, (4), 256 (1972).

172. Hurrle, A., and M. Schulz, in *Lattice Defects in Semiconductors*, 1974, Inst. Phys. Conf. Ser., 23, London, (1975), Ch. 7, p. 474.

173. Hurrle, A., and G. Sixt, *Appl. Phys.*, **8**, 293 (1975).

173a. Inoue, K., T. Hirao, Y. Yaegashi, and S. Takayanagi, *Jpn. J. Appl. Phys.*, **18**, 367 (1979).

173b. Ishitani, T., R. Shimizu, and K. Murata, *Phys. Status Solidi B*, **50**, 681 (1972); *Jpn. J. Appl. Phys.*, **11**, 125 (1972).

173c. Ishitani, T., and R. Shimizu, *Phys. Lett. A*, **46**, 487 (1974).

174. Ishitani, T., and R. Shimizu, *Appl. Phys.*, **6**, 241 (1975).

175. Ishitani, T., R. Shimizu, and H. Tamura, *Appl. Phys.*, **6**, 277 (1975).

176. Iwaki, M., S. Nambo, K. Yoshida, and N. Soda, *Proc. of the 7th Internl. Vac. Congr. and the 3rd Internl. Conf. on Solid Surfaces*, Vienna, 1977, p. 1429.

177. Iwaki, M., S. Nambo, K. Yoshida, N. Soda, K. Yukawa, and T. Sato, *Jpn. J. Appl. Phys.*, **16**, 1475 (1977).

177a. Jackson, D. P., *Can. J. Phys.*, **53**, 1513 (1975).

178. Jeantet, A. Y., R. Martoja, and M. Truchet, *C.R. Acad. Sc. Paris, Ser. D*, **278**, 1441 (1974).

179. Johnson, C. E., and D. V. Steidl, *Adv. Chem. Ser.*, **158**, 349 (1976).

180. Jones, R. H., *Metall. Trans.*, **8A**, 378 (1977).

181. Joshi, A., L. E. Davis, and P. W. Palmberg, in A. W. Czanderna, Ed., *Methods of Surface Analysis*, Elsevier, Amsterdam, 1975, p. 159.

181a. Joyes, P., *J. Physique*, **29**, 774 (1968).

181b. Joyes, P., *J. Physique*, **30**, 243 (1969); **30**, 365 (1969).

181c. Joyes, P., *J. Phys. Chem. Solids*, **32**, 1269 (1971).

181d. Joyes, P., *J. Phys. B; Atom. Molec. Phys.*, **4**, L15 (1971).

181e. Joyes, P., and G. Toulouse, *Phys. Lett. A*, **39**, 267 (1972).

182. Joyes, P., *Radiat. Eff.*, **19**, 235 (1973).

182a. Joyes, P., *C.R. Acad. Sc. Paris, Ser. B*, **288**, 155 (1979).

183. Jurela, Z., and B. Perovic, *Can. J. Phys.*, **46**, 773 (1968).

184. Jurela, Z., in D. W. Palmer, M. W. Thompson, and D. P. Townsend, Eds., *Atomic Collision Phenomena in Solids*, North-Holland, Amsterdam, 1970, p. 339.

185. Jurela, Z., *Int. J. Mass Spectrom. Ion Phys.*, **12**, 33 (1973).

185a. Kalinowski, L., and R. Seguin, *Appl. Phys. Lett.*, **35**, 211 (1979).

186. Kaminsky, M., *Atomic and Ionic Impact Phenomena on Metal Surfaces*, Springer-Verlag, Berlin, 1965.

187. Kato, Y., Y. Tajima, H. Hayakawa, A. Shibata, and K. Nishiwaki, *Mass Spectrosc., (Jpn.)*, **23**, 271 (1975).

187a. Kazmerski, L. L., *Thin Solid Films*, **57**, 99 (1979).

188. Keefer, D. W., and A. G. Pard, *J. Nucl. Mater.*, **47**, 97 (1973).

189. Kelly, R., and N. Q. Lam, *Radiat. Eff.*, **19**, 39 (1973).

189a. Kelly, R., and C. B. Kerkdijk, *Surf. Sci.*, **46**, 537 (1974).

190. Kempf, J., and G. Kaus, *Appl. Phys.*, **13**, 261 (1977).

191. Kempf, J., and G. Kaus, Presented at the Internl. SIMS Conf. Muenster, Germany, 1977.

191a. Kempf, J., in A. Benninghoven, C. A. Evans Jr., R. A. Powell, R. Shimizu, and H. A. Storms, Eds., *Secondary Ion Mass Spectrometry (SIMS-II), Proc. of the Secondary Internl. Conf. on Secondary Ion Mass Spectrometry (SIMS-II)*, Stanford University, Stanford, California, USA, August 27-31, 1979, Springer-Verlag, Berlin-Heidelberg—New York, 1979, p. 97.

191b. Kerkdijk, C. B., and R. Kelly, *Radiat. Eff.*, **38**, 73 (1978).

191c. Kienitz, H., Ed., *Massenspektrometrie*, Verlag Chemie, GmbH, Weinheim/Bergstrasse, 1968.

192. Kiko, J., H. W. Mueller, T. Kirsten, S. Kalbitzer, M. Warhaut, and K. Buechler, Presented at the Internl. SIMS Conf., Muenster, Germany, 1977.

193. Kim, H. B., G. G. Sweeney, and T. M. S. Heng, *Inst. Phys. Conf. Ser. No. 24*, 307 (1975).

194. Kingery, W. D., W. L. Robbins, A. F. Henriksen, and C. E. Johnson, *J. Am. Ceram. Soc.*, **59**, 239 (1976).

194a. Kirschner, J., and H. W. Etzkorn, *Appl. Surf. Sci.*, **3**, 251 (1979).

195. Kisieleski, W., and R. Ringo, *J. Microsc.*, **104**, pt. 2, 199 (1975).

196. Kitada, A., and H. Tamura, *Mass Spectrosc.*, **25**, 85 (1977).

196a. Koennen, G. P., A. Tip, and A. E. de Vries, *Radiat. Eff.*, **21**, 269 (1974); **26**, 23 (1975).

196b. Kolonits, V. P., M. Koltai, and D. Marton, *Thin Solid Films*, **57**, 221 (1979).

197. Kolot, V. Ya., V. I. Tatus, V. V. Vodolazhchenko, V. F. Rybalko, A. E. Grodshtein, N. D. Kirsanov, and Ya. M. Fogel, *Sov. Phys. Tech. Phys.*, **17**, 111 (1972).

198. Komiya, S., T. Narusawa, and T. Satake, *J. Vac. Sci. Technol.*, **12**, 361 (1975).

199. Krauss, A. R., and D. M. Gruen, *Appl. Phys.*, **14**, 89 (1977); *Nucl. Instrum. Methods*, **149**, 547 (1978).

199a. Krimmel, E. F., and H. Pfleiderer, *Radiat. Eff.*, **19**, 83 (1973).

200. Krohn, V. E., *J. Appl. Phys.*, **33**, 3523 (1962).

201. Krohn, V. E., and G. R. Ringo, *Rev. Sci. Instrum.*, **43**, 1771 (1972).

202. Krohn, V. E., and G. R. Ringo, *Appl. Phys. Lett.*, **27**, 479 (1975); *Int. J. Mass Spectrom. Ion Phys.*, **22**, 307 (1976).

202a. Krohn, V. E., *Int. J. Mass Spectrom. Ion Phys.*, **22**, 43 (1976).

203. Kuehl, Ch., M. Druminski, and K. Witmaack, *Thin Solid Films*, **37**, 317 (1976).

204. Kusao, K., Y. Yoshioka, N. Nakamura, and F. Konishi, *Proc. of the 23rd Annual Conf. on Mass Spectrom. and Allied Topics*, Houston, Texas, May 25-30, 1975, p. 368.

205. Laegreid, N., and G. K. Wehner, *J. Appl. Phys.*, **32**, 365 (1961).

206. Lane, W. C., and N. C. Yew, in O. Johari and I. Corvin, Eds., *Scanning Electron Microscopy/1973*, IIT Res. Inst., Chicago, 1973, p. 81.

207. Larsson, S. J., A. Lodding, H. Odelius, and L. G. Petersson, *Calc. Tiss. Res.*, **24**, 179 (1977).

208. Larsson, S. J., A. Lodding, H. Odelius, and G. J. Flim, *Adv. Mass Spectrom.*, **7A**, 797 (1978).

208a. Laurent, R., and G. Slodzian, *Radiat. Eff.*, **19**, 181 (1973).

209. Lee, D. H., and R. M. Malbon, *Appl. Phys. Lett.*, **30**, 327 (1977).

209a. Lefevre, R., *J. Microscopic Biol. Cell.*, **22**, 335 (1975).

209b. Lefevre, R., R. M. Frank, and J. C. Voegel, *Calcif. Tiss. Res.*, **19**, 251 (1976).

210. Lefevre, R., and M. T. Venec-Peyre, *C.R. Acad. Sc. Paris, Ser. D*, **285**, 23 (1977).

211. Lefevre, R., *8th Internl. ConF. on X-Ray Optics and Microanalysis*, Boston, U.S.A., 1977, paper no. 143.

211a. Leleyter, M., and P. Joyes, *Radiat. Eff.*, **18**, 105 (1973).

212. Leroy, V., J-P. Servais, and L. Habraken, *Metall. Rep. CRM*, **35**, 69 (1973).

213. Leroy, V., J-P. Servais, and J. Richelmi, *J. Microsc. Spectrosc. Electron*, **1**, 81 (1976).

214. Lewis, R. K., J. M. Morabito, and J. C. C. Tsai, *Appl. Phys. Lett.*, **23**, 260 (1973).

215. Leys, J. A., and J. T. McKinney, in O. Johari and I. Corvin, Eds., *Scanning Electron Microscopy/1976*, IIT Res. Inst., Chicago, 1976, p. 231.

215a. Liau, Z. L., B. Y. Tsaur, and J. W. Mayer, *J. Vac. Sci. Technol.*, **16**, 121 (1979).

216. Lidow, A., J. F. Gibbons, V. R. Deline, and C. A. Evans, *Appl. Phys. Lett.*, **32**, 15 (1978); **32**, 149 (1978).

216a. Lidow, A., J. F. Gibbons, V. R. Deline, and C. A. Evans, Jr., *Appl. Phys. Lett.*, **32**, 572 (1978).

217. Liebl, H., and R. F. K. Herzog, *J. Appl. Phys.*, **34**, 2893 (1963).

218. Liebl, H., *J. Appl. Phys.*, **38**, 5277 (1967).

219. Liebl, H., *Int. J. Mass Spectrom. Ion Phys.*, **6**, 401 (1971).

220. Liebl, H., *Anal. Chem.*, **46**, 22A (1974).

221. Liebl, H., *J. Phys. E*, **8**, 797 (1975).

222. Liebl, H., *J. Vac. Sci. Technol.*, **12**, 385 (1975).

223. Liebl, H., *Natl. Bur. Std. (U.S.) Spec. Publ.*, **427**, 1 (1975).

224. Liebl, H., *Adv. Mass Spectrom.*, **7A**, 751 (1978).

225. Liebl, H., *Adv. Mass Spectrom.*, **7A**, 807 (1978).

226. Liebl, H., and W. W. Harrison, *Int. J. Mass Spectrom. Ion Phys.*, **22**, 237 (1976).

226a. Lindhard, J., M. Scharff, and H. E. Schiott, *Mat. Fys. Medd. Dan. Vid. Selsk.*, **33**, nr. 14 (1963).

226b. Lindhard, J., V. Nielsen, M. Scharff, and P. V. Thomson, *Mat. Fys. Medd. Dan. Vid. Selsk.*, **33**, nr. 10 (1963).

226c. Lindhard, J., V. Nielsen, and M. Scharff, *Mat. Fys. Medd. Dan. Vid. Selsk.*, **36**, nr. 10 (1968).

227. Linton, R. W., P. Williams, C. A. Evans, and D. F. S. Natusch, *Anal. Chem.*, **49**, 1514 (1977).

228. Litovchenko, V. G., G. Ph. Romanova, and R. I. Marchenko, *Proc. of the 7th Internl. Vac. Congr. and the 3rd Internl. Conf. on Solid Surfaces*, Vienna, 1977, p. 2047.

228a. Litovchenko, V. G., R. I. Marchenko, G. Romanova, and V. N. Vasilevskaya, *Thin Solid Films*, **44**, 295 (1977).

228b. Littmark, U., G. Maderlechner, B. Berisch, B. M. U. Scherzer, and M. T. Robinson, *Nucl. Instrum. Methods*, **132**, 661 (1976).

229. Lodding, A., J-M. Gourgout, L. G. Petersson, and G. Frostell, *Z. Naturforsch.*, **29A**, 897 (1974).

230. Lodding, A., and L. Lundkvist, *Thin Solid Films*, **25**, 491 (1975).

231. Lodding, A., H. Odelius, and L. Ekbom, *Proc. of the 7th Internl. Vac. Congr. and the 3rd Internl. Conf. on Solid Surfaces*, Vienna, 1977, p. 2531.

232. Lodding, A., and H. Odelius, in P. Echlin et al., Eds., *Proc. Internl. Conf. on Microprobe Analysis in Biology and Medicine*, Muenster, Sept. 1977.

233. Long, J. V. P., D. M. Astill, J. N. Coles, and S. J. B. Reed, *8th Internl. Conf. on X-Ray Optics and Microanalysis*, Boston, U.S.A., 1977, paper 132.

234. Lovering, J. F., *Comments on Earth Sciences: Geophysics*, **3**, 153 (1973).

235. Lovering, J. F., *Natl. Bur. Std. (U.S.) Spec. Publ.*, **427**, 135 (1975).

236. Lyon, O., G. Blaise, C. Roques-Carmes, and G. Slodzian, *C.R. Hebd. Seances Acad. Sc., Ser. B*, **281**, 313 (1975).

236a. MacDonald, R. J., and D. Haneman, *J. Appl. Phys.*, **37**, 1609 (1966).

237. MacDonald, R. J., and P. J. Martin, *Surf. Sci.*, **66**, 423 (1977).

238. Magee, C. W., and W. L. Harrington, *Proc. 25th Annual Conf. on Mass Spectrom. and Allied Topics*, Washington, D.C., May 29-June 3, 1977, p. 316.

239. Magee, C. W., and C. P. Wu, *Nucl. Instrum. Methods*, **149**, 529 (1978).

240. Magee, C. W., W. L. Harrington, and R. E. Honig, *Rev. Sci. Instrum.*, **49**, 477 (1978).

240a. Magee, C. W., and W. L. Harrington, *Appl. Phys.*, **33**, 193 (1978).

240b. Magee, T. J., J. Peng, J. D. Hong, C. A. Evans, Jr., V. R. Deline, and R. M. Malborn, *Appl. Phys. Lett.*, **35**, 277 (1979).

240c. Magee, T. J., J. Peng, J. D. Hong, V. R. Deline, and C. A. Evans, Jr., *Appl. Phys. Lett.*, **35**, 615 (1979).

241. Mai, H., and H. Wagner, *J. Sci. Instrum.*, **44**, 883 (1967).

242. Malm, D. L., M. J. Vasile, F. J. Padden, D. B. Dove, and C. G. Patano, Jr., *J. Vac. Sci. Technol.*, **15**, 35 (1978).

242a. Martoja, R., A. Szoelloesi, and M. Truchet, *J. Microsc. Biol. Cell.*, **22**, 247 (1975).

242b. Martoja, R., J. Alibert, C. Ballan-Dufrancais, A. Y. Jeantet, D. Lhonore, and M. Truchet, *J. Microsc. Biol. Cell.*, **22**, 441 (1975).

243. Mathieu, H. J., D. E. McClure, and D. Landolt, *Thin Solid Films*, **38**, 281 (1976).

244. Matsumoto, R., K. Sato, and K. Suzuki, Proc. 6th Internl. Vac. Congr. 1974, *Jpn. J. Appl. Phys.*, Suppl. 2, pt. 1, (1974), p. 387.

245. Matsushima, Y., and S. Gonda, *Jpn. J. Appl. Phys.*, **15**, 2093 (1976).

246. Mayer, J. W., and A. Turos, *Thin Solid Films*, **19**, 1 (1973).

246a. Mazur, P., and J. B. Sanders, *Physica*, **44**, 444 (1969).

247. McCaughan, D. V., and R. A. Kushner, in R. F. Kane and G. B. Larrabee, Eds., *Characterization of Solid Surfaces*, Plenum, New York, 1974, p. 627.

248. McCrea, J. M., in R. F. Kane and G. B. Larrabee, Eds., *Characterization of Solid Surfaces*, Plenum, New York, 1974, p. 577.

248a. McDowell, C. A., Ed., *Mass Spectrometry*, McGraw-Hill, New York, 1963.

249. McHugh, J. A., and J. F. Stevens, *Anal. Chem.*, **44**, 2187 (1972).

250. McHugh, J. A., *Radiat. Eff.*, **21**, 209 (1974).

251. McHugh, J. A., in A. W. Czanderna, Ed., *Methods of Surface Analysis*, Elsevier, Amsterdam, 1975, p. 223.

252. McHugh, J. A., *Natl. Bur. Std. (U.S.) Spec. Publ.*, **427**, 129 (1975).

253. McHugh, J. A., J. C. Sheffield, L. R. Hanrahan, and D. S. Simons, *Proc. of the 25th Annual Conf. on Mass Spectrom. and Allied Topics*, Washington, D.C., May 29-June 3, 1977, p. 706.

254. McCune, R. C., *J. Vac. Sci. Technol.*, **15**, 31 (1977).

254a. McLevige, W. V., K. V. Vaidyanathan, B. G. Streetman, M. Legems, J. Comas, and L. Plew, *Appl. Phys. Lett.*, **32**, 127 (1978).

255. Meisenheimer, R. G., *J. Nucl. Mater.*, **63**, 429 (1976).

255a. Melton, C. E., *Principles of Mass Spectrometry and Negative Ions*, Marcel Dekker, New York, 1970.

256. Menzinger, M., and L. Whalin, *Rev. Sci. Instrum.*, **40**, 102 (1969).

257. Meriaux, J. P., R. Goutte, and C. Guillaud, *J. Radioanal. Chem.*, **12**, 53 (1972).

258. Meyer, C., D. S. McKay, D. H. Andersen, and P. Butler, *Geochim. Cosmochim. Acta.* (Suppl.), **6**, 1673 (1975).

259. Miyagawa, S., *J. Appl. Phys.*, **44**, 5617 (1973).

260. Miyake, M., T. Ishiki, H. Takeshita, Y. Yamamoto, and T. Sano, *Thin Solid Films*, **40**, 149 (1977).

261. Monnier, J., H. Hilleret, E. Ligeon, and J. B. Quoirin, *J. Radioanal. Chem.*, **12**, 353 (1972).

262. Moon, D. M., and R. C. Koo, *Metall. Trans.*, **2A**, 2115 (1971).
262a. Morabito, J. M., and R. K. Lewis, *Anal. Chem.*, **45**, 869 (1973).
263. Morgan, A. E., and H. W. Werner, *Anal. Chem.*, **48**, 699 (1976).
264. Morgan, A. E., and H. W. Werner, *Appl. Phys.*, **11**, 193 (1976).
265. Morgan, A. E., and H. W. Werner, *Anal. Chem.*, **49**, 927 (1977).
266. Morgan, A. E., H. W. Werner, and J. M. Gourgout, *Appl. Phys.*, **12**, 283 (1977).
267. Morgan, A. E., and H. W. Werner, *J. Microsc. Spectrosc. Electron.*, **2**, 285 (1977).
268. Morgan, A. E., and H. W. Werner, *Surf. Sci.*, **65**, 687 (1977).
269. Morgan, A. E., and H. W. Werner, *J. Chem. Phys.*, **68**, 3900 (1978).
270. Morgan, A. E., and H. W. Werner, *Mikrochimica Acta.*, **II**, 31 (1978).
271. Morgan, A. E., and H. W. Werner, *Physica Scripta*, **18**, 451 (1978).
272. Morrison, G. H., and G. Slodzian, *Anal. Chem.*, **47**, 932A (1975).
272a. Mueller, A., and A. Benninghoven, *Surf. Sci.*, **39**, 427 (1973).
272b. Mueller, A., and A. Benninghoven, *Surf. Sci.*, **41**, 493 (1974).
273. Mueller, G., *Appl. Phys.*, **10**, 317 (1976).
274. Mueller, G., M. Haubold, R. Schimko, C-E. Richter, and G. Schwarz, *Phys. Status Solidi A*, **42**, 579 (1977).
275. Mueller, G., Presented at the Internl. SIMS Conf., Muenster, Germany, 1977.
275a. Mueller, G., M. Haubold, R. Schimko, M. Trapp, and G. Schwarz, *Phys. Status Solidi A*, **49**, 279 (1978).
275b. Mueller, G., M. Trapp, R. Schimko, and C. E. Richter, *Phys. Status Solidi A*, **51**, 87 (1979).
275c. Mukherjee, S. D., D. V. Morgan, M. J. Howes, J. G. Smith, and P. Brook, *J. Vac. Sci. Technol.*, **16**, 138 (1979).
276. Nakamura, K., S. Aoki, Y. Nakajima, H. Doi, and H. Tamura, *Mass Spectrom.*, **20**, 1 (1972).
276a. Nakamura, K., H. Hirose, A. Shibata, and H. Tamura, *Jpn. J. Appl. Phys.*, **16**, 1307 (1977).
277. Namdar-Irani, R., *J. Microsc. Spectrosc. Electron.*, **2**, 293 (1977).
278. Narusawa, T., and S. Komiya, *J. Vac. Sci. Technol.*, **11**, 312 (1974).
278a. Navincek, B., *Proc. Surf. Sci.*, **7**, 49 (1976).
278b. Nelson, R. S., *Phil. Mag.*, **11**, 291 (1965).
278c. Newbury, D. E., Paper Nr. 234, Pittsburgh Meeting on Anal. Chem. Cleveland (1977).
279. Nier, A. O., *Rev. Sci. Instrum.*, **18**, 398 (1947).
280. Nishimura, H., and J. Okano, Proc. 6th Internl. Vac. Congr. 1974, *Jpn. J. Appl. Phys. Suppl. 2*, pt. 1 (1974), p. 399; *Adv. Mass Spectrom.*, **7A**, 569 (1978).
281. Nolen, R. L., and D. E. Solomon, *J. Non-Cryst. Solids*, **19**, 377 (1975).
281a. Noya, A., S. Kuriki, G. Matsumoto, and M. Hirano, *Thin Solid Films*, **59**, 143 (1979).
282. Oechsner, H., and W. Gerhard, *Phys. Lett. A*, **40**, 211 (1972); *Surf. Sci.*, **44**, 480 (1974).
283. Oechsner, H., *Appl. Phys.*, **8**, 185 (1975).
283a. Oechsner, H., and E. Stumpe, *Appl. Phys.*, **14**, 43 (1977).
283b. Oechsner, H., W. Ruehe, and E. Stumpe, *Surf. Sci.*, **85**, 289 (1979).
284. Okazima, Y., and Y. Aizawa, *Mass Spectrom.*, **25**, 91 (1977).
285. Palmstrom, C. J., D. V. Morgan, and M. J. Howes, *Nucl. Instrum. Methods*, **150**, 305 (1978).
286. Panitz, J. A., and J. A. Foesch, *Rev. Sci. Instrum.*, **47**, 44 (1976).
287. Paul, W., and U. von Zahn, *Z. Physik*, **152**, 143 (1958).
288. Pavlyak, F., L. Bori, J. Giber, and R. Buhl, *Jpn. J. Appl. Phys.*, **16**, 335 (1977).
289. Pebler, A., G. G. Sweeney, and P. M. Castle, *Metall. Trans.*, **6A**, 991 (1975).
290. Petersson, L. G., G. Frostell, and A. Lodding, *Z. Naturforsch.*, **29C**, 417 (1974).

291. Petersson, L. G., H. Odelius, A. Lodding, S. J. Larsson, and G. Frostell, *J. Dent. Res.*, **55**, 980 (1976).

292. Phillips, B. F., *J. Vac. Sci. Technol.*, **11**, 1093 (1974).

293. Phillips, B. F., A. E. Austin, and H. L. Hughes, *Natl. Bur. Std., (U.S.) Spec. Publ.*, **400-23**, 65 (1976).

293a. Phillips, B. F., and M. Jackson, in R. Bakish, Ed., *Proc. 7th Internl. Conf. Electron-Ion Beam Sci. Technol.*, 1976.

294. Phillips, D. H., *Natl. Bur. Std., (U.S.) Spec. Publ.*, **400-23**, 73 (1976).

295. Pichlmayer, F., *Vakuum-Technik*, **23**, 97 (1974).

296. Pivin, J. C., C. Roques-Carmes, and P. Lacombe, *J. Microsc. Spectrosc. Electron.*, **2**, 90 (1977).

297. Pivin, J. C., C. Roques-Carmes, and P. Lacombe, *Mem. Sci. Rev. Metall.*, **74**, 161 (1977).

298. Pivin, J. C., C. Roques-Carmes, and G. Slodzian, *Int. J. Mass Spectrom. Ion Phys.*, **26**, 219 (1978).

298a. Plog, C., Thesis University Muenster, W-Germany, 1974.

298b. Plog, C., L. Wiedmann, and A. Benninghoven., *Surf. Sci.*, **67**, 565 (1977).

299. Ploog, K., and A. Fischer, *Appl. Phys.*, **13**, 111 (1977).

300. Ploog, K., A. Fischer, and A. Diebold, Presented at the Internl. SIMS Conf., Muenster, Germany, 1977.

300a. Ploog, K. and A. Fischer, *J. Vac. Sci. Technol.*, **15**, 255 (1978).

301. Ponpon, J. P., J. J. Grob, A. Grob, R. Stuck, and P. Siffert, *Nucl. Instrum. Methods*, **149**, 647 (1978).

302. Prager, M., *Appl. Phys.*, **8**, 361 (1975).

303. Pregerbauer, F., and F. G. Ruedenauer, *Proc. of the 7th Internl. Vac. Congr. and the 3rd Internl. Conf. on Solid Surfaces*, Vienna, 1977, p. 2539.

303a. Price, D, and J. E. Williams, Eds., *Time-of-Flight Mass Spectrometry*, Pergamon, Oxford, 1967: Based on the Proceedings of the First European Symposium on Time-of-flight Mass Spectrometry, held at the University of Salford, July 3–5, 1967.

303b. Prival, H. G., *Surf. Sci.*, **76**, 443 (1978).

304. Purdes, A. J., B. F. T. Bolker, J. D. Bucci, and T. S. Tisone, *J. Vac. Sci. Technol.*, **14**, 981 (1977).

305. Quatert, D., and F. Coen-Porisini, *J. Nucl. Mater.*, **36**, 20 (1970).

306. Rachidi, I., J. Monte, J. Pelletier, C. Pomot, and F. Rinchet, *Appl. Phys. Lett.*, **28**, 292 (1976).

307. Radermacher, L., and H. E. Beske, *Int. J. Mass Spectrom. Ion Phys.*, **20**, 333 (1976).

308. Randtke, P., and J. Warren, Presented at the Pittsburgh Conf. on Anal. Chem. and Appl. Spectrosc., Cleveland, Ohio, Febr. 1977.

309. Reed, S. J. B., J. V. P. Long, J. N. Coles, and D. M. Astill, *Int. J. Mass Spectrom. Ion Phys.*, **22**, 333 (1976).

310. Remond, G., *J. Lumin.*, **15**, 121 (1977).

311. Revesz, A. G., and T. D. Kirkendall, *J. Electrochem. Soc.*, **123**, 1514 (1976).

311a. Robinson, M. T., and I. M. Torrens, *Phys. Rev. B*, **9**, 5008 (1974).

312. Roden, H. J., and R. D. Fralick, paper 70, 26th Pittsburgh Conf. on Anal. Chem. and Appl. Spectrosc., Cleveland, Ohio, March 3–7, 1975.

313. Roques-Carmes, C., G. Slodzian, and P. Lacombe, *Can. Metall. Q.*, **13**, 99 (1974).

314. Rosenstiel, A. P. von, J. D. Brown, and D. J. Gras, Presented at the Internl. SIMS Conf., Muenster, Germany, 1977.

315. Roth, J., J. Bohdansky, W. O. Hofer, and J. Kirschner, *Plasma Wall Interaction*, Pergammon, Oxford, 1977, p. 309.

316. Rouberol, J. M., Ph. Basseville, and J-P. Lenoir, *J. Radioanal. Chem.*, **12**, 59 (1972).

317. Rouberol, J. M., M. Lepareur, B. Autier, and J-M. Gourgout, 8th Internl. Conf. on X-Ray Optics and Microanalysis, Boston, U.S.A., (1977), paper 133.

318. Ruedenauer, F. G., and W. Steiger, Proc. 6th Internl. Vac. Congr. 1974, *Jpn. J. Appl. Phys. Suppl. 2*, pt. 1 (1974), p. 383.

319. Ruedenauer, F. G., *Vacuum*, **25**, 409 (1975).

320. Rudenauer, F. G., and W. Steiger, *Vacuum*, **26**, 537 (1976).

320a. Ruedenauer, F. G., W. Steiger, and H. W. Werner, *Surf. Sci.*, **54**, 553 (1976).

321. Ruedenauer, F. G., and W. Steiger, *Proc. of the 7th Internl. Vac. Congr. and the 3rd Internl. Conf. on Solid Surfaces*, Vienna, 1977, p. 2535.

322. Ruedenauer, F. G., Presented at the Internl. SIMS Conf., Muenster, Germany, 1977.

322a. Sanders, J. B., *Can. J. Phys.*, **46**, 455 (1968).

322b. Sanders, J. B., *Proc. 5th. Intern. Conf. Atomic Coll. in Solids*, Gatlinburg, 1973, p. 125.

323. Santhanam, A. T., and J. R. Gavaler, *J. Appl. Phys.*, **46**, 3633 (1975).

324. Sapin, M., R. Sequin, and F. Maurice, *J. Microsc. Spectrosc. Electron.*, **2**, 87 (1977).

325. Satake, T., T. Narusawa, O. Tsukakoshi, and S. Komiya, *Jpn. J. Appl. Phys.*, **15**, 1359 (1976).

326. Satkiewicz, F. G., *Proc. of the 19th Annual Conf. on Mass Spectrom. and Allied Topics*, Atlanta, Georgia, May 2-7, 1971, p. 276.

327. Satkiewicz, F. G., and H. K. Charles, Jr., *Proc. of the 24th Annual Conf. on Mass Spectrom. and Allied Topics*, San Diego, California, May 9–13, 1976, p. 307.

328. Sato, K., K. Suzuki, R. Matsumoto, and S. Nagashima, *Trans. Jpn. Inst. Met.*, **18**, 61 (1977).

329. Schemmer, M., and A. Benninghoven, Presented at the Internl. SIMS Conf., Muenster, Germany, 1977.

330. Scherrer, S., and F. Naudin, *Proc. of the 11th Internl. Congr. on Glass*, Prague, July 1977, Vol. 3, p. 301.

331. Schillalies, H., and H. D. Polaschegg, Presented at LH-Symp. in Bajkowa Institute, Moscow, Jan. 1975.

332. Schilling, J. H., and P. A. Bueger, *Appl. Phys.*, **15**, 115 (1978).

332a. Schiott, H. E., *Mat. Fys. Medd. Dan Vid. Selsk.*, **35**, nr. 9 (1966).

333. Schiott, H. E., *Radiat. Eff.*, **6**, 107 (1970).

334. Schroeer, J. M., T. N. Rodin, and R. C. Bradley, *Surf. Sci.*, **34**, 571 (1973).

335. Schubert, R., and J. Tracy, *Rev. Sci. Instrum.*, **44**, 487 (1973).

336. Schubert, R., *J. Vac. Sci. Technol.*, **12**, 505 (1975).

337. Schulz, M., E. Klausmann, and A. Hurrle, *Crit. Rev. Solid State Sci.*, **5**, 319 (1975).

337a. Schwarz, S. A., and C. R. Helms, *J. Vac. Sci. Technol.*, **16**, 781 (1979).

338. Schwarzer, R., and K. H. Gaukler, *Proc. of the 7th Internl. Vac. Congr. and the 3rd Internl. Conf. on Solid Surfaces*, Vienna, 1977, p. 2547.

339. Scilla, G. J., and G. H. Morrison, *Anal. Chem.*, **49**, 1529 (1977).

340. Sell, H. G., D. F. Stein, R. Stickler, A. Joshi, and E. Berkey, *J. Inst. Metals*, **100**, 275 (1972).

341. Seran, J. L., *Acta Metall.*, **24**, 627 (1976).

342. Servais, J. P., H. Graas, V. Leroy, and L. Habraken, *Le Vide*, No. **181**, 27 (1976).

343. Servais, J. P., and V. Leroy, Presented at the Apeldoorn SIMS Meeting, Organized by P. von Rosentiel (TNO) 1976, unpublished.

344. Shafranovskii, E. A., S. Yu. Semenov, V. D. Grishin, and V. L. Tal'rose, *Instr. Exp. Technol.*, **17**, 177 (1974).

345. Shimizu, R., T. Ishitani, and Y. Ueshima, *Jpn. J. Appl. Phys.*, **13**, 249 (1974).

346. Shimizu, R., T. Ishitani, T. Kondo, and H. Tamura, *Anal. Chem.*, **47**, 1020 (1975).

347. Shimizu, R., T. Okutani, T. Ishitani, and H. Tamura, *Surf. Sci.*, **69**, 349 (1977).

347a. Shimizu, R., *Proc. 7th Intern. Vac. Congr. & 3rd Intern. Conf. Solid Surf.*, Vienna, (1977), p. 1417.

348. Shimojo, T., and T. Nagasako, *Proc. of the 7th Internl. Vac. Congr. and the 3rd Internl. Conf. on Solid Surfaces*, Vienna, 1977, p. 2311.

349. Shinoki, F., and A. Itoh, *J. Appl. Phys.*, **46**, 3381 (1975).

349a. Sigmund, P., *Phys. Rev.*, **184**, 383 (1969).

350. Sigmund, P., *Rev. Roum. Phys.*, **17**, 823 (1972); **17**, 969 (1972); **17**, 1079 (1972).

351. Sigmund, P., *Appl. Phys. Lett.*, **25**, 169 (1974); **27**, 52 (1975).

351a. Sigmund, P., in N. H. Tolk, J. C. Tuly, W. Heiland, and C. W. White, Eds., *Inelastic Ion Surface Collisions*, Academic, New York, 1977, p. 212.

351b. Sixt, G., K. H. Ziegler, and W. R. Fahrner, *Thin Solid Films*, **56**, 107 (1979).

352. Simons, D. S., J. E. Baker, and C. A. Evans Jr., *Anal. Chem.*, **48**, 1341 (1976).

353. Slevin, P. J., W. W. Harrison, *Appl. Spectrosc.*, **10**, 201 (1975).

354. Slodzian, G., *Ann. Phys.*, (Paris), **9**, 591 (1964).

354a. Slodzian, G., and J. F. Hennequin, *C.R. Acad. Sc. Paris, Ser. B*, **263**, 1246 (1966).

354b. Slodzian, G., *Rev. Phys. Appl.*, **3**, 360 (1968).

355. Slodzian, G., and A. Havette, *Adv. Mass Spectrom.*, **6**, 629 (1974).

356. Slodzian, G., *Surf. Sci.*, **48**, 161 (1975).

357. Smit, A. L. C., M. A. J. Rossetto, and F. H. Field, *Anal. Chem.*, **48**, 2042 (1976).

358. Smith, D. H., and W. H. Christie, *Int. J. Mass Spectrom. Ion Phys.*, **26**, 61 (1978).

359. Snowdon, K., *Proc. of the 7th Internl. Vac. Congr. and the 3rd Internl. Conf. on Solid Surfaces*, Vienna, 1977, p. 2557.

360. Socha, A. J., *Surf. Sci.*, **25**, 147 (1971).

361. Someno, M., M. Kobayashi, and H. Saito, *Proc. of the 7th Internl. Vac. Congr. and the 3rd Internl. Conf. on Solid Surfaces*, Vienna, 1977, p. 2593.

361a. Southern, A. L., W. R. Willis, and M. T. Robinson, *J. Appl. Phys.*, **34**, 153 (1963).

362. Sparrow, G. R., *Proc. of the 24th Annual Conf. on Mass Spectrom. and Allied Topics*, San Diego, California, May 9 - 14, 1976, p. 298.

362a. Sparrow, G. R., Paper Nr. 348, Pittsburgh Meeting on Anal. Chem., Cleveland, 1977.

363. Spitzer-Aronson, M., *Proc. of the 11th Internl. Congr. on Glass*, Prague, July 1977, Vol. 3, p. 323.

364. Sroubek, Z., *Rev. Sci. Instrum.*, **44**, 487 (1973).

365. Sroubek, Z., *Surf. Sci.*, **44**, 47 (1974).

365a. Sroubek, Z., J. Zavadil, F. Kubec, and K. Zdansky, *Surf. Sci.*, **77**, 603 (1978).

366. Staib, P., *J. Phys. E*, **5**, 484 (1972).

367. Staib, P., and G. Staudenmaier, *J. Nucl. Mater.*, **63**, 37 (1976).

368. Stanton, H. E., *J. Appl. Phys.*, **31**, 678 (1960).

368a. Staudenmeier, G., *Radiat. Eff.*, **13**, 87 (1972).

368b. Staudenmeier, W., *Radiat. Eff.*, **18**, 181 (1973).

369. Steckelmacher, W., *J. Phys. E*, **6**, 1061 (1973).

370. Steiger, W., and F. G. Ruedenauer, *Adv. Mass Spectrom.*, **7A**, 770 (1978).

370a. Steiger, W., and F. G. Ruedenauer, *Anal. Chem.*, **51**, 2107 (1979).

371. Storms, H. A., K. F. Brown, and J. D. Stein, *Anal. Chem.*, **49**, 2023 (1977).

372. Summers, A. J., N. J. Freeman, and N. R. Daly, *Rev. Sci. Instrum.*, **42**, 1353 (1971).

373. Sweeney, G., *Proc. of the 24th Annual Conf. on Mass Spectrom. and Allied Topics*, San Diego, California, May 9-13, 1976, p. 319.

373a. Szymonski, M., and A. E. de Vries, *Phys. Lett. A*, **63**, 359 (1977).

373b. Szymonski, M., H. Overeijnder, and A. E. de Vries, *Radiat. Eff.*, **36**, 189 (1978).

373c. Szymonski, M., R. S. Bhattacharya, H. Overeijnder, and A. E. de Vries, *J. Phys. D: Appl. Phys.*, **11**, 751 (1978).

374. Tamura, H., T. Kondo, I. Kanomata, K. Nakamura, and Y. Nakajima, Proc. 6th Internl. Vac. Congr. 1974, *Jpn. J. Appl. Phys. Suppl. 2*, pt. 1 (1974), p. 379.

375. Tarng, M. L., and G. K. Wehner, *J. Appl. Phys.*, **43**, 2268 (1972).

375a. Taylor, J. A., and J. W. Rabalais, *Surf. Sci.*, **74**, 229 (1978).

376. Thauvin, G., C. Roques-Carmes, G. Blaise, G. Slodzian, and R. Pichoir, *La Recherche Aerospatiale*, **No. 4**, 217 (1976).

376a. Thomas, G., *Radiat. Eff.*, **31**, 185 (1977).

376b. Thompson, M. W., and R. S. Nelson, *Phil. Mag.*, **7**, 2015 (1962).

376c. Thompson, M. W., *Phil. Mag.*, **18**, 377 (1968).

377. Tolk, N. H., I. S. T. Tong, and C. W. White, *Anal. Chem.*, **49**, 16A (1977).

378. Tongson, L. L., and C. B. Cooper, *J. Phys. E*, **10**, 1245 (1977).

378a. Tousimis, A. J., in C. J. Arceneaux, Ed., *30th Ann. Proc. Electron Microscopy Society of America*, (1972).

378b. Trilhe, J., J. Borel, and J. P. Duchemin, *J. Crystal Growth*, **45**, 439 (1978).

378c. Truchet, M., Thesis, University Paris (1974).

379. Truchet, M., *J. Microsc. Biol. Cell.*, **22**, 465 (1975); **24**, 1 (1975).

379a. Truchet, M., *C.R. Acad. Sc. Paris, Ser. D*, **282**, 1785 (1976).

380. Truchet, M., and J. Vovelle, *Calcif. Tiss. Res.*, **24**, 231 (1977).

381. Tsai, J. C. C., and J. M. Morabito, *Surf. Sci.*, **44**, 247 (1974).

382. Tsai, J. C. C., and J. M. Morabito, in S. Namba, Ed., *Ion Implantation in Semiconductors*, Plenum, New York, 1975, p. 155.

383. Tsai, M. Y., B. G. Streetman, P. Williams, and C. A. Evans, *Appl. Phys. Lett.*, **32**, 144 (1978).

383a. Tsukamoto, K., Y. Akasaka, and K. Horie, *Jpn. J. Appl. Phys.*, **16**, 663 (1977).

383b. Tsukamoto, K., Y. Akasaka, and K. Horie, *J. Appl. Phys.*, **48**, 1815 (1977).

384. Tsunoyama, K., Y. Ohashi, T. Suzuki, and K. Tsuruoka, *Jpn. J. Appl. Phys.*, **13**, 1683 (1974).

385. Tsunoyama, K., Y. Ohashi, and T. Suzuki, *Anal. Chem.*, **48**, 832 (1976).

386. Tsuruoka, K., K. Tsunoyama, Y. Ohashi, and T. Suzuki, Proc. 6th Internl., Vac. Congr. 1974, *Jpn. J. Appl. Phys. Suppl. 2*, pt. 1 (1974), p. 391.

386a. Tuck, B., K. T. Ip, and L. F. Eastman, *Thin Solid Films*, **55**, 41 (1978).

387. Tuithof, H. H., A. J. H. Boerboom, and H. L. C. Meuzelaar, *Int. J. Mass Spectrom. Ion Phys.*, **15**, 437 (1974).

388. Ueda, Y., and J. Okano, *Mass Spectrosc.*, **20**, 185 (1972).

389. Valencourt, L. R., C. E. Johnson, D. V. Steidl, and H. T. Davis, *J. Nucl. Mater.*, **58**, 293 (1975).

390. Vallerand, P., and M. Baril, *Can. J. Phys.*, **52**, 482 (1974).

391. Vallerand, P., and M. Baril, *Int. J. Mass Spectrom. Ion Phys.*, **24**, 241 (1977).

392. Vasile, M. J., and D. L. Malm, *Int. J. Mass Spectrom. Ion Phys.*, **21**, 145 (1976).

393. Viel, L., C. Benazeth, and N. Benazeth, *Surf. Sci.*, **54**, 635 (1976).

394. Voll, H., J-P. Servais, V. Leroy, and J. Lueckers, *Arch. Eisenhuettenwes.*, **48**, 13 (1977).

394a. Vyas, P. D., and B. L. Sharma, *Thin Solid Films*, **51**, L21 (1978).

395. Walsh, J. M., *Metall. Trans.*, **5**, 2104 (1974).

396. Walsh, J. M., and B. H. Kear, *Metall. Trans. A*, **6**, 226 (1975).

397. Wang, K. L., and H. A. Storms, *J. Appl. Phys.*, **47**, 2539 (1976).

397a. Watanabe, K., M. Hashiba, Y. Hirohata, M. Nishino, and T. Yamashina, *Thin Solid Films*, **56**, 63 (1979).

398. Weg, W. F. van der, and D. J. Bierman, *Physica*, **44**, 177 (1969); **44**, 206 (1969).

398a. Wehner, G. K., *Phys. Rev.*, **108**, 108 (1957); **112**, 1120 (1958).

399. Wehner, G. K., in A. W. Czanderna, Ed., *Methods of Surface Analysis*, Elsevier, Amsterdam, 1975, p. 5.

399a. Weijsenfeld, C. H., Thesis, University Utrecht, The Netherlands, 1966, Philips Res. Repts., Suppl. 1967, Nr. 2.

399b. Weinke, H. H., W. Kiesl, R. Saelens, and R. Gijbels, *Meteoritia*, **13**, 665 (1978).

399c. Weinke, H. H., W. Kiesl, and R. Gijbels, *Mikrochimica Acta, Suppl.*, **8**, 87 (1979).

400. Werner, H. W., and H. A. M. de Grefte, *Vakuum-Technik*, **2**, 37 (1968).

401. Werner, H. W., in E. L. Grove and A. J. Perkins, Eds., *Developments in Applied Spectroscopy*, Vol. 7A, Plenum, New York, 1969, p. 239.

402. Werner, H. W., H. A. M. de Grefte, and J. van den Berg, *Int. J. Mass Spectrom. Ion Phys.*, **8**, 459 (1972).

403. Werner, H. W., H. A. M. de Grefte, and J. van den Berg, *Radiat. Eff.*, **18**, 269 (1973).

403a. Werner, H. W., and H. A. M. de Grefte, *Surf. Sci.*, **35**, 458 (1973).

404. Werner, H. W., H. A. M. de Grefte, and J. van den Berg, in A. R. West, Ed., *Advances in Mass Spectrometry*, Vol. 6, Applied Science Publishers, Barking, Essex (1974), p. 673.

405. Werner, H. W., Proc. 6th Internl. Vac. Congr. 1974, *Jpn. J. Appl. Phys. Suppl. 2*, pt. 1 (1974), p. 367.

406. Werner, H. W., *Vacuum*, **24**, 493 (1974).

407. Werner, H. W., *Acta Electronica*, **18**, 51 (1975).

408. Werner, H. W., A. E. Morgan, and H. A. M. de Grefte, *Appl. Phys.*, **7**, 65 (1975).

409. Werner, H. W., *Surf. Sci.*, **47**, 301 (1975).

409a. Werner, H. W., Paper Presented at the joint Japan-US Seminar on Quantitative SIMS, Honolulu, Oct. 13-17, (1975).

410. Werner, H. W., *Acta Electronica*, **19**, 53 (1976).

411. Werner, H. W., and A. E. Morgan, *J. Appl. Phys.*, **47**, 1232 (1976).

412. Werner, H. W., *Sci. Ceram.*, **8**, 55 (1976).

413. Werner, H. W., *Mikrochimica Acta (Wien)*, Suppl. **7**, 63 (1977).

414. Werner, H. W., *Proc. of the 7th Internl. Vac. Congr. and the 3rd Internl. Conf. on Solid Surfaces*, Vienna, 1977, p. 2135.

415. Werner, H. W., and H. A. M. de Grefte, unpublished.

415a. Werner, H. W., and A. E. Morgan, in N. Daly, Ed., *Advances in Mass Spectrometry*, Vol. 7A, The Institute of Petroleum, London, 1977, p. 764.

416. White, C. W., D. L. Simms, and N. H. Tolk, *Science*, **177**, 481 (1972).

416a. White, C. W., W. H. Christie, B. R. Appleton, S. R. Wilson, P. P. Pronko, and C. W. Magee, *Appl. Phys. Lett.*, **33**, 662 (1978).

416b. Whitehouse, D. J., in P. F. Kane and G. B. Larrabee, Eds., *Characterization of Solid Surfaces*, Plenum, New York, 1974, Chapter 3.

417. Wichner, R., and E. J. Charlson, *J. Electron. Mater.*, **5**, 513 (1976).

418. Williams, P., and C. A. Evans, *Natl. Bur. Std. (U.S.) Spec. Publ.*, **427**, 63 (1975).

419. Williams, P., and C. A. Evans, M. L. Grossbeck, and H. K. Birnbaum, *Anal. Chem.*, **48**, 964 (1976).

420. Williams, P., R. K. Lewis, C. A. Evans, and P. R. Hanley, *Anal. Chem.*, **49**, 139 (1977).

421. Williams, P., and C. A. Evans, *Appl. Phys. Lett.*, **30**, 559 (1977).

422. Williams, P., and C. A. Evans, Presented at the Internl. SIMS Conf., Muenster, Germany, 1977.

422a. Williams, P., and C. A. Evans, Jr., *Surf. Sci.*, **78**, 324 (1978).

422b. Williams, P., *Surf. Sci.*, **90**, 588 (1979).

422c. Winograd, N., D. E. Harrison, Jr., and B. J. Garrison, *Surf. Sci.*, **78**, 467 (1978).

422d. Winograd, N., K. E. Foley, B. J. Garrison, and D. E. Harrison, Jr., *Phys. Lett. A*, **73**, 253 (1979).

422e. Winograd, N., in A. Benninghoven, C. A. Evans Jr., R. A. Powell, R. Shimizu, and H. A. Storms, Eds., *Secondary Ion Mass Spectrometry (SIMS-II)*, Proc. of the Secondary Internl. Conf. on Secondary Ion Mass Spectrometry (SIMS-II), Stanford University, Stanford, California, August 27 - 31, 1979, Springer-Verlag, Berlin - Heidelberg - New York, 1979, p. 2.

422f. Winterbon, K. B., Range-Energy data for keV ions in amorphous materials, Chalk River report AECL-3194, 1968, Ontario, Canada.

422g. Winterbon, K. B., P. Sigmund, and J. B. Sanders, *Mat. Fys. Medd. Dan. Vid. Selsk.*, **37**, Nr. 14 (1970).

422h. Winterbon, K. B., *Ion Implantation Range and Energy Deposition Distributions*, Vol. 2, *Low Incident Ion Energies*, IFI/Plenum, New York, Washington, London, (1975).

423. Wittmaack, K., J. Maul, and F. Schulz, *Int. J. Mass Spectrom. Ion Phys.*, **11**, 23 (1973).

424. Wittmaack, K., J. Maul, and F. Schulz, in R. Bakish, Ed., *Proc. of the 6th Internl. Conf. on Electron and Ion Beam Sci. and Technol.*, The Electrochemical Society, Princeton, 1974, p. 164.

425. Wittmaack, K., *Surf. Sci.*, **53**, 626 (1975).

426. Wittmaack, K., *Appl. Phys. Lett.*, **29**, 552 (1976).

427. Wittmaack, K., in N. H. Tolk, J. C. Tully, W. Heiland, and C. W. White, Eds., *Inelastic Ion-Surface Collisions*, Academic, New York, 1977, p. 153.

427a. Wittmaack, K., *Phys. Lett. A*, **69**, 322 (1979).

427b. Wittmaack, K., *Phys. Rev. Lett.*, **43**, 872 (1979).

427c. Wu, C. P., and C. W. Magee, *Appl. Phys. Lett.*, **34**, 737 (1979).

427d. Yamaguchi, N., K. Suzuki, K. Sato, H. Tamura, *Anal. Chem.*, **51**, 695 (1979).

427e. Yamamura, Y., and Y. Kitazoe, *Radiat. Eff.*, **39**, 251 (1978).

428. Yao, T., S. Amano, Y. Makita, and S. Maekawa, *Jpn. J. Appl. Phys.*, **15**, 1001 (1976).

429. Yiou, F., M. Baril, J. Dufaure De Citres, P. Fontes, E. Gradsztajn, and R. Bernan, *Phys. Rev.*, **166**, 968 (1968).

430. Yu, M. L., *Phys. Rev. Lett.*, **40**, 574 (1978).

431. Yurasova, V. E., A. A. Sysoev, G. A. Samsonov, V. M. Bukhanov, L. N. Nevzorova, and L. B. Shelyakin, *Radiat. Eff.*, **20**, 89 (1973).

432. Zinner, E., and R. M. Walker, *Geochim. Cosmochim. Acta, Suppl.*, **6**, 3601 (1975).

433. Zinner, E., R. M. Walker, J. Chaumont, and J. C. Dran, *Geochim. Cosmochim. Acta, Suppl.*, **7**, 953 (1976); **8**, 3859 (1977).

Part 1
Section I

Chapter 6

MÖSSBAUER SPECTROSCOPY

BY JOHN G. STEVENS AND MICHAEL J. RUIZ, *University of North Carolina at Asheville*

Contents

I. INTRODUCTION

Spectroscopic methods are important analytic tools in the study of various microscopic systems such as nuclei, atoms, molecules, and crystal lattices. The knowledge gained from the study of the various absorption, emission, or reemission processes in microscopic systems is generally of interest either to the nuclear physicist or to the solid-state physicist and the chemist, but seldom to both groups simultaneously. Mössbauer spectroscopy, however, makes very enriching contributions to both fields. This spectroscopy is based on observing recoilless emission and resonance absorption of gamma rays by nuclei in solids. Resonance absorption of gamma rays had been predicted since the beginning of this century; however, experimental observation of this behavior was difficult due to the amount of energy lost in the recoil of a nucleus emitting and/or absorbing a gamma ray. Below, we briefly describe this problem.

From a classical point of view, a free atom of mass m, moving in the x direction with a given velocity v, has a linear momentum of mv. If this free atom is in an excited nuclear state and undergoes an energy transition to the corresponding ground state by emitting a gamma ray, the momentum of the system must be conserved. To conserve momentum, the momentum of the emitted gamma ray must be balanced by a change in the velocity of the nucleus. This change in the velocity imparts to the nucleus an energy associated with its recoil after emission. In most optical spectroscopic studies, this loss of energy due to recoil is insignificant because it is much less than the experimental spectral line width. However, in the study of the recoilless emission and resonance absorption of gamma rays, the energy lost due to recoil is much greater than the line width and thus becomes an important factor.

In pre-Mössbauer time, several experimental methods were utilized to either broaden the line width or add to the energy of the gamma ray to make up for the loss in recoil. One such method was the utilization of the Doppler effect by accelerating the source toward the absorber in an attempt to compensate for the energy lost in recoil. Temperature broadening and the use of recoil momentum imparted by a preceding transition were also successfully employed in some cases in restoring energy lost due to recoil. However, these techniques did not afford the opportunity to observe hyperfine interactions due to the large broadening of the lines which were necessarily incurred.

In 1958 Rudolph L. Mössbauer discovered that recoilless emission and resonance absorption of gamma rays could be observed if solid substances were used (28). Without the need to compensate for energy loss due to recoil, line widths were greatly

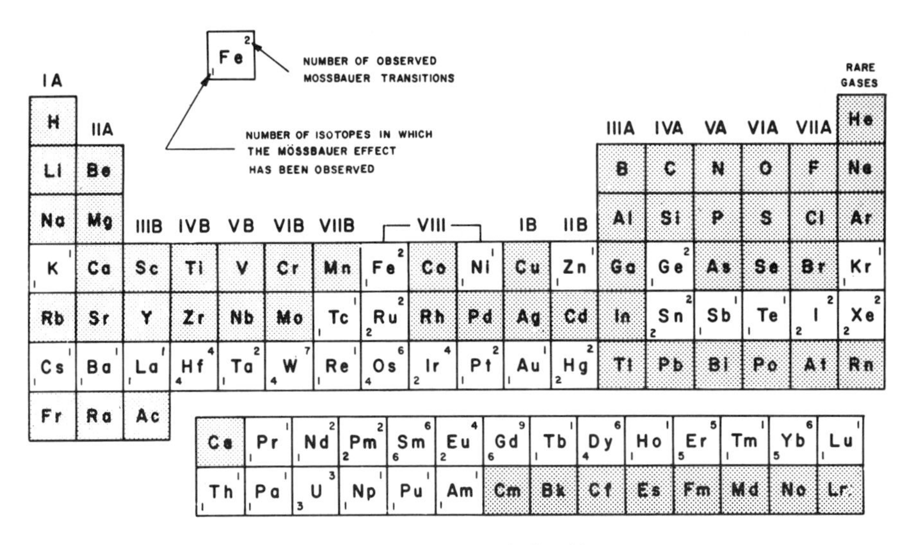

Fig. 6.1. Mössbauer periodic table.

reduced to the order of 10^{-7} eV. A high-resolution technique of this type enables one to observe hyperfine interactions and thus study events affecting the nucleus and its immediate environment.

This recoilless resonance phenomenon, which is referred to as the Mössbauer effect or nuclear gamma resonance (NGR), has been reported for over 100 nuclear gamma ray transitions. The elements in which these transitions are observable are shown in Fig. 6.1.

II. BASIC PRINCIPLES

NGR is the process whereby a gamma ray emitted during the transition of a nucleus from an excited to a ground level excites an identical nucleus in the reverse manner. This process is the basis of the Mössbauer effect. However, the unique characteristic of the Mössbauer effect is the ability to observe hyperfine interactions. This can largely be attributed to emission and absorption of the gamma ray by the nucleus without any loss of energy or line broadening due to recoil.

To fully appreciate the significance of recoilless emission, we will examine gamma emission from a classical point of view. Consider a nucleus of mass m at rest in an excited state. If during a transition from this excited state to the ground state, the nucleus emits a photon of energy E_γ, then, the conservation of energy principle states that the change in energy of the nucleus (E_0) due to the transition must be equal to the quantum of energy carried away by the photon plus the recoil kinetic energy (E_r) gained by the nucleus due to the emission process. Therefore, conservation of energy requires that

$$E_e - E_g = E_0 = E_\gamma + E_r \tag{1}$$

where E_e is the energy of the excited state and E_g is the energy of the ground state. For there to be appreciable resonance absorption, the energy of the transition must be approximately equal to the energy of the radiation emitted, i.e., $E_0 \approx E_\gamma$. Conservation of momentum requires that the momentum imparted to the nucleus (p) from the transition be equal to the momentum of the gamma ray (p_γ), i.e.,

$$p = m\text{v} = p_\gamma = \frac{E_\gamma}{c} \tag{2}$$

where v is the recoil velocity and c is the speed of light. The recoil energy can be written as

$$E_r = \tfrac{1}{2} m\text{v}^2 = \frac{E_\gamma^2}{2mc^2} \tag{3}$$

The problem that results when the recoil energy is greater than the uncertainty in the energy of the photon (Γ) is shown in Fig. 6.2. When $E_r \approx \Gamma$ there is almost no overlap of the emission and absorption energies, resulting in virtually no resonance. Even for cases where $2E_r \approx \Gamma$, there is no appreciable overlap. Only when $3E_r \leqslant \Gamma$ is there significant overlap. For most other spectroscopies, the recoil energy is usually much smaller than the line width and resonance occurs frequently.

Now consider an atom bound inside a solid. If for low-energy nuclear gamma rays, the atom containing the nucleus of interest is bound strongly enough to its nearest neighbors, there is a probability that the entire solid will recoil instead of the individual atom. In this case, the mass in equation 3 must be replaced by the mass of the bulk material. The larger mass of the solid results in an extremely small recoil energy, virtually allowing for complete overlapping of the emission and absorption

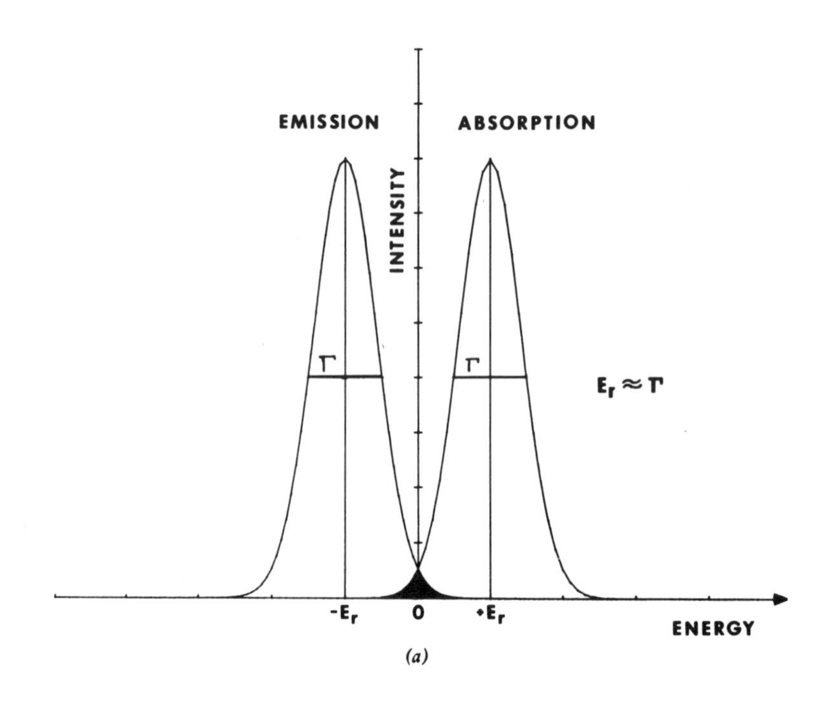

(a)

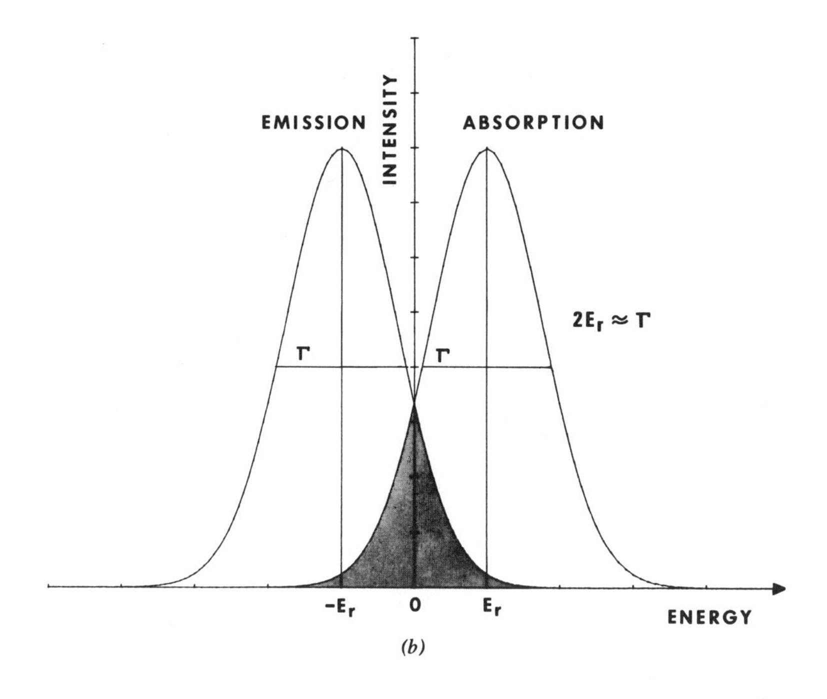

(b)

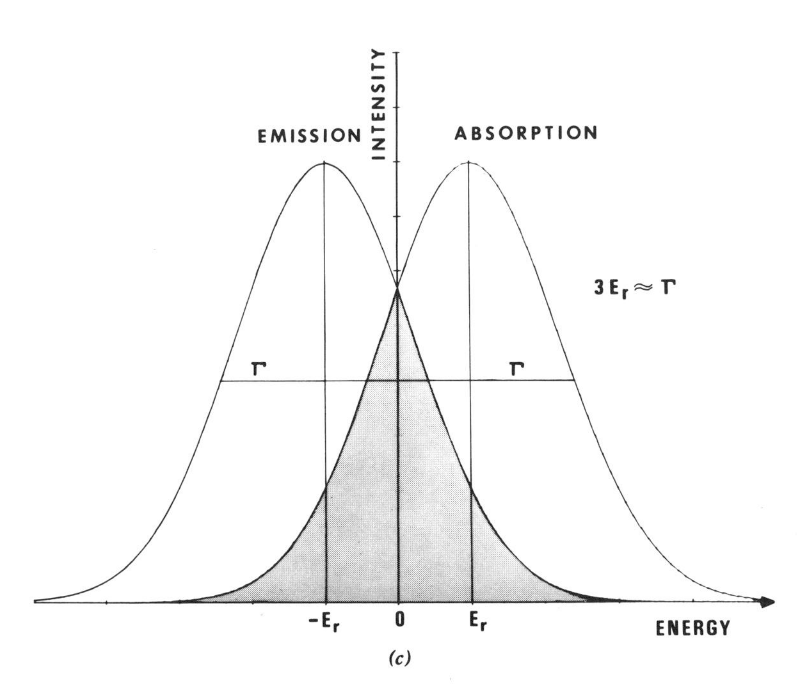

(c)

Fig. 6.2. Emission and absorption Lorentzian lines as a function of relative values of the recoil energy (E_r) and the line width (Γ). (*a*) $E_r \approx \Gamma$, (*b*) $2E_r \approx \Gamma$, and (*c*) $3E_r \approx \Gamma$.

lines. Resonance can now occur easily with the natural line widths preserved, a very important phenomenon since it allows for measurement of several types of hyperfine interactions. These will be discussed in the following sections.

The general line shapes of absorption spectra in Mössbauer spectroscopy are Lorentzian "for infinitely thin" absorbers, i.e.,

$$I(E_\gamma) = I_0 \frac{(\Gamma/2)^2}{(E_\gamma - E_0)^2 + (\Gamma/2)^2} \tag{4}$$

where I is the intensity of radiation for a particular gamma energy (E_γ) and E_0 is the resonance energy giving an intensity of I_0. The line width, Γ, (full width at half-maximum) is generally twice the natural line width (Γ_{NLW}) of the source. Using the uncertainty principle, one finds that

$$\Gamma = 2\,\Gamma_{NLW} = \frac{2\,\hbar\,\ln 2}{t_{1/2}} \tag{5}$$

where $t_{1/2}$ is the half-life of the excited nuclear level.

A comparison of the shape of the Lorentzian with the more common Gaussian is made in Fig. 6.3. As sample thickness is increased, the intensity of absorption increases, the line width broadens, and the line shape goes from a Lorentzian to that of a Gaussian. Before we define a "thin" or "thick" absorber, resonance cross sections and Mössbauer fractions will be discussed.

The general laws of quantum mechanical scattering give the cross-section for resonance, assuming a single line, no internal conversion and 100% zero phonon absorption (no loss of recoil energy to the lattice) as

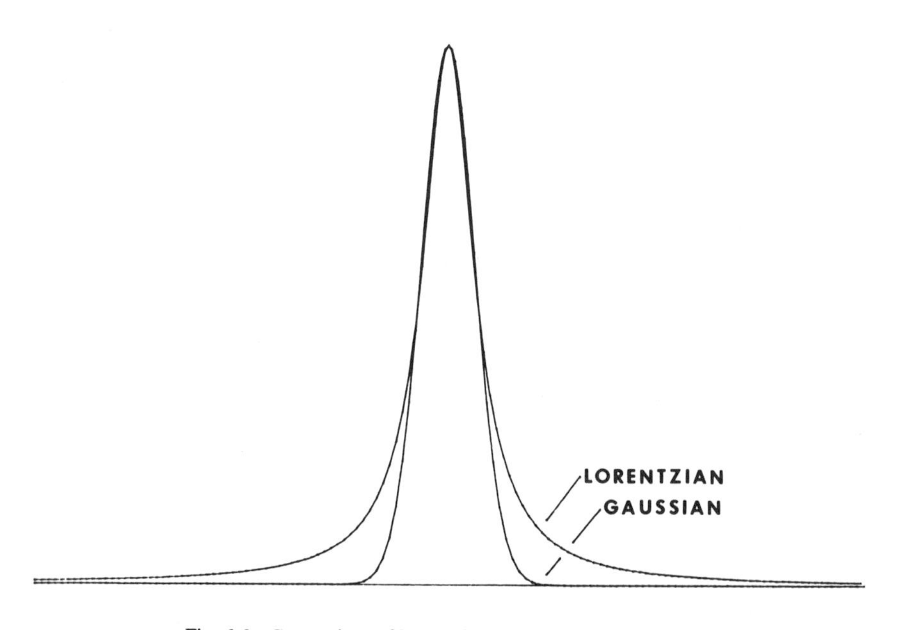

Fig. 6.3. Comparison of Lorentzian and Gaussian line shapes.

$$\sigma_0 = 2\pi \lambda\!\!\!^{-2} \frac{(2I_e + 1)}{(2I_g + 1)} \tag{6}$$

where $\lambda\!\!\!^{-}$ is the wavelength of the photon divided by 2π, I_e is the nuclear spin of the excited level, and I_g is the nuclear spin of the ground level. For most Mössbauer transitions internal conversion (IC) must be considered. Internal conversion is the deexcitation of an excited nucleus in which energy is carried away by the emission of an atomic electron. Inclusion of internal conversion modifies equation 6, giving for the resonance cross-section

$$\sigma_0 = 2\pi \lambda\!\!\!^{-2} \frac{(2I_e + 1)}{(2I_g + 1)} \cdot \frac{1}{(\alpha + 1)} \tag{7}$$

where α, the internal conversion coefficient, is the ratio of the probability of internal conversion to that of gamma emission.

To appreciate how large this cross section is, a comparison can be made between the actual geometric cross section of the nucleus and the calculated cross section from equation 7. For the case of the 14.4 keV transition in ^{57}Fe, the ratio of the Mössbauer resonance cross section to the nuclear geometric cross section is approximately 2.5×10^6.

This is the most dramatic of the Mössbauer transitions and is the main reason that the ^{57}Fe transition is the one most often used in Mössbauer spectroscopy. This and other more common Mössbauer transitions are listed in Table 6.I with useful

TABLE 6.I

Parameters for Selected Mössbauer Transitions

Isotope	Isotope abundance (%)	E_γ (keV)	I_e I_g	Half life $t_{1/2}$ (ns)	Internal conversion coefficient	Mössbauer cross section (10^{-20} cm^2)	Mössbauer line width (mm/s)
^{57}Fe	2.14	14.41	3/2 1/2	97.8	8.21	256	0.194
^{61}Ni	1.19	67.41	5/2 3/2	5.27	0.135	71.2	0.770
^{99}Ru	12.72	89.36	3/2 5/2	20.5	1.54	8.0	0.149
^{119}Sn	8.58	23.87	3/2 1/2	17.8	5.12	140	0.647
^{121}Sb	57.25	37.15	7/2 5/2	3.5	11.1	19.5	2.10
^{125}Te	6.99	35.46	3/2 1/2	1.48	13.6	26.6	5.21
^{127}I	100	57.60	7/2 5/2	1.91	3.78	20.6	2.49
^{129}I	[a]	27.77	5/2 7/2	16.8	5.1	39.0	0.586
^{133}Cs	100	81.00	5/2 7/2	6.31	1.72	10.3	0.535
^{151}Eu	47.82	21.53	7/2 5/2	9.7	28.6	23.8	1.31
^{153}Eu	52.18	103.18	3/2 5/2	3.9	1.78	5.46	0.68
^{155}Gd	14.73	86.54	5/2 3/2	6.33	0.43	34	0.499
^{161}Dy	18.88	25.66	5/2 5/2	28.2	2.9	95	0.378
^{166}Er	33.41	80.56	2 0	1.87	6.93	23.8	1.82
^{170}Yb	3.03	84.25	2 0	1.61	8.05	19.0	2.02
^{181}Ta	99.99	6.24	9/2 7/2	6800	46	167	0.0064
^{193}Ir	62.70	73.04	1/2 3/2	6.3	6.5	3.06	0.594
^{197}Au	100	77.34	1/2 3/2	1.88	4.30	3.86	1.88
^{237}Np	[b]	59.54	5/2 5/2	68.3	1.12	31	0.0673

[a] Radioactive, $t_{1/2} = 1.57 \times 10^7 y$.
[b] Radioactive, $t_{1/2} = 2.14 \times 10^6 y$.

information such as nuclear spin states, gamma energies, nuclear lifetimes, line widths, internal conversion coefficients, and cross sections.

The intensity of absorption can be related to two other parameters; these are the sample thickness and the fraction of recoilless emissions. This latter parameter is called the Mössbauer fraction or recoil-free fraction and will be discussed first.

For an emitting atom bound in a lattice, recoil energy can be given to the lattice in the form of vibrational energy (phonons). When this occurs, the emitted gamma has less energy than the transition energy due to the recoil energy gained by the emitting atom. If no energy is given to vibrational excitations in the lattice, the whole lattice recoils. Then the recoil energy is exceedingly small since the recoiling mass is very large, being the mass of the whole crystal. The emitted gamma energy is therefore essentially equal to the transition energy. This is called recoil-free or zero-phonon absorption. When both of these processes take place, there is nuclear gamma resonance.

A simplified description of the above processes can be found in the Einstein model of a lattice, where there is a single vibrational frequency (ν_E) for the atoms. If the recoil energy given by equation 3 is greater than $h\nu_E$, there will not be zero-phonon interactions as the atoms will absorb vibrational energy. However, if the recoil energy is less than $h\nu_E$, then the Mössbauer fraction, i.e., the fraction of recoilless emissions is

$$f = \mathrm{e}^{-\frac{E_r}{k\theta_E}} \tag{8}$$

where $\theta_E = \dfrac{h\nu_E}{k}$ is the Einstein temperature and k is the Boltzmann constant.

This factor can also be written as

$$f = \mathrm{e}^{\frac{-\langle \overline{x^2} \rangle}{\lambda\!\!\!^{-2}}} \tag{9}$$

where $\langle \overline{x^2} \rangle$ is the thermal average of the mean square displacement of the emitting or absorbing atom and $2\pi\lambda\!\!\!^-$ is the wavelength of the radiation. This factor was used in earlier X-ray diffraction studies and known as the Debye-Waller factor.

From equation 8, it is evident that the larger the recoil (corresponding to higher gamma energies), the smaller the Mössbauer fraction. Smaller atomic masses give smaller Mössbauer fractions while stronger lattice forces support larger recoilless fractions.

The Einstein model is oversimplified as it considers only one natural frequency for the oscillators. The Debye model is an improved model where a distribution of oscillator frequencies is incorporated into the calculation of the Mössbauer fraction. The distribution in the model is proportional to ν where ν goes from zero to a maximum called the Debye frequency (ν_D). The final result for the Mössbauer fraction is

$$f = \mathrm{e}^{-2W} \tag{10}$$

where

$$W = \frac{3E_r}{k\theta_D}\left[\frac{1}{4} + \left(\frac{T}{\theta_D}\right)^2 \int_0^{\theta_D/T} \frac{x\,dx}{e^x - 1}\right]$$

and $\theta_D = \dfrac{h\nu_D}{k}$. Plots of equation 10 are given for ^{57}Fe (14.4 keV), ^{121}Sb (37.15 keV), and ^{197}Au (77.34 keV) in Fig. 6.4.

In addition to the effect of the gamma energy and lattice force discussed above, note the additional effect of temperature in the Debye model. Increasing the temperature decreases the recoil-free fraction. Serious consideration must be given to the operating temperature for the experiment. This will be discussed in more detail in the experimental section.

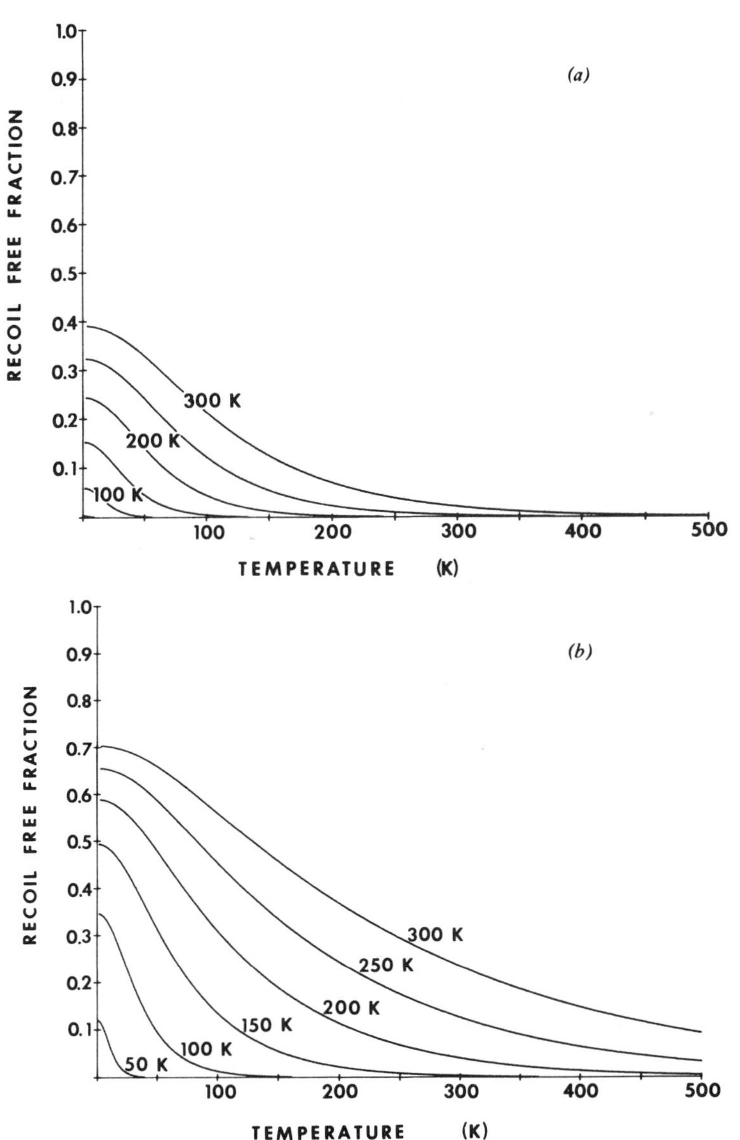

Fig. 6.4. Plots of the Mössbauer fraction versus temperature for various Debye temperatures. (*a*) ^{197}Au, (*b*) ^{121}Sb, and (*c*) ^{57}Fe.

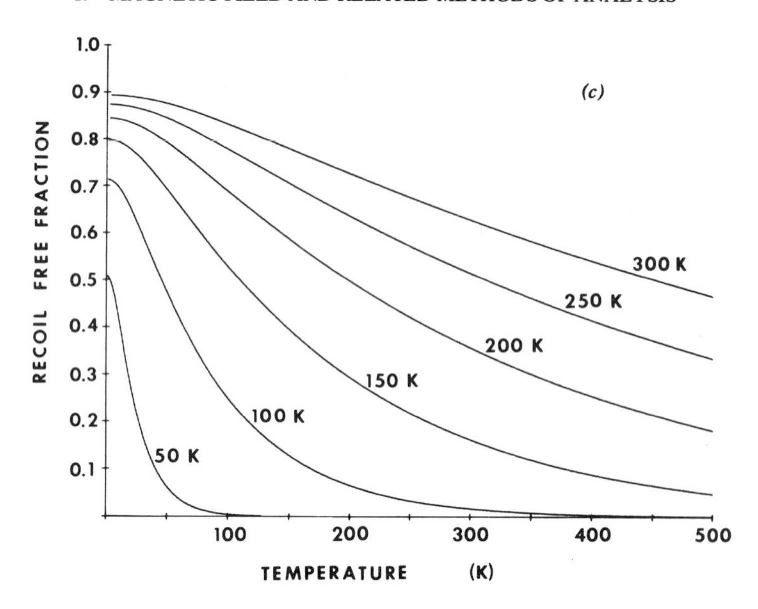

Fig. 6.4. (*Continued*).

The effective thickness for an absorber (t_a) is another very useful Mössbauer parameter. This parameter is directly related to the observed intensity of a Mössbauer absorption peak. The effective thickness is defined as

$$t_a = f_a \cdot n_a \cdot d_a \cdot IA \cdot \sigma_0 \tag{11}$$

where f_a is the Mössbauer fraction in the absorber, n_a the number of atoms per cubic centimeter of the element, d_a the thickness (cm) of the absorber, IA the isotopic abundance, and σ_0 the resonance cross section. The above equation can be expressed in more convenient terms by replacing n_a and d_a with the surface density σ_a expressed in mg/cm^2 of the element of interest. Therefore,

$$t_a = \frac{f_a \cdot N_0 \cdot IA \cdot \sigma_0}{1000 \cdot A_a} \tag{12}$$

where N_0 is Avogadro's number and A_a is the atomic weight for the absorber.

Now a more general expression can be given for the line shape resulting from the nuclear resonance for an absorber with uniform finite thickness as

$$I(E) = I_0 \left\{ 1 - f_s \left[1 - \frac{2}{\Gamma_s \Gamma_a} \int_{-\infty}^{+\infty} \frac{(\Gamma/2)^2}{(E - E_0)^2 + (\Gamma/2)^2} \right] \right.$$

$$\left. \times \exp \left[-t_a \frac{(\Gamma/2)^2}{E^2 + (\Gamma/2)^2} \right] dE \right\} \tag{13}$$

where f_s is the Mössbauer fraction for the source (26).

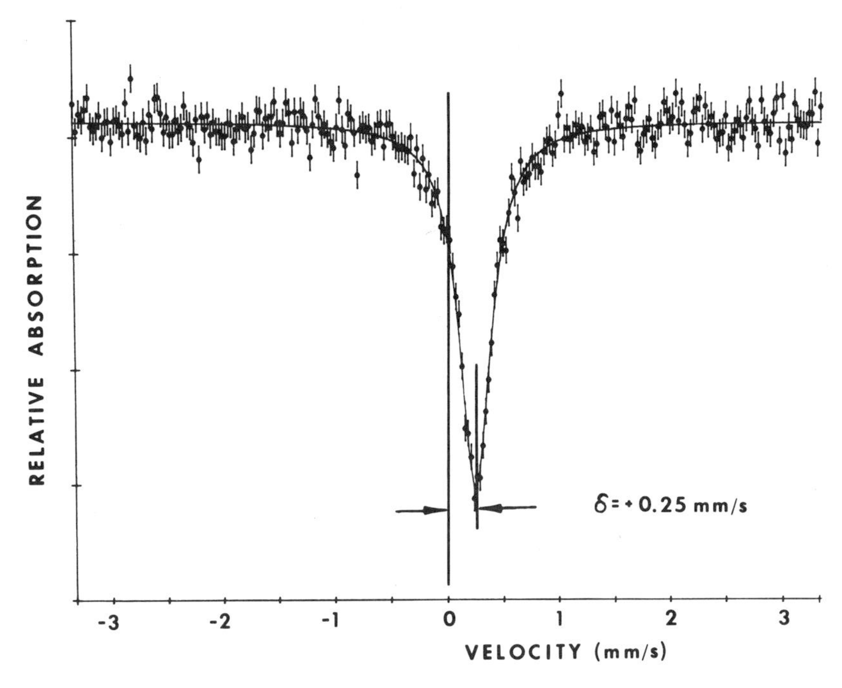

Fig. 6.5. ^{57}Fe Mössbauer spectrum of potassium ferrocyanide.

A Mössbauer spectrum of potassium ferrocyanide is shown in Fig. 6.5 in which the data points are fitted to a Lorentzian. Note that the energy axis is expressed in terms of velocity (mm/s). The reason for the choice of units has to do with the way in which the energy is varied to obtain thc spectrum. The observable line widths are about one part in $10^{12}-10^{13}$ of the actual energy of the photon. Such resolution far exceeds any of the ordinary forms of spectroscopy and requires a unique method of varying the energy, one method of which utilizes the first-order Doppler shift (6). The absorber and source are moved relative to each other with velocity v resulting in a Doppler energy shift for the gamma ray of

$$\varepsilon = \frac{\mathrm{v}}{c} E_{\gamma} \tag{14}$$

Conversions from the velocity units (mm/s) to several corresponding energy units are given in Table 6.II for a number of the more common transitions. Positive velocity refers to the case when the absorber and source are approaching each other. Methods for obtaining the velocities are discussed in the Experimental Methods section.

Observing nuclear gamma resonance is interesting, but the major usefulness comes as a result of the extremely high resolution that can be achieved. One of the main areas that can be investigated are nuclear hyperfine interactions. These are interactions between a property of the nucleus (e.g., magnetic dipole moment) and a feature of the environment of the nucleus (e.g., magnetic field). There are three such interactions that are extremely important to Mössbauer spectroscopy. These are the

TABLE 6.II
Energy Factors for Mössbauer Transitions

Isotope	Transition energy (keV)	1 mm/s = (10^{-8} eV)	1 mm/s = (MHz)	1 mm/s = (10^{-27} J/molecule)	1 mm/s = (mK)[a]
^{57}Fe	14.4	4.808	11.62	7.703	0.5579
^{61}Ni	67.4	22.48	54.37	36.02	2.609
^{99}Ru	89.4	29.81	72.07	47.76	3.459
^{119}Sn	23.9	7.963	19.25	12.76	0.9240
^{121}Sb	37.2	12.39	29.96	19.85	1.438
^{125}Te	35.5	11.83	28.60	18.95	1.373
^{127}I	57.6	19.21	46.46	30.78	2.230
^{129}I	27.8	9.263	22.40	14.84	1.075
^{133}Cs	81.0	27.02	65.33	43.29	3.135
^{151}Eu	21.5	7.182	17.37	11.51	0.8335
^{153}Eu	103.2	34.42	83.22	55.14	3.994
^{155}Gd	86.5	28.87	69.80	46.25	3.350
^{161}Dy	25.7	8.558	20.69	13.71	0.9931
^{166}Er	80.6	26.87	64.97	43.05	3.118
^{170}Yb	84.2	28.10	67.95	45.03	3.261
^{181}Ta	6.2	2.081	5.031	3.334	0.2415
^{193}Ir	73.0	24.36	58.91	39.03	2.827
^{197}Au	77.3	25.80	62.38	41.34	2.994
^{237}Np	59.5	19.86	48.02	31.82	2.305

[a] mK = milli Kelvin.

electric monopole interaction (E0), the magnetic dipole interaction (M1), and the electric quadrupole interaction (E2). We discuss these hyperfine interactions in the next section.

III. THE PRINCIPAL INTERACTIONS

A. ELECTRIC MONOPOLE (E0)—ISOMER SHIFT

An electrostatic interaction occurs between the nuclear charge of the nucleus and the atomic electrons that penetrate the nucleus. At the nucleus, the electronic charge density is given by $-e|\Psi(0)|^2$ and is approximately constant over the nuclear volume. Usually only the s electrons can penetrate the nucleus due to their wavefunction symmetries. The effect of this interaction is to raise the nuclear energy level slightly as shown in Fig. 6.6. Approximating the nucleus as a uniform sphere of radius R, one finds for this shift in energy

$$\delta E = \frac{2\pi}{5} Ze^2|\Psi(0)|^2 R^2 \tag{15}$$

where Z is the nuclear charge. A net change in the energy of the transition (ΔE) will occur depending on the shifts for both the excited (δE_e) and ground (δE_g) levels, i.e.,

$$\Delta E = \delta E_e - \delta E_g = \frac{2\pi}{5} Ze^2|\Psi(0)|^2 (R_e^2 - R_g^2) \tag{16}$$

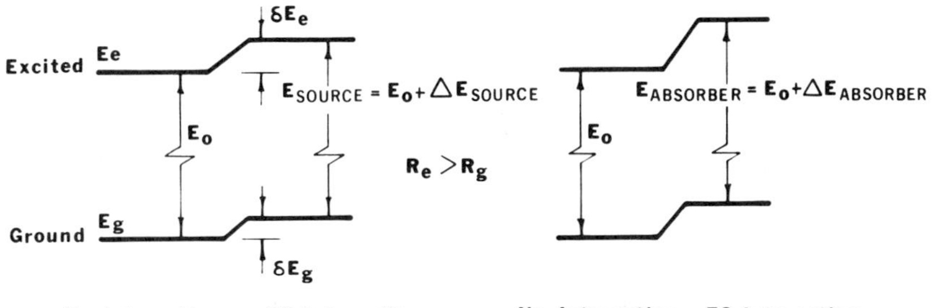

Fig. 6.6. Energy level diagrams for identifying the isomer shift.

This change in the transition energy will occur in both the source (emission) and the absorber (absorption). The difference between these in velocity units is called the isomer shift (δ) and can be measured as a shift in the absorption line as shown in Fig. 6.5. The expression for the isomer shift in velocity units using equation 16 becomes

$$\begin{aligned}\delta &= [\Delta E_{\text{absorber}} - \Delta E_{\text{source}}] \frac{c}{E_\gamma} \\ &= \frac{2\pi Z e^2 c}{5E_\gamma} (R_e^2 - R_g^2) [|\Psi(0)|^2 \text{absorber} - |\Psi(0)|^2 \text{source}] \\ &= \frac{4\pi Z e^2 c R^2}{5E_\gamma} \left(\frac{\Delta R}{R}\right) [|\Psi_a(0)|^2 - |\Psi_s(0)|^2] \qquad (17)\end{aligned}$$

where $\Delta R = R_e - R_g$. The electron density terms in this equation are nonrelativistic; however, a relativistic model calculation (36) indicates that the only modification of equation 17 necessary is the addition of an overall factor $S(Z)$. Therefore

$$\delta = \frac{4\pi Z e^2 c R^2}{5E_\gamma} S(Z) \frac{\Delta R}{R} [|\Psi_a(0)|^2 - |\Psi_s(0)|^2] \qquad (18)$$

is the relativistic generalization of equation 17.

Equation 18 can be simply written as

$$\delta = \alpha \Delta |\Psi(0)|^2 \qquad (19)$$

where α is called the isomer shift calibration constant and $\Delta|\Psi(0)|^2$ is the difference in electron densities at the nuclei in the two substances. Values for $\Delta R/R$, $S(Z)$, and α are given in Table 6.III. It is evident from these equations that the isomer shift is a function of the electron density at the nucleus. This electron density will be very much dependent on electronic structure of the Mössbauer atom and the bonding between this atom and its ligands. For α less than zero (e.g., ^{57}Fe), if iron compound

TABLE 6.III

Parameters for the Isomer Shift (34)

Isotope	E_γ (keV)	Relativistic correction $S(Z)$	α ($10^{-3}\ a_0^3$) (a_0 mm/s)[a]	$\Delta R/R$ (10^{-4})
^{57}Fe	14.41	1.294	−0.157	−5.6
^{61}Ni	67.41	1.343	−0.0085	−1.25
^{99}Ru	89.36	1.927	+0.060	+5.4
^{119}Sn	23.87	2.306	+0.042	+0.79
^{121}Sb	37.15	2.381	−0.21	−5.8
^{125}Te	35.46	2.438	+0.024	+0.62
^{127}I	57.60	2.530	−0.081	−3.3
^{129}I	27.77	2.530	+0.21	+4.2
^{133}Cs	81.00	2.685	+0.0156	+0.84
^{151}Eu	21.53	3.511	+0.34	+3.9
^{153}Eu	103.18	3.511	−0.40	−22
^{155}Gd	86.54	3.678	−0.025	−1.12
^{161}Dy	25.66	3.993	+0.115	+1.44
^{170}Yb	84.25	4.667	+0.0060	+0.22
^{181}Ta	6.24	5.196	−3.1	−7.8
^{193}Ir	73.04	6.213	+0.035	+0.95
^{197}Au	77.34	6.840	+0.053	+1.46
^{237}Np	59.54	13.580	−0.26	−4.1

[a] a_0 = Bohr radius.

A has a δ greater than that of iron compound B, then the electron density at nucleus B is greater than that at A.

Usually isomer shifts are given relative to the source used in the experiment or relative to a standard reference material. To compare literature data it is necessary to have all δ's relative to the same substance. Conversions relative to one material can be obtained relative to another by using the evaluated data given in Table 6.IV. This table also gives recommended standard reference materials. All δ data are usually reported relative to these materials.

B. MAGNETIC DIPOLE (M1)—MAGNETIC HYPERFINE SPLITTING

Energy levels in nuclei having spin quantum numbers (I) greater than zero will have a nonzero magnetic dipole moment ($\vec{\mu}$). In the presence of a magnetic field ($\vec{H}$), there will be an interaction that results in the splitting of nuclear energy levels removing degeneracies. The Hamiltonian describing this interaction is simply

$$\mathcal{H} = -\vec{\mu} \cdot \vec{H} \tag{20}$$

The magnetic moment can be expressed as

$$\vec{\mu} = g_N \beta_N \vec{I} \tag{21}$$

where g_N is the nuclear Landé factor (sometimes called the nuclear g factor) and β_N is the nuclear magneton ($\beta_N = 5.051 \times 10^{-27}$ Joule/Tesla). Substituting equation 21 into equation 20 gives

$$\mathcal{H} = -g_N\beta_N\vec{I}\cdot\vec{H} \tag{22}$$

The diagonalization of the first-order perturbation matrix results in the following eigenvalues (E_M) for the Hamiltonian:

$$E_M(m_I) = -\mu H m_I/I = -g_N\beta_N H m_I \tag{23}$$

where m_I is the nuclear magnetic quantum number, having the $(2I+1)$ values: $-I$, $-I+1$, ... $I-1$, $+I$. As an example, the resulting splitting and transitions for ^{57}Fe are shown in Fig. 6.7. However, two of the transitions, $m_I = +3/2$ to $m_I = -1/2$ and $m_I = -3/2$ to $m_I = +1/2$, are forbidden since the selection rule is $\Delta m_I = 0, \pm 1$. Spectra that result are often quite complex. In the spectrum for metallic iron, shown in Fig. 6.8, the magnetic field is an internal field of 33 Tesla.

As for the isomer shift, the term that is of most interest is the environmental parameter, in this case the magnetic field. This field can either exist internally or be applied. There are three principal contributions to the internal magnetic field, each being generated by unpaired electrons (45). Usually the dominant contribution is the Fermi contact field (H_c) which results from a spin density (either spin up or spin down) at the nucleus. These are the s-electrons that can be spin-polarized by the electrons in the outer shells. The other two fields are the orbital field (H_L), which results from the orbital motion of the valence electrons, and the dipolar field (H_D),

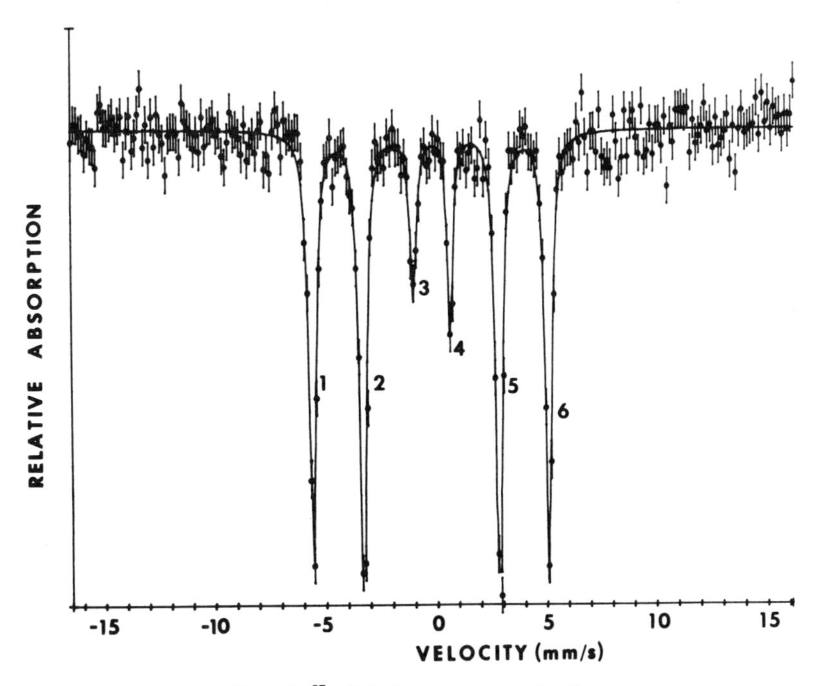

Fig. 6.7. ^{57}Fe Mössbauer spectrum of α-Fe.

TABLE 6.IV
Isomer Shift Reference Scales,
Uncertainty in Last Digit(s) Given in Parenthesis

Isotope	Transition (keV)	Reference material	Absorber and source materials (isomer shift relative to reference material in mm/s)		
^{57}Fe	14.4	α-Fe	$Na_2Fe(CN)_5NO \cdot 2H_2O$	Cr	Stainless steel
		(T=300 K)	−0.2576(14)	−0.146(3)	−0.086(3)
			$Na_4Fe(CN)_6 \cdot 10H_2O$	$K_4Fe(CN)_6 \cdot 3H_2O$	Rh
			−0.0553(21)	−0.042(3)	+0.1209(22)
			Pd	Cu	Pt
			+0.1798(12)	+0.2242(10)	+0.3484(24)
			α-Fe_2O_3		
			+0.365(3)		
		$Na_2Fe(CN)_5NO \cdot 2H_2O$	Cr	Stainless steel	$Na_4Fe(CN)_6 \cdot 10H_2O$
		(T=300 K)	+0.111(3)	+0.171(3)	+0.2024(22)
			$K_4Fe(CN)_6 \cdot 3H_2O$	α-Fe	Rh
			+0.2152(24)	+0.2576(14)	+0.3786(24)
			Pd	Cu	Pt
			+0.4374(13)	+0.4819(13)	+0.606(3)
			α-Fe_2O_3		
			+0.623(3)		
^{99}Ru	89.4	Ru	RuO_2	$K_4Ru(CN)_6 \cdot 3H_2O$	Ru(Rh)
			−0.249(8)	−0.224(10)	0.000(4)
^{119}Sn	23.9	$BaSnO_3$	SnO_2	$CaSnO_3$	Me_2SnF_2
		(T=77 K)	0.000	0.000	+1.301(16)
			Pd(Sn)	Pd_3Sn	V(Sn)
			+1.505(13)	+1.571(14)	+1.577(6)
			Mg_2Sn	α-Sn	β-Sn
			+1.908(12)	+1.998(15)	+2.559(8)

			SnTe +3.446(15)		
		$BaSnO_3$ (T=300 K)	SnO_2 0.000	$CaSnO_3$ 0.000	Me_2SnF_2 +1.291(16)
			Pd(Sn) +1.507(13)	V(Sn) +1.624(14)	Pd_3Sn +1.579(7)
			Mg_2Sn +1.905(12)	α-Sn 1.995(15)	β-Sn +2.555(8)
			SnTe +3.441(15)		
^{121}Sb	37.2	InSb	β-Sn −2.70(4)	$Ni_{21}Sn_2B_6$ +1.648(19)	$BaSnO_3$ +8.47(4)
			SnO_2 +8.51(2)	$CaSnO_3$ +8.53(3)	
^{125}Te	35.5	ZnTe	β-TeO_3 −1.16(3)	PbTe +0.00(6)	Cu(I) +0.01(2)
			Rh(Sb) +0.05(10)	Cu(Sb) +0.08(3)	SnTe +0.23(5)
			Te +0.57(3)	TeO_2 +0.78(4)	
^{127}I	57.6	CuI	NaI −0.024(8)	KI −0.01(3)	CsI +0.00(3)
			ZnTe +0.12(2)		
^{129}I	27.8	CuI	NaI −0.076(18)	KI −0.062(14)	CsI +0.007(17)
			ZnTe +0.384(11)	SnTe +0.81(5)	
^{149}Sm	22.5	SmF_3	SmF_2 −0.90(8)	Eu −0.02(11)	Eu_2O_3 −0.01(6)

TABLE 6.IV (Cont.)
Isomer Shift Reference Scales,
Uncertainty in Last Digit(s) Given in Parenthesis

Isotope	Transition (keV)	Reference material	Absorber and source materials (isomer shift relative to reference material in mm/s)		
			EuF_3	Sm_2O_3	$SmAl_2$
			0.00(2)	+0.04(3)	+0.15(12)
^{151}Eu	21.5	EuF_3	EuS	$EuF_3 \cdot 2H_2O$	$SmF_3 \cdot 2H_2O$
			−11.65(4)	−0.046(9)	−0.003(13)
			SmF_3	Sm_2O_3	Eu_2O_3
			+0.05(3)	+0.85(4)	+1.017(8)
^{153}Eu	103.2	EuF_3	Eu_2O_3	Sm_2O_3	EuS
			−1.18(14)	−0.94(14)	+14.0(10)
^{155}Gd	86.5	GdF_3	Pd(Eu)	Gd	Sm(Eu)
			−0.684(9)	−0.678(9)	−0.53(4)
			EuF_2	$GdAl_2$	$SmAl_3$(Eu)
			−0.51(3)	−0.234(13)	−0.169(13)
			$GdAl_3$	$Sm_2Sn_2O_7$(Eu)	SmF_3(Eu)
			−0.159(12)	−0.12(2)	−0.02(4)
^{161}Dy	25.6	DyF_3	GdF_3(Tb)	Gd_2O_3(Tb)	Dy_2O_3
		(T=300 K)	+0.12(14)	+0.1(5)	+0.62(6)
			Gd(Tb)	Dy	
			+2.25(5)	+2.82(10)	
^{170}Yb	84.3	$YbAl_2$	$YbSO_4$	YbB_6	Yb
			−0.34(4)	−0.202(16)	0.00(2)
			TmB_{12}	$TmAl_2$	$YbAl_3$
			0.00(2)	+0.060(12)	+0.09(2)

			Tm +0.12(3)		
^{181}Ta	6.2	Ta	Mo(W) −22.56(8)	W −0.835(3)	Ta(W) −0.074(4)
			Pt(W) +2.71(8)		
^{193}Ir	73.0	Ir	Pt(Os) −0.644(6)	Os +0.539(7)	Nb(Os) +1.0(2)
			V(Os) +1.71(3)		
^{197}Au	77.3	Au	Pt +1.22(2)		
^{237}Np	59.5	$NpAl_2$	Th(Am) −8.9(3)	NpO_2 −6.10(4)	UO_2 −5.17(6)

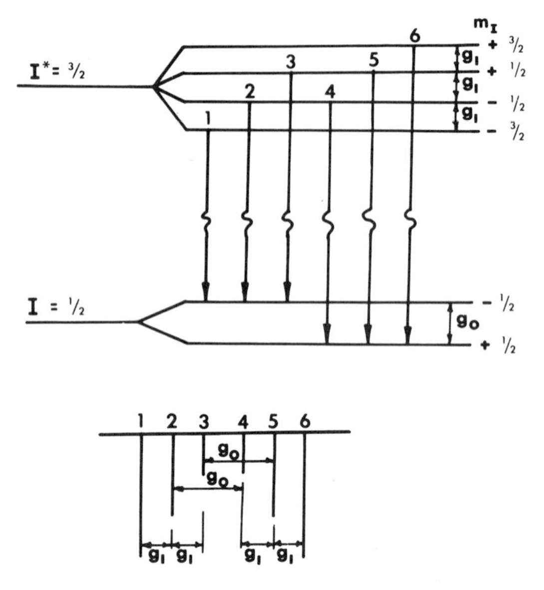

Fig. 6.8. Energy level diagram and line spectrum due to the magnetic hyperfine interaction in ^{57}Fe.

due to the coupling between the nucleus and outer electrons. Therefore, the total internal field is

$$H = H_C + H_L + H_D \tag{24}$$

C. ELECTRIC QUADRUPOLE (E2)—QUADRUPOLE COUPLING CONSTANT

When the nuclear-spin quantum number is greater than $\frac{1}{2}$ there is a nuclear quadrupole moment (Q). This moment can interact with the electric field gradient (EFG) to result in the splitting of nuclear energy levels. The EFG is $(-\vec{\nabla}\,\vec{\nabla}\,V)$, i.e.,

$$\text{EFG} = \vec{\nabla}\,\vec{E} = -\vec{\nabla}\,\vec{\nabla}\,V = \begin{bmatrix} V_{xx} & V_{xy} & V_{xz} \\ V_{yx} & V_{yy} & V_{yz} \\ V_{zx} & V_{zy} & V_{zz} \end{bmatrix} \tag{25}$$

where the components are given as

$$-V_{ij} = \frac{\partial^2 V}{\partial x_i \partial x_j}$$

The above symmetric tensor can be diagonalized by the appropriate choice of axes. The resulting tensor has three non-zero elements, which are the diagonal elements. Only two of these diagonal elements are independent due to Laplace's equation, which states

$$V_{xx} + V_{yy} + V_{zz} = 0$$

These two independent elements give rise to two experimentally observable parameters. One of these is the z component of the electric field gradient defined as

$$q = \frac{V_{zz}}{e} \tag{26a}$$

while the other, the asymmetry parameter, is

$$\eta = \frac{V_{xx} - V_{yy}}{V_{zz}} \tag{26b}$$

where $|V_{zz}| \geqslant |V_{yy}| \geqslant |V_{xx}|$. This constraint restricts η to lie between 0 and 1.

The Hamiltonian for the interaction between the nuclear quadrupole moment and the EFG is

$$\mathcal{H}_Q = \frac{e^2Qq}{4I(2I-1)} [3I_z^2 - I^2 + \eta (I_+^2 + I_-^2)/2] \tag{27}$$

where I is the nuclear spin operator, I_z the operator for the nuclear spin projected in the z direction, and $I_\pm$ are shift operators. For cases of axial symmetry, i.e., $\eta = 0$, the energy eigenvalue equation is

$$E_Q = \frac{e^2Qq}{4I(2I-1)} [3m_I^2 - I(I+1)] \tag{28}$$

For the case where $I = 3/2$ (applicable to ^{119}Sn and ^{57}Fe) equation 27 becomes

$$E_Q(m_I) = \frac{e^2Qq}{12} \left[3m_I^2 - \frac{15}{4}\right] [1 + \eta^2/3]^{1/2} \tag{29}$$

This gives two levels:

$$E_Q(\pm 1/2) = -1/4e^2Qq\,(1 + \eta^2/3)^{1/2}$$

$$E_Q(\pm 3/2) = +1/4e^2Qq\,(1 + \eta^2/3)^{1/2} \tag{30}$$

The resulting energy-level diagram for this interaction is shown in Fig. 6.9. Instead of a single absorption line there are two. The observed splitting of the single line into

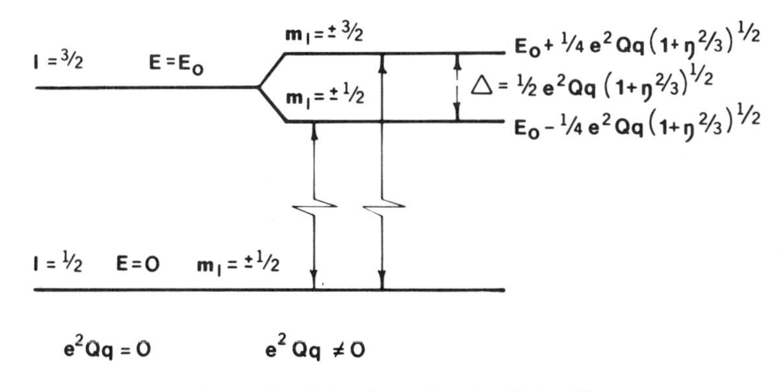

Fig. 6.9. Energy level diagram identifying the quadrupole splitting (Δ).

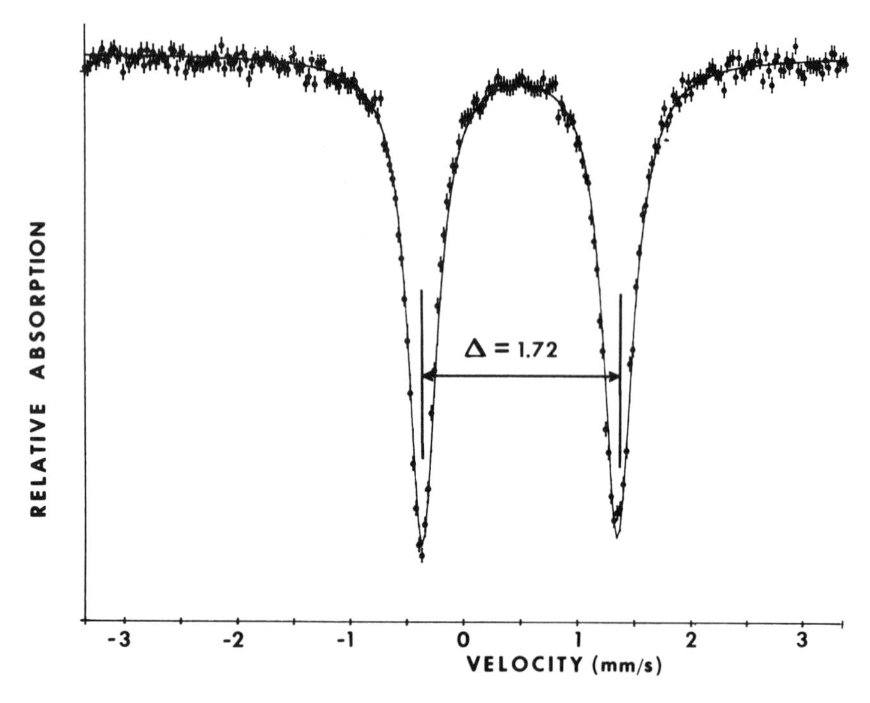

Fig. 6.10. ^{57}Fe Mössbauer spectrum of sodium nitroprusside.

two lines (see Fig. 6.10) is called the quadrupole splitting (Δ) and from equation 30

$$\Delta = 1/2e^2Qq\,(1 + \eta^2/3)^{1/2} \tag{31}$$

Often $\eta = 0$ and consequently the quadrupole splitting for a particular transition will be dependent only on V_{zz}.

When the nuclear spins are different from those of the common transitions in iron and tin, a much more complicated situation arises. For example, ^{121}Sb has a ground nuclear spin of 5/2 while the excited level is 7/2. If $\eta = 0$, there will be eight allowed transitions resulting in a more complex spectrum. It is unfortunate that for ^{121}Sb the eight absorption peaks overlap each other and none of the lines can be resolved. A typical ^{121}Sb Mössbauer spectrum is shown in Fig. 6.11. Because both the relative positions and intensities are known, it is not too difficult with the use of a digital computer to determine the quadrupole coupling. More details on this will be provided in the sections on Experimental Methods, (Section IV) and Quadrupole Coupling (Section VII).

IV. EXPERIMENTAL METHODS

A. SPECTROMETERS

A Mössbauer spectrum is a plot of intensity (of gamma rays) versus Doppler velocity. While gamma rays are detected and counted by using normal nuclear-counting instrumentation methods, velocity-modulation techniques are relatively

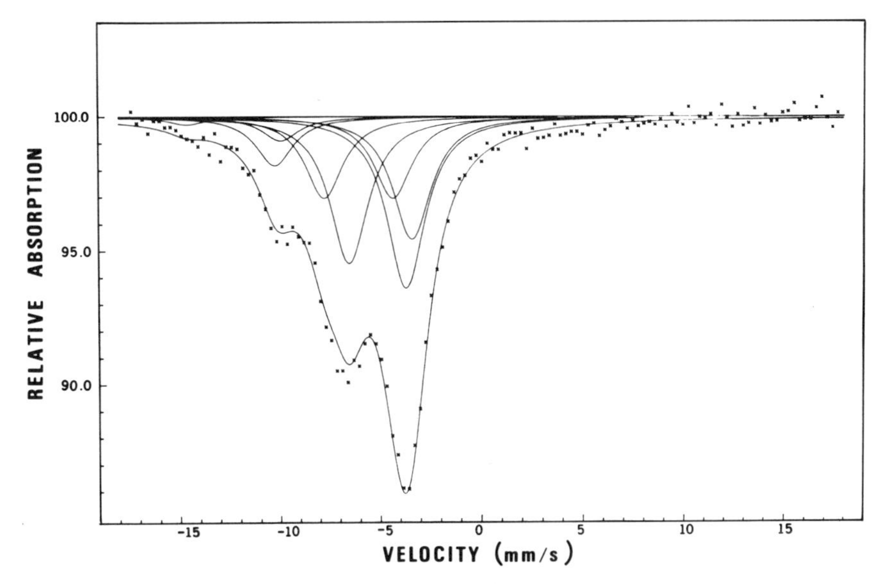

Fig. 6.11. ^{121}Sb Mössbauer spectrum of $PhSb(Et_2dtc)_2$.

unique to Mössbauer spectroscopy and are central to any Mössbauer spectrometer. Many of the earlier spectrometers were mechanical devices in which the source or absorber moved at a constant velocity relative to one another. With these devices, the spectrum is obtained by counting for a specific period of time at one velocity, recording the number of gammas counted, and then selecting another velocity, etc. Such a procedure is obviously very time-consuming and requires much effort. Programming techniques can be used to reduce some of this effort. However, current spectrometers use electromechanical devices which sweep a range of velocities with a frequency of the order of 10 Hz. This allows the complete range of velocities (i.e., energies) to be counted almost simultaneously, although a spectrum does not begin to appear until after many scans.

The primary element of a Mössbauer spectrometer is an electromagnetic transducer, often referred to as the Kankeleit drive. This has one basic design consisting of a drive coil, which is located in the field of a permanent cylindrical magnet, and a velocity-monitoring coil, both of which are attached to a center rod. The coils can be either specifically designed for the spectrometer or obtained from a commercially produced loudspeaker. A cross-section diagram of a drive is shown in Fig. 6.12.

The rod is driven by a current running through the drive coil. The current can be varied to produce several different periodic motions, shown in Fig. 6.13. The triangle, which is the most common waveform, and the sawtooth, sometimes called "flyback," both give velocities which vary linearly in time. The triangular waveform gives a true spectrum along with its mirror image. The sawtooth does not give a mirror spectrum. The sinusoidal waveform is especially suited for large velocities and fine precision.

A schematic block diagram of a typical Mössbauer spectrometer is shown in Fig. 6.14. It illustrates how the electromagnetic transducer, discussed above, is

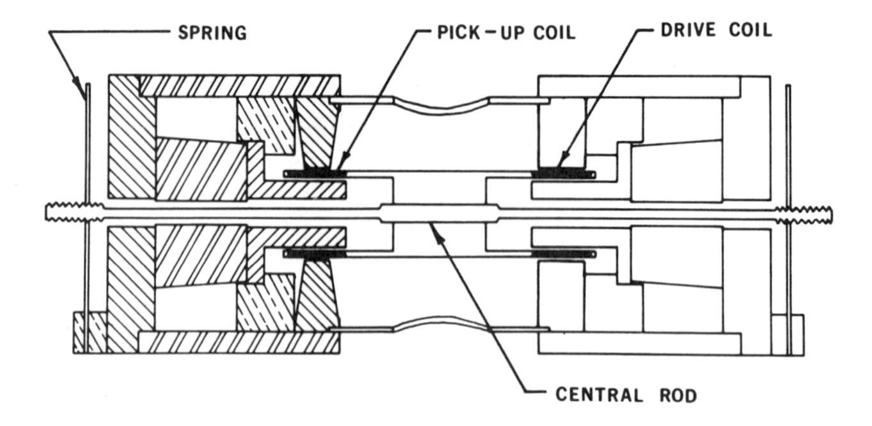

Fig. 6.12. Mössbauer drive motor (courtesy of Austin Science Associates).

related to the other major components of the spectrometer. Central to the electronic part of a Mössbauer spectrometer is a multichannel analyzer (MCA), an on-line computer system, or a microprocessor which stores the collected data in the form of the number of gammas counted at each velocity.

B. SOURCES

The usual sources are radioactive isotopes that first decay by electron capture or alpha, beta, or gamma ray emission. These radioactive isotopes subsequently undergo Mössbauer transitions. Simplified nuclear energy level diagrams illustrating typical decays are given in Fig. 6.15 for four typical Mössbauer transitions. It is

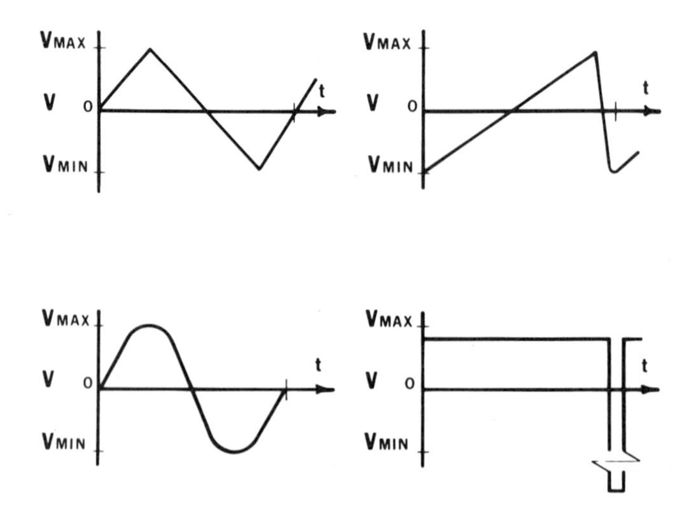

Fig. 6.13. Periodic motions of a Mössbauer spectrometer (triangle, sawtooth, sinusoidal, and flyback).

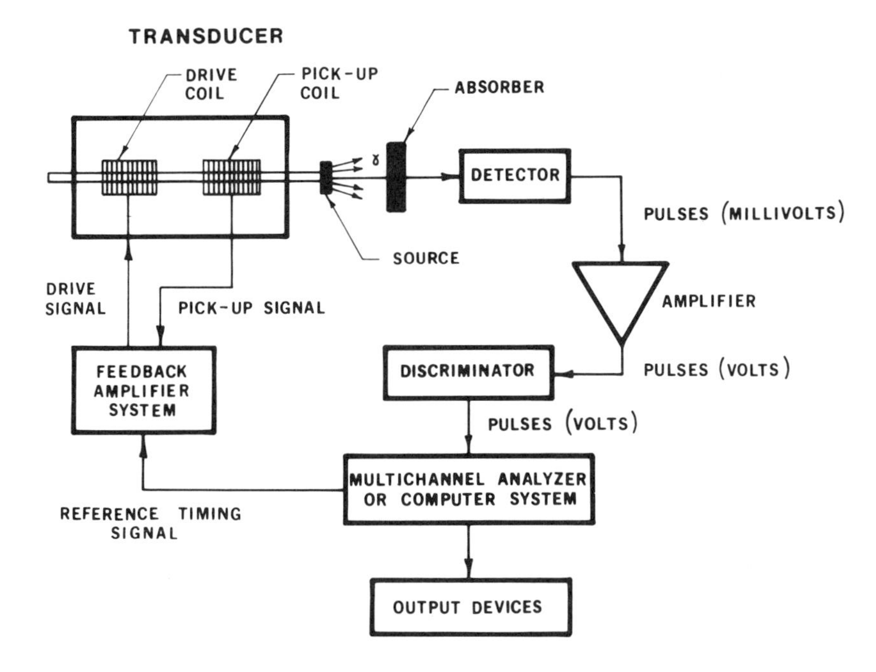

Fig. 6.14. Block diagram of a Mössbauer spectrometer.

desirable for the gamma ray to be emitted with zero recoil during the Mössbauer transition. From equations 8 and 10, it is apparent that the Mössbauer fraction, which gives the probability for a recoilless emission, is significant provided that the energy of the gamma is not too large. In fact, all the observable Mössbauer transitions have gamma energies below 200 keV. See Table 6.I for the energies of the more common transitions.

The lifetimes of any excited nuclear level used in Mössbauer spectroscopy must have natural line widths (see equation 5) that can be observed by the Doppler velocity-scan method of varying the energy. Consequently these lifetimes are usually in the range of 1–100 ns. If they are shorter, the line width will be too broad and, if longer, the line width will normally be too narrow to be observed. It is important to select source materials that give a large Mössbauer fraction (f_s) and have single, narrow lines. Table 6.V contains a list of such materials with f_s values at those temperatures normally used for the spectroscopy.

C. DETECTORS

Since Mössbauer gamma rays are quite low in energy, the detectors employed are those that normally detect X-rays. Basically there are three different types of detectors. These are the scintillation detectors, the proportional counters and the semiconductor devices. Scintillation detectors are usually NaI(Tl) crystals and are excellent for the higher-energy Mössbauer transitions due to their counting efficiency.

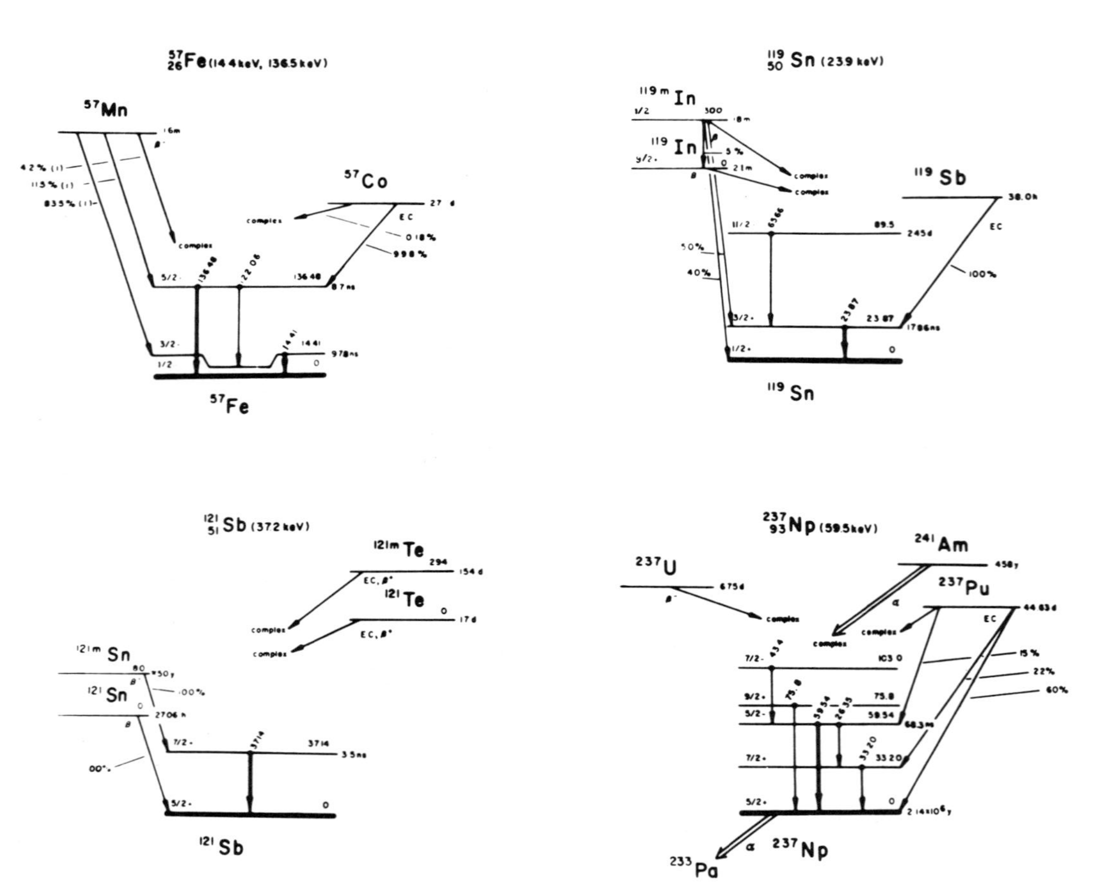

Fig. 6.15. Simplified nuclear energy level diagrams for ^{57}Fe, ^{119}Sn, ^{121}Sb and ^{237}Np.

They are also relatively inexpensive, but have poor resolution. This makes them inappropriate for many transitions. Proportional counters, also inexpensive, have resolutions that are much improved over the scintillation detectors, but have poor efficiency at the higher energies. The semiconductor detectors are usually Ge(Li) or Si(Li). More recently intrinsic Ge detectors have been employed. The efficiency is excellent for these devices at all gamma energies of interest and they have resolutions that are even better than proportional detectors. An example of their improvement in resolution can be seen in Fig. 6.16 where a comparison is made with results obtained from a proportional detector. However, their cost ranges from $5,000 to $15,000. In addition, they must be maintained at liquid nitrogen temperatures.

TABLE 6.V
Properties of Mössbauer Sources (37)

Isotope	Transition energy (keV)	$W_o{}^a$ (mm/s)	Source material	Source temperature (K)	f_s
^{57}Fe	14.41	0.194	Cr	300	0.784
			SS310	300	0.604
			SS	300	0.678
			Rh	300	0.784
				4.2	0.875
			Pd	300	0.660
				77	0.863
				4.2	0.813
			Cu	300	0.708
				4.2	0.910
			Pt	300	0.724
				77	0.890
				4.2	0.851
			CoO	300	0.735
^{61}Ni	67.41	0.770	Ni–Cr alloy	4.2	
			Ni–V alloy	4.2	0.162
^{99}Ru	89.36	0.149	Ru(Rh)	4.2	0.140
^{119}Sn	23.87	0.646	SnO_2	300	0.471
				77	0.585
				4.2	0.885
			$CaSnO_3$	300	0.574
			$BaSnO_3$	300	0.623
			Pd(Sn)	300	0.383
			Pd_3Sn	300	0.340
				4.2	0.750
			V(Sn)	300	0.460
				77	0.780
			Mg_2Sn	300	0.280
				77	0.770
			α-Sn	77	
			β-Sn	300	0.046
				77	0.446
				4.2	0.716

TABLE 6.V (Cont.)
Properties of Mössbauer Sources (37)

Isotope	Transition energy (keV)	W_o [a] (mm/s)	Source material	Source temperature (K)	f_s
^{121}Sb	37.15	2.10	SnO_2	300	0.212
				77	0.320
			$BaSnO_3$	77	0.450
			β-Sn	77	0.160
			$Ni_{21}Sn_2B_6$	300	0.070
				77	0.290
^{125}Te	35.46	5.209	β-TeO_3	300	0.320
				77	0.531
			PbTe	300	<0.029
				77	0.250
			Cu(I)	77	0.143
				4.2	0.400
			Rh(Sb)	4.2–77	
			Cu(Sb)	300	<0.029
				77	0.442
^{127}I	57.60	2.49	ZnTe	4.2	0.120
^{129}I	27.77	0.586	ZnTe	77	0.232
^{151}Eu	21.53	1.31	SmF_3	300	0.275
			$SmF_3 \cdot 2H_2O$	300	
			Sm_2O_3	300	0.440
^{153}Eu	103.2	0.68	Sm_2O_3	20	0.050
^{155}Gd	86.55	0.499	Pd(Eu)	4.2	0.110
			Sm(Eu)	4.2	
			$Sm_2Sn_2O_7$	4.2	
^{161}Dy	26.66	0.378	Gd_2O_3	300	0.230
			GdF_3	300	
			Gd(Tb)	300	
^{166}Er	80.56	1.816	$HoAl_2$	25–30	
^{170}Yb	84.25	2.019	TmB_{12}	4.2	0.340
			$TmAl_2$	4.2	0.180
			Tm	4.2	
^{181}Ta	6.24	0.0064	Mo(W)	300	
			W	300	
			Ta(W)	300	
			Pt(W)	300	
^{193}Ir	73.04	0.595	Pt(Os)	4.2	
			Os	4.2	
			Nb(Os)	4.2	
			V(Os)	4.2	
^{197}Au	77.35	1.882	Pt	77	0.069
				4.2	0.272
^{237}Np	59.54	0.067	Th(Am)	4.2–77	
			VO_2	4.2–77	

[a] Natural line width in units of mm/s.

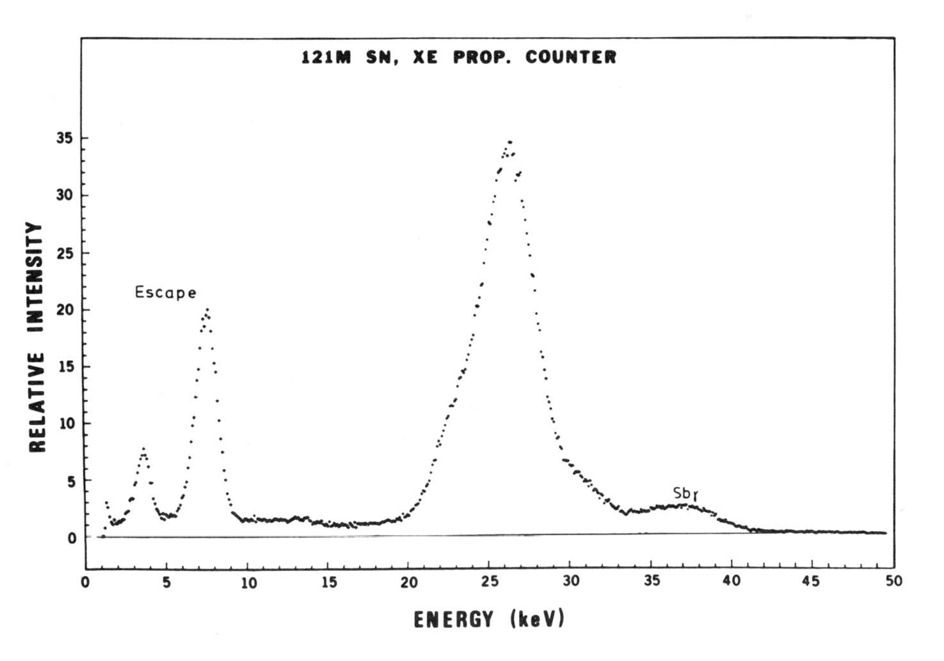

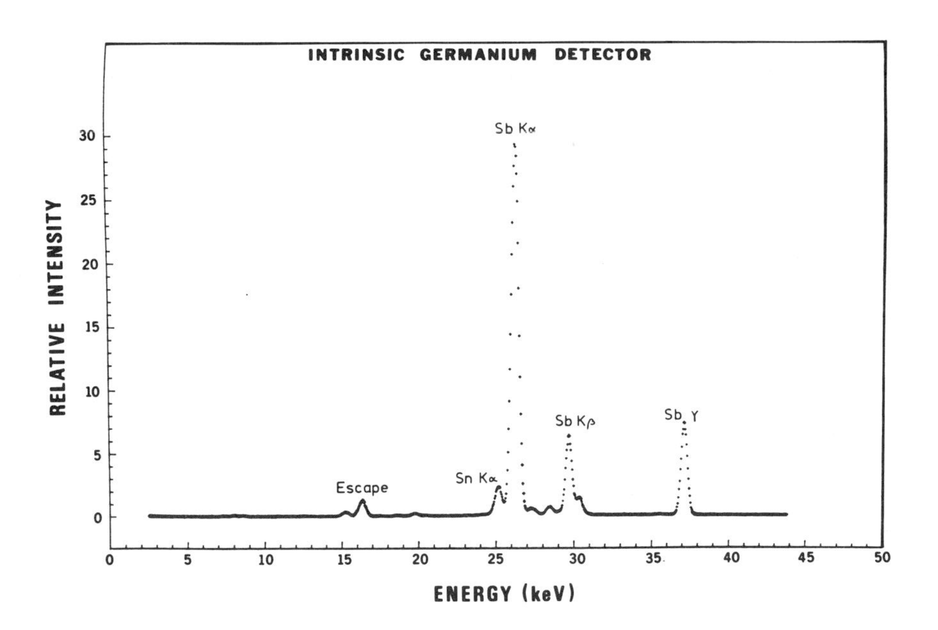

Fig. 6.16. Comparison of the emission spectra using a $Xe(CH_4)$ proportion detector and an intrinsic germanium detector.

TABLE 6.VI
Effective Absorber Thickness Constant (40)

Isotope	Transition (keV)	t_e (cm^2/mg)	Isotope	Transition (keV)	t_e (cm^2/mg)
^{57}Fe	14.4	0.578	^{155}Gd	86.5	0.194
^{61}Ni	67.4	0.0836	^{161}Dy	43.8	0.225
^{99}Ru	89.4	0.0622	^{166}Er	80.6	0.288
^{119}Sn	23.9	0.609	^{169}Tm	8.4	0.918
^{121}Sb	37.1	0.557	^{170}Yb	3.0	0.0204
^{125}Te	35.5	0.0895	^{181}Ta	6.2	5.56
^{127}I	57.6	0.975	^{182}W	100.1	0.219
^{129}I[a]	27.8	1.82	^{193}Ir	73.0	0.0598
^{133}Cs	81.0	0.466	^{195}Pt	98.8	0.0637
^{149}Sm	22.5	0.0397	^{197}Au	77.3	0.118
^{151}Eu	21.5	0.453	^{237}Np[a]	59.5	0.778
^{153}Eu	83.4	0.200			

[a] Isotopic abundance assumed to be one.

D. ABSORBERS

Special care needs to be taken in preparing an absorber with particular attention given to optimizing the thickness. A sample too thin will result in little observable absorption while one too thick will absorb most of the gammas nonresonantly, thus washing out the resonant absorption spectrum. In most other spectroscopies adjusting the thickness by trial and error is usually the most efficient procedure. However, in Mössbauer spectroscopy it often takes hours (and sometimes even days) to obtain a spectrum. Therefore it is important, if at all possible, to prepare the sample correctly on the first attempt. The thickness of the sample needed to give a good spectrum can be determined by using equation 12. This expression can be simplified by defining an "effective absorber thickness constant" (t_e) as

$$t_e = \frac{N_0 \cdot IA \cdot \sigma_0}{1000 \cdot A_a} \tag{32}$$

Using equation 32 with equation 12 we find

$$\sigma_a = \frac{t_a}{f_a \cdot t_e} \tag{33}$$

which is an expression for sample thickness in units of mg of the atom of interest per cm^2. Values for the constant t_e are given in Table 6.VI for the common transitions and values for f_a can be estimated using plots like those in Fig. 6.4. To use these plots one must assume an approximate Debye temperature for the material. Usually organic substances are 50–150K, inorganic substances 100–300 K, and metals and alloys 200–300 K. The Debye temperature is related to the strength of the bonds between the atom of interest and its neighbors. For selecting the sample thickness, a rough value of θ_D will suffice. Using equation 33 and letting $t_a = 1$, an appropriate sample thickness can be determined. If a spectrum is expected to consist of more than a single line, then one should use a larger value for t_a up to five, depending on the anticipated complexity. Some values of sample thicknesses using $t_a = 1$ are listed in Table 6.VII.

It is quite important to have a sample that is approximately uniform in thickness. This is quite difficult to achieve when the sample needs to be as thin as indicated for some cases in the table. The usual procedure is to mix the material with a filler substance that is relatively transparent to the gamma ray. These substances should have atoms of low atomic weights (low Z) and be nonreactive with the sample. Examples are fine powders of boron nitride, sugar, polymethylmethacrylate and glass.

The absorber is placed in a container also made of a low Z material. Common materials include Plexiglass, beryllium, aluminum, and Teflon. If the absorber material is a metal, then it can be rolled into a thin foil.

It is important to maximize the count rate by minimizing the source detector distance. However, at some point minimizing can begin to add a serious error in the velocity scale. This is often referred to as the "cosine effect" because the error in the Doppler energy (ΔE) is

$$\Delta E = \frac{v}{c} E_\gamma \cos\theta \tag{34}$$

where θ is the angle between the direction of the photon and the normal direction between the source and the absorber. As a general rule, it is desirable that the ratio between the detector-window radius and the source-detector distance be less (assuming the source radius is equal to the window radius) than 0.1. Sometimes when a particular source may be quite weak, closer distances are necessary to get any sort of spectrum in a reasonable amount of time.

E. TEMPERATURE CONSIDERATIONS

An examination of the plots in Fig. 6.4 reveal the importance of temperature in obtaining observable spectra. For most Mössbauer transitions, it is necessary that the experiment be done at low temperature, often down to 4.2 K, which can be achieved using liquid helium. Some spectra can be obtained using liquid nitrogen for cooling (77 K). Commercial Dewars readily available which are not already suitable can be easily modified for Mössbauer spectroscopy. The Dewars are constructed either from stainless steel or glass, the latter being less expensive but more easily broken.

Many different absorber-source-detector geometries and configurations are possible. The most common has the absorber and the source at the same temperature inside the cryostat, and the detector outside. Mylar windows (usually aluminum coated) are most commonly used to minimize the nonresonant absorption of the gammas.

Often it is quite important to gather Mössbauer data as a function of temperature. These can be obtained using feedback heating devices that give temperatures from 4.2 K to well above room temperature. For higher temperatures, specially con-

TABLE 6.VII

Calculated Values of Absorber Thickness (σ_a) in mg/cm^2 of the Natural Isotopic Abundance (2)

Isotope	Transition (keV)	$S(\theta_D = 100$ K)			$S(\theta_D = 150$ K)			$S(\theta_D = 200$ K)			$S(\theta_D = 250$ K)		
		300	77	4.2	300	77	4.2	300	77	4.2	300	77	4.2
^{57}Fe	14.4	261	11	5	24	5.9	4.4	10.3	4.8	4.1	7	4.3	4
^{61}Ni	67.4			40268			3281		18483	951		2341	454
^{99}Ru	89.4						6635			1735		4604	776
^{119}Sn	23.8	965	15	5.3	41	6.7	4.5	14	5	4.2	8.3	4.4	4
^{121}Sb	37.1		129	11	1447	19	7.3	105	9.5	6.1	31	7	5.4
^{125}Te	35.5		546	61		100	43	455	55	37	156	41	33
^{127}I	57.6			28		100	11		21	7.5	317	10	5.8
^{129}I	27.8		7.3	2	26	2.7	1.6	6.6	1.9	1.5	3.5	1.6	1.4
^{133}Cs	80.9			575			110		343	48		87	30
^{151}Eu	21.5	169	12	6	22	7	5.4	11	5.8	5.1	8	5.3	5
^{153}Eu	103						1765		8878	555		1288	278
^{155}Gd	86.5			1317			259		795	116		208	72
^{161}Dy	25.6	380[a]	11[a]	4.5[a]	26[a]	5.5[a]	3.9[a]	10[a]	4.3[a]	3.6[a]	6.5[a]	3.8[a]	3.5[a]
^{170}Yb	84.2			5014			1230		3240	614		1017	406
^{193}Ir	73			527		2040	208		395	131		183	100
^{197}Au	77.3			351		1552	126		255	76		110	56
^{237}Np	59.5		298[a]	11.4[a]	7084[a]	24[a]	6.9[a]	227[a]	9.8[a]	5.4[a]	46[a]	6.5[a]	4.6[a]

[a] Should be multiplied by a factor of 10 because of line broadening.

structed vacuum chambers are used. They are constructed with materials that are nonreactive with the sample at temperatures as high as 2000 K. On the other end of the temperature scale it is possible to carry out Mössbauer experiments well below 4.2 K (to the order of 10^{-4}) by using $^3He-^4He$ dilution refrigerators, now commercially available. More details of these cooling and heating devices are described in several good review articles (6,12,22).

F. APPLIED MAGNETIC FIELDS

The common types of electromagnets are usually not adequate for use in Mössbauer spectroscopy because they produce fields that are too small to resolve any interesting information. Most of the studies use superconducting solenoids capable of giving fields as high as 10 Tesla. These magnetic devices are constructed inside the Dewar to give large fields parallel (longitudinal) and perpendicular (transverse) to the source-absorber direction. These magnets are commercially available for Mössbauer spectroscopy.

G. VELOCITY CALIBRATION

Calibrating the energy function of a Mössbauer spectrometer is a nontrivial procedure. In most instances one of two types of procedures is used. The simplest and most common is the use of standard reference materials whose Mössbauer spectra have peaks that are well defined in velocity units. The other method is an optical one which uses either a Michelson interferometer or a Moiré fringe device.

Several standard calibration references are available. The most common reference is the ^{57}Fe Mössbauer spectrum of α-Fe. The splittings for various materials are given in Fig. 6.17. The α-Fe has several advantages, including multiple peaks that not only allow for the determination of the velocity calibration scale constant, but also enable a check of the linearity of the spectrometer. Sodium nitroprusside ($Na_2Fe(CN)_5NO \cdot 2H_2O$) is another common material but no check can be made on the linearity because there are only two peaks. This is usually employed when the velocity scale is small in a particular experiment, i.e., a maximum velocity of less than 3 mm/s. Although both of these materials are the most widely used, there are several other substances that are used. These give multiple line spectra for larger velocity scales than those for which α-Fe is suitable. α-Fe_2O_3 can be employed, but care must be taken because of the possibility of other phases. To achieve lines at larger velocities one can use a source of ^{57}Co in α-Fe and an α-Fe absorber, which will give lines over a range of velocities of 20 mm/s. Finally the largest practical splitting currently used is the ^{161}Dy Mössbauer spectrum of Dy metal, which gives peaks over a range of velocities exceeding 400 mm/s.

Recently two optical devices for calibration have been gaining wide support (9,13,14). These are now available on most commercial spectrometers and are more precise than the reference calibration discussed above. Both optical devices can use either a lamp or a laser, but the latter is preferred.

With a Michelson interferometer, one can measure distance and time very precisely to determine velocity. There are two basic mirrors: one is fixed and the other

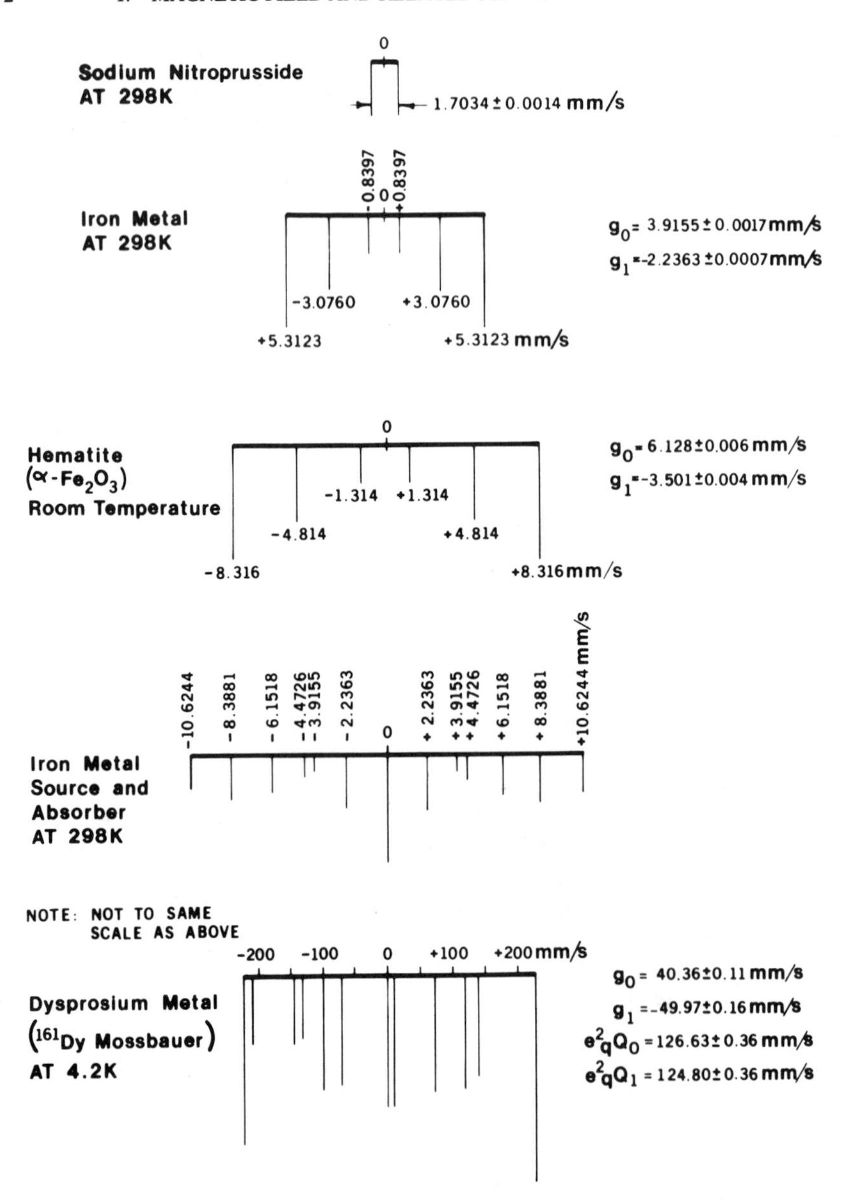

Fig. 6.17. Standard splittings for Mössbauer velocity calibrations.

is connected to the moving rod of the transducer. The intensity of light detected at the photodiode depends on the position (x) of the moving mirror (see Fig. 6.18), such that

$$I \sim \sin \frac{2\pi x}{\lambda/2} \tag{35}$$

where λ is the wavelength of the laser. When the mirror moves $\lambda/4$, the intensity of the laser beam at the photodiode will change from a maximum to zero. The photodiode is used to count the number (n_i) of the times there is a change from dark to bright to dark. The calibration of a particular velocity channel is achieved using the following relation for the average velocity:

$$\bar{v}_i = \frac{n_i \lambda}{2N \Delta t_i} \tag{36}$$

where Δt_i is the time spent in the channel and N is the number of times the channel has been opened for counting.

Similar is the Moiré fringe method, which is also shown in Fig. 6.18. The average velocity of the i channel is given by

$$\bar{v}_i = \frac{n_i d}{4N \Delta t_i} \tag{37}$$

where d is the grating distance. The Moiré method does not require the sometimes difficult aligning and focusing necessary when the interferometer is used, but it is an order of magnitude less precise. The interferometer gives a "direct measurement," while the Moiré method requires a knowledge of the spacing between lines in the grating. However, an advantage of the Moiré devices is their compactness.

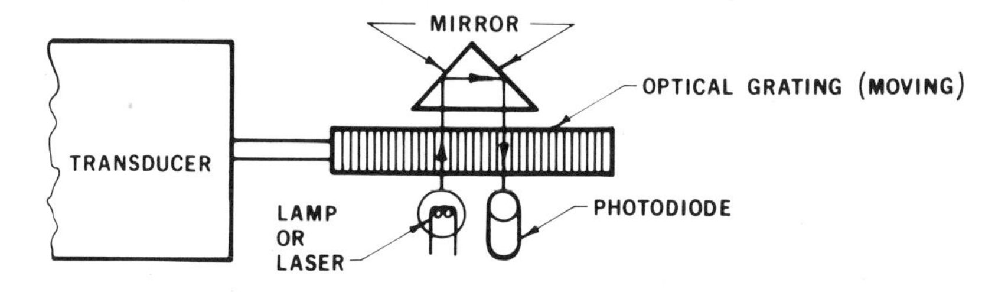

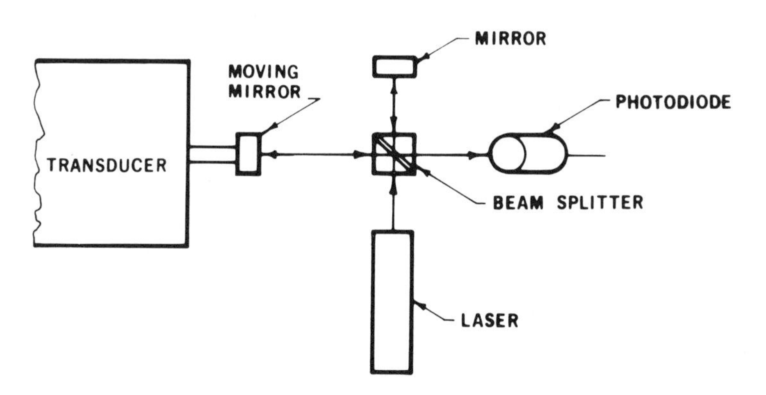

Fig. 6.18. Schematic diagram for a Moiré fringe device (upper diagram) and a Michelson interferometer (lower diagram).

H. CURVE FITTING

Mössbauer spectra are collected and stored in the form of digital data. This allows spectra to be examined very carefully, usually by attempting to fit the data to a theoretical model. The most important data frequently are the positions of each of the spectral lines. In fitting the data to the model, it is normally important to know the intensity and the line width of each absorption peak. As discussed earlier, the shape of absorption peaks for zero-thickness absorbers is Lorentzian. Even though samples are of finite thickness, the Lorentzian shape is a good approximation and can be treated with computers quite easily. A more general line shape is the "transmission integral" but the computer time necessary for the fitting is long and prohibitively expensive. However, when there are two or more Lorentzians overlapping, serious consideration must be given to using a transmission integral fit as opposed to a simple sum of Lorentzians.

Most of the computer programs used contain subroutines that perform least-square fits for the data. Before this is done, a certain amount of preprocessing of the data is required. The digital data is in the form of channel numbers (representing velocity) versus counts (representing intensity). If the spectrometer is not linear in velocity, velocity values are assigned to each data point using the data obtained by one of the calibration techniques described earlier. If a transmission integral fit is desired for the least squares, then the velocity scales must be adjusted to allow for a constant velocity increment between data points. If the asymmetric velocity waveform is used, then the spectrum can also be folded since the first half of the data is a mirror image of the second half. This procedure will remove some of the unwanted features of the data due to the geometry.

After the preprocessing, the least-squares computation can be performed. Comparison is made between the experimental data (Y_i) and the theoretical data (A_i) for a particular model. In particular, a function (χ^2) is minimized:

$$\chi^2 = \sum_i \frac{(Y_i - A_i)^2}{\sigma_i^2} \tag{38}$$

where σ_i is the standard deviation of Y_i. Since nuclear decay data is described by Poisson statistics, i.e., the standard deviation is nothing more than the square root of counts in the channel of interest, equation 38 becomes

$$\chi^2 = \sum_i \frac{(Y_i - A_i)^2}{Y_i} \tag{39}$$

Care must be taken in proposing a certain model and getting a good fit for the data, since it is often possible to have two models give almost identical values for χ^2.

Some laboratories do not perform a least-squares fit but merely estimate the peak positions, widths, and intensity from a plot. However, digital computation gives experimental parameters that are approximately an order of magnitude more accurate and precise. The data from a storage device can be outputted by one of several devices such as *x-y* plotters, strip chart recorders, paper computer tapes, oscilloscopes, teletypes, and even on-line devices.

V. ISOMER SHIFT AND ITS APPLICATION

The basic interaction which results in the isomer shift (δ) has already been described in Section III. The isomer shift can be used to measure electron densities at the nucleus although primarily one measures changes in it when going from one state to another (e.g., changes in the chemical species, the physical phases, or reference frames). Of the various Mössbauer parameters, δ is certainly the most unique since the information it provides cannot easily be obtained by other means. The first report of δ data was made in 1960 by Kistner and Sunyar (23). Since then there has been a voluminous amount of experimental and theoretical work reported in the literature. One of the major contributions has been an entire book on the subject (35). This approximately 1000-page volume covers a comprehensive range of topics discussed by leading Mössbauer spectroscopists.

For chemists and solid-state physicists the isomer shift can be correlated to a number of factors. These include the number of ligands, the geometric arrangement of the ligands about the Mössbauer nucleus, the electronegativity of the ligands, the bonding characteristics between the Mössbauer atom and the ligands, and the electronic state of the Mössbauer atom. Most fruitful Mössbauer isomer shift data is obtained when a series is considered in which all variables are held constant except one.

Most of the δ data is interpreted in the context of empirical relations, i.e., the isomer shift is correlated either with those factors mentioned above or with data from other experimental methods (e.g., NMR, IR, ESCA, and powder X-ray diffraction). Theoretical development has been gradual but shows promise as quantum methods are continuously being refined.

A. ELECTRON DENSITY CALCULATIONS

Since the isomer shift is a measurement of electron density in the vicinity of the nucleus, quantum determinations in the form of $\Psi^2(r = 0)$ have offered much insight into those species studied. For example, various self-consistent field (SCF) calculations have been used. In these calculations the Mössbauer atom is treated independently of any ligands, i.e., as a free ion. One of the first treatments was the Hartree-Fock calculations for iron by Walker, Wertheim, and Jaccarino (43). Their results are given in Fig. 6.19. This plot can be used to interpret isomer shifts of ionic materials. Similar calculations have been done for a number of other Mössbauer atoms. As an example, the results for antimony are shown in Fig. 6.20 (32). In both of these figures of plots of electron density versus electronic configuration, the isomer-shift scale has been superimposed.

While the SCF results are instructional in understanding factors that affect the isomer shift, they neglect covalency. This can be incorporated into the model by using some type of molecular orbital (MO) method, which usually considers only the valence atomic orbitals. Specifically, molecular orbitals are assumed to be made up of a linear combination of atomic orbitals (LCAO), i.e.,

$$\Psi_{MO} = a_M\phi_M + a_L\phi_L \tag{40}$$

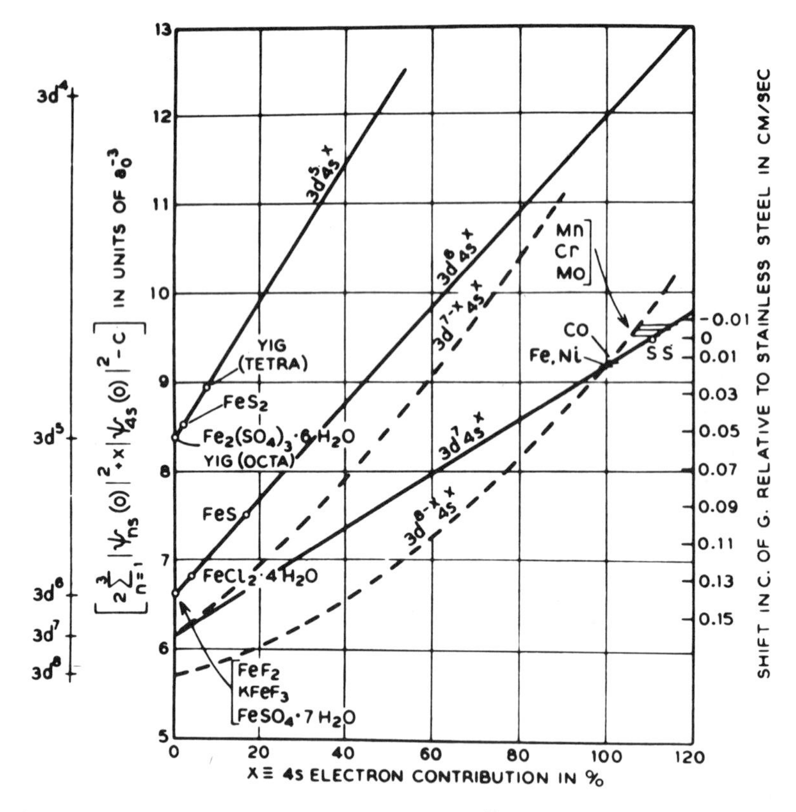

Fig. 6.19. Hartree-Fock calculations for the isomer shift of ^{57}Fe as a function of 3d- and 4s- electron charge density. Reproduced by permission from *Phys. Rev. Letters* (43).

where ϕ_M and ϕ_L are the atomic orbitals of the metals and its ligands. The coefficients a_M and a_L control the amount of mixing of each atomic orbital which gives the molecular orbital. An example of an energy diagram which results from these considerations is given in Fig. 6.21 for transition octahedral metal complexes containing ligands with bonding (17). The relative spacing of the energy of the molecular orbitals is related to the ligands and the geometric structure. The filling of these levels and values for the coefficients (a_M and a_L) are quite important in the interpretation of isomer shift data since they allow the determination of electron populations for the Mössbauer atom.

Extended Hückel MO theory provides a fairly simple procedure for obtaining the needed electron population to interpret Mössbauer parameters. In this procedure all atoms in the molecular system are considered. The basis set is usually taken from Slater-type atomic orbitals in which all the valence orbitals of each atom are considered. Overlap integrals are calculated, but Coulomb integrals are set equal to the proper valence state ionization energies (42). The Wolfsberg-Helmholtz approximation is used to obtain numerical results for the Hamiltonian matrix (47). This model has been successful as a semiempirical approach.

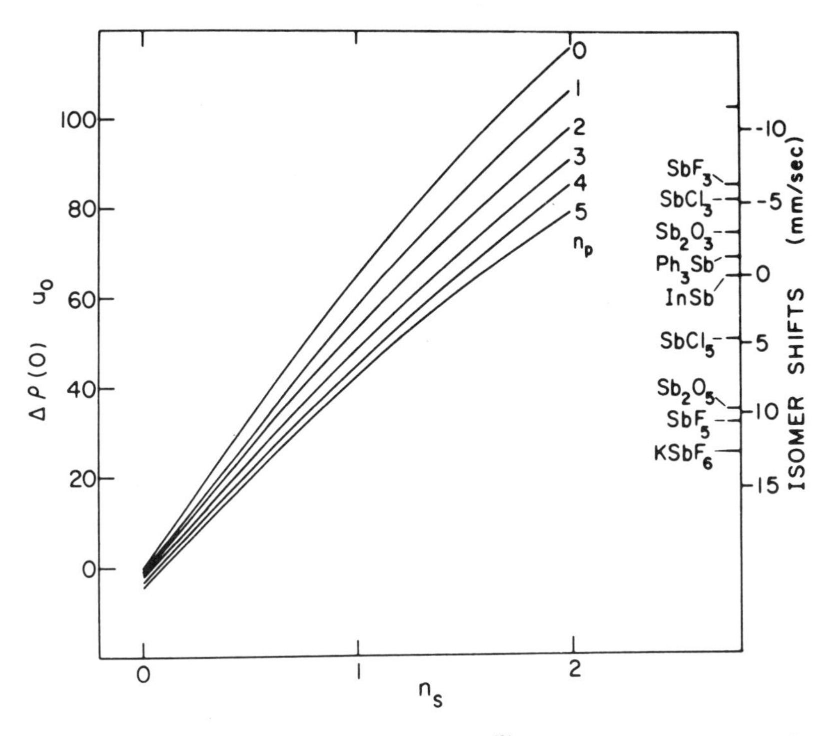

Fig. 6.20. Hartree-Fock calculations for the isomer shift of ^{121}Sb as a function of 5p- and 5s- electron charge density. Reproduced by permission from *Phys. Rev.* (32).

B. OXIDATION STATES

From Figs. 6.19 and 6.20, it can be seen that isomer shifts can easily be used in many cases to differentiate between oxidation states. Even for iron where the difference between Fe(II) and Fe(III) is only a $3d$ electron, it is usually fairly simple to differentiate between the two δ's for many compounds. While $3d$ electrons have essentially no direct effect on the electron density at the nucleus because they do not penetrate it, they do shield the $4s$ electron of an iron atom from the nucleus. Fig. 6.22 contains several examples of collected data showing the relation between δ and the oxidation state of the Mössbauer atom. In many cases, there is a distinct range of isomer shifts for a particular oxidation that does not overlap with the range of another.

C. ELECTRONEGATIVITY

Within the range of isomer shifts for a particular oxidation state, the second factor that affects these values is the electronegativities of the ligands. Generally, as the electronegativity of the ligand increases, there is a corresponding decrease in electron density at the nucleus. There are many cases of linear relations between the isomer shift and electronegativity or a related parameter. For example, many iodine-containing molecules are made of bonds which are only pure p. One such empirical

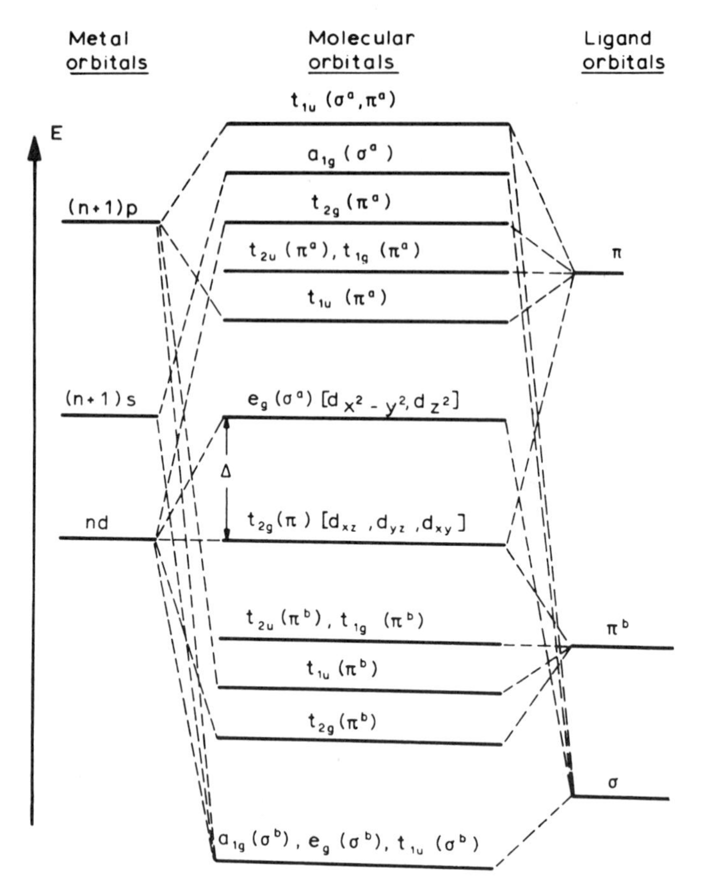

Fig. 6.21. Molecular orbital energy level diagram for octahedral metal complexes containing ligands with π^b (bonding) and relatively stable π^a (antibonding) orbitals. Reproduced from (17).

result for this class of compounds is

$$\delta = 0.136h_p - 0.54 \tag{41}$$

where h_p is the number of vacant "p" orbitals at the iodine (30). The values for h_p are directly related to electronegativity.

More commonly, the linear empirical relation exists directly between the isomer shift and the ligand electronegativity. Such an example is shown in Fig. 6.23. The resulting linear equations for these two series are $\delta = 2.6 + 0.019 \cdot$ (Ionicity in percent) for R_4SbX and $\delta = 1.4 + 0.034 \cdot$ (Ionicity in percent) for R_3SbX_2 (10).

D. PARTIAL CHEMICAL SHIFTS

Numerous cases of linear relations between ligand electronegativity and isomer shift have led to the concept of partial chemical shift (pcs) (18), i.e.,

$$\delta = \text{Constant} + \sum_{i=1}^{n} (\text{pcs})_i \tag{42}$$

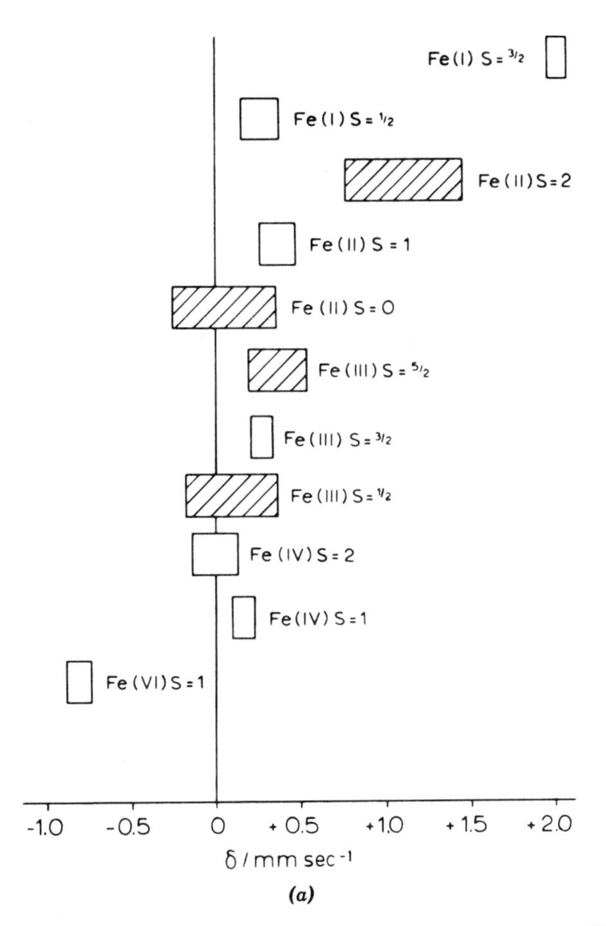

(a)

Fig. 6.22. Relation between isomer shift and oxidation state of several elements. (*a*) ^{57}Fe, reproduced from (29); (*b*) ^{99}Ru, reproduced from (19); (*c*) ^{119}Sn, reproduced from (20); (*d*) ^{121}Sb, reproduced from (16); (*e*) ^{170}Yb, reproduced from (21); (*f*) ^{197}Au, reproduced from (16); and (*g*) ^{237}Np, reproduced from (24).

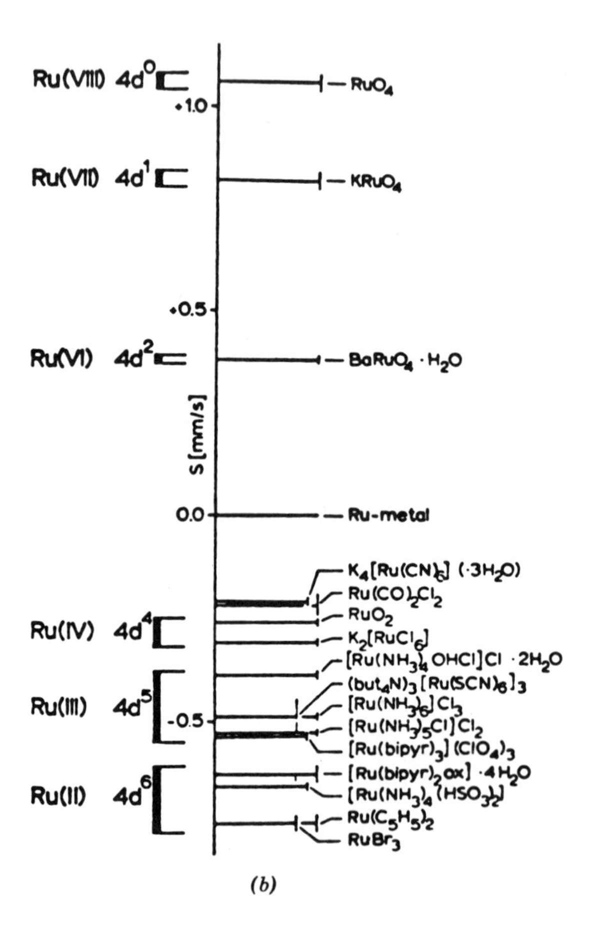

(b)

Fig. 6.22. (*Continued*).

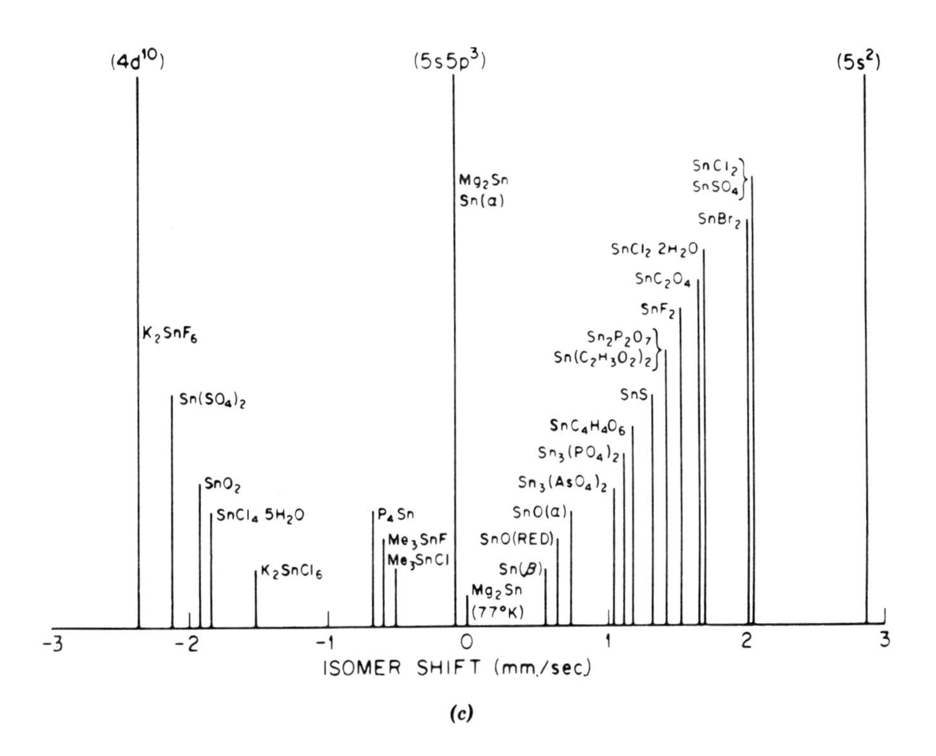

(c)

Fig. 6.22. (*Continued*).

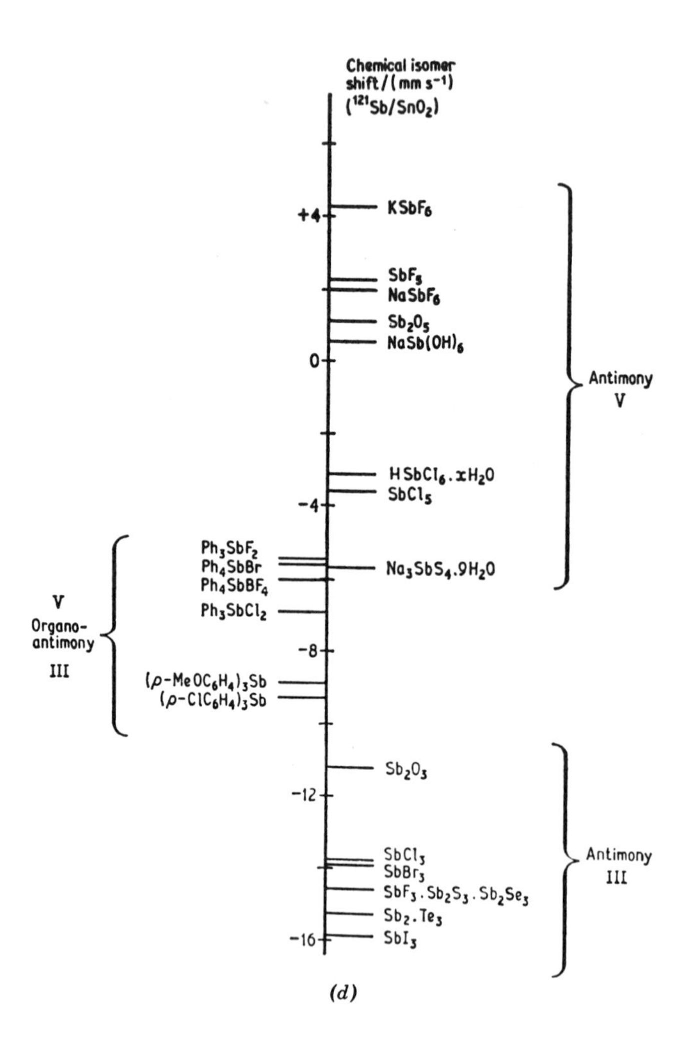

Fig. 6.22. (*Continued*).

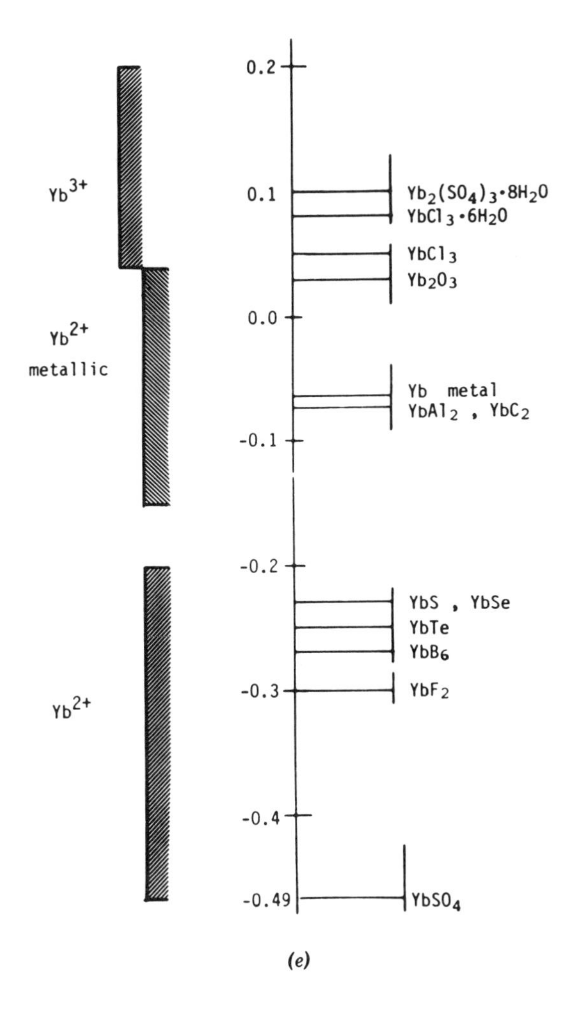

Fig. 6.22. (*Continued*).

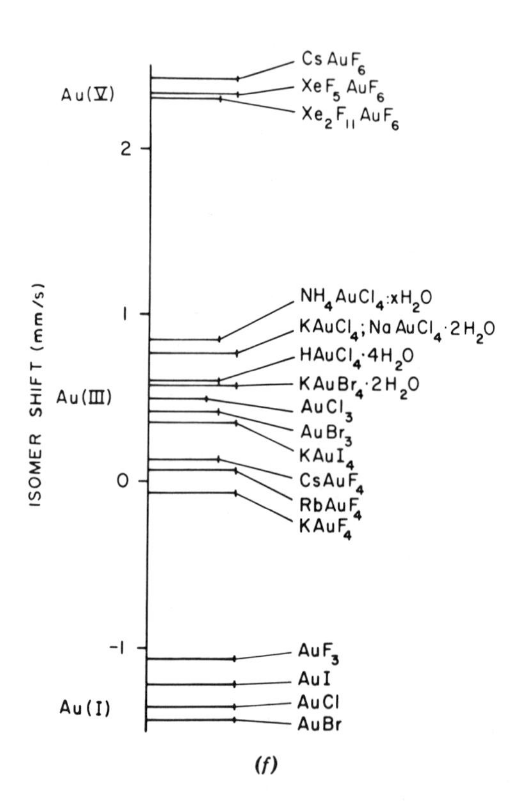

(f)

Fig. 6.22. (*Continued*).

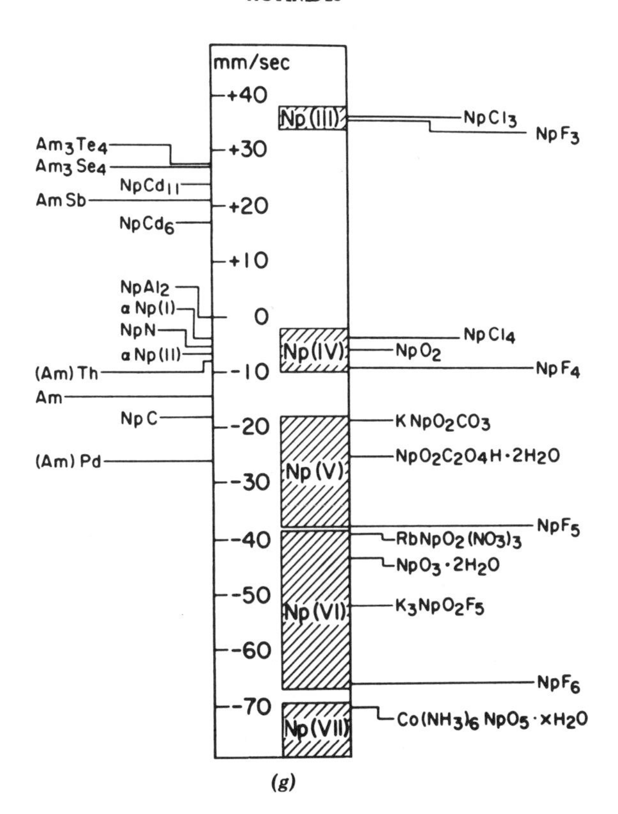

(g)

Fig. 6.22. (*Continued*).

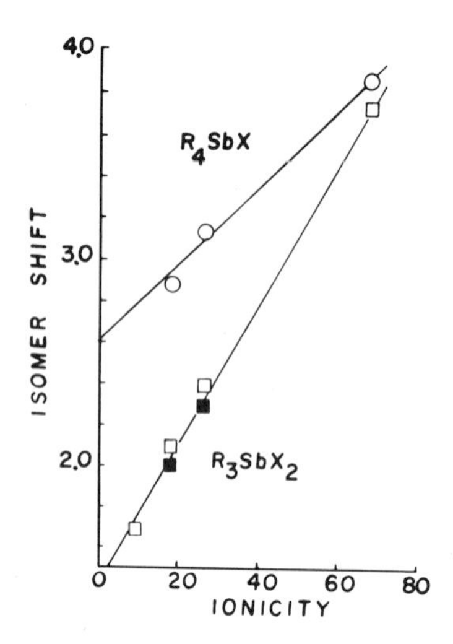

Fig. 6.23. Isomer shift vs. Pauling bond ionicity (of the SbX bond) from some organoantimony compounds. □, O(R = C_6H_5); ■ R = CH_3. Reproduced from (10).

where n is the coordination number. Using pcs tables, it is possible to predict isomer shift values for various compounds, or conversely, isomer-shift data can possibly be used to determine geometric structures. For example, see the recent ^{121}Sb Mössbauer data summarized in Table 6.VIII for $Me_xSbCl_{(3-x)}$. X-ray single-crystal data is only available for $SbCl_3$, which indicates that the Cl's are trigonally coordinated to the antimony and the Cl–Sb–Cl bond angles are about 95° (25). It is unlikely that X-ray structural data can be obtained for the other three compounds due to the difficulty of growing single crystals and their reactivity. Mössbauer spectroscopy is then of special value because it gives clues to structure. In this particular example, it is apparent that the structures of all four species are quite similar since there is the constant change in δ's, i.e., [pcs(CH_3^-) − pcs(Cl^-)] = 1.9 mm/s and the model of partial chemical shifts requires that the structures in a series be the same.

The values of the pcs, as has been mentioned above, depend on a number of factors; but if all of these are kept constant and only the ligands are allowed to vary, then regardless of the particular Mössbauer isotope or the structure being considered,

TABLE 6.VIII

^{121}Sb Mössbauer Data for $Me_xSbCl_{(3-x)}$ (39)

x	δ (mm/s)	e^2qQ | exp (mm/s) (±1.0)	e^2qQ | theory (mm/s)	η exp (±0.1)	η theory
0	−5.9	+13.3	+15	0.2	0.0
1	−4.2	+31.0	+29	0.4	0.3
2	−2.5	−30.8	−28	0.8	0.6
3	−0.1	+15.8	+15	0.0	0.0

a special ordering is observed. That is, if the ligands are ordered according to their pcs values, the resulting series will be nearly identical to the spectrochemical series.

E. SECOND-ORDER DOPPLER SHIFT

Care must be taken when interpreting isomer-shift data because included with all isomer shift values is a very small contribution due to the second-order Doppler (SOD) shift. This contribution needs to be considered and corrected for in those few cases when it is significant enough to measurably change an isomer-shift value.

SOD shift results from the relativistic emission or absorption energy shift of a stationary system seen by another system that is moving; in this case the systems are the source and the absorber. This energy shift in the emitting (or absorbing) gamma is given by

$$\delta_{SOD} = E_\gamma (1 + \mathrm{v}/c + 1/2\,(\mathrm{v}/c)^2 + \ldots) \tag{43}$$

The second term will average out since $\langle \mathrm{v} \rangle$ for a vibrating nulceus is zero; however $\langle \mathrm{v}^2 \rangle$ is nonzero. Therefore, the second-order term will contribute to an energy change in the gamma referred to as the second-order Doppler shift. The Debye model can be used to evaluate this shift (δ_{SOD}). One finds

$$\delta_{SOD} = \frac{-3kT}{2mc}\left[\frac{3}{8}\frac{\theta_D}{T} + 3\left(\frac{T}{\theta_D}\right)^3 \int_0^{\theta_D/T} \frac{x^3 dx}{(e^x - 1)}\right] \tag{44}$$

Values of these shifts have been calculated for ^{57}Fe, ^{119}Sn, and ^{121}Sb and are in Table 6.IX.

TABLE 6.IX

Second-Order Doppler Shifts in mm/s

	Debye temperature	Experiment temperature (K)							
Isotope	(K)	4.2	40	77	100	150	200	250	300
^{57}Fe	20	0.0059	0.0296	0.0564	0.0731	0.1096	0.1461	0.1826	0.2191
	40	0.0110	0.0306	0.0570	0.0736	0.1099	0.1463	0.1827	0.2192
	60	0.0164	0.0324	0.0579	0.0743	0.1104	0.1467	0.1830	0.2194
	80	0.0219	0.0348	0.0592	0.0753	0.1111	0.1472	0.1834	0.2198
	100	0.0274	0.0377	0.0609	0.0766	0.1119	0.1478	0.1840	0.2202
	120	0.0329	0.0411	0.0629	0.0782	0.1130	0.1486	0.1846	0.2208
	140	0.0383	0.0450	0.0652	0.0800	0.1142	0.1496	0.1854	0.2214
	160	0.0438	0.0491	0.0678	0.0821	0.1157	0.1506	0.1862	0.2221
	180	0.0493	0.0535	0.0707	0.0844	0.1173	0.1519	0.1872	0.2229
	200	0.0548	0.0582	0.0738	0.0870	0.1190	0.1532	0.1883	0.2239
	220	0.0602	0.0630	0.0772	0.0897	0.1210	0.1547	0.1895	0.2249
	240	0.0657	0.0680	0.0808	0.0927	0.1231	0.1563	0.1908	0.2260
	260	0.0712	0.0730	0.0847	0.0959	0.1254	0.1581	0.1923	0.2272
	280	0.0767	0.0782	0.0887	0.0993	0.1278	0.1600	0.1938	0.2285

TABLE 6.IX (Cont.)

Second-Order Doppler Shifts in mm/s

Isotope	Debye temperature (K)	Experiment temperature (K)							
		4.2	40	77	100	150	200	250	300
	300	0.0821	0.0834	0.0928	0.1028	0.1304	0.1620	0.1954	0.2298
	320	0.0876	0.0887	0.0971	0.1065	0.1332	0.1642	0.1972	0.2313
	340	0.0931	0.0940	0.1016	0.1104	0.1361	0.1664	0.1990	0.2329
	360	0.0986	0.0993	0.1062	0.1144	0.1391	0.1688	0.2010	0.2345
	380	0.1040	0.1047	0.1108	0.1185	0.1422	0.1713	0.2030	0.2363
	400	0.1095	0.1101	0.1156	0.1228	0.1455	0.1739	0.2052	0.2381
^{119}Sn	20	0.0028	0.0142	0.0270	0.0350	0.0525	0.0700	0.0874	0.1049
	40	0.0053	0.0147	0.0273	0.0352	0.0526	0.0701	0.0875	0.1050
	60	0.0079	0.0155	0.0277	0.0356	0.0529	0.0703	0.0877	0.1051
	80	0.0105	0.0167	0.0284	0.0361	0.0532	0.0705	0.0879	0.1053
	100	0.0131	0.0181	0.0292	0.0367	0.0536	0.0708	0.0881	0.1055
	120	0.0157	0.0197	0.0301	0.0374	0.0541	0.0712	0.0884	0.1057
	140	0.0184	0.0215	0.0312	0.0383	0.0547	0.0716	0.0888	0.1060
	160	0.0210	0.0235	0.0325	0.0393	0.0554	0.0722	0.0892	0.1064
	180	0.0236	0.0256	0.0338	0.0404	0.0562	0.0727	0.0897	0.1068
	200	0.0262	0.0279	0.0354	0.0417	0.0570	0.0734	0.0902	0.1072
	220	0.0288	0.0302	0.0370	0.0430	0.0580	0.0741	0.0908	0.1077
	240	0.0315	0.0326	0.0387	0.0444	0.0590	0.0749	0.0914	0.1082
	260	0.0341	0.0350	0.0405	0.0459	0.0601	0.0757	0.0921	0.1088
	280	0.0367	0.0375	0.0425	0.0476	0.0612	0.0766	0.0928	0.1094
	300	0.0393	0.0399	0.0445	0.0493	0.0625	0.0776	0.0936	0.1101
	320	0.0420	0.0425	0.0465	0.0510	0.0638	0.0786	0.0944	0.1108
	340	0.0446	0.0450	0.0487	0.0529	0.0652	0.0797	0.0953	0.1115
	360	0.0472	0.0476	0.0508	0.0548	0.0666	0.0809	0.0963	0.1123
	380	0.0498	0.0501	0.0531	0.0568	0.0681	0.0821	0.0973	0.1132
	400	0.0525	0.0527	0.0554	0.0588	0.0697	0.0833	0.0983	0.1140
^{121}Sb	20	0.0028	0.0139	0.0266	0.0345	0.0516	0.0688	0.0860	0.1032
	40	0.0052	0.0144	0.0268	0.0347	0.0518	0.0689	0.0861	0.1033
	60	0.0077	0.0153	0.0273	0.0350	0.0520	0.0691	0.0862	0.1034
	80	0.0103	0.0164	0.0279	0.0355	0.0523	0.0693	0.0864	0.1035
	100	0.0129	0.0178	0.0287	0.0361	0.0527	0.0696	0.0867	0.1037
	120	0.0155	0.0194	0.0296	0.0368	0.0532	0.0700	0.0870	0.1040
	140	0.0181	0.0212	0.0307	0.0377	0.0538	0.0705	0.0873	0.1043
	160	0.0206	0.0231	0.0319	0.0387	0.0545	0.0710	0.0877	0.1046
	180	0.0232	0.0252	0.0333	0.0398	0.0552	0.0715	0.0882	0.1050
	200	0.0258	0.0274	0.0348	0.0410	0.0561	0.0722	0.0887	0.1055
	220	0.0284	0.0297	0.0364	0.0423	0.0570	0.0729	0.0893	0.1059
	240	0.0310	0.0320	0.0381	0.0437	0.0580	0.0736	0.0899	0.1064
	260	0.0335	0.0344	0.0399	0.0452	0.0591	0.0745	0.0906	0.1070
	280	0.0361	0.0368	0.0418	0.0468	0.0602	0.0754	0.0913	0.1076
	300	0.0387	0.0393	0.0437	0.0484	0.0614	0.0763	0.0921	0.1083
	320	0.0413	0.0418	0.0458	0.0502	0.0627	0.0773	0.0929	0.1090
	340	0.0438	0.0443	0.0479	0.0520	0.0641	0.0784	0.0938	0.1097
	360	0.0464	0.0468	0.0500	0.0539	0.0655	0.0795	0.0947	0.1105
	380	0.0490	0.0493	0.0522	0.0558	0.0670	0.0807	0.0956	0.1113
	400	0.0516	0.0519	0.0545	0.0578	0.0686	0.0819	0.0967	0.1122

The isomer shift measured is a sum of the chemical isomer shifts, the primary contribution, and the second-order Doppler shift (δ_{SOD}), which in most cases can be ignored. Therefore, in general one has

$$\delta_{\text{measure}} = \delta_{\text{SOD}} + \delta \tag{45}$$

F. PHASE ANALYSIS

1. Phase Transitions

In most cases, the electronic structure is different enough in two different phases to allow the use of the isomer shift to determine if and where a phase transition has occurred. In these cases of first-order transitions, there will be a discontinuity in the isomer-shift value. The usual parameter varied is temperature.

Some spectrometers have been designed that operate at a single velocity set on an absorption peak. Then the temperature is varied and the number of gammas detected for each temperature increment is recorded. When a phase change occurs, the count rate will increase due to a decrease in resonance absorption. The resulting plot of counts versus temperature is called a thermal scan and is considerably less complicated than obtaining a complete Mössbauer spectrum at each temperature.

2. High Pressure

Besides temperature, the pressure can also be varied. This has been successfully accomplished for about 10 Mössbauer transitions. Studies have usually been centered around phase transitions and/or the effect of pressure on the isomer shift. There is a definite volume dependence of the isomer shift which can be expressed as

$$\left(\frac{\partial \delta}{\partial \ln V}\right)_T = \text{constant} \tag{46}$$

As for temperature, it is possible to vary the pressure as a function of the number of counts at a constant velocity.

3. Chemical Identification

Mössbauer spectroscopy can be used for assisting in the identification of particular chemical substances. For example, it has been used to identify those iron minerals found in samples brought back from the moon. In these identifications, the isomer shift is usually used along with possible quadrupole coupling and magnetic hyperfine data when attempting to identify unknown materials. Other active scientific areas of interest in which chemical identification is important include corrosion processes, mechanisms in catalysts, biological activity of iron-containing systems, and even the history of ancient artifacts, primarily pottery. Using Mössbauer spectroscopy as a fingerprint is discussed quite thoroughly in several chapters of a book by Bancroft (3). Numerous detailed examples can be found in Stevens and Shenoy (39a).

VI. MAGNETISM

A. LINE INTENSITIES

As discussed in an earlier section, there are six Mössbauer transitions observable for ^{57}Fe due to the interaction of the nuclear magnetic moment and a magnetic field. For these transitions $\Delta m_I = 0, \pm 1$. The intensities of these transitions have an angular dependence which is related to Δm_I. A convenient angle (θ) is defined as the angle between the directions of the magnetic field and the gamma-ray emission. A number of the angular relations are listed in Table 6.X. Usually there is no preferred direction, and thus the relative intensities of the peaks can be determined by averaging the angular dependence over all angles. For ^{57}Fe this gives a 3:2:1:1:2:3 ratio of line intensities. If a ^{57}Fe magnetic Mössbauer spectra does not give this ratio, then there is a preferred orientation in the material. For example, in the two extreme cases, $\theta = 0°$ gives 3:0:1:1:0:3 and $\theta = 90°$ gives 3:4:1:1:4:3. The spectrum shown in Fig. 6.8 is, in fact, closer to this latter ratio, indicating that the aligned fields in the foil are mainly in a direction approximately 90° relative to the gamma direction. It is useful to know if there are preferred magnetic field directions in materials and a determination of this is possible with Mössbauer spectroscopy. When preferred directions do exist these materials are said to have "texture."

B. CONTRIBUTIONS TO THE MAGNETIC FIELD INTERACTIONS

Besides magnetic fields due to external sources, there are three primary internal magnetic field interactions (44) which have been discussed previously in Section III.B. One of these, the Fermi contact interaction, results from a direct coupling between the spin density of s electrons at the nucleus and the nuclear spin. It can be expressed as

$$H_c = -\frac{16}{3}\pi\beta\sum_i \delta(\vec{r}_i)\,\vec{S}_i \cdot \vec{I} \tag{47}$$

where β is the Bohr magneton ($\beta = 9.274\text{x}/10^{-24}$ Joule/Tesla) and S_i is the core electron spin. The summation represents an imbalance of electron density at the nucleus. This polarization comes about via the unpaired electrons in the outer electron shells of the atom. The Fermi contact is usually the largest of the magnetic field interaction terms.

A second contribution to the internal magnetic field interaction is the orbital term

$$H_L = 2\beta\frac{1}{r^3}\vec{L} \cdot \vec{I} \tag{48}$$

There are several cases when this term is zero. These include those when an outer electronic shell is either half full and is high spin or completely full. For example, high-spin Fe(III) compounds do not have this magnetic field contribution.

While the Fermi contact considers the interaction of the nucleus and the spin density of the electrons at the nucleus, a third contribution considers the interaction between the nuclear spin and the spin of the electrons outside of the nucleus. This is called the dipole interaction and the resulting field can be expressed as

$$H_D = -2\beta\left[3\vec{r}\frac{\vec{S}\cdot\vec{r}}{r^5} - \frac{\vec{S}}{r^3}\right]\cdot\vec{I} \tag{49}$$

The dipole contribution is usually quite small compared to the others and is zero in cases of cubic symmetry.

Other possible sources for an effective magnetic field include the conduction electrons that are polarized by neighboring atoms or other electrons in the atom. Also possibly contributing are fields produced by neighboring atoms either by overlap distortion of the core s orbitals or dipole fields of localized moments (41).

C. MAGNETIC HYPERFINE FIELD SPECTRA

The common difficulty encountered when interpreting quantitatively the Mössbauer data (δ, e^2qQ, magnetic hyperfine interaction) is not knowing the value of the nuclear components of the interaction equations. In the case of the magnetic hyperfine field interactions, however, the nuclear term (the nuclear moment or nuclear g factor) is often known well enough. Values of these are given in Table 6.XI. The nuclear g factors in this table are given in units such that the values in mm/s of observed spectra splittings can be then directly converted to units of Tesla. For example, (see Figs. 6.7 and 6.8 for α-Fe) using the g_0 and g_1 splittings in mm/s and the values 0.1188 mm/(S · T) and 0.06790 mm/(S · T), respectively, the value of 33 Tesla is obtained for the effective internal field of metallic iron.

VII. QUADRUPOLE INTERACTION AND ITS APPLICATION

A great deal of our understanding about the nuclear quadrupole interaction has come from NQR (nuclear quadrupole resonance) spectroscopy, which was already fairly well established when Mössbauer spectroscopy began to be used for making quadrupole measurements. While Mössbauer spectroscopy does not have the precision of NQR, it does add to our knowledge of quadrupole interactions because it enables observations of quadrupole coupling in many materials not possible with NQR. These are materials that have no quadrupole interaction in their ground states since their spins are either 0 or 1/2. However, excited nuclear states usually have spins of 1 or greater. As nuclei undergo Mössbauer transitions, their spectra reveal information about the quadrupole coupling in their excited states. Primary examples of such cases are ^{57}Fe and ^{119}Sn. Another contribution of Mössbauer spectroscopy to the study of quadrupole interactions is the easy determination of the sign of the quadrupole coupling constant. Examples include ^{121}Sb and ^{129}I.

A. ELECTRIC FIELD GRADIENTS

Quadrupole measurements give information about the electric field gradient (EFG). Although the EFG in general contains nine elements (see equation 25), the information of interest is condensed into two parameters: the principal diagonal

TABLE 6.X

Relative Line Intensities of Magnetic Hyperfine Interactions (15)

I_1	I_2	Multipolarity	Examples	$\vert m_I(1)\rangle$	$\vert m_I(2)\rangle$	$\vert \Delta m_I \vert$ [a]	A^2	$A^2(\theta = 0°)$ [b]	$A^2(\theta = 90°)$ [b]
3/2	1/2	M1	^{57}Fe, ^{119}Sn, ^{125}Te	±3/2	±1/2	1	3	3	3
				±1/2	±1/2	0	2	0	4
				∓1/2	±1/2	1	1	1	1
2	0	E2	^{166}Er, ^{170}Yb	±2	0	2	1	0	1
				±1	0	1	1	1	1
				0	0	0	1	0	0
5/2	3/2	M1	^{61}Ni, ^{99}Ru, ^{155}Gd	±5/2	±3/2	1	10	10	10
				±3/2	±3/2	0	4	0	8
				±1/2	±3/2	1	1	1	1
				±3/2	±1/2	1	6	6	6
				±1/2	±1/2	0	6	0	12
				∓1/2	±1/2	1	3	3	3
5/2	5/2	M1	^{161}Dy, ^{237}Np	±5/2	±3/2	0	25	0	25
				±3/2	±5/2	1	10	10	5
				±5/2	±3/2	1	10	10	5
				±3/2	±3/2	0	9	0	9
				±1/2	±3/2	1	16	16	8
				±3/2	±1/2	1	16	16	8
				±1/2	±1/2	0	1	0	1
				∓1/2	±1/2	1	18	18	9
7/2	5/2	M1	^{121}Sb, ^{127}I, ^{129}I, ^{151}Eu	±7/2	±5/2	1	21	21	21
				±5/2	±5/2	0	6	0	12
				±3/2	±5/2	1	1	1	1
				±5/2	±3/2	1	15	15	15
				±3/2	±3/2	0	10	0	20
				±1/2	±3/2	1	3	3	3
				±3/2	±1/2	1	10	10	10

				$\pm 1/2$	$\pm 1/2$	0	12	0	24
				$\mp 1/2$	$\pm 1/2$	1	6	6	6
9/2	7/2	M1	^{191}Ta	$\pm 9/2$	$\pm 7/2$	1	36	36	36
				$\pm 7/2$	$\pm 7/2$	0	8	0	16
				$\pm 5/2$	$\pm 7/2$	1	1	1	1
				$\pm 7/2$	$\pm 5/2$	1	28	28	28
				$\pm 5/2$	$\pm 5/2$	0	14	0	28
				$\pm 3/2$	$\pm 5/2$	1	3	3	3
				$\pm 5/2$	$\pm 3/2$	1	21	21	21
				$\pm 3/2$	$\pm 3/2$	0	18	0	36
				$\pm 1/2$	$\pm 3/2$	1	6	6	6
				$\pm 3/2$	$\pm 1/2$	1	15	15	15
				$\pm 1/2$	$\pm 1/2$	0	20	0	40
				$\mp 1/2$	$\pm 1/2$	1	10	10	10

[a] The angular dependence is related to the multipolarity and the $|\Delta m_I|$ value, i.e., or dipole radiation (M1)

$|\Delta m_I| = 0 : \sin^2\theta$
$|\Delta m_I| = 1 : (1 + \cos^2\theta)^{\frac{1}{2}}$

and for quadrupole radiation (E2)

$|\Delta m_I| = 0 : 3/4 \sin^2 2\theta$
$|\Delta m_I| = 1 : 1/2 (\cos^2\theta + \cos^2 2\theta)$
$|\Delta m_I| = 2 : 1/8 (4 \sin^2\theta + \sin^2 2\theta)$

where θ is the angle between the direction of the magnetic field and the direction of the emission of the gamma.

[b] A^2 are relative intensities.

TABLE 6.XI

Nuclear Magnetic Moment Data (38)[a]

				Magnetic moment			Gyromagnetic ratio	
Isotope	E_γ (keV)	I_e	I_g	μ_0 Ground (nuclear magnetons)	μ Excited (nuclear magneton)	Ratio of magnetic moments	g_0 Ground mm/(S · T)	g_1 Excited mm/(S · T)
^{57}Fe	14.4	3/2	1/2	0.090604(9)	−0.15532(4)	−1.7142(4)	0.118821(12)	−0.067897(17)
^{61}Ni	67.4	5/2	3/2	0.74980(10)	0.478(7)	−0.637(11)	0.070083(12)	0.0268(4)
^{99}Ru	89.4	3/2	5/2	−0.626(13)	−0.285(5)	0.456(2)	−0.0265(6)	−0.0201(4)
^{119}Sn	23.9	3/2	1/2	−1.0461(3)	0.633(18)	−0.605(17)	−0.8283(3)	0.167(5)
^{121}Sb	37.2	7/2	5/2	3.3591(6)	2.47(3)	0.735(9)	0.3418(2)	0.180(2)
^{125}Te	35.5	3/2	1/2	−0.8872(3)	0.604(6)	−0.681(4)	−0.4729(4)	0.1073(11)
^{127}I	57.6	7/2	5/2	2.8091(4)	2.54(4)	0.905(16)	0.18436(7)	0.1191(19)
^{129}I	27.8	5/2	7/2	2.6174(8)	2.797(3)	1.0687(11)	0.2545(2)	0.3808(5)
^{133}Cs	81.0	5/2	7/2	−2.5786(8)	3.443(21)	1.335(8)	−0.08596(3)	0.1607(10)
^{151}Eu	21.5	7/2	5/2	3.465(2)	2.587(3)	0.7465(6)	0.6083(4)	0.3244(3)
^{153}Eu	103.2	3/2	5/2	1.5294(7)	2.043(5)	1.336(3)	0.05604(3)	0.1248(3)
^{155}Gd	86.5	5/2	3/2	−0.2584(5)	−0.529(5)	2.05(2)	−0.01881(4)	−0.0231(2)
^{161}Dy	25.7	5/2	5/2	−0.479(5)	0.592(6)	−1.2368(14)	−0.0706(7)	0.0872(9)
^{166}Er	80.6	2	0	0.0	0.629(10)	—	0.0	0.0369(6)
^{170}Yb	84.3	2	0	0.0	0.669(8)	—	0.0	0.0375(4)
^{181}Ta	6.2	9/2	7/2	2.356(7)	5.24(7)	2.23(3)	1.020(4)	1.764(24)
^{193}Ir	73.0	1/2	3/2	0.1583(6)	0.4683(20)	2.958(6)	0.1366(5)	0.1212(5)
^{197}Au	77.3	1/2	3/2	0.1448(7)	0.416(3)	2.875(22)	0.01180(6)	0.1017(7)
^{237}Np	59.5	5/2	5/2	2.5(3)	1.34(12)	0.535(4)	0.159(19)	0.085(8)

[a] Uncertainties in the last digit(s) are given in the parenthesis.

TABLE 6.XII

The Elements of the Electric Field Gradient in Spherical Coordinates for a Point Charge q

$V_{xx} = qr^{-3}(3\sin^2\theta\cos^2\phi - 1)$
$V_{yy} = qr^{-3}(3\sin^2\theta\sin^2\phi - 1)$
$V_{zz} = qr^{-3}(3\cos^2\theta - 1)$
$V_{xy} = V_{yx} = qr^{-3}(3\sin^2\theta\sin\phi\cos\phi)$
$V_{xz} = V_{zx} = qr^{-3}(3\sin\theta\cos\theta\cos\phi)$
$V_{yz} = V_{zy} = qr^{-3}(3\sin\theta\cos\theta\sin\phi)$

component of the diagonalized EFG (see equation 26a) and the asymmetry parameter (see equation 26b).

The elements of the EFG for a single point charge q are given in Table 6.XII. In generalized coordinates these elements are

$$V_{ij} = q(3x_ix_j - r^2\delta_{ij}).$$

When there are several point charges, the contributions from each must be added to obtain the EFG elements of the configuration. The source of the point charges are either valence electrons or ligands. The contributions to the EFG can often be obtained from theoretical calculations using molecular orbital-wave functions. For convenience, the EFG should be diagonalized. By choosing the most apparent symmetry axis of the system as the z axis, the EFG matrix will usually be diagonalized with V_{zz} as the maximum valued element.

An immediate application of the terms in Table 6.XII can be made by comparing the ligand contribution of the *cis*- and *trans*-octahedral complexes of type MA_2B_4. The diagonalized matrix elements for the *cis*-ligand complex are

$$\begin{aligned} V_{xx} = V_{yy} &= (A - B)e \\ V_{zz} &= (-2A + 2B)e \end{aligned} \tag{51}$$

where $A = Z_A/r_A^2$ and $B = Z_B/r_B^2$. The asymmetry parameter, $\eta = 0$. Likewise, for the *trans*-ligand complex, the diagonal elements are

$$\begin{aligned} V_{xx} = V_{yy} &= -2(A - B)e \\ V_{zz} &= -2(-2A + 2B)e \end{aligned} \tag{52}$$

and the asymmetry parameter $\eta = 0$ once again. Note the difference in sign and a factor of 2 when comparing equation 51 with equation 52. Mössbauer spectroscopy allows for easy differentiation between these two structures. The success of the Mössbauer results is demonstrated in Table 6.XIII.

As mentioned above, valence electrons can also contribute to the EFG. S electron wave functions are spherical and therefore do not contribute to the EFG. Likewise, if the valence p or d shells are half filled with no spin-pairing or completely filled, there is no contribution to the EFG. In all other cases, valence electrons contribute to the components of the diagonalized EFG. The q values (see equation 26a) for each of the p and d electrons are listed in Table 6.XIV.

TABLE 6.XIII
Comparison of Quadrupole Splittings for *cis-trans* Isomers of Low-Spin Fe(II) (4,7)

Compounds	Δ (mm/s)
trans-$FeCl_2(ArNC)_4$[a]	+1.55
cis-$FeCl_2(ArNC)_4$	−0.78
trans-$Fe(SnCL_3)_2(ArNC)_4$	(+)1.05
cis-Fe $(SnCl_3)_2(ArNC)_4$	(−)0.50
trans-Fe $(CN)_2(EtNC)_4$	−0.60
cis-Fe $(CN)_2(EtNC)_4$	(+)0.30

[a] ArNC = *p*-methoxyphenylisocyanide.

A general expression for the total q value can be written by summing the contributions from the ligands (lattice) and the valence electrons, i.e.,

$$q = (1 - \gamma_\infty)\, q_{\text{lattice}} + (1 - R)\, q_{\text{valence}}$$

where γ_∞ and R are Sternheimer antishielding factors. These factors correct for the polarization of the core electrons by the ligands (γ_∞) and the valence electrons (R). They can be calculated from self-consistent field methods.

The valence term can be further subdivided into the contributions from the crystal field and from the electrons in the molecular orbitals that are created by the metal and its ligands. The crystal field term will be important when considering nontransition metal complexes. Using the values in Table 6.XIV, an expression for the p (q_p) and d (q_d) contributions can be written based on the populations of the atomic orbitals, i.e.,

$$q_p = \frac{-4/5\,(1 - R_p)\,[N_z - 1/2\,(N_x + N_y)]}{\langle r_p^{-3} \rangle} \tag{53}$$

TABLE 6.XIV
Magnitude of the Diagonal Electric Field Gradient Tensor Elements for p and d Electrons

Wavefunction	V_{xx} $(e\langle r^{-3}\rangle)$	V_{yy} $(e\langle r^{-3}\rangle)$	V_{zz} $(e\langle r^{-3}\rangle)$
p_x	−4/5	+2/5	+2/5
p_y	+2/5	−4/5	+2/5
p_z	+2/5	+2/5	−4/5
d_{xy}	−2/7	−2/7	+4/7
d_{xz}	−2/7	+4/7	−2/7
d_{yz}	+4/7	−2/7	−2/7
d_{x2-y2}	−2/7	−2/7	+4/7
d_{z2}	+2/7	+2/7	−4/7

$$q_d = \frac{-4/7\,(1 - R_d)\,[N_{z^2} + 1/2\,(N_{xz} + N_{yz}) - (N_{x^2-y^2} + N_{xy})]}{\langle r_d^{-3} \rangle} \tag{54}$$

The lattice term can be evaluated if it is assumed that contributions come from only the nearest neighbor. When the particular geometry is known, the components of the EFG can be found from Table 6.XV. The remaining quantities to be determined are the charges (Z_A, Z_B, . . .) and the radii ($1/r_A^3$, $1/r_B^3$, . . .). It is possible to use MO calculations for the Z values and structural data for the $1/r^3$ values. Another method for evaluating this data which is qualitative, will be described in the Section "Additive Models."

Recall from equation 26b that the values of the diagonalized elements of the EFG all contribute to the asymmetry parameter, η. If $V_{xx} = V_{yy}$ (true for the many complexes that have cylindrical symmetry), then $\eta = 0$. The other extreme occurs when either V_{xx} or V_{yy} is equal to zero; then $\eta = 1$. It is unfortunate that for ^{57}Fe and ^{119}Sn Mössbauer spectroscopy the value for η cannot be determined from data for the pure quadrupole interaction. This is due to the fact that the measured quadrupole splitting (Δ) is a function of both q and η (see equation 31), therefore not allowing for the independent determination of these two parameters. However, these quantities can be determined from Mössbauer spectra of higher-spin nuclei. For ^{57}Fe and ^{119}Sn, it is possible to determine η by removing the remaining degeneracy in nuclear levels through the application of a magnetic field.

B. SPECTRA

As discussed in Section III.C, when a nucleus has a nuclear quadrupole moment and an electric field gradient is present, the nuclear level splits into $(2I - 1)$ levels. The energies of the two resulting levels for $I = 3/2$ are given in equation 30. It is not possible to express the energies of the split levels for most cases of I in closed form. A series approximation has been worked out for the various I states (33). These can be obtained by using the series expression for the split energy levels:

$$E_q(I, m_I) = e^2 q\, Q \sum_{n=0}^{4} a_n(I, m_I)\eta^n \tag{55}$$

The values for a_n are given in Table 6.XVI. The Mössbauer spectra will indicate quadrupole splittings, if they are present, in both the excited and the ground nuclear levels. The shift for each possible Mössbauer line (ΔE_Q) can then be expressed as

$$\Delta E_Q = E_Q^* - E_Q \tag{56}$$

where the * represents the excited nuclear level. When $I \geqslant 1$ for both excited and ground nuclear levels, one may substitute equation 56 into equation 55 giving the energy shift as

$$\Delta E(I, m_I) = e^2 q\, Q\, [R_Q E_Q^*(I^*, m_{I^*}) - E_Q(I, m_I)] \tag{57}$$

TABLE 6.XV.

Point Charge Model Expressions for the Components of the EFG Tensor at a Nucleus M Due to Ligands A,B,C,D for Common Structures (5)

Structure	Components of the EFG [a,b]	Structure	Compound of the EFG
Z ⋮ B | M·······x B B B	$V_{zz} = V_{xx} = V_{yy} = 0$	Z ⋮ A [d] | M B C C	$V_{zz} = \{2[A] - 2/3([B] + 2[C])\}e$ $V_{yy} = \{-[A] - [B] + 2[C]\}e$ $V_{xx} = \{-[A] + 5/3\,[B] - 2/3[C]\}e$ $V_{xz} = V_{zx} = \{\sqrt{2}/3(-2[B] + 2[C])\}e$ $V_{xy} = V_{yx} = V_{yz} = V_{zy} = 0$ $\eta \neq 0$
Z ⋮ A | M·······x B B B	$V_{zz} = \{2[A] - 2[B]\}e$ $V_{yy} = \{-[B] + [A]\}e$ $V_{xx} = \{+[B] + [A]\}e$ $\eta = 0$		$\Delta = 1/2e^2Q(4/3p^2 + 8/3\,Q^2)^{1/2}$ $P = [A] + [B] - 2[C]$ $Q = [A] - [B]$ Sign = sign of P
A | M [c] A B B	$V_{zz} = \{2[A] - 2[B]\}e$ $V_{yy} = \{2[B] - 2[A]\}e$ $V_{xx} = 0$ $\eta = 1$	Z ⋮ A [e] | M D B C	$V_{zz} = \{2[A] - 2/3([B] + [C] + [D])\}e$ $V_{yy} = \{-[A] - [B] + [C] + [D]\}e$ $V_{xx} = \{-[A] + 5/3[B] - 1/3([C] + [D])\}e$ $V_{yz} = V_{zy} = \{\sqrt{2}/\sqrt{3}([C] - [D])\}e$ $V_{xz} = V_{zx} = \{\sqrt{2}/3(-2[B] + [C] + [D])\}e$ $V_{xy} = V_{yx} = \{2/\sqrt{3}([C] - [B])\}e$ $\eta \neq 0$
Z ⋮ B B [f] M—B·····x B B	$V_{zz} = \{4[B]^{tba} - 3[B]^{tbe}\}e$ $V_{yy} = \{3/2[B]^{tbe} - 2[B]^{tba}\}e$ $V_{xx} = \{3/2[B]^{tbe} - 2[B]^{tba}\}e$ $\eta = 0$	Z ⋮ A B M—B · · · x A B	$V_{zz} = \{-2[A]^{tbe} - [B]^{tbe} + 4[B]^{tba}\}e$ $V_{yy} = \{5/2[A]^{tbe} - [B]^{tbe} - 2[B]^{tba}\}e$ $V_{xx} = \{-1/2[A]^{tbe} + 2[B]^{tbe} - 2[B]^{tba}\}e$ $\eta = 0$

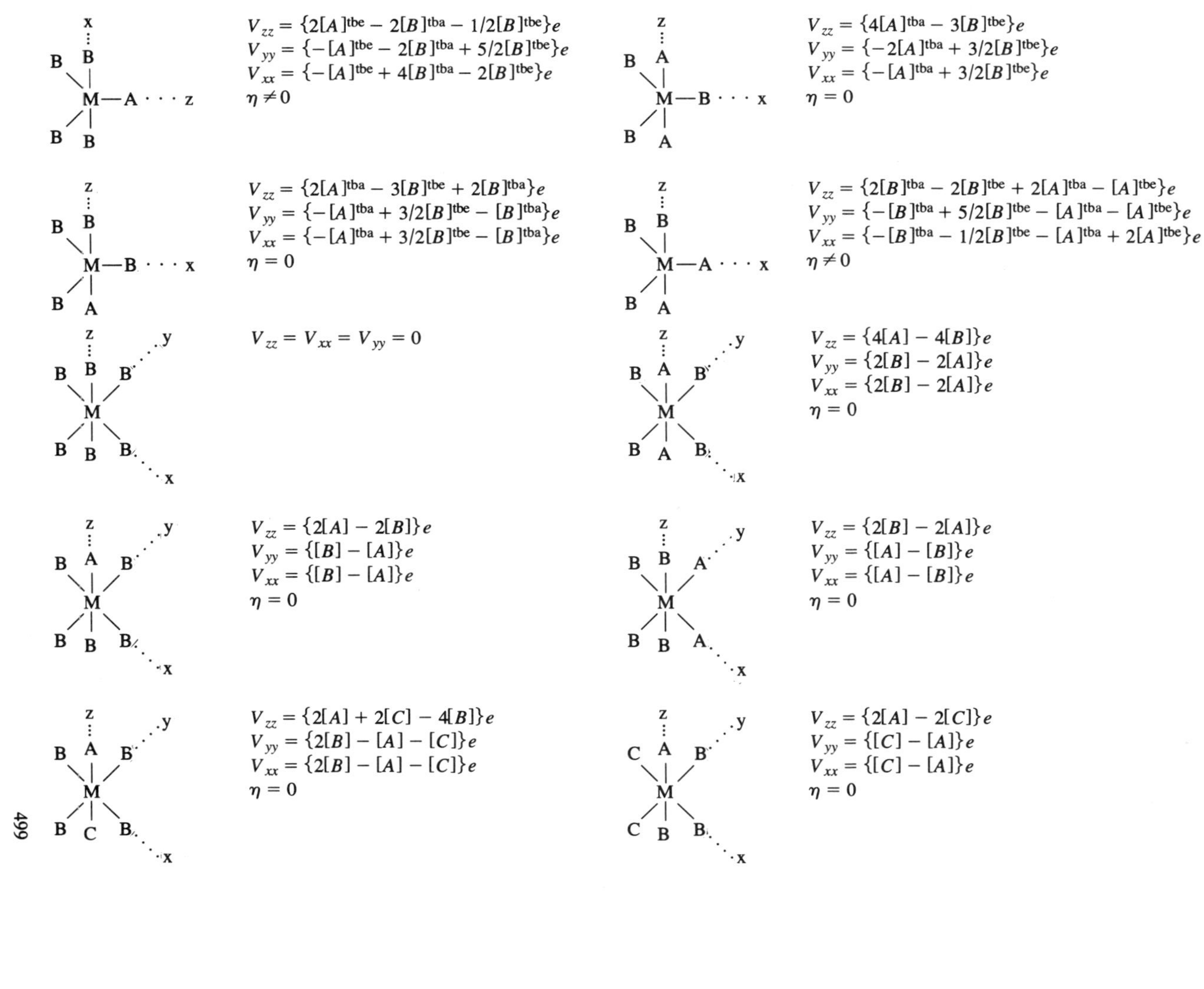

$$V_{zz} = \{2[A]^{tbe} - 2[B]^{tba} - 1/2[B]^{tbe}\}e$$
$$V_{yy} = \{-[A]^{tbe} - 2[B]^{tba} + 5/2[B]^{tbe}\}e$$
$$V_{xx} = \{-[A]^{tbe} + 4[B]^{tba} - 2[B]^{tbe}\}e$$
$$\eta \neq 0$$

$$V_{zz} = \{4[A]^{tba} - 3[B]^{tbe}\}e$$
$$V_{yy} = \{-2[A]^{tba} + 3/2[B]^{tbe}\}e$$
$$V_{xx} = \{-[A]^{tba} + 3/2[B]^{tbe}\}e$$
$$\eta = 0$$

$$V_{zz} = \{2[A]^{tba} - 3[B]^{tbe} + 2[B]^{tba}\}e$$
$$V_{yy} = \{-[A]^{tba} + 3/2[B]^{tbe} - [B]^{tba}\}e$$
$$V_{xx} = \{-[A]^{tba} + 3/2[B]^{tbe} - [B]^{tba}\}e$$
$$\eta = 0$$

$$V_{zz} = \{2[B]^{tba} - 2[B]^{tbe} + 2[A]^{tba} - [A]^{tbe}\}e$$
$$V_{yy} = \{-[B]^{tba} + 5/2[B]^{tbe} - [A]^{tba} - [A]^{tbe}\}e$$
$$V_{xx} = \{-[B]^{tba} - 1/2[B]^{tbe} - [A]^{tba} + 2[A]^{tbe}\}e$$
$$\eta \neq 0$$

$$V_{zz} = V_{xx} = V_{yy} = 0$$

$$V_{zz} = \{4[A] - 4[B]\}e$$
$$V_{yy} = \{2[B] - 2[A]\}e$$
$$V_{xx} = \{2[B] - 2[A]\}e$$
$$\eta = 0$$

$$V_{zz} = \{2[A] - 2[B]\}e$$
$$V_{yy} = \{[B] - [A]\}e$$
$$V_{xx} = \{[B] - [A]\}e$$
$$\eta = 0$$

$$V_{zz} = \{2[B] - 2[A]\}e$$
$$V_{yy} = \{[A] - [B]\}e$$
$$V_{xx} = \{[A] - [B]\}e$$
$$\eta = 0$$

$$V_{zz} = \{2[A] + 2[C] - 4[B]\}e$$
$$V_{yy} = \{2[B] - [A] - [C]\}e$$
$$V_{xx} = \{2[B] - [A] - [C]\}e$$
$$\eta = 0$$

$$V_{zz} = \{2[A] - 2[C]\}e$$
$$V_{yy} = \{[C] - [A]\}e$$
$$V_{xx} = \{[C] - [A]\}e$$
$$\eta = 0$$

TABLE 6.XV. *(Cont.)*

Point Charge Model Expressions for the Components of the EFG Tensor at a Nucleus M Due to Ligands A,B,C,D for Common Structures (5).

Structure	Components of the EFG [a,b]	Structure	Compound of the EFG
z ⋮ y C A B M B B B x	$V_{zz} = \{2[A] - [B] - [C]\}e$ $V_{zz} = \{-[B] + 2[C] - [A]\}e$ $V_{yy} = \{2[B] - [C] - [A]\}e$ $\eta \neq 0$	z ⋮ y C A B M B B C x	$V_{zz} = \{2[A] - 2[C]\}e$ $V_{xx} = \{4[C] - 3[B] - [A]\}e$ $V_{yy} = \{3[B] - 2[C] - [A]\}e$ $\eta \neq 0$
z ⋮ y A A B M A B B x	$V_{zz} = V_{yy} = V_{xx} = 0$	z ⋮ y B A C M B A C x	$V_{zz} = \{4[A] - 2[B] - 2[C]\}e$ $V_{yy} = \{[C] + [B] - 2[A]\}e$ $V_{xx} = \{[C] + [B] - 2[A]\}e$ $\eta = 0$
z ⋮ y B A B M B A A x	$V_{zz} = \{3[A] - 3[B]\}e$ $V_{yy} = \{3[B] - 3[A]\}e$ $V_{xx} = 0$ $\eta = 1$	z ⋮ y B A C M C A B x	$V_{zz} = \{4[A] - 2[B] - 2[C]\}e$ $V_{xx} = \{4[B] - 2[C] - 2[A]\}e$ $V_{yy} = \{4[C] - 2[B] - 2[A]\}e$ $\eta \neq 0$

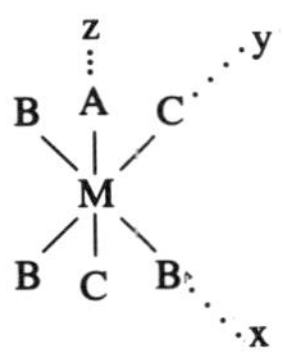

z B A C ·y M B C B ·x	$V_{zz} = \{2[A] + [C] - 3[B]\}e$ $V_{xx} = \{3[B] - 2[C] - [A]\}e$ $V_{yy} = \{[C] - [A]\}e$ $\eta \neq 0$
z C A B ·y M C B A ·x	$V_{zz} = \{[A] + [B] - 2[C]\}e$ $V_{yy} = \{[B] + [C] - 2[A]\}e$ $V_{xx} = \{[A] + [C] - 2[B]\}e$ $\eta \neq 0$
z C B A ·y M C B A ·x	$V_{zz} = \{4[B] - 2[A] - 2[C]\}e$ $V_{yy} = \{[A] + [C] - 2[B]\}e$ $V_{xx} = \{[A] + [C] - 2[B]\}e$ $\eta = 0$
z B C A ·y M B C A ·x	$V_{zz} = \{4[C] - 2[A] - 2[B]\}e$ $V_{yy} = \{[A] + [B] - 2[C]\}e$ $V_{xx} = \{[A] + [B] - 2[C]\}e$ $\eta = 0$

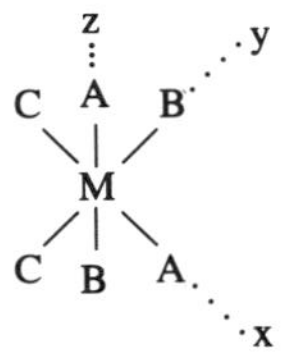

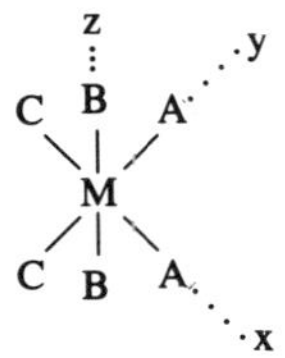

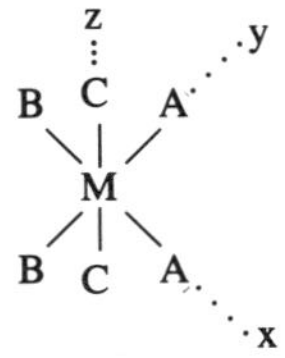

[a] The choice of EFG axes is usually indicated on the diagram of the structure or in a footnote. In all cases, except the four-coordinate $MABC_2$ and MABCD structures, this choice of axes serves to diagonalize the EFG tensor. The ordering of the axes to preserve the convention $|V_{zz}| \geq |V_{yy}| \geq |V_{xx}|$ will depend on the $[L]$ values. Thus, the final choice of axes may not be the same as given here; i.e., V_{zz} may become V_{xx} or V_{yy}, etc.

[b] Whenever η is not 0 or 1, it is easily calculated (after diagonalizing the tensor), taking $|V_{xx}| \leq |V_{yy}| \leq |V_{zz}|$, and using $\eta = (V_{xx} - V_{yy})/V_{zz}$. For example, take the third last structure in the table, $MA_2B_2C_2$: $\eta = 3[C] - 3[B]/[B] + [C] - 2[A]$.

[c] The x axes coincide with the C_2 symmetry axis, and the y and z axes lie in the symmetry planes.

[d] The y axis is perpendicular to the symmetry plane, while the x and z axes lie in the plane. The orientation of the x and z axes depends on the relative magnitudes of $[A]$, $[B]$, and $[C]$, and the tensor must be diagonalized separately for each case considered. The P and Q expressions lead to the magnitude of the quadrupole splitting and is obtained from the symmetrized parameters of Clark (2).

[e] The EFG tensor must be diagonalized for each example considered.

[f] The superscripts tbe and tba refer to trigonal-bipyramidal equatorial and trigonal-bipyramidal axial bonds, respectively.

TABLE 6.XVI

Eigenvalue Coefficients (A_n) for the Nuclear Quadrupole Interaction (33)

I	m_I	a_0	a_1	a_2	a_3	a_4
3/2	±3/2	0.2500	−0.0001	0.0425	−0.0020	−0.0017
	±1/2	−0.2500	−.0001	−0.0425	0.0020	0.0017
2	2	0.2500	0	0	0	0
	1	−0.1250	0.1250	0	0	0
	0	−0.2500	0.0001	−0.0425	0.0020	0.0017
	−1	−0.1250	−0.1250	0	0	0
	−2	0.2500	−0.0001	0.0425	−0.0020	−0.0017
5/2	±5/2	0.2500	0.0003	0.0125	0.0020	−0.0003
	±3/2	−0.0500	−0.0001	0.0880	−0.0431	0.0006
	±1/2	−0.2000	0.0001	−0.1005	0.0410	−0.0006
7/2	±7/2	0.2500	0.0001	0.0063	0.0048	−0.0026
	±5/2	0.0357	0.0008	0.0242	0.0124	−0.0058
	±3/2	−0.1071	−0.0009	0.1318	−0.1327	0.0417
	±1/2	−0.1785	0.0003	−0.1659	0.1239	−0.0385
9/2	±9/2	0.2500	−0.0004	0.0077	−0.0020	0.0009
	±7/2	0.0833	0.0001	0.0201	−0.0014	0.0017
	±5/2	−0.0417	0.0009	0.0359	0.0253	−0.0204
	±3/2	−0.1250	0.0052	0.1404	−0.2085	0.0842
	±1/2	−0.1667	−0.0048	−0.2066	0.1898	−0.0679

R_Q is the ratio of the excited nuclear quadrupole moment to that of the ground nuclear quadrupole moment. Values of R_Q are listed in Table 6.XVII. If $I = 0$ or 1/2 for one of the nuclear levels, then simply

$$E_Q = E_Q(I, m_I) \text{ or}$$
$$E_Q = E_Q^*(I^*, m_{I^*}) \tag{58}$$

Table 6.XVII also gives values for the nuclear quadrupole moments and for the quadrupole coupling constant e^2qQ, which are needed for evaluating and interpreting quadrupole Mössbauer spectra.

The only remaining items that need to be evaluated are the transition intensities for each Mössbauer line. Again, in most cases, these are not known in closed form, but have been evaluated as a series (33). The intensities can be expressed as

$$A(I^*, I, m_{I^*}, m_I) = \sum_{n=0}^{4} b_n(I^*, I, m_{I^*}, m_I)\eta^n \tag{59}$$

where the values of b_n are given in Table 6.XVIII.

In Fig. 6.24 ^{121}Sb is considered in which plots are shown of the energy splittings, the relative intensity of each transition, and the effect of η on the relative energy positions of each peak. An example spectrum in which $\eta = 0$ is shown in Fig. 6.11.

For many substances $\eta = 0$, simplifying a number of factors. For ^{57}Fe and ^{119}Sn, the spectra reduce to doublets (see Fig. 6.10). If the absorber is powdered and there is

TABLE 6.XVII

Nuclear Quadrupole Moment Data (38)

Isotope	E_γ	I_e	I_g	Quadrupole Moments Q_0 ground[a]	Q excited[a]	Ratio of quadrupole moments[a]	$\frac{eQ}{4I(2I-1)}$ $[10^{-22}\,(\text{mm/s})(V\cdot m^2)]$ Ground[a]	Excited[a]
^{57}Fe	14.4	3/2	1/2	0.0	0.21(1)[s]		0.0	0.364(17)
^{61}Ni	67.4	5/2	3/2	0.162(15)[s]	−1.21(13)[s]	−1.21(13)	0.060(6)	−0.134(15)
^{99}Ru	89.4	3/2	5/2	0.12(3)	0.35(9)	2.93(5)	0.010(3)	0.10(3)
^{119}Sn	23.9	3/2	1/2	0.0	−0.06		0.0	−0.063(11)
^{121}Sb	37.2	7/2	5/2	0.28(6)	−0.38(8)	1.340(10)	−0.057(17)	−0.37(8)
^{125}Te	35.5	3/2	1/2	0.0	−0.20(2)		0.0	−0.141(14)
^{127}I	57.6	7/2	5/2	−0.79(10)	−0.71(9)	−0.896(2)	−0.103(13)	−0.044(6)
^{129}I	27.8	5/2	7/2	−0.55(7)	−0.68(6)	1.2380(16)	−0.071(9)	−0.184(16)
^{133}Cs	81.0	5/2	7/2	−0.0030(7)			−0.00013(3)	
^{151}Eu	21.5	7/2	5/2	1.14(5)	1.50(7)	1.312(9)	0.397(17)	0.249(12)
^{153}Eu	103.2	3/2	5/2	2.90(12)	1.51(6)	0.520(3)	0.211(9)	0.366(15)
^{155}Gd	86.5	5/2	3/2	1.59(16)	0.32(8)	0.20(5)	0.46(5)	0.028(7)
^{161}Dy	25.7	5/2	5/2	2.35(16)[s]	2.34(16)[s]	0.9996(4)	0.69(5)	0.68(5)
^{166}Er	80.6	2	0	0.0	−1.59(15)		0.0	−0.247(23)
^{170}Yb	84.3	2	0	0.0	−2.11(11)		0.0	−0.313(16)
^{181}Ta	6.2	9/2	7/2	3.9(4)	4.4(5)	1.133(10)	2.23(23)	1.47(17)
^{193}Ir	73.0	1/2	3/2	0.70(18)	0.0		0.24(6)	0.0
^{197}Au	77.3	1/2	3/2	0.594(10)	0.0		0.192(3)	0.0
^{237}Np	59.5	5/2	5/2	4.1(7)	4.1(7)	0.990(10)	0.52(9)	0.52(9)

[a] Uncertainties in the last digit(s) are given in the parenthesis.

[s] Sternheimer corrected

TABLE 6.XVIII

Intensity Coefficients (b_n) for the Nuclear Quadrupole Interaction (33)

I^*	I	m_I^*	m_I	b_0	b_1	b_2	b_3	b_4
7/2	9/2	±7/2	±9/2	0.2000	0	−0.0021	0.0003	−0.0005
		±7/2	±7/2	0.0444	−0.0001	0.0014	−0.0009	0.0007
		±7/2	±5/2	0.0056	0.0001	−0.0003	−0.0001	0.0007
		±7/2	±3/2	0	−0.0001	0.0007	0.0011	0.0011
		±7/2	±1/2	0	0.0001	0.0002	−0.0003	0.0002
		±5/2	±9/2	0	0	0.0019	−0.0007	0.0006
		±5/2	±7/2	0.1555	−0.0007	−0.0027	−0.0168	0.0093
		±5/2	±5/2	0.0778	0.0028	−0.0245	0.0725	−0.0346
		±5/2	±3/2	0.0167	−0.0073	0.0321	−0.0546	0.0224
		±5/2	±1/2	0	0.0051	−0.0058	−0.0020	0.0031
		±3/2	±9/2	0	0	0.0004	0	0
		±3/2	±7/2	0	0.0007	−0.0025	0.0229	−0.0118
		±3/2	±5/2	0.1167	−0.0135	−0.0190	0.0275	−0.0178

TABLE 6.XVIII (Cont.)

Intensity Coefficients (b_n) for the Nuclear Quadrupole Interaction (33)

I^*	I	m_I^*	m_I	b_0	b_1	b_2	b_3	b_4
		±3/2	±3/2	0.1000	0.0640	0.1008	−0.2557	0.1359
		±3/2	±1/2	0.033	−0.0513	−0.0790	0.2043	−0.1058
		±1/2	±9/2	0	0	0	0.0001	0
		±1/2	±7/2	0	0	0.0041	−0.0055	0.0020
		±1/2	±5/2	0	0.0105	0.0441	−0.1003	0.0519
		±1/2	±3/2	0.0833	−0.0566	−0.1327	0.3076	−0.1564
		±1/2	±1/2	0.1667	0.0461	0.0847	−0.2021	0.1026
5/2	7/2	±5/2	±7/2	0.2500	0.0007	−0.0097	0.0100	−0.0061
		±5/2	±5/2	0.0715	0.0001	0.0011	−0.0033	0.0064
		±5/2	±3/2	0.0120	−0.0027	0.0037	0.0029	−0.0041
		±5/2	±1/2	0	0.0013	0.0021	−0.0045	0.0017
		±3/2	±7/2	0	−0.0004	0.0040	−0.0001	0.0009
		±3/2	±5/2	0.1784	0.0014	−0.0677	0.0839	−0.0338
		±3/2	±3/2	0.1189	0.0043	0.2327	−0.3327	0.1393
		±3/2	±1/2	0.0356	−0.0033	−0.1743	0.2543	−0.1081
		±1/2	±7/2	0	−0.0012	0.0056	−0.0058	0.0020
		±1/2	±5/2	0	−0.0017	0.0672	−0.0819	0.0286
		±1/2	±3/2	0.1191	−0.0011	−0.2379	0.3294	−0.1340
		±1/2	±1/2	0.2140	0.0099	0.1344	−0.1907	0.0775
5/2	5/2	±5/2	±5/2	0.2381	0	−0.0110	−0.0022	0.0007
		±5/2	±3/2	0.0952	0.0009	−0.0251	0.0454	−0.0154
		±5/2	±1/2	0	−0.0008	0.0361	−0.0435	0.0149
		±3/2	±5/2	0.0952	0.0009	−0.0251	0.0454	−0.0154
		±3/2	±3/2	0.0857	−0.0035	0.2519	−0.3138	0.1109
		±3/2	±1/2	0.1524	0.0025	−0.2265	0.2679	−0.0952
		±1/2	±5/2	0	−0.0008	0.0361	−0.0435	0.0149
		±1/2	±3/2	0.1524	0.0025	−0.2265	0.2679	−0.0952
		±1/2	±1/2	0.1810	−0.0017	0.1903	−0.2242	0.0803
3/2	5/2	±3/2	±5/2	0.3333	0.0003	−0.0205	0.0049	0.0013
		±3/2	±3/2	0.1333	−0.0019	0.0707	−0.0436	0.0081
		±3/2	±1/2	0.0333	0.0015	−0.0498	0.0380	−0.0092
		±1/2	±5/2	0	−0.0003	0.0205	−0.0049	−0.0013
		±1/2	±3/2	0.2000	0.0018	−0.0704	0.0430	−0.0078
		±1/2	±1/2	0.3000	−0.0015	0.0498	−0.0380	0.0092
3/2	1/2	±3/2	±1/2	0.5000	0	0	0	0
		±1/2	±1/2	0.5000	0	0	0	0
2	0	All five transitions		0.2000	0	0	0	0
5/2	1/2	All three transitions		0.3333	0	0	0	0

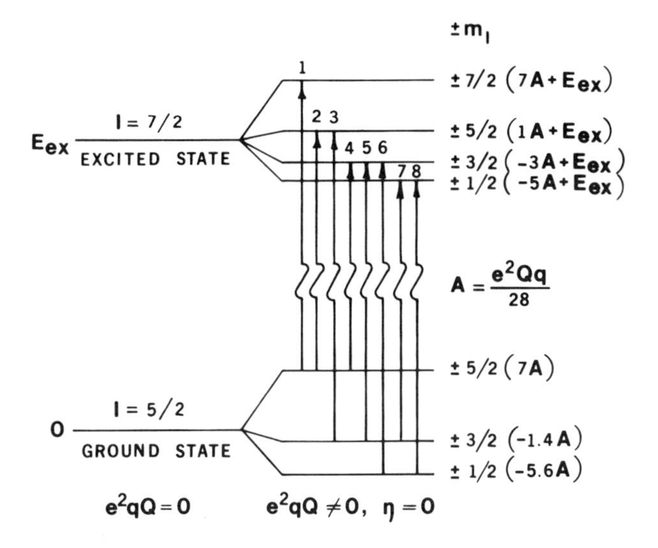

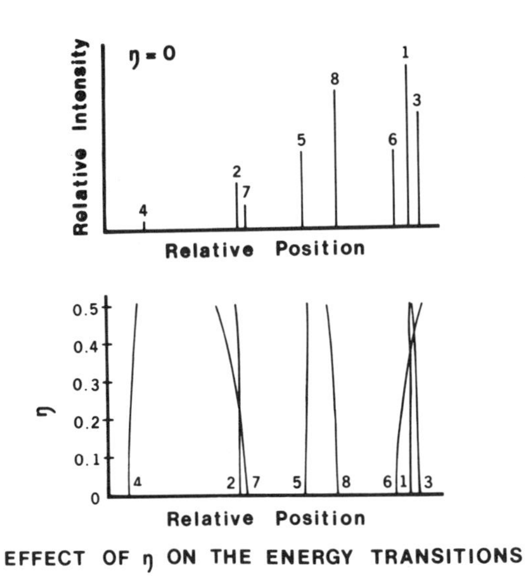

Fig. 6.24. Quadrupole splittings for ^{121}Sb.

no preferred direction for the small crystals, the intensities of the two peaks will be the same. An additional requirement, however, is that there be no lattice anisotropy. If there are preferred directions, the intensities will be unequal. This is illustrated in Fig. 6.25 where the intensities are plotted as a function of the angle between the direction of the gamma and the crystallographic z direction.

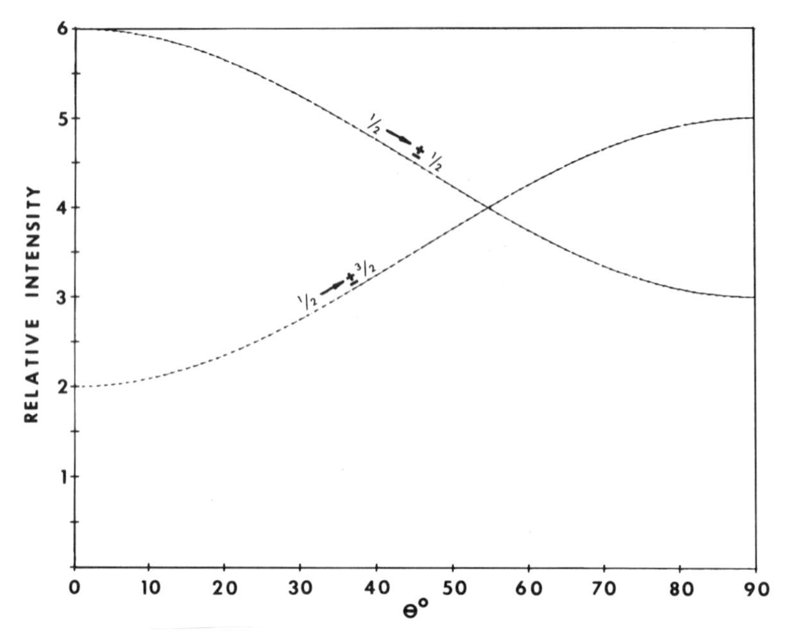

Fig. 6.25. Relative intensities of the two quadrupolar split absorption peaks as a function of the angle between the direction of the gamma and the crystallographic direction.

C. ADDITIVE MODEL

The quadrupole coupling data can be used quite successfully with the additive model. This model has similarities to the partial chemical shifts discussed in Section V.D. Using the equations in Table 6.XV and the experimental data, A's, B's. . . can be determined. These values are called partial quadrupole splittings (pqs). To determine the rest of the pqs values, one must assign a value of the pqs for one ligand. These values can then be used to assist in the interpretation of Mössbauer data on substances about which little is known. The model works quite well in most instances because, although the equations in Table 6.XV consider only the ligand contribution to q, they are also applicable to the populations of the molecular orbitals when the orbitals are in the direction of the ligands, as they usually are.

Examples of the usefulness of the additive model are apparent in Table 6.VIII. Using pqs values for chloride as zero, for methyl as 13.0 mm/s, and for the lone pair as 7.5 mm/s, one gets the calculated values for the quadrupole splitting shown there. The relative values of the pqs for these ligands indicate, as expected, that the methyls are electron-donating while the chlorides are electron-withdrawing.

Many other examples have been reported by Bancroft and are summarized in a book (3). This model has also been extensively discussed by M.G. Clark (11).

VIII. SPIN HAMILTONIAN AND RELAXATION

The principal interactions discussed in Section III can be collected and the interaction Hamiltonian written as

$$\mathcal{H}_{\text{Int}} = \mathcal{H}_{(\text{E0})} + \mathcal{H}_{(\text{M1})} + \mathcal{H}_{(\text{E2})} \quad (60)$$

The magnetic interaction $\mathcal{H}_{(\text{M1})}$ consists of the nuclear Zeeman interaction (Z), the orbital term (L), and the magnetic or spin hyperfine interaction (M):

$$\mathcal{H}_{(\text{M1})} = \mathcal{H}_Z + \mathcal{H}_L + \mathcal{H}_M \quad (61)$$

The nuclear Zeeman interaction occurs when there is an externally applied magnetic field, while the magnetic hyperfine interaction is due to internal magnetic fields produced by the electronic spins. These internal fields produce the Fermi contact and dipole interactions which have been discussed in Section VI along with the orbital term in equation 61.

The magnetic hyperfine interaction is analogous to the electric quadrupole interaction, which is an electric hyperfine interaction. A magnetic hyperfine tensor $\overleftrightarrow{A}$, similar to the electric field gradient tensor of equation 25, can be defined and a spin Hamiltonian written as

$$\mathcal{H}_M = \vec{I} \cdot \overleftrightarrow{A} \cdot \vec{S} \quad (62)$$

where $\vec{I}$ is the instrinsic nuclear spin and $\vec{S}$ is the effective electronic spin. For many applications $\overleftrightarrow{A}$ can be taken to be diagonal, i.e.,

$$\overleftrightarrow{A} = \begin{pmatrix} A_x & 0 & 0 \\ 0 & A_y & 0 \\ 0 & 0 & A_z \end{pmatrix}$$

Substituting the above tensor in equation 62 gives

$$\mathcal{H}_M = A_x I_x S_x + A_y I_y S_y + A_z I_z S_z \quad (63)$$

Following Wickman, Klein, and Shirley (45) the effects of equation 63 on the Mössbauer spectrum for iron are investigated below. The effective electronic-spin case considered is $S = 1/2$.

For the Mössbauer transition in ^{57}Fe, $I_e = 3/2$ and $I_g = 1/2$ for the excited and ground states, respectively. The excited state has $(2I_e + 1)(2S + 1) = 8$ degenerate levels while the ground state has $(2I_g + 1)(2S + 1) = 4$ degenerate levels. Each of the $2I + 1$ nuclear spin states can occur with an effective electronic spin $S_z = \pm 1/2$. With a magnetic field in the z direction only, equation 63 reduces to

$$\mathcal{H}_M = A_z I_z S_z \quad (64)$$

Since the degenerate states are not mixed by the Hamiltonian of equation 64, the energies are simply

$$E_{mM} = \langle m M \mid \mathcal{H}_M \mid m M \rangle \quad (65)$$

where m and M are the z quantum numbers of the nuclear and electron spins, respectively. Equation 65 implies four energy levels for the excited state and two for the ground state. However, each of these levels have two-fold degeneracy (see Fig. 6.26). The dipole selection rule allows for six pairs of transitions, one (see

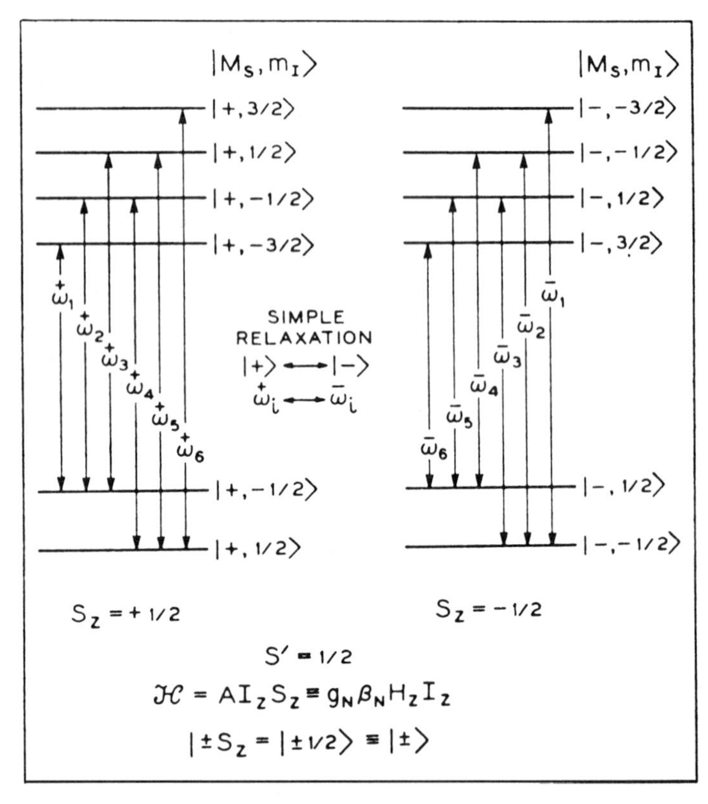

Fig. 6.26. Twofold degeneracy of energy levels in ^{57}Fe where the z component of the effective electron spin can have values $S_z = \pm\frac{1}{2}$. Reproduced by permission from *Phys. Rev.* (45).

Fig. 6.7) for each of the two electronic spin states. Since the Lorentzian pairs overlap, the resulting Mössbauer spectrum consists of six line shapes.

Maintaining axial symmetry, but allowing small transverse magnetic fields such that $A_x = A_y \ll A_z$, the Mössbauer spectrum becomes more complicated. Note that the components of the hyperfine tensor are different for the excited and the ground states. The spin Hamiltonian equation 63 in terms of raising and lowering operators is

$$\mathcal{H}_M = A_z I_z S_z + \left(\frac{A_x + A_y}{4}\right)(I_+S_- + I_-S_+) + \left(\frac{A_x - A_y}{4}\right)(I_+S_+ + I_-S_-) \quad (66)$$

The last term in equation 66 is zero when $A_x = A_y$. However, the second term removes the degeneracy of the $|\pm 1/2, \mp 1/2\rangle$ states in both the excited and ground nuclear levels. The splitting is illustrated in Fig. 6.27 which also indicates the second-order nondegenerate perturbation shifts on the adjacent pairs in the excited nucleus. Two transitions between the split levels are forbidden and the resulting spectrum consists of ten line shapes.

For the case $A_z \gg A_x > A_y$, the axial symmetry is destroyed. The third term in equation 66 splits the degenerate levels $|\pm 1/2, \pm 1/2\rangle$ in both excited and ground-

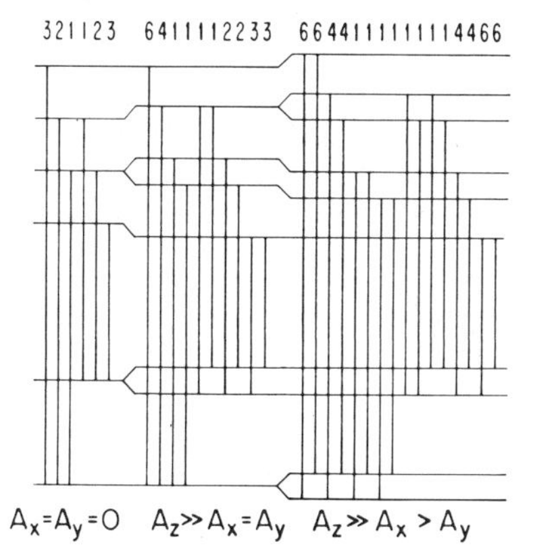

Fig. 6.27. Energy levels and transitions with relative intensities for ^{57}Fe (effective electronic spin case $S=\frac{1}{2}$) under various magnetic hyperfine interactions. Reproduced by permission from *Phys. Rev.* (45).

state nuclei. Adjacent levels in the excited nuclear state are shifted in second order (see Fig. 6.27). The Mössbauer spectrum now has 16 different transition energies.

The above considerations assume static magnetic fields. However, the internal fields at the nuclear sites are time dependent. The simple case of a random or randomly fluctuating magnetic field in the z direction is considered below with particular attention given to rapid fluctuations (fast relaxation) and slow fluctuations (slow relaxation). To determine whether the magnetic field is changing rapidly, one compares the time for the field to change polarity (τ) with the period (τ_L) of the Larmor precession. A magnetic moment $\vec{\mu} = \gamma \vec{I}$ in a constant magnetic field $\vec{H} = (0,0,H)$ rotates about the z axis with the Larmor frequency $\omega_L = \gamma H$. The discussion below is patterned after Blume and Tjon (8).

The line shape for gamma emission is determined by the transition probability (27) (for an exponential decay law)

$$P(E_\gamma) = \frac{|\langle f \mid \mathcal{H}^{(+)} \mid i \rangle|^2}{(E_\gamma - E_f + E_i)^2 + \Gamma^2/4} \tag{67}$$

In equation 67 i and f represent the initial (excited) and final (ground) states respectively, E_γ is the gamma energy, Γ is the line width, and $\mathcal{H}^{(+)}$ contains the creation operator for the photon.

The energies of the nuclear states $\mid I\ m \rangle$ are found from the unperturbed and the spin Hamiltonians

$$\mathcal{H} = \mathcal{H}_0 + A_z I_z (1/2) \tag{68}$$

where the effective spin is assumed to be constant ($S_z = 1/2$) for the moment. It is the effective spin that produces the internal magnetic field at the nucleus. There is a finite probability for the effective spin to flip to $S_z = -1/2$; such transitions induce the time-dependent magnetic fields. The energies calculated from equation 68 are

$$E_i = \langle I_e\ m_e \mid \mathcal{H} \mid I_e\ m_e \rangle = E_e + \frac{A_e}{2} m_e \equiv E_e + g\beta H\ m_e \tag{69a}$$

$$E_f = \langle I_g\, m_g \mid \mathcal{H} \mid I_g\, m_g \rangle = E_g + \frac{A_g}{2} m_g \equiv E_g + g'\beta H\, m_g \qquad (69b)$$

where H is the internal magnetic field, g and g' are Landé factors, and β is the Bohr magneton. The probability for the emission of a photon with energy E_γ is obtained by averaging over the four initial states and summing over final states:

$$P(E_\gamma) = 1/4 \sum_{m_e m_g} \frac{\mid \langle I_g\, m_g \mid \mathcal{H}^{(+)} \mid I_e\, m_e \rangle \mid^2}{[E_\gamma - E_o - (gm_e - g'm_g)\beta H]^2 + \Gamma^2/4} \qquad (70)$$

where $E_o = E_e - E_g$. If $H = 0$, there is one Lorentian peek at $E_\gamma = E_o$; however, equation 70 implies six Lorentzians when $H \neq 0$ with peaks $E_\gamma = E_o + (gm_e - g'm_g)$.

The fluctuations of the magnetic field due to the spin of the electronic spins can be easily incorporated in the calculations by letting

$$H \rightarrow H f(t) \qquad (71)$$

where $f(t)$ randomly takes on values of ± 1.

The generalization of equation 70 for time dependent fields is (8,46)

$$P(E_\gamma) = \frac{2}{\Gamma} Re \int_0^\infty e^{\,iE_\gamma t/\hbar - \frac{1}{2}\Gamma t}\, G(t)\, dt \qquad (72)$$

where

$$G(t) = \overline{\begin{array}{c} 1/4 \sum_{m_e m_g} \mid \langle I_g m_g \mid \mathcal{H}^{(+)} \mid I_e m_e \rangle \mid^2 e^{-iE_o t/\hbar} \\ X \exp[-i(gm_e - g'm_o)\, \beta\, H \int_0^t f(t')dt'] \end{array}} \qquad (73)$$

The bar in equation 73 designates averaging. When the magnetic field is constant, i.e., $f(t) = 1$, one finds

$$G(t) = 1/4 \sum_{m_e m_g} \mid \langle I_g m_g \mid \mathcal{H}^{(+)} \mid I_e m_e \rangle \mid^2 e^{-i[E_o + (gm_e - g'm_o)\beta Ht]/\hbar} \qquad (74)$$

If equation 74 is substituted into equation 72, the real part of the resulting integral gives equation 70.

Defining $\alpha = g'm_o - gm_e$, the average needed in equation 73 is (1,8)

$$\overline{e^{i\alpha \int_0^t f(t')\,dt'}} = \left(\cos xWt + \frac{\sin xWt}{x}\right) e^{-Wt} \qquad (75)$$

where $W \approx 1/\tau$ is the probability for the effective spin to flip and $x = (\alpha^2/W^2 - 1)^{\frac{1}{2}}$. For slow relaxation $\tau \gg \tau_L \approx \frac{1}{\alpha}$, i.e., $W \ll \alpha$ and

$$\overline{e^{i\alpha \int_0^t f(t')dt'}} \approx \cos \alpha\, t = 1/2(e^{\,i\alpha t} + e^{-i\alpha t}) \qquad (76)$$

using equation 75. Equation 76 implies two sets of six Lorentzians (see Fig. 6.27); there will be six line shapes in the Mössbauer spectrum as pairs overlap.

For fast relaxation $W \gg \alpha$ and $x = \sqrt{-1} = i$. The right side of equation 75 becomes

$$\left(\cos iWt + \frac{\sin iWt}{i}\right)e^{-Wt} = e^{-i(iWt)}e^{-Wt} = 1 \quad (77)$$

using the Euler formula

$$\cos\theta + \sin\theta/i = \cos\theta - i\sin\theta = e^{-i\theta} \quad (78)$$

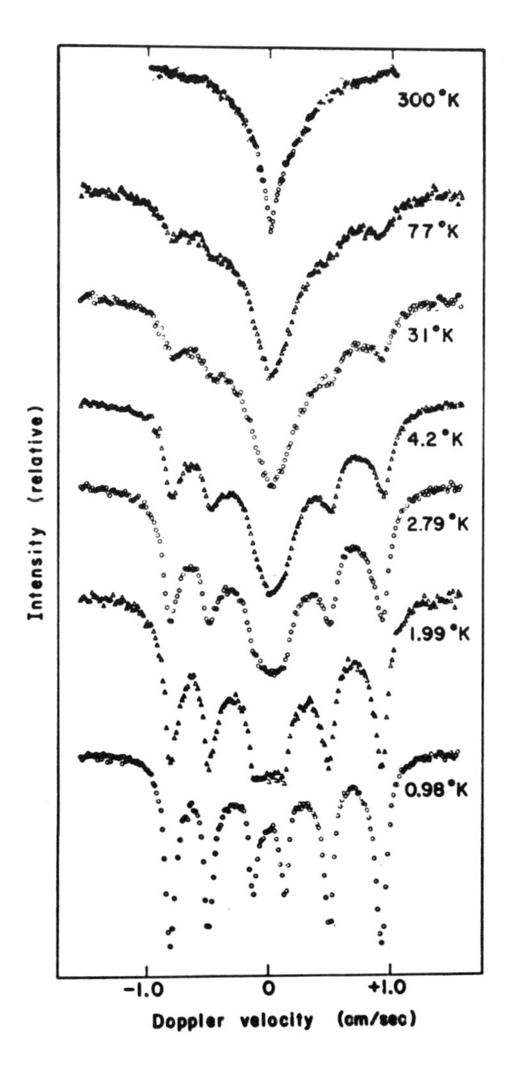

Fig. 6.28. ^{57}Fe Mössbauer spectra of Ferrichrome A at various temperatures. Reproduced by permission from *Phys. Rev.* (45).

Equation 73 is then simply

$$G(t) = 1/4 \sum_{m_e m_g} | \langle I_g m_g \; | \mathcal{H}^{(+)} | \; I_e \, m_{ei} \rangle \; |^2 e^{-iE_o t} \tag{79}$$

and equation 70 reduces to

$$P(E_\gamma) = 1/4 \sum_{m_e m_g} \frac{| \langle I_g m_g \,|\, \mathcal{H}^{(+)} \,|\, I_e m_e \rangle \; |^2}{(E_\gamma - E_o)^2 + \Gamma^2/4} \tag{80}$$

The Mössbauer spectrum consists of one line shape. The fluctuations of the magnetic field are so rapid that the field at the nucleus averages to zero during times of the order of the Larmor period. Fig. 6.28 indicates how the six line shapes reduce to one as temperature is increased. Increasing temperature induces rapid spin flips and therefore fast relaxation. However, due to other factors the resulting single line is non-Lorentzian.

An alternative approach to relaxation employs the Bloch equations (1,31). A theoretical spectrum using this approach is given for iron (45,46) in Figure 6.29. Once again, the six line shapes collapse to one as relaxation times become short.

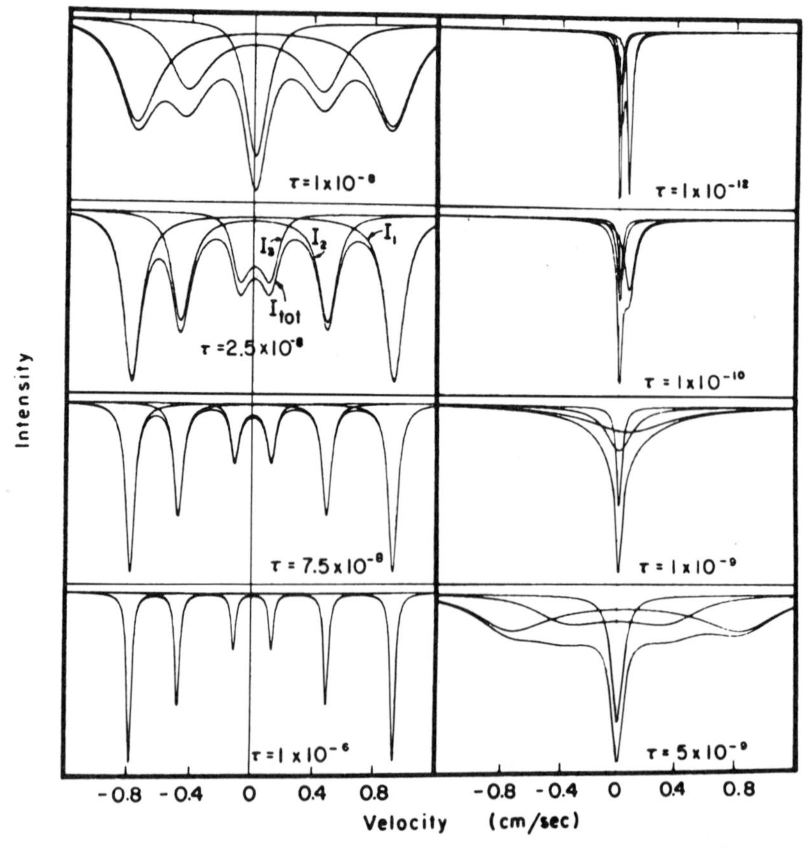

Fig. 6.29. ^{57}Fe iron relaxation spectra for Ferrichrome A with various relaxation times. Reproduced by permission from *Phys. Rev.* (45).

Acknowledgments

We express our special appreciation to Virginia Stevens and Robert Detjen who have laboriously proofread the text. We thank Karen Ball and Mary Jane Winfrey for typing this manuscript and Julie Blankenship for preparation of the figures. This chapter has been adapted from a set of notes for a course on Mössbauer spectroscopy and we wish to thank the students for their suggestions and comments.

APPENDIX

Nomenclature and Conventions for Reporting Mössbauer Spectroscopic Data

I. INTRODUCTION

The guidelines that follow are based considerably on the reports of several groups; most notably the Mössbauer Spectroscopy Task Group of Committee E-4 (Metallography) of the American Society for Testing and Materials, the *ad hoc* Panel on Mössbauer Data of the Numerical Data Advisory Board of the Division of Chemistry and Chemical Technology of the National Research Council (USA), and the Commission on Molecular Structure and Spectroscopy (Physical Chemistry Division) of the International Union of Pure and Applied Chemistry.

II. CONVENTIONS FOR THE REPORTING OF MÖSSBAUER DATA

A. TEXT

The text should include information about the following:

1. Method of sample mounting, sample thickness control, sample confinement, and appropriate composition data for alloys, solid solutions, or frozen solution samples;

2. Absorber form (single crystal, polycrystalline powder, inert matrix if used, evaporated film, rolled foil, isotopic enrichment, etc.);

3. Apparatus and detector used and comments about associated electronics (e.g., single-channel window, escape-peak measurements, solid-state detector characteristics, etc.) if appropriate or unusual; data acquisition time if unusual;

4. Geometry of the experiment (transmission, scattering, in-beam, angular dependence, etc.); direction and strength of applied magnetic field if used;

5. Critical absorbers of filters if used;

6. Method of data reduction (e.g., visual, computer, etc.) and curve-fitting procedure; (1)

7. Isomer-shift convention used or the isomer shift of a standard (reference) absorber. Positive velocities are defined as source approaching absorber. Sufficient details concerning the isomer shift standard should be included to facilitate interlaboratory comparison of data;

8. Method of energy calibration (e.g., calibrated absorber, Michelson interferometer, Moiré fringes, etc.);

9. An estimate of systematic and statistical errors of the quoted parameters.

B. NUMERICAL OR TABULATED DATA

Information collected and summarized in tabular form should include:

1. Chemical state of source matrix and absorber;
2. Temperature of source and absorber and the constancy of these parameters over the length of the data acquisition period;
3. Values of the parameters required to characterize the features in the Mössbauer spectrum (given in mm/s or other appropriate units) with estimated errors (see below the section on "Terminology, Symbols, and Units" for parameters used);
4. Isomer shift reference point with respect to which the positional parameter is reported;
5. Observed line widths defined as the full-width at half-maximum peak-height or other appropriate line width (e.g., line widths calculated by a transmission integral computer fit, line widths of the single Lorentzian peaks when the spectrum is the result of a sum of overlapping Lorentzian peaks, etc);
6. Line intensities or (relative) area of each component of the hyperfine interaction spectrum observed, if pertinent.

C. FIGURES ILLUSTRATING SPECTRA

Scientific communications in which Mössbauer effect measurements constitute a primary or significant source of experimental information should include an illustration of one spectrum (i.e., percent transmission or absorption or counting rate versus an appropriate energy parameter) to indicate the quality of the data. Such figures should include the following information:

1. A horizontal axis normally scaled in velocity (mm/s; channel number or analyzer address values should not be used for this purpose);
2. A vertical axis normally scaled in terms of the effect magnitude, transmission per scattering intensity, counts per channel, or related units; (2)
3. Statistical counting error limits indicated for at least one data point; (3)
4. Individual data points (rather than a smoothed curve alone) should be shown. Computed fits should be indicated in such a way that they are clearly distinguished from the experimental points.

III. TERMINOLOGY, SYMBOLS, AND UNITS

If the units selected by the experimenter are not SI units, they should be defined in the text.

Name	Symbol	SI Unit	Suggested decimal multiple of submultiple SI Units for Mössbauer data	Definition and comments
Isomer Shift	δ	m/s	mm/s ($=10^{-3}$m/s)	Measure of the energy difference between the source (E_s) and the absorber (E_a) transition. The measured Doppler velocity shift, δ, is related to the energy difference by $E_a - E_s = E_\gamma v/c$ (where E_γ is the Mössbauer gamma energy and c is the speed of light in a vacuum). [4],[5]
Nuclear/gyromagnetic ratio	γ	$T^{-1}s^{-1}$		The parameter that is the proportionality constant between the nuclear moment and the angular momentum. [6]

III. TERMINOLOGY, SYMBOLS, AND UNITS *(Cont.)*

Name	Symbol	SI Unit	Suggested decimal multiple of submultiple SI Units for Mössbauer data	Definition and comments
Magnetic flux density	B	T		Magnetic flux density at the nucleus (from experiment) in those cases in which the magnetic hyperfine interaction can be described by an effective field. In other cases the tensor components of the magnetic hyperfine interaction should be reported if possible.
Magnetic hyperfine splitting	$\gamma \hbar B$	J	mm/s($=10^{-3}$m/s)	The energy difference between two adjacent levels that are the results of the interaction of the nuclear magnetic dipole moment and the magnetic flux density. (5)
Components of the magnetic hyperfine interaction tensor	A_x, A_y, A_z	J	mm/s($=10^{-3}$m/s)	Used when the magnetic hyperfine interaction is to be described by the Hamiltonian $\vec{I} \cdot \overrightarrow{\overrightarrow{A}} \cdot \vec{S}$.(7)
Nuclear quadrupole moment (spectroscopic)	eQ	Cm^2		A parameter that describes the effective shape of the equivalent

				ellipsoid of nuclear charge distribution $Q>O$ for a prolate (e.g., ^{57}Fe, ^{197}Au) and $Q < O$ for an oblate (e.g., ^{119}Sn, ^{129}I) nucleus.
Electric field gradient (EFG) tensor		V/m^2		A second-rank tensor describing the electric field gradient specified by η and V_{zz}.[8]
Principal component of EFG	$-V_{zz}$	V/m^2		$\partial^2 V/\partial z^2 = eq$ (e is the proton charge, V_{zz} is the largest component of the diagonalized EFG).
Quadrupole coupling constant	e^2qQ/h	Hz	MHz($=10^6$Hz)	Product of V_{zz}/h and the nuclear quadrupole moment, eQ.[5]
Asymmetry	η			$= (V_{xx} - V_{yy})/V_{zz}$.
Line width	W	m/s	mm/s($=10^{-3}$m/s)	Full width at half maximum of the observed resonance line(s).[5]
Natural line width	W_0	m/s	mm/s($=10^{-3}$m/s)	Theoretical value of the full width at half maximum, usually calculated from lifetime data.

III. TERMINOLOGY, SYMBOLS, AND UNITS *(Cont.)*

Name	Symbol	SI Unit	Suggested decimal multiple of submultiple SI Units for Mössbauer data	Definition and comments
Resonance effect magnitude	I_0			The difference in the transmitted or scattered intensity at resonance maximum and off-resonance, relative to the intensity off resonance.[9]
Recoil-free fraction	f			The fraction of all gamma rays of the Mössbauer transition which are emitted (f_s) or absorbed (f_a) without recoil energy loss.[10]
Mössbauer thickness	t			The effective thickness of a source (t_s) or an absorber (t_a) in the optical path.[11]
Resonance/cross section	σ_0	m^2		The cross section for resonant absorption of the Mössbauer gamma ray.[12]
Vibrational anisotropy	ε_m			When the vibrational anisotropy tensor $(\langle x^2_{ij} \rangle)$ is axially

symmetric $\varepsilon_m = (1/\lambda\!\!\!^{-2})$ $(\langle x_{\parallel}^2 \rangle - \langle x_{\perp}^2 \rangle)$ where $\langle x_{\parallel}^2 \rangle$ and $\langle x_{\perp}^2 \rangle$ are the mean square vibrational amplitudes of the Mössbauer nucleus parallel and perpendicular to the cylindrical symmetry axis through the Mössbauer atom and $\lambda\!\!\!^{-}$ is the wavelength of the Mössbauer radiation divided by 2π.

FOOTNOTES

(1) If data are analyzed by computer, a brief description of the program should be given to identify the algorithm used. The number of constraints should be specified (e.g., equal line widths or intensities, etc.) and a measure of the goodness of fit should be indicated.

If measurements of very high accuracy are reported and the discussion of the reality of small effects is an important part of the work, then the following items should be included:

a. the functional form and all parameters used in fitting (i.e., the constraints should be clearly stated);

b. the treatment of the background (e.g., assumed energy independent, experimentally subtracted, etc.);

c. the relative weighting of abscissa and ordinate (e.g., equal weighting);

d. a measure of the statistical reliability;

e. the number of replications and the agreement between these if applicable;

f. an estimate of systematic errors as primary results.

(2) It has become customary to display data obtained in transmission geometry with the resonance maximum "down" and scattering data with the resonance maximum "up." In either case, sufficient data should be shown far enough from the resonance peaks to establish the nonresonant base line.

(3) In most instances (where the data are uncorrected counting results), the standard deviation (i.e., the square root of the second moment of the distribution) is given by $\pm N^{\frac{1}{2}}$, where N is the number of counts scaled per velocity point. For corrected data (i.e., when background or other nonresonant effects are subtracted from the raw data), the propagated error should be computed by normal statistical methods which are briefly described in the figure legend. Fiducial marks bracketing the data points to show the magnitude of the standard deviation are often used in indicating the spread of the data.

(4) The center of a Mössbauer spectrum is defined as the Doppler velocity at which the resonance maximum is (or would be) observed when all magnetic dipole, electric quadrupole, etc. hyperfine interactions are (or would be) absent. The contribution of the second-order Doppler shift should be indicated, if possible. The isomer shift is the sum of this term and the chemical isomer shift.

(5) The SI unit of energy for isomer shift, quadrupole-coupling constant, quadrupole splitting, and line width is the Joule. The measured quantity is the velocity (m/s) which can be converted to the desired energy units.

(6) The nuclear gyromagnetic ratio can be expressed in terms of the nuclear g factor (Landé factor) as $\gamma\hbar = g\mu_N$ where μ_N is called the nuclear magneton and is defined as $\mu_N = eh/2m_pc$ (m_p is the mass of the proton). Another quantity that is often used is the nuclear magnetic moment (μ) which is related to γ and g by $\mu = \gamma hI = g\mu_N I$. The usual unit for μ is μ_N, i.e., nuclear magnetons. Note that the units of γh are JT^{-1}.

(7) In the case of an isotropic interaction the symbol "a" is used (i.e., $a = A_x = A_y = A_z$).

(8) The sum $V_{xx} + V_{yy} + V_{zz} = 0$ regardless of the choice of axes. In the absence of magnetic hyperfine interaction, principal axes are chosen so that the off-diagonal matrix elements vanish, $V_{ij} = 0 (i,j = x,z; i \neq j)$ and are defined such that $| V_{zz} | \geqslant | V_{yy} | \geqslant | V_{xx} |$, so that $0 \leqslant \eta \leqslant 1$. $(\text{EFG})_{ij} = -(\partial^2 V/\partial x_i \partial x_j)$, where $x_i, x_j = x, y,$ or z.

(9) This parameter is calculated from the relationship $I_0 = [N(\infty) - N(0)]/N(\infty)$, where $N(0)$ is the counting (or transmission or scattering intensity) at the resonance maximum, and $N(\infty)$ is the corresponding rate at a velocity at which the resonance effect is negligible. If corrections for nonresonant gamma- or X-rays, or other base line corrections have been made in evaluating I_0, these should be listed.

(10) The recoil-free fraction can be related to the expectation value of the mean square displacement of the Mössbauer atom by the relationship $f = \exp(-k^2\langle x^2 \rangle)$ where k is the wave number of the Mössbauer gamma ray and x is the displacement taken along the optical axis.

(11) The t parameter is usually calculated from the relationship $t = n \cdot \sigma_0 \cdot IA$, in which n is the number of Mössbauer element atoms per unit area in the optical path, σ_0 is the cross section for recoilless scattering, IA is the fractional abundance of the Mössbauer transition active nuclides, and f is the recoil-free fraction (*vide supra*).

(12) This parameter is usually calculated from the relationship
$\sigma_0 = (h^2c^2/2\pi)E_\gamma^{-2}(1 + 2I_e)(1 + 2I_g)^{-1}(1 + \alpha_T)^{-1}(W_0/2W_a)$,
where E_γ is the transition energy; W_a is the absorber line width; I_e and I_g are the excited and ground-state spins, respectively, and α_T is the total internal conversion coefficient of the Mössbauer transition.

REFERENCES

1. Abragam, A., *The Theory of Nuclear Magnetism*, Oxford University, London (1961).
2. Asch, L., and G. M. Kalvius, "Calculated Sample Thicknesses for Absorbers," in J. G. Stevens, Ed., "Mössbauer Spectroscopy," in J. W. Robinson, Ed., *CRC Handbook of Spectroscopy, Volume 3*, CRC Press, Boca Raton, Florida, 1981, pp. 404-416.
3. Bancroft, G. M., *Mössbauer Spectroscopy, An Introduction for Inorganic Chemists and Geochemists*, McGraw Hill, New York (1973).
4. Bancroft, G. M., M. J. Mays, and B. E. Prater, *J. Chem. Soc. (A)*, 956 (1970).
5. Bancroft, G.M., "Point Charge Model Tables," in J. G. Stevens, Ed., "Mössbauer Spectroscopy," in J. W. Robinson, Ed., *CRC Handbook of Spectroscopy, Volume 3*, CRC Press, Boca Raton, Florida, 1981, pp. 480-490.
6. Benczer-Koller, N., and R. H. Herber, "Experimental Methods," in V. I. Gol'danskii and R. H. Herber, Eds., *Chemical Applications of Mössbauer Spectroscopy*, Academic, New York, 1968, pp. 114-158.
7. Berrett, R. R., and B. W. Fitzsimmons, *J. Chem. Soc. (A)*, 525 (1967).
8. Blume, M., and J. A. Tjon, *Phys. Rev.*, **165**, 446 (1968).
9. Biscar, J. P., W. Kundig, H. Bommel, and R. S. Hargrove, *Nucl. Instr. Methods*, **75**, 165 (1969).
10. Bowen, L. H., "^{121}Sb Mössbauer Spectroscopy," in J. G. Stevens and V. E. Stevens, Eds., *Mössbauer Effect Data Index—1972*, Plenum, 1973, pp. 71-110.
11. Clark, M. G., "Hyperfine Interactions and Molecular Structure," in A. D. Buckingham, Ed., *Molecular Structure and Properties*, Butterworths, London, 1975, pp. 239-297.
12. Cohen, R. L., and G. K. Wertheim, "Experimental Methods in Mössbauer Spectroscopy," in R. V. Coleman, Ed., *Methods of Experimental Physics, 11*, Academic, New York, 1974, pp. 307-369.
13. Cosgrove, J. G., and R. L. Collins, *Nucl. Instr. Methods*, **95**, 269 (1971).
14. De Waard, H., *Rev. Sci. Instr.*, **36**, 1728 (1965).
15. Fischer, H., and U. Gonser, "Relative Line Intensities of Magnetic Hyperfine Interactions," in J. G. Stevens, Ed., "Mössbauer Spectroscopy," in J. W. Robinson, Ed., *CRC Handbook of Spectroscopy, Volume 3*, CRC Press, Baca Raton, Florida, 1981, pp. 451-465.
16. Greenwood, N. N., and T. C. Gibb, *Mössbauer Spectroscopy*, Chapman and Hall, London (1971).
17. Gütlich, P. "Mössbauer Spectroscopy in Chemistry," in U. Gonser, Ed., *Mössbauer Spectroscopy*, Springer-Verlag, New York, 1975, pp. 53-96.
18. Herber, R. H., *Inorg. Chem.*, **3**, 101 (1964).
19. Kaindl, G., K. Leary, and N. Bartlett, *J. Chem. Phys.*, **59** 5050 (1973).
20. Kaindl, G., W. Potzel, F. Wagner, U. Zahn, and R. L. Mössbauer, *Z. Phys.*, **226**, 103 (1969).
21. Kalvius, G. M., "Mössbauer Spectroscopy in the Actinides," in *Mössbauer Spectroscopy and Its Applications*, International Atomic Energy Agency, 1972, pp. 169-196.
22. Kalvius, G. M., and E. Kankeleit, "Recent Improvements in Instrumentation and Methods of Mössbauer Spectroscopy," in *Mössbauer Spectroscopy and Its Applications*, International Atomic Energy Agency, Vienna, 1972, pp. 9-88.
23. Kistner, O. C., and A. W. Sunyar, *Phys. Rev. Lett*, **4**, 412 (1960).
24. Lees, J. K., Ph.D. Thesis, Carnegie Institute of Technology (1966).

25. Lindqvist, I., and Aiggli, *J. Inorg. Nucl. Chem.*, **2**, 345 (1956).
26. Margulies, S., and J. R. Ehrman, *Nucl. Instr. Methods*, **12**, 131 (1961).
27. Merzbacher, E., *Quantum Mechanics*, John Wiley, New York (1961).
28. Mössbauer, R. L., *Z. Physik*, **151**, 124 (1958).
29. Nowik, I., and E. R. Bauminger, "^{170}Yb Mössbauer Spectroscopy," in J. G. Stevens and V. G. Stevens, Ed., *Mössbauer Effect Data Index—1975*, Plenum, 1976, pp. 407–442.
30. Pasternak, M., A. Simopoulos, and Y. Hazony, *Phys. Rev.*, **140A**, 1892 (1965).
31. Pople, J. A., W. G. Schneider, and H. J. Bernstein, *High-Resolution Nuclear Magnetic Resonance*, McGraw Hill, New York (1959).
32. Ruby, S. L., G. M. Kalvius, G. B. Beard, and R. E. Snyder, *Phys. Rev.*, **159**, 239 (1967)
33. Shenoy, G. K., and B. D. Dunlap, *Nucl. Instr. Meth.*, **71**, 285 (1969).
34. Shenoy, G. K., and B. D. Dunlap, "Constants in Isomer Shift Expression," in G. K. Shenoy, and F. E. Wagner, Eds., *Mössbauer Isomer Shifts*, North-Holland, New York, 1978, pp. 889–894.
35. Shenoy, G. K., and F. E. Wagner, *Mössbauer Isomer Shifts*, North-Holland, New York, 1978.
36. Shirley, D. A., *Rev. Mod. Phys.*, **36**, 339 (1964).
37. Stadnik, Z. M., *Mössbauer Effect Ref. Data J.*, **1**, 217 (1978).
38. Stevens, J. G., and B. D. Dunlap, *J. Phys. Chem. Ref. Data*, **5**, 1093 (1976).
39. Stevens, J. G., J. M. Trooster, and H. A. Meinema, *Inorg. Chem.*, **20**, 801 (1981).
39a. Stevens, J. G., and G. K. Shenoy, Eds., *Mössbauer Spectroscopy and Its Chemical Applications*, American Chemical Society, Washington, 1981.
40. Trooster, J. M., and M. P. A. Viegers, *Mössbauer Effect Ref. Data J.*, **1**, 154 (1978).
41. Van der Woude, F., and G. A. Sawatzky, *Phys. Rep.*, **12**, 335 (1974).
42. Viste, A., and H. B. Gray, *Inorg. Chem.*, **3**, 1113 (1964).
43. Walker, L. R., G. K. Wertheim and V. Jaccarino, *Phys. Rev. Lett.*, **6**, 98 (1961).
44. Watson, R. E., and A. J. Freeman, *Phys. Rev.*, **123**, 2027 (1961).
45. Wickman, H. H., M. P. Klein, and D. A. Shirley, *Phys. Rev.*, **152**, 345 (1966).
46. Wickman, H. H., and G. K. Wertheim, "Spin Relaxation in Solids and Aftereffects of Nuclear Transformations," in V. I. Gol'danskii and R. H. Herber, Eds., *Chemical Applications of Mössbauer Spectroscopy*, Academic, New York, 1968, pp. 548–621.
47. Wolfsberg, M., and L. Helmholtz, *J. Chem. Phys.*, **20**, 837 (1952).

SUBJECT INDEX